STATISTICS

The Jones and Bartlett Series in Health Sciences

AIDS Smartbook, Kopec

Anatomy and Physiology: An Easy Learner, Sloane

Aquatic Exercise, Sova

Aquatics: The Complete Reference Guide for Aquatic Fitness Professionals, Sova

Aquatics Activities Handbook, Sova

An Athlete's Guide to Agents, Ruxin

Basic Epidemiological Methods and Biostatistics, Page, et al.

Basic Law for the Allied Health Professions, Second Edition, Cowdrey/Drew

The Biology of AIDS, Third Edition, Fan, et al.

Bloodborne Pathogens, National Safety Council

A Challenge for Living: Dying, Death, and Bereavement, Corless, et al.

CPR Manual, National Safety Council

Dawn Brown's Complete Guide to Step Aerobics, Brown

Drugs and Society, Fourth Edition, Hanson/Venturelli

Drug Use in America, Venturelli

Dying, Death, and Bereavement: Theoretical Perspectives and Other Ways of Knowing, Corless, et al.

Emergency Encounters: EMTs and Their Work, Mannon

Essential Medical Terminology, Stanfield

Ethics Consultation: A Practical Guide, La Puma/Schiedermayer

First Aid, National Safety Council

First Aid and CPR, Second Edition, National Safety Council

First Aid and CPR, Infants and Children, National Safety Counicl

Fitness and Health: Life-Style Strategies, Thygerson

Fostering Emotional Well-Being in the Classroom, Page/Page

Golf: Your Turn for Success, Fisher/Geertsen

Grant Application Writer's Handbook, Reif-Lehrer

Health and Wellness, Fourth Edition, Edlin/Golanty

Health Eucation: Creating Strategies for School and Community Health, Gilbert/Sawyer

Healthy Children 2000, U.S. Department of Health and Human Services

Healthy People 2000, U.S. Department of Health and Human Services

Healthy People 2000--Summary Report, U.S. Department of Health and Human Services

Human Aging and Chronic Disease, Kart, et al.

Human Anatomy and Physiology Coloring Workbook and Study Guide, Anderson

An Introduction to Epidemiology, Timmreck

Introduction to Human Disease, Third Edition, Crowley

Introduction to the Health Professions, Second Edition, Stanfield/Hui

Managing Stress: Principles and Strategies for Health and Wellbeing, Seaward

Medical Terminology, Second Edition, Stanfield/Hui

Medical Terminology with Vikki Wetle, RN, MA, Video Series, Wetle

National Pool and Waterpark Lifeguard/CPR Training, Ellis & Associates

The Nation's Health, Fourth Edition, Lee/Estes

Omaha Orange: A Popular History of EMS in America, Post

Oxygen Administration, National Safety Council

Planning, Program Development, and Evaluation: A Handbook for Health Promotion, Aging, and Health Services, Timmreck

Perspectives on Death and Dying, Fulton/Metress

Primeros Auxilios y RCP, National Safety Council

Safety, Second Edition, Thygerson

Skill-Building Activities for Alcohol and Drug Eduction, Bates/Wigtil

Sports Equipment Management, Walker/Seidler

Sports Injury Care, Abdenour/Thygerson

Statistics: An Interactive Text for the Health and Life Sciences, Krishnamurty et al.

Stress Management, National Safety Council

Teaching Elementary Health Science, Third Edition, Bender/Sorochan

Weight Training for Strength and Fitness, Silvester

STATISTICS

An Interactive Text for the Health and Life Sciences

Goteti Bala Krishnamurty

Professor of Health Sciences
California State University
Northridge, California

Patricia Kasovia-Schmitt

Consultant in Nursing and Epidemiology

David J. Ostroff

Professor of Mathematics
California State University
Northridge, California

Jones and Bartlett Publishers
Sudbury, Massachusetts
Boston London Singapore

Editorial, Sales, and Customer Service Offices
Jones and Bartlett Publishers
40 Tall Pine Drive
Sudbury, MA 01776
(508) 443-5000
(800) 832-0034

Jones and Bartlett Publishers International
Barb House, Barb Mews
London W6 7PA
UK

Library of Congress Cataloging-in-Publication Data

Krishnamurty, Goteti Bala.
Statistics : an interactive text for the health and life sciences / Goteti Bala Krishnamurty, Patricia Kasovia-Schmitt, David J. Ostroff.
p. cm.
Includes index.
ISBN 0-86720-868-6
1. Statistics. I. Kasovia-Schmitt, Patricia. II. Ostroff, David J. III. Title.
QA276. 12. K75 1994
001.4'22--dc20 93-49765
CIP

Acquisitions Editor: Arthur C. Bartlett
Production Editor: Judy Songdahl
Manufacturing Buyer: Dana L. Cerrito
Design: George H. McLean
Typesetting: Scanway Graphics International
Cover Design: Marshall Hendrichs
Printing and Binding: Braun-Brumfield

Acknowledgment is made to each of the following:
Sterling Lord Literistic, Inc. for permission to reprint excerpts from *The Phantom Tollbooth*, Copyright © 1961 by Norton Juster.
The American Public Health Association for permission to reprint excerpts from the following articles from the *Amerian Journal of Public Health*:
"Adolescents and AIDS: A Survey of Knowledge, Attitudes and Beliefs about AIDS in San Francisco," by R. J. DiClemente, J. Zorn, and L. Temoshok; 1986, 76:1443–1445.
"Formaldehyde-Related Health Complaints of Residents Living in Mobile and Conventional Homes," by I. M. Ritchie and R. G. Lehnen; 1987, 77:323–328.
Additional credits appear on pages 705–709, which are a continuation of the copyright page.

Printed in the United States of America
98 97 96 10 9 8 7 6 5 4 3 2

Contents

17 Designs Using Analysis of Variance 633

LITERATURE EXAMPLES

Preface

This book is intended for undergraduate students in the health sciences and related fields. Parts have also been used in a graduate course in experimental design. The book is intended to provide the student with a working knowledge of basic statistical techniques and their application to various experimental designs. As far as we know, the book is unique in its experiential approach and in the large number of examples taken from professional journals.

We purposely paced this text for the vast majority of our students, many of whom have had bad experiences with mathematics in the past and hence approach this subject with anxiety. We have provided special experiential exercises for those students to help them overcome their math and computer anxiety. We find that once our students gain confidence in their ability to master the material, they can accept larger doses of technical material. The later chapters are therefore considerably more demanding than the earlier chapters, and they move at a significantly accelerated pace.

We have tried to describe the statistical and research tools presented accurately and completely enough to help the reader understand their use. We have not tried to give mathematical justifications for the various techniques, nor have we given complete details, in some cases, of the mathematical assumptions on which they are based. Our intention was to supply the reader with enough knowledge of statistics that she or he can read and assess research articles. We did not intend this book to be a complete handbook of all commonly used statistical techniques, nor did we intend that our readers would become skilled users of statistical tools simply by mastering its contents. Rather, we have tried to provide basic tools that will serve the readers as a foundation for further study and help them become discerning readers of scientific literature containing statistical arguments.

Almost all of our students have been successful in mastering the material in this book. Many were delighted to discover that they could do and understand statistics. We hope that our readers will be similarly successful.

We intended this text to be self-contained so that a student could use the book without additional instruction. Our own teaching experience has shown that we are most successful in communicating the material when we give students little, if any, lecture instruction. Instead, we encourage students to read the text and do the exercises on their own. Then we provide an opportunity in class for them to discuss in small groups what they've learned and what they've been puzzled by. We spend most of our class time answering questions and listening to the group discussions, in some cases asking questions to guide those discussions in fruitful directions. We highly recommend this approach to instructors using this book.

The authors are firmly committed to methods of experiential learning. Reading about methods of research is a very small step toward learning how to do research. To truly learn the methods presented here, one must participate. We have provided data sets (in the Appendixes) and a wide variety of exercises and activities throughout the book. Our students tell us that they benefit greatly from doing these exercises. Professional practitioners who have taken our course, such as nurses, have told us that as a result of experience gained from it they are now able to read and understand the supporting arguments in their professional journals. Previously, they say, they only read the conclusions.

Organization

The chapters of the book cover the following material. Chapter 1 discusses the structure of a journal article. Chapter 2 tells how data are generated and provides a questionnaire that may be used to collect data from the class to use as examples. Chapter 3 discusses the nature and interpretation of various uses of numbers and also introduces the data sets used in the book. Chapter 4 introduces various sampling methods and gives examples of their use. Chapters 5 and 6 introduce typical types of tables and graphs, as well as summary statistics such as mean, median, mode, percentiles, etc. Summary statistics are further explored in Chapter 8.

Chapter 7 introduces most of the mathematical formulas used later in the book. Our intention was to introduce the formulas in isolation so that students could learn how to substitute values into them and compute from them with no regard for the meaning or application of the computational results. We find that separating the computational details from the meaning and application of formulas helps to overcome one aspect of math anxiety. When students meet the formulas in context later, they can concentrate on the statistical concepts; the mathematics has already been mastered.

Chapter 9 presents some common designs for experiments and observational studies. Chapter 10 is a very brief introduction to the laws of probability. Chapter 11 presents some standard mathematical tables and explains how to read them. Like the material on formulas in Chapter 7, this material is presented out of context to isolate the technical skill—in this case, reading tables—from the more demanding skill of knowing when and how to use those tables.

Chapter 12 introduces the subject of formulating and testing hypotheses, a subject that is developed in later chapters. Chapter 13 discusses the distribution of sample means and statistical inference using the Central Limit Theorem. Chapters 14 and 15 discuss confidence intervals and hypothesis tests for means, proportions, correlation coefficients, and regression coefficients. Estimating required sample size is also discussed. The chapter also gives a brief introduction to analysis of variance.

Chapter 16 presents several distribution-free tests including chi-square tests of independence. Relative risk, odds ratio, and the phi coefficient are also covered. A very brief introduction to resampling methods is also presented. We conclude with Chapter 17 on analysis of variance, which gives overviews of fixed and random effects models, block designs, and general linear models and briefly introduces repeated measures and Latin square designs. If time is short, the instructor may choose to omit topics from the later chapters.

Acknowledgments

A great many people have contributed to this book. Without their contributions it would never have reached its current form. We wish to express our gratitude to Art Bartlett, who nurtured this project from its infancy. We also thank our colleagues, Jan Adams, Lawrence Clevenson, Steve Gabany, Ernest Scheuer, and Mark Schilling (all of California State University, Northridge) as well as Valery L. Parillo of Kaiser-Permanente, for their many valuable suggestions. We also appreciate the suggestions we received from Donna Brogan of Emory University School of Public Health, William E. Bros of San Jose State University, and Sandra J. Eyres, who reviewed earlier versions of the manuscript. Their suggestions motivated changes that significantly improved the book. We want to thank D. P. Brown, S. F. Loy, S. J. Hall, M. Heng, G. J. Holland, and S. R. Shaw for providing us with the Faculty Exercise Data, which is used for numerous examples throughout the book. Thanks to Janet Morgenstern who read most of the manuscript and spotted a number of errors that we had missed. Thanks also to Cathy Nyfack for her help with the supplementary materials, to Gail Magin for the splendid graphics, and to Jeannie Yost for correcting the page proofs.

Our editor, Quica Ostrander, made major improvements in the text for which we are extremely grateful. Not only did she correct rhetorical and grammatical errors (as one would expect an editor to do), she also vastly improved the pedagogy. For example, in many cases her suggested answers to the exercises were much more precise than were our original answers.

We profoundly appreciate the support, encouragement, and direct assistance provided by our spouses, Shari Klein, Mary Krishnamurty, and John Alan Schmitt. They helped in numerous ways to make this book a reality.

We wish to thank the many journals, publishers, and researchers who graciously granted us permission to reproduce materials from their publications. We especially want to thank Norton Juster and Sterling Lord Literistic, Inc. for permission to reprint the extensive excerpts from *The Phantom Tollbooth* that introduce each chapter. We hope the excerpts will engage your curiosity enough to read the entire book, assuming you haven't already done so.

Finally, we wish to give special thanks to our students, who taught us how to teach, and from whom we have learned at least as much as we have taught them.

Introduction

Scientific research contributes to the betterment of humanity in many ways. Some of you will do research in the future, and we hope you will apply the tools of statistics correctly and appropriately to strengthen your research conclusions. We also hope that those of you who do not do research will acquire deeper understanding of the articles you read in technical journals as well as popular magazine and newspaper articles that describe research results.

We have provided many examples of research studies and their results from technical journals. Some of these studies produced results that will change the lives of many people for the better. Indeed, some concern life-and-death matters. Other studies are of limited significance. We hope you will be inspired by the examples of excellent research and empowered by the tools acquired from this text to make significant contributions to the betterment of humanity.

The tools presented in this book are an introduction to proven experimental designs and techniques of statistical analysis. Statistics deals with probabilities, not certainties. Much scientific conjecture presents plausible solutions to scientific problems, and statistical analysis of experimental results allows us to measure how likely it is that these conjectures actually are the true solutions. We hope you will be rationally skeptical of research results, accepting what is likely, rejecting what is highly unlikely, and remaining open-minded about the rest.

Statistics has its place in research, but we must be careful not to invest it with more persuasive power than it warrants. Statistical tools are just that: tools. Just as one should not use a hammer to drive screws, one should not use statistical tools, such as hypothesis tests, in situations where they do not apply. This text provides an introduction to a vast and varied field. Statistical tests are often quite sensitive to the assumptions made in the analysis and design of an experiment. In many cases we have only presented the simplest examples, and as complexities are introduced, the analysis must be modified accordingly. In addition, even when a result has statistical significance, it may not have practical significance.

We have found that researchers frequently assume that their experimental designs match common design patterns, whereas, in fact, there are (perhaps subtle) differences that profoundly affect the analysis. For this reason, we strongly recommend that researchers discuss their designs with a statistician before collecting data. We know from experience that a statistician can often save valuable time and money for the researcher, and with minor adjustments in experimental design, more useful information can often be gathered.

Furthermore, research does not always address the most important problems or the most interesting ones. No matter how many questionnaires a sociologist administers, for instance, the essence of the homeless problem will probably elude him or her if the problem is viewed only through glasses tinted by theory. As scientists, we must learn to use our hearts and our moral and ethical instincts as well as our intellects to help us see the whole picture. The answers to the problems that plague us do not lie in statistics, but statistics can be used as a persuasive tool for convincing other scientists and laypersons that our observations and solutions are correct.

Following are some things to keep in mind as you read:

Answers to the exercises and activities can be found at the end of each chapter.

We use an asterisk (*) as a multiplication sign. Most computer languages have adopted this convention, and we find it preferable to other common notations because it is less ambiguous. We hope that those who are unfamiliar with this usage will find it easy to adapt to this convention.

We use SMALL CAPS for the names of variables in the two data sets supplied with the text (see Appendixes). We found that setting the variable names off in this way makes some passages easier to read. Other variables, such as those in the Literature Examples, are not given a special typeface.

The Examples from the Literature do not always use the techniques exactly as they are described in the text. You will find, for example, that some authors give graphs in different formats from the ones described. The text introduces and describes some common representations, but there are many variations on these forms and many new forms being developed. Some authors invent their own methods to display their data or to reinforce the points they want to make. Our text introduces the basic concepts, which are "seeds" for further development. The examples presented show what can grow from these seeds.

CHAPTER

1

Research and Reading Abstracts

Prelude

"It *has* been a long trip," said Milo, climbing on to the couch where the princesses sat; "but we would have been here much sooner if I hadn't made so many mistakes. I'm afraid it's all my fault."

"You must never feel badly about making mistakes," explained Reason quietly, "as long as you take the trouble to learn from them. For you often learn more by being wrong for right reasons than you do by being right for the wrong reasons."

"But there's so *much* to learn," he said, with a thoughtful frown.

"Yes, that's true," admitted Rhyme, "but it's not just learning things that's important. It's learning what to do with what you learn and learning why you learn things at all that matters. . . ."

"Many of the things I'm supposed to know seem so useless that I can't see the purpose of learning them at all."

"You may not see it now," said the Princess of Pure Reason, looking knowingly at Milo's puzzled face, "but whatever we learn has a purpose, and whatever we do affects everything and everyone else, if even in the tiniest way. Why, when a housefly flaps his wings, a breeze goes round the world; when a speck of dust falls to the ground, the entire planet weighs a little more; and when you stamp your foot, the earth moves slightly off its course. Whenever you laugh, gladness spreads like the ripples in a pond; and whenever you're sad, no one anywhere can be really happy. And it's much the same thing with knowledge, for whenever you learn something new, the whole world becomes that much richer."

"And remember, also," added the Princess of Sweet Rhyme, "that many places you would like to see are just off the map and many things you want to know are just out of sight or a little beyond your reach. But someday you'll reach them all, for what you learn today, for no reason at all, will help you discover all the wonderful secrets of tomorrow."

— Norton Juster
The Phantom Toll Booth, pp. 233–234

Chapter Outline

- Why Do the Self-Assessment?
- Self-Assessment
- Research
- The Sections of a Research Paper
- Reading Abstracts, Figures, and Tables
- Self-Assessment

Why Do the Self-Assessment?

We start each chapter by having you assess your ability to do the tasks listed in the table. This serves four purposes:

1. The table gives a list of major concepts in the chapter, which you can use as a guide in reading the chapter; you can plan to spend more time on the concepts that you are least familiar with.
2. The table will help you locate the main concepts introduced in the chapter by giving their page numbers.
3. Doing the self-assessment will help you preview your own competency before reading the chapter.
4. The self-assessment allows you to measure your progress by comparing your competency after you finish the chapter with your assessment before you started. This will give you a tangible record of what you have learned.

Self-Assessment

Please assess your competency.

Task:	**How well can you do it?**					**Page**
Explain each of the following:	*Poorly*				*Very well*	
1. Abstract	1	2	3	4	5	8
2. Methods	1	2	3	4	5	9
3. Target group	1	2	3	4	5	9
4. Measuring instrument	1	2	3	4	5	9
5. Results	1	2	3	4	5	10
6. Discussion	1	2	3	4	5	11
7. Medline search	1	2	3	4	5	8
8. References	1	2	3	4	5	11

Research

In an attempt to make sense out of the world, children ask questions all the time. From the answers they receive, they learn a great deal about the world, about language, and about how adults acquire information. Adults, too, ask questions. We ask questions of ourselves and of other people we know. When we want accurate or detailed information we often question people who have special or expert knowledge of a subject. Often we cannot address our questions directly to the experts, so we seek answers by reviewing the literature in print on the subject. As we delve deeper into the subject, more questions occur to us. Questioning that seeks answers in a systematic and organized fashion is called **research**.

Research questions are often thrust upon us by changes in situations or conditions in the world around us. Currently there is much interest in the AIDS issue, because to date there is no known cure; to the best of our knowledge, the disease is always fatal. Furthermore, our experience with other sexually transmitted diseases has taught us that disease transmitted through sexual contact is extremely difficult to control.

Our first activity takes a look at knowledge, attitudes, and beliefs held by adolescents about AIDS. This sort of research is important, because currently the only weapon available to stem the spread of this dread disease is education. In order to educate effectively, it is helpful to know what information and misinformation the target group already has.

The Sections of a Research Paper

ACTIVITY 1.1

Read the following article about AIDS. We selected it from many that we located by conducting a literature review. As you read, note the sections of the research paper: Abstract, Introduction, Methods, Results, Discussion, and References.

LITERATURE EXAMPLE 1.1

Adolescents and AIDS: A survey of knowledge, attitudes and beliefs about AIDS in San Francisco

Ralph J. DiClemente, PhD, Jim Zorn, BA, and Lydia Temoshok, PhD

American Journal of Public Health, December 1986, vol. 76, no. 12

Abstract

To assess adolescents' knowledge, attitudes, and beliefs about AIDS in San Francisco, data were obtained from 1,326 adolescents. There was marked variability in knowledge across informational items, particularly about the precautionary measures to be taken during sexual intercourse which may reduce the risk of infection. We conclude that development and implementation of school health education programs on AIDS and other sexually transmitted diseases are needed in this population. (*Am J Public Health* 1986; 76:1443–1445.)

Introduction

Although Acquired Immune Deficiency Syndrome (AIDS) is rare among adolescents,[1] this should not be grounds for neglecting preventive health education in school systems. Epidemiologic data on the use of drugs and the spread of other sexually transmitted diseases (STD) among this population suggest that the rate of disease transmission may far exceed its reported rate,[2–6] and that STDs and adolescent drug use are the most pervasive, destructive, and costly health problems confronting adolescents today in the United States. Knowledge about high-risk behaviors associated with AIDS virus infection could help prevent the spread of disease in this population.

One published report suggests that the high school students did not possess a great deal of information about AIDS nor were many concerned about the threat of AIDS.[7] The findings, however, are based on a relatively small sample size (n = 250), and reflect a geographic area of the United States which has a low incidence of AIDS. The present study reports a survey of adolescents' knowledge, attitudes, and beliefs about AIDS in an AIDS epicenter (San Francisco).

Methods

Students enrolled in Family Life Education classes at 10 high schools in the San Francisco Unified School District (SFUSD) were eligible to participate in this project. The schools selected represent those with the largest enrollment in the SFUSD. Family Life Education classes are mandatory in the SFUSD.

In cooperation with the Division of Health Education Programs for the SFUSD, all Family Life Education teachers were personally contacted by the investigators. The aims of the project were explained and their cooperation was solicited.

This project involves administering a newly developed self-report questionnaire, consisting of 30 questions which evaluated students' knowledge about the cause, transmission, and treatment of AIDS. A second group of 11 questions tapped students' attitudes and beliefs regarding personal susceptibility, disease severity, and the need for AIDS instruction to be included in high school curricula. Students were requested to give "True," "False" or "Don't Know" responses to all questions. All questionnaires were completely anonymous; they were distributed and returned within a one-week period in may 1985. Of the 1,332 questionnaires

TABLE 1 Subject responses for each knowledge statement

Statement	True %	False %	Don't Know %
AIDS is a medical condition in which your body cannot fight off diseases.	73.9 (964)*	12.4 (161)	13.7 (179)
AIDS is caused by a virus.	60.3 (790)*	16.6 (218)	23.1 (303)
AIDS is a condition you are born with	6.9 (90)	84.4 (1105)*	8.8 (115)
Stress causes AIDS.	6.7 (87)	76.4 (998)*	17.0 (222)
If you kiss someone with AIDS you will get the disease.	41.9 (553)	41.0 (541)*	17.1 (225)
If you touch someone with AIDS you can get AIDS.	17.1 (226)	67.9 (896)*	14.9 (197)
All gay men have AIDS.	11.6 (152)	80.5 (1057)*	7.9 (104)
What you eat can give you AIDS.	9.6 (126)	75.0 (967)*	15.4 (203)
Anybody can get AIDS.	84.6 (1114)*	10.3 (136)	5.1 (67)
AIDS can be cured.	13.4 (175)	60.5 (790)*	26.1 (341)
Women are more likely to get AIDS during their period.	7.8 (102)	56.9 (747)*	35.4 (465)
AIDS can be spread by using someone's personal belongings like a comb or hairbrush.	15.7 (206)	66.3 (869)*	17.9 (235)
AIDS is not at all serious, it is like having a cold.	3.3 (43)	89.9 (1179)*	6.9 (90)
AIDS is caused by the same virus that causes VD.	20.0 (261)	41.4 (541)*	38.6 (504)
The cause of AIDS is unknown.	46.5 (608)	33.8 (442)*	19.7 (257)
Just being around someone with AIDS can give you the disease.	15.3 (200)	70.4 (926)*	14.0 (183)
Having sex with someone who has AIDS is one way of getting it.	92.4 (1213)*	3.9 (51)	3.7 (49)
If a pregnant woman has AIDS, there is a chance it may harm her unborn baby.	85.7 (1124)*	3.0 (39)	11.3 (148)
Most people who get AIDS usually die from the disease.	79.6 (1036)*	9.7 (126)	10.8 (140)
Using a condom during sex can lower the risk of getting AIDS.	60.0 (782)*	14.2 (185)	25.8 (336)
You can get AIDS by shaking hands with someone who has it.	10.0 (131)	74.7 (975)*	15.3 (200)
Receiving a blood transfusion with infected blood can give a person AIDS.	84.4 (1099)*	5.8 (75)	9.8 (128)
You can get AIDS by sharing a needle with a drug user who has the disease.	81.1 (1055)*	6.3 (82)	12.6 (164)
AIDS is a life-threatening disease.	83.8 (1087)*	5.9 (77)	10.3 (134)
People with AIDS usually have lots of other diseases as a result of AIDS.	36.6 (475)*	27.1 (351)	36.3 (471)
All gay women have AIDS.	8.3 (108)	73.4 (950)*	18.3 (237)
There is no cure for AIDS.	60.5 (783)*	16.9 (219)	22.6 (293)
I can avoid getting AIDS by exercising regularly.	6.8 (88)	77.4 (1003)*	15.8 (205)
AIDS can be cured if treated early.	30.1 (389)	36.8 (476)*	33.1 (428)
A new vaccine has recently been developed for the treatment of AIDS.	31.7 (411)	25.3 (328)*	42.9 (556)

NOTE: Actual frequencies in parentheses. An asterisk denotes correct responses.

TABLE 2 Subject responses for attitudes and beliefs about AIDS

Statement	True %	False %	Don't Know %
AIDS is not as big a problem as the media suggests.	8.2 (106)	76.5 (995)	15.4 (200)
I am afraid of getting AIDS.	78.7 (1003)	15.9 (202)	5.4 (69)
Living in the Bay Area increases my chances of getting AIDS.	41.8 (535)	41.1 (526)	17.0 (218)
I am not worried about getting AIDS.	26.3 (336)	66.3 (847)	7.36 (94)
I am not the kind of person who is likely to get AIDS.	61.5 (784)	20.0 (255)	18.5 (236)
I am less likely than most people to get AIDS.	52.5 (662)	20.7 (261)	26.9 (339)
I'd rather get any other disease than AIDS.	50.6 (641)	18.1 (229)	31.3 (397)
If a free blood test was available to see if you have the AIDS virus, would you take it?	51.5 (653)	25.3 (321)	23.2 (294)
I've heard enough about AIDS and I don't want to hear any more about it.	28.9 (367)	58.8 (745)	12.3 (156)
It is important that students learn about AIDS in Family Life Education classes.	87.6 (1119)	5.8 (74)	6.6 (84)
Have you had any instruction about AIDS in your school curriculum?	35.3 (447)	54.4 (689)	10.3 (131)

NOTE: Actual frequencies in parentheses.

distributed, a total of 1,326 (more than 99 per cent) usable questionnaires were returned and represent the sample upon which all analyses are based. Students ranged in age from 14–18 years. All major ethnic groups were represented.

RESULTS

The findings suggest that students possess some knowledge of AIDS—although this knowledge is uneven (see Table 1). With respect to disease transmission, 92 per cent of the students correctly indicated that "sexual intercourse was one mode of contracting AIDS," however, only 60 per cent were aware that "use of a condom during sexual intercourse may lower the risk of getting the disease." This large discrepancy suggests that many adolescents, while knowing a major route of disease transmission, nonetheless, will be engaging in unprotected sexual activity. Most adolescents were aware that receiving infected blood from a transfusion (84 per cent) or sharing intravenous drug needles (81 per cent) were also identified routes of disease transmission.

On the other hand, only 66 per cent of students surveyed were aware that AIDS could not be spread by using someone's personal belongings and only 68 per cent knew that engaging in casual contact (i.e., shaking hands) would not lead to contracting the disease. Moreover, less than half the students (41 per cent) correctly reported that kissing was not a route of AIDS transmission. Adolescents were less informed with respect to the treatment of AIDS. Surprisingly, only 25.3 per cent and 36.8 per cent, respectively, were aware that "no new vaccine was available for treating AIDS" nor "could AIDS be cured if treated early," and only 60.5 per cent reported that "AIDS could not be cured."

With respect to adolescents' attitudes and beliefs about AIDS (Table 2), 78.7 per cent of the adolescents report "being afraid of getting AIDS" and 73.7 per cent report being "worried about contracting the disease." Over half the adolescents surveyed (50.6 per cent) would rather contract "any other disease than AIDS." One attitude, which was most pervasive (87.6 per cent agreeing), was that it is important for students to receive AIDS instruction in the school curriculum.

DISCUSSION

Our findings are in stark contrast to those reported by Price, *et al.*,[7] suggesting that geographic proximity to a high-density AIDS epicenter has a great deal of saliency for what students know and the attitudes and beliefs they possess about AIDS. Furthermore, these findings indicate the need for a teaching module which would be included in the school curricula to overcome misconceptions about AIDS. Preferably AIDS education would not be provided in isolation, but as part of the teaching plan to instruct students about communicable diseases, in general. Equally important as understanding the cause and transmission of AIDS, as Yankauer points out, is providing education on the role that social values play in the control of sexually transmitted diseases.[8]

As the projected epidemic of AIDS cases and HTLV-III infection continue to escalate,[9] increased attention must be directed at the adolescent population, which has heretofore been neglected, if in the long run we are to curtail the spread of disease in this population.

REFERENCES

1. Centers for Disease Control: AIDS Weekly Surveillance Report, May 12, 1986.
2. Robins LN: The natural history of adolescent drug use. Am J Public Health 1984; 74:656–657.
3. National Institute of Allergies and Infectious Disease Study Group: Sexually Transmitted Diseases-Summary and Recommendations, 1980. US Department HEW, National Institutes of Health.
4. Kroger F, Wiesner PJ: STD Education: Challenge for the 80s. J Sch Health 1981; 51:242–246.
5. Shafer M, Beck A, Blain B, Dole P, Irwin C, Sweet R, Schacter J: Chlamydia trachomatis: Important relationships to race, contraception, lower genital tract infection, and Papanicolaou smear. J Pediatr 1984; 104:141–146.
6. Bell TA, Holmes KK: Age-specific risks of syphilis, gonorrhea, and hospitalized pelvic inflammatory disease in sexually experienced US women. Sex Transm Dis 1984; 11:291–295.
7. Price JH, Desmond S, Kukulka G: High school students' perceptions and misperceptions of AIDS. J Sch Health 1985; 55:107–109.
8. Yankauer A: The persistence of public health problems: SF, STD, and AIDS. (editorial) Am J Public Health 1986; 76:494–495.
9. Curran JW, Morgan WM, Hardy AM, Jaffe HW, Darrow WW, Dowdle WR: The epidemiology of AIDS: Current status and future prospects. Science 1985; 229:1352–1357.

Note what appears under each heading of the article. Then answer the following questions about the Abstract, Introduction, Methods, Results, and Discussion sections of the AIDS article.

Abstract

The **abstract** briefly and succinctly states the problem, the methods employed, and the results of the research project. It is usually the last part of the paper to be written. See "Reading Abstracts, Figures, and Tables" later in this chapter for more information.

The **problem statement** is usually given in one sentence and is often stated as a question. However, in this abstract and in the rest of the abstracts given in this chapter, the problem is not stated as a question.

1. Find the problem statement in the paper and copy it below.

   ____

   ____

2. The authors refer to two sources to substantiate the following claims:

   AIDS is rare among adolescents.

   High school students had little knowledge of or concern about AIDS.

   What sources do the authors use to support these claims?

   ____

   ____

ACTIVITY 1.2

This activity is highly recommended for people who are not experienced in using the library and/or computerized searches. Answers are not given because the information requested is specific to your particular college library.

1. Are the sources referred to in question 2 above available in your college library?

   ____

2. If so, where exactly will you find them? If not, can they be gotten from another library?

   ____

3. Consult with the reference librarian about running a computerized search of the *periodical literature*. An example is a **Medline** search.
4. Can your library conduct a computerized search for texts? If it can, look up a subject that interests you and record the following.
   Book #1:

   Title: ______________________________

   Author(s): ______________________________

Book #2:

Title: ______________________________

Author(s): ______________________________

Introduction

The **introduction** provides the context and the rationale for the research project. It should also clearly and succinctly state the purpose of the study and let the reader know what to expect from the paper.

Methods

The **methods** section of a research paper is written so that any other investigator wishing to repeat the study will be able to do so. In other words, it is a detailed account of how the research was conducted. If you wish to repeat a study, it is customary to write to the authors for details and copies of any instruments that they may have used. (An instrument is any device used to measure the variables of interest, for example, a survey or questionnaire.) It is essential that you do this if you would like to compare your results with those of other investigators.

1. The methods section has a description of the **target group** (the group that was chosen for the study) and the relevant variables. A **variable** is any characteristic in a sample or population that can change its **value** from case to case. Below are some examples of variables and possible values for each. (Note that a "value" is not necessarily a number.)

Variable	**Variable Values**
"AIDS is caused by a virus."	"True," "False," or "Don't Know"
Weight	135 pounds (for example)
Sex	Female, Male

Who was in the target group?

____

____

What was measured in this study? State two of the variables.

____

____

2. The methods section has a description of the instrument used to obtain the data. What was the instrument?

____

Was it new? ____________

What are the problems with using a new questionnaire?

____

____

3. How was the instrument administered?

____

____

4. Was the instrument tested? ____________
5. What were the oral or written instructions to the students?

____

____

6. Was the instrument revised? ____________
7. Would you be able to repeat this study using the methods described in this article? ____________
8. No scientific finding is considered an established fact until several researchers substantiate it. Why not?

____

____

Results

After the data have been analyzed using statistical techniques, the investigators present their findings in the Results section. In this section the selected characteristics of the population are described.

It is customary to present important findings in table form. Tables must be self-explanatory. In other words, an informed reader should be able to make sense out of a table without referring to the text.

Statistical methods fall into two broad categories: **Descriptive statistics** describes or summarizes data in the form of either tables, charts, and graphs or single-number statistics such as mean, median, and mode. **Inferential statistics** is used to generalize about entire populations, using data obtained from a sample from that population.

1. Did the authors use descriptive statistics (describing data) or inferential statistics (inferring about a population from the sample)? ____________
2. Examine the titles of the tables presented with the research article: Table 1, "Subject Responses for Each Knowledge Statement" and Table 2, "Subject Responses for Attitudes and Beliefs about AIDS." How do these two tables differ in what they describe?

____

____

3. Look at Table 1, "Subject Responses for Each Knowledge Statement." Note that each statement (variable) is assigned one of three values (can have one of three responses).

Table 1 has thirty variables. How many variables are there in Table 2? ____________

60.3 percent of the subjects thought that "AIDS is caused by a virus." What percent of subjects said they "Don't Know"? ____________

Which statement got the smallest percentage of "Don't Know" answers? ____________

The statement "AIDS is caused by a virus" is true. What percent of the subjects either answered incorrectly or did not know the answer? ____________

Discussion

The main component of the **discussion** portion of a research paper is the comparison of the authors' results with those of others as reported in the literature. This comparison must be based on the data presented in the paper. The authors must clearly indicate where their results differ and where they are substantiated by others' findings. The authors should clearly distinguish between opinion and fact, that is, between an impression, hunch, or informed guess and a conclusion arrived at through applied logic or knowledge from related fields. Be aware that authors sometimes base their conclusions on unstated assumptions.

1. When the authors mention a need for adding AIDS education to the school curriculum, what assumption are they making?

   ____

   ____

2. Does the assumption come from the data collected in this article?

   ____

3. If not, where did it come from?

   ____

   ____

4. What information (data) was collected?

   ____

   ____

5. Should the authors' recommendation have been made on the basis of the data that were collected? ____________

6. The data indicated a lack of knowledge. Do the data say anything about the effectiveness of including AIDS education in the curriculum?

   ____

The discussion portion of a paper also includes recommendations. Note how the authors substantiate their concern about the escalation of AIDS.

References

The information in the **references** section should help you, the reader, find additional information on the subject, should you wish to learn more.

Each journal establishes its own format for references. If you are interested in publishing a paper, you can obtain format instructions by writing to the journal you plan to submit your paper to.

Abstracts, computerized literature searches, computerized book searches by author, title, and subject, and computerized periodical searches are available at most educational institutions. It is important for researchers to learn how to use these resources. The reference librarians provide invaluable help, both to beginners in learning how to use the library's resources and to experienced researchers in tracking down references that are obscure or difficult to find.

Reading Abstracts, Figures, and Tables

Abstracts describe the following:

1. The problem under investigation
2. The research method
3. The population of interest
4. The specific sample size
5. The statistical analysis performed
6. The salient results
7. The implications of the research
8. The reference to the journal

Before reading an article, the reader can scan the abstract to decide if the article will be of interest. To get more information about the article's content, one can then scan the tables, figures, and graphs. Their titles and legends should be self-explanatory.

Abstracts provide the context for the interpretation of tables and figures, which contain the detailed information about the subject.

One of the purposes of this book is to illustrate how researchers use statistics to describe their findings and how they draw inferences from these findings. We accomplish this by giving abstracts and tables from published articles and having you interpret them. This chapter contains abstracts, tables, and figures from three such articles to get you started. The first abstract is taken from the article on AIDS that we have already presented.

LITERATURE EXAMPLE 1.2

Adolescents and AIDS: A survey of knowledge, attitudes and beliefs about AIDS in San Francisco

Ralph J. DiClemente, Ph.D., Jim Zorn, B.A., and Lydia Temoshok, Ph.D.

American Journal of Public Health, December 1986, Vol. 76, No. 12, pp. 1443–1445 [*Am J Public Health,1986; 76 (12):1443–1445*]

To assess adolescents' knowledge, attitudes, and beliefs about AIDS in San Francisco, data were obtained from 1,326 adolescents. There was marked variability in knowledge across informational items, particularly about the precautionary measures to be taken during sexual intercourse which may reduce the risk of infection. We conclude that development and implementation of school health education programs on AIDS and other sexually transmitted diseases are needed in this population. (*Am J Public Health* 1986; 76:1443–1445.)

Component of the abstract	**Content of the abstract**
1. Problem under investigation	To assess adolescents' knowledge, attitudes, and beliefs about AIDS in San Francisco
2. Research method	Implied to be a survey
3. Population of interest (target group)	All adolescents
4. Specific sample size	1,326 adolescents
5. Statistical analysis performed	This abstract does not indicate the analysis. See Table 1; note that percentages are used.
6. Salient results	Marked variability in knowledge across informational items
7. Implications of the research	Need for the development and implementation of school health education programs on AIDS
8. Reference to the journal	(*Am J Public Health* 1986; 76: 1443–1445)

1. Table 1 gives responses of adolescents to each of the knowledge statements. There are three possible responses to each statement. What are they? ____________ ____________ ____________

   Besides the percentage, actual numbers are given in the parentheses. Read the results for the following statements: "Having sex with someone

TABLE 1 Subject responses for each knowledge statement

Statement	True %	False %	Don't Know %
AIDS is a medical condition in which your body cannot fight off diseases.	73.9 (964)*	12.4 (161)	13.7 (179)
AIDS is caused by a virus.	60.3 (790)*	16.6 (218)	23.1 (303)
AIDS is a condition you are born with	6.9 (90)	84.4 (1105)*	8.8 (115)
Stress causes AIDS.	6.7 (87)	76.4 (998)*	17.0 (222)
If you kiss someone with AIDS you will get the disease.	41.9 (553)	41.0 (541)*	17.1 (225)
If you touch someone with AIDS you can get AIDS.	17.1 (226)	67.9 (896)*	14.9 (197)
All gay men have AIDS.	11.6 (152)	80.5 (1057)*	7.9 (104)
What you eat can give you AIDS.	9.6 (126)	75.0 (967)*	15.4 (203)
Anybody can get AIDS.	84.6 (1114)*	10.3 (136)	5.1 (67)
AIDS can be cured.	13.4 (175)	60.5 (790)*	26.1 (341)
Women are more likely to get AIDS during their period.	7.8 (102)	56.9 (747)*	35.4 (465)
AIDS can be spread by using someone's personal belongings like a comb or hairbrush.	15.7 (206)	66.3 (869)*	17.9 (235)
AIDS is not at all serious, it is like having a cold.	3.3 (43)	89.9 (1179)*	6.9 (90)
AIDS is caused by the same virus that causes VD.	20.0 (261)	41.4 (541)*	38.6 (504)
The cause of AIDS is unknown.	46.5 (608)	33.8 (442)*	19.7 (257)
Just being around someone with AIDS can give you the disease.	15.3 (200)	70.4 (926)*	14.0 (183)
Having sex with someone who has AIDS is one way of getting it.	92.4 (1213)*	3.9 (51)	3.7 (49)
If a pregnant woman has AIDS, there is a chance it may harm her unborn baby.	85.7 (1124)*	3.0 (39)	11.3 (148)
Most people who get AIDS usually die from the disease.	79.6 (1036)*	9.7 (126)	10.8 (140)
Using a condom during sex can lower the risk of getting AIDS.	60.0 (782)*	14.2 (185)	25.8 (336)
You can get AIDS by shaking hands with someone who has it.	10.0 (131)	74.7 (975)*	15.3 (200)
Receiving a blood transfusion with infected blood can give a person AIDS.	84.4 (1099)*	5.8 (75)	9.8 (126)
You can get AIDS by sharing a needle with a drug user who has the disease.	81.1 (1055)*	6.3 (82)	12.6 (164)
AIDS is a life-threatening disease.	83.8 (1087)*	5.9 (77)	10.3 (134)
People with AIDS usually have lots of other diseases as a result of AIDS.	36.6 (475)*	27.1 (351)	36.3 (471)
All gay women have AIDS.	8.3 (108)	73.4 (950)*	18.3 (237)
There is no cure for AIDS.	60.5 (783)*	16.9 (219)	22.6 (293)
I can avoid getting AIDS by exercising regularly.	6.8 (88)	77.4 (1003)*	15.8 (205)
AIDS can be cured if treated early.	30.1 (389)	36.8 (476)*	33.1 (428)
A new vaccine has recently been developed for the treatment of AIDS.	31.7 (411)	25.3 (328)*	42.9 (556)

who has AIDS is one way of getting it." "Using a condom during sex can lower the risk of getting AIDS."

2. What practical implications do these results have for the prevention of AIDS?

Literature Example 1.3 is from an article entitled "Formaldehyde-related Health Complaints of Residents Living in Mobile and Conventional Homes" by Ingrid M. Ritchie, PhD and Robert G. Lehnen, PhD.

LITERATURE EXAMPLE 1.3

This paper explores the dose-response relation between formaldehyde (HCHO) concentration and reported health complaints (eye irritation, nose/throat irritation, headaches and skin rash) of nearly 2,000 residents living in 397 mobile and 494 conventional homes. The study analyzes the effects of HCHO concentrations, age and sex of respondent, and smoking behavior on each of the four health effects. The results demonstrate a positive dose-response relation between HCHO concentration and reported health complaints, with reported health complaints demonstrated at HCHO concentrations of 0.1 ppm and above. Concentrations of 0.4 ppm in manufactured homes as targeted by the Department of Housing and Urban Development (HUD), may not be adequate to protect occupants from discomfort and from acute effects of HCHO exposure. (*Am J Public Health* 1987; 77:323–328.)

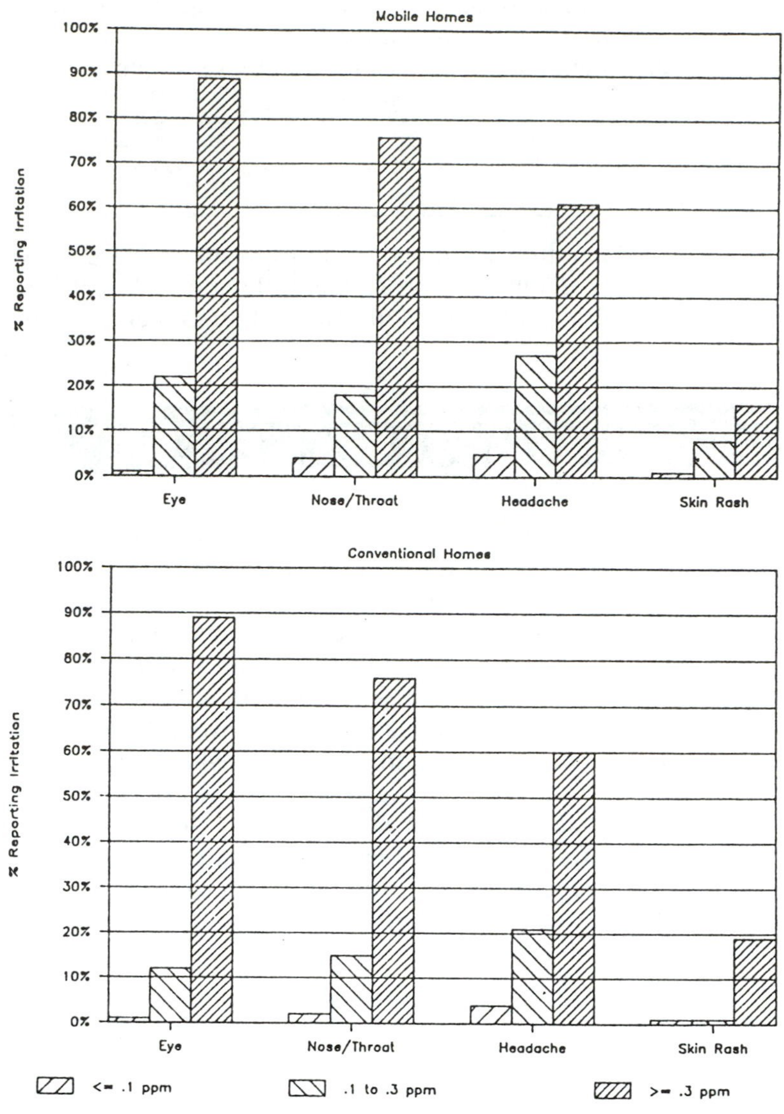

FIGURE 1 Health Effects by HCHO in Mobile Homes (top) and Conventional Homes (bottom).

1. Look at the abstract and copy the journal reference here:

____

____

2. What is the volume number? ______________

3. What are the page numbers? ______________

Components of the abstract	**Fill in information** from abstract here. (You may wish to enter the line number to save time.)
1. Problem under investigation	1 ______________________
2. Research method	2 ______________________
3. Population of interest (target group)	3 ______________________
4. Specific sample size	4 ______________________
5. Statistical analysis performed	5 ______________________
6. Salient results	6 ______________________
7. Implications of the research	7 ______________________
8. Reference to the journal	8 ______________________

The graphs in Figure 1 show some of the results of the investigation. Refer to the graphs and answer the following questions:

1. Read the title. What does Figure 1 portray?

   ____

   ____

2. Read the labels under the horizontal axis. What health effects did the authors study?

   ____

3. Read the legend given with the lower graph. What were the concentrations of HCHO?

   ____

4. Look for the tallest bars. Which HCHO concentration was associated with the greatest irritation?

   ____

5. From which health effects did the subjects suffer most?

   ____

Literature Example 1.4 is from an article entitled "Progress toward Meeting the 1990 Nutrition Objectives for the Nation: Nutrition Services and Data Collection in State/Territorial Health Agencies" by Mildred Kaufman, RD, MS; Jerianne Heimendinger, DSc, MPH; Susan Foerster, RD, MPH; and Mary Ann Carroll, RD, MPH.

LITERATURE EXAMPLE 1.4

Promoting Health, Preventing Disease, Objectives for the Nation specifies nutrition as a priority area for improving the health of Americans by 1990. Eleven of the 15 nutrition objectives target adults or the public, while four address pregnant and lactating women, infants, and children. To determine progress by states toward achieving the nutrition objectives, the Association of State and Territorial Public Health Nutrition Directors conducted a survey in 1955 to which 54 state and territorial nutrition directors responded. Three-fourths of the nutrition personnel focus efforts on maternal and child populations and were supported largely by federal funds for the special Supplemental Food Program for Women, Infants, and Children (WIC) and the Maternal and Child Health Block Grant. Only 15 percent were funded by state and local sources. While most agencies reported using nutrition consultation in adult health programs, only 25 percent paid for these personnel. Data to monitor progress were commonly available for only five of the 15 objectives. Achievement of the nutrition objectives by states will require a more comprehensive approach to nutrition programming with increased allocation of appropriate resources and expansion of health data systems. (*Am J Public Health* 1987; 77:299–303.)

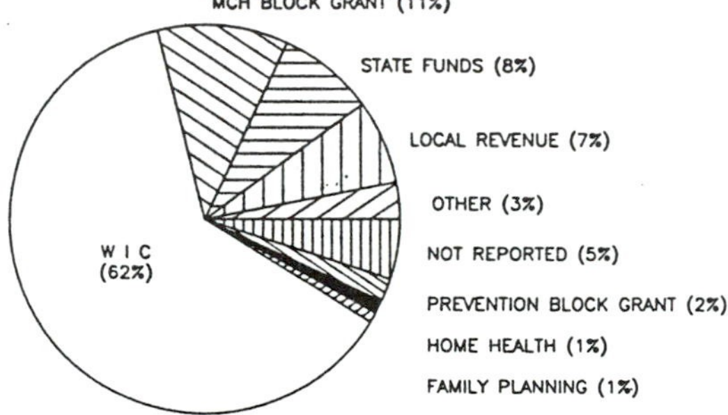

FIGURE 1 Funding Sources for Public Health Nutrition Positions in State and Local Official Health Agencies in the United States, Spring 1985

Look at the abstract and copy the journal reference here:

____

____

Components of the abstract	**Fill in information** from abstract here. (You may wish to enter the line number to save time.)
1. Problem under investigation	1 ______________
2. Research method	2 ______________
3. Population of interest (target group)	3 ______________
4. Specific sample size	4 ______________
5. Statistical analysis performed	5 ______________
6. Salient results	6 ______________
7. Implications of the research	7 ______________
8. Reference to the journal	8 ______________

Figure 1 shows some of the results of the investigation. Refer to the graph and answer the following questions.

1. Read the title. What does this figure portray?

   ____

2. Which funding source offers the most public health nutrition positions?

   ____

3. What percent of positions are funded this way? _______
4. What percent of public health nutrition positions are funded by state and local sources? _______
5. What observation does the abstract make concerning the data in questions 3 and 4?

   ____

   ____

Self-Assessment

Now that you have studied the chapter, please assess your competency again.

Task:	How well can you do it?					Page
Explain each of the following:	*Poorly*				*Very well*	
1. Abstract	1	2	3	4	5	8
2. Methods	1	2	3	4	5	9
3. Target group	1	2	3	4	5	9
4. Measuring instrument	1	2	3	4	5	10
5. Results	1	2	3	4	5	10
6. Discussion	1	2	3	4	5	11
7. Medline search	1	2	3	4	5	8
8. References	1	2	3	4	5	11

Self-assessment is subjective. This assessment device is used to alert you to the main concepts you need to master. If you are not sure you can explain a concept, you should review the section of the text in which it is discussed.

Answers for Chapter 1

Answer for question 1, p. 8: The first sentence in the abstract states: "To assess adolescents' knowledge, attitudes, and beliefs about AIDS in San Francisco . . . ." The last sentence in the abstract implies, however, that the authors' purpose may have been to develop a mechanism to change these attitudes.

Answers for question 2, p. 8: Reference 1; Reference 7

Answers for questions 1–8, pp. 9–10: 1) Adolescents; adolescents' knowledge, attitudes, and beliefs about AIDS; any statement from Table 1 or Table 2 is a variable. 2) A questionnaire; yes; results cannot be compared with those from other studies. 3) Students were requested to complete the questionnaire anonymously. 4) The article gives no information. 5) Students were asked to give "True," "False," or "Don't know" responses to all questions. 6) The article gives no information. 7) We think there is not enough information given in the article to repeat the study. A researcher who wanted to repeat the study would have to contact the authors for more details. 8) There are many reasons why a scientific finding must be substantiated by several researchers. Among them are the facts that researchers make errors and they can be led to observe what they want to observe; samples can be atypical; unknown factors can produce irreproducible results; or a researcher may have been dishonest.

Answers to questions 1–3, pp. 10–11: 1) Descriptive statistics. 2) Table 1 is a tabulation of responses to knowledge questions and Table 2 is dealing with attitudes and beliefs. 3) 11; 23.1; "Having sex with someone . . ."; 39.7

Answers to questions 1–6, p. 11: 1) Assumption: Adding AIDS to the curriculum helps adolescents learn the essential concepts and it will lead to appropriate preventive behaviors. 2) No. However, one item in Table 2 indicates that 87.6 percent of the respondents believed "It is important that students learn about AIDS in Family Life Education Classes." The data indicate that many do not have the knowledge. 3) Perhaps the conclusion that adding AIDS education to the curriculum comes from other studies not cited here. 4) Information was collected on students' knowledge about the cause, transmission, and treatment of AIDS. Also surveyed were students' attitudes and beliefs regarding personal susceptibility, disease severity, and the need for AIDS instruction in high schools. 5) No. Strictly speaking, one needs to teach about AIDS and evaluate the results before such a recommendation can be made. 6) No. See the answer to question 5.

Answers for Literature Example 1.2, pp. 13 and 15: 1) True; False; Don't Know 2) "The findings suggest that students possess some knowledge of AIDS—although this knowledge is uneven." Therefore more effective educational activities are needed.

Answers for Literature Example 1.3, p. 16: 1) *Am J Public Health* 1987; 77:323–328 2) 77 3) 323–328

Answers for Components of the Abstract, p. 17: 1) Lines 1–5 2) Survey is implied. 3) Persons who reside in mobile and conventional homes 4) Lines 3–4 5) Lines 7–10; but the statistical tool used is not specified. 6) Lines 7–12 7) Lines 9–12 8) Lines 12–13

Answers for questions with Figure 1, p. 17: 1) Health effects by HCHO in mobile homes and conventional homes 2) Irritation of eye, nose/throat, headache, and skin rash 3) $\leq$ .1 ppm, .1 to .3 ppm and $\geq$ .3 ppm 4) $\geq$ .3 ppm 5) Eye irritation

Answer for Literature Example 1.4, p. 19: *Am J Public Health* 1987; 77: 299–303

Answers for Components of the Abstract, p. 19: 1) Lines 5–8 2) Lines 7–8, survey 3) Line 3: indirectly, adults or the public; Lines 6–8: specifically, nutrition directors 4) 54 nutrition directors 5) Calculation of percents 6) Lines 10–17 7) Lines 16–19 8) Lines 18–19

Answers for Figure 1, p. 19: 1) Funding sources for public health nutrition positions 2) WIC 3) 62 percent 4) 8 percent + 7 percent = 15 percent 5) Lines 12–14

CHAPTER

2

Generation of Data

Prelude

"What do you want our names for?" asked Milo, looking anxiously over his shoulder. "We're in a bit of a hurry."

"Oh, this won't take a minute," the man assured them. "I'm the official Senses Taker, and I must have some information before I can take your senses. Now, if you'll just tell me when you were born, where you were born, why you were born, how old you are now, how old you were then, how old you'll be in a little while, your mother's name, your father's name, your aunt's name, your uncle's name, your cousin's name, where you live, how long you've lived there, the schools you've attended, the schools you haven't attended, your hobbies, your telephone number, your shoe size, shirt size, collar size, hat size, and the names and addresses of six people who can verify all this information, we'll get started. . . ."

— Norton Juster
The Phantom Toll Booth, p. 226

Chapter Outline

Self-Assessment

Please assess your competency.

Task:	How well can you do it?					Page
Explain each of the following:	*Poorly*				*Very well*	
1. Variable	1	2	3	4	5	23
2. Case	1	2	3	4	5	28
3. Structured data generation	1	2	3	4	5	24
4. Unstructured data generation	1	2	3	4	5	24
5. Coding a questionnaire	1	2	3	4	5	28
6. Decoding the numbers into phrases	1	2	3	4	5	28
7. Designing a recording sheet for data	1	2	3	4	5	29

Introduction

Statistics helps us consolidate and bring order into the bewildering inputs (data) we experience (gather) unwittingly or systematically. It also helps us make decisions.

A **variable** is any characteristic of a person or thing that may take on different values for different people or things. We collect the values of different variables in order to describe a person or a group. These values form the data set. Each individual whose characteristics are recorded is called a **case** or a **subject**. For example, people who respond to a questionnaire are cases or subjects. Characteristics that we measure and record, such as height and weight, are variables. The values of the variables make up the data set that furnishes us with the characteristics of that group.

We may collect data three ways: 1) from existing records (for example, census data or hospital medical records); 2) by observing and recording the measurements of characteristics (for exampie, by administering a questionnaire or actually weighing people; 3) by conducting experiments and recording the results. We will now focus on data collection using a questionnaire.

Data can be gathered in an **unstructured** or in a **structured** way. The unstructured way allows the respondent to articulate unique responses to specific questions, in contrast to choosing from given alternatives. The structured format establishes the possible values of each variable by specifying alternatives from which the respondent chooses.

Ways of Collecting Data Using a Questionnaire

ACTIVITY 2.1

Generation of unstructured data

By doing the following exercise, you will gain insights into your own attitudes toward statistics and recognize the value and limitations of unstructured data generation.

1. Take out a blank sheet of paper and set it on the desk in front of you.
2. Close your eyes and draw a representational diagram of your feelings about the statistics class. (2 minutes)
3. Now look at your drawing and write five words that describe your feelings about statistics. (1 minute)

____

____

ACTIVITY 2.2

Generation of structured data

By doing the following exercise, you will recognize the value and limitations of structured data generation.

Respond to each question by placing the number of your answer in the space provided: (1) Agree, (2) Disagree, (3) No comment. (1 minute)

I think learning statistics will be an enjoyable experience. _______

I fear statistics class. _______

Statistics is a necessary class. _______

ACTIVITY 2.3

Dyad interaction: Taking timed turns

In this kind of exercise, you choose a partner and divide the allocated time evenly between partners. During the first half of the time, one person speaks while the partner listens without comment. During the second half of the time, you reverse roles.

Use your time as speaker to talk about the concepts you have learned. It is also often useful to articulate feelings that you have about the particular material or about learning it.

Avoid turning the dyad interaction into a conversation. When it's your turn to listen, you should listen with the fullest attention that you can muster; try not to interrupt your partner by giving your interpretation, even if you think that what your partner is saying is wrong.

Rationale for the Dyad Interaction

Intention shapes our attention. Your intention during the dyad interaction should be to summarize the new concepts you have learned in your own words. Because you know that you will be listened to, you will automatically pay better attention to the process of summarizing the concepts. Summarizing serves the purpose of review, and review is known to facilitate understanding and move concepts from short-term to long-term memory.

Next you will listen to another version of the concept from your partner. During the dyad interaction you will experience the same concept three or four times from different view points.

The dyad interaction may also help you formulate questions about the material. If you are doing this exercise in class, the instructor will allow time immediately after the interaction for students to raise any questions that they have. If you are doing the interaction as a study aid outside of class (and we urge you to do this activity frequently with your study partners), write down questions that arise so that you will remember to ask them at the next class meeting.

Another thing that you can do during the dyad interaction is to articulate any negative feelings you may have about the material or about studying or learning the material. This is helpful because negative feelings (such as fear, anger, annoyance, irritation) interfere with the learning process. Acknowledging that you have these feelings is an important step toward eliminating them and thereby removing their detrimental effect on your ability to learn the material.

If your instructor does not assign the following as an exercise to be done in class, you should do it on your own. Choose a dyad partner and share with your partner your thoughts about the following questions. Take timed turns. (Suggested time: 2 minutes each)

1. What is the value of unstructured data generation?
2. What are the limitations of unstructured data generation?
3. What is the value of structured data generation?
4. What are the limitations of structured data generation?

ACTIVITY 2.4

Filling out a questionnaire

A questionnaire can be used as an instrument to generate data about an individual or a group of people. A questionnaire is presented here to generate data from your class. After the data have been generated, values can be consolidated into tables and graphs. Each variable in the questionnaire has a data-generation method implicit in it. For example, when we ask for your weight, we assume you know your weight because you weighed yourself at some time in the past. The instrument of measurement you used when you originally obtained your weight was a scale, and now the instrument for collecting the data is a questionnaire. The variable is weight. Its value is the number you give. When all the completed questionnaires are collected and data are consolidated, you will have some idea about the values of the variables for your class.

We designed this questionnaire (also referred to as the instrument of measurement) to serve the following purposes:

1. to illustrate an instrument of structured (see questions 1–14) and unstructured (see question 15) data collection;
2. to illustrate an experiment producing before and after data (see questions 6 and 7);
3. to illustrate two kinds of variables: categorical variables (those that help us place different cases into categories such as male and female) and measured variables (height, weight, and pulse). We will elaborate on different kinds of variables in Chapter 3.
4. to yield observations that lend themselves to different kinds of statistical analyses.

Data obtained with a questionnaire that does not involve any experimental intervention are said to be obtained through observation, in contrast to data obtained through an experiment.

STUDENT QUESTIONNAIRE

Please complete the following questionnaire by filling in the blank or circling the most appropriate answer.

1. Sex: (1) Female (2) Male

2. Age: _______

3. Class status:
   (1) Freshman (2) Sophomore (3) Junior (4) Senior (5) Graduate

4. Height in inches: _______

5. Weight in pounds: _______

6. Pulse (take your pulse for 15 seconds, multiply by 4, and record the number): _______

7. Run in place for one minute and then take your pulse again.

   Pulse after running in place for one minute: _______

8. Smoking cigarettes:
   (1) I never smoked. (2) I quit smoking. (3) I smoke. (4) Other

9. Daily physical activity level:
   (1) Minimal (2) Moderate (3) Vigorous

10. Daily stress level:
    (1) Minimal (2) Moderate (3) Heavy

11. I have at least three friends or relatives with whom I can discuss the positive and negative events of my day.
    (1) True (2) False

12. I take at least one five-minute "time out" per day to be quiet.
    (1) True (2) False

13. I would rate my overall health as:
    (1) Excellent (2) Good (3) Average (4) Below average

14. I generally keep the thermostat at my residence set at _______ °F.

15. What is your opinion about national health insurance?

    ____

    ____

Sources of Data When You Fill Out a Questionnaire

To complete the above questionnaire you needed to obtain information from various sources. Your height may have been measured some time ago, perhaps at a physical examination; maybe your weight was taken in a doctor's office as well. You measured and recorded PULSE1 yourself; you measured and recorded PULSE2 after running in place for one minute. Your opinion about national health insurance may be a result of your experience with the health delivery system.

Data Obtained Through Observation

Coding and Decoding

Notice that some answers in the questionnaire you just completed have numbers preceding them. These are called **codes**, and the process of creating numbered categories in a questionnaire is called **coding**. Notice that answers to the fifteenth question have not been coded. This is because it is an unstructured question, and you are to respond in your own words. Refer to the questionnaire and list the codes for the following variables:

1. Daily stress level _______
2. Health status _______
3. Daily "time out" _______

Each code corresponds to a particular word or phrase. Coding consists of assigning numbers to words or phrases, whereas **decoding** assigns words to the numbers. Some statistical computer programs allow the user to input the actual names of the categories into the data base. If you use that kind of software, assigning numbers to the categories (coding) may not be necessary.

4. Refer to question 8 in the questionnaire and decode the numbers of the answers.

Q8 Code	Decoded Phrase
1	______________________
2	______________________
3	______________________
4	______________________

Recording Data

One usually records data by transcribing them into rows and columns. Each **row** stands for a case, or record (the data from one person), and each **column** represents a variable. A questionnaire data recording form, or coding sheet, may be structured as follows.

Question number		1	2	3	4	5	6	7	8	9	10	11	12	13	14
Abbreviated names of variables	case	SEX	AGE	CLA	HT	WT	PUL1	PUL2	SM	ACT	STR	DIS	TOU	HLT	TEM

Each row in the recording sheet lists measurements from a different case (i.e., each row is the set of all answers from a particular person). Please record your data in the row of blanks. By entering your data, you have created one case. Each class member will record his or her data similarly. Collectively, the data from all class members will make up a data set.

Each time we transcribe data it is possible to make mistakes. If you are using a computer you can minimize this source of error by inputting the data directly from the questionnaire to the computer, rather than transcribing it first by hand to a table or a chart.

Data Obtained by Performing an Experiment

Measuring Variables

Listed below are the variables used in the experiment described in Literature Example 2.3 and a note on how they have been measured. A researcher would give details of how the variables are measured in the methods section of a research paper. The following example will be examined again in more detail in Activity 2.7 (p. 35).

Variable	**How the variable was measured**
Initial heart rate (PREHR)	Number of ventricular beats per minute. Initial heart rate was recorded using a 3-channel (12-lead) electrocardiograph.
Maximum heart rate (MAXHR)	Maximum heart rate. The subjects pedaled a stationary cycle to their maximum capacity and then the heart rate was recorded by the same method as above.
Volume of oxygen consumed (VO2)	Volume of oxygen consumed was determined by inserting a tube in the mouth of the participant (whose nose was plugged) and linking the tube to a computer. VO2 was measured while the participant bicycled.
Body fat percent (FAT%)	Skin folds from different parts of the body were measured and the percent fat was computed using a formula.

Resting systolic blood pressure (RESSYS)	The resting systolic blood pressure was measured with a sphygmomanometer after a rest period which came before each exercise workout.

The same variables were measured for each individual *before* and *after* the experimental period.

Other instruments were also used to obtain data in this exercise study. These data were also recorded on a coding sheet.

(Fill in the blanks.)

1. ____________ are recorded in the rows.
2. ____________ are recorded in the columns.

Recording Variables

You can maintain a laboratory book to record measurements as you obtain data. The exact structure of the records is simply a question of convenience. The following is an example of the record of one subject:

Subject's name: ______________________________

Variable name	Measurement	Measurement after the experiment

Examples from the Literature

Activity 2.5

Collection of data using a questionnaire

Read the following abstract, "Adolescent Smoking, Weight Changes, and Binge-Purge Behavior: Associations with Secondary Amenorrhea" and study the table that accompanies this research article.

1. Name the variables measured in this study.

____

____

2. Write one structured question that would elicit one piece of data in the table.

____

____

LITERATURE EXAMPLE 2.1

Adolescent Smoking, Weight Changes, and Binge-Purge Behavior: Associations with Secondary Amenorrhea

Jim Johnson, PhD, and Agnes H. Whitaker, MD

Background. The association of secondary amenorrhea with extreme forms of substance use, weight control, and exercise in nonrepresentative samples raises questions as to whether adolescents in the general population who engage in these behaviors are at increased risk for secondary amenorrhea. We examined the prevalence and behavioral correlates of secondary amenorrhea in a county-wide high school population of 2544 girls aged 13 to 18.

Methods. A survey questionnaire, which elicited menstrual history as well as weight history, weight control practices, level of exercise, and use of cigarettes, wine, and beer, was administered during school hours; absentees were also surveyed. The completion rate was 91%.

Results. The 1-year prevalence of secondary amenorrhea was 8.5%. Secondary amenorrhea was associated with smoking one or more packs of cigarettes per day (adjusted relative risk [RRa] = 1.96, 1.21–3.10), with multiple binge-eating behaviors in combination with laxative use or self-induced vomiting (RRa = 4.17, 2.54–6.32), and with weight fluctuation due to weight control (RRa = 2.59, 1.33–4.79). There was no association between amenorrhea and alcohol consumption or exercise level.

Conclusions. Estimates of attributable risk are provided and indicate that bulimic behaviors and cigarette smoking may result in a considerable excess of cases of secondary amenorrhea in an adolescent population. (*Am J Public Health.* 1992;82:47–54)

TABLE 3—Rate/100 of Adolescent Females Who Reported Missing Three Consecutive Menstrual Periods in the Past Year, by Background Characteristics (n = 2156)

	Rate/100 Secondary Amenorrhea
Socioeconomic status (SES) quintiles	
I (low SES)	10.4
II	10.2
III	7.5
IV	8.5
V (high SES)	8.5
Body mass index (BMI) quartile	
I (thin girls)	8.9
II	9.3
III	8.1
IV (heavy girls)	7.8
Chronological age	
13	10.8
14	7.8
15	9.4
16	7.9
17	8.1
18	9.8
Gynecologic age	
0	12.5
1	13.5
2	9.7
3	9.6
4	6.1
5	8.5
6	5.3
7	5.4

ACTIVITY 2.6

Collection of data from an existing data bank

At some institutions, data are collected routinely and stored in a data bank. Researchers collect data from the data bank. They begin with a predetermined protocol (which may consist of a set of questions or variable names) and copy the required data. Some institutions store data in a computer. A researcher can selectively copy the needed data into a file and analyze them. Manual entry of data may not be required.

Read Literature Example 2.2, "The Effect of Legal Drinking Age on Fatal Injuries of Adolescents and Young Adults" and study the table that accompanies it.

1. Name the three variables selected from the data bank.

   ____

   ____

2. Name the source for the mortality data (see the Methods section of the Abstract).

   ____

   ____

3. What is the source for legal drinking age data?

   ____

   ____

LITERATURE EXAMPLE 2.2

The Effect of Legal Drinking Age on Fatal Injuries of Adolescents and Young Adults

Nancy E. Jones, DrPH, CPNP, Carl F. Pieper, DrPH, and Leon S. Robertson, PhD

This study examined the effect of legal drinking age (LDA) on fatal injuries in persons aged 15 to 24 years in the United States between 1979 and 1984. Effects on pre-LDA teens, adolescents targeted by LDA, initiation at LDA, and post-LDA drinking experience were assessed. A higher LDA was also associated with reduced death rates for motor vehicle drivers, pedestrians, unintentional injuries excluding motor vehicle injuries, and suicide. An initiation effect on homicides was identified. Reductions in injury deaths related to drinking experience were not found. In general, a higher LDA reduced deaths among adolescents and young adults for various categories of violent death. (*Am J Public Health.* 1992;82:112–115)

TABLE 1—Legal Drinking Age (LDA) Changes, United States, 1979–1984[a]

State	LDA	Date of Change	State	LDA	Date of Change
Ala	19		Mont	19	
Alaska[b]	19		Neb	19–20	7/19/80
Ariz[c]	19		Nev	21	
Ark	21		NH[i]	20	
Calif	21		NJ[j]	18–19	1/02/80
Colo[d]	21			19–21	1/01/83
Conn	18–19	7/01/82	NM	21	
	19–20	10/01/83	NY	18–19	12/04/82
Del[e]	20–21	1/01/84	NC[k]	18–19	10/01/83
DC	18		ND	21	
Fla	18–19	10/01/80	Ohio[d]	21	
Ga	18–19	9/01/80	Okla[d]	21	
Hawaii	18		Ore	21	
Idaho	19		Pa	21	
Ill[f]	19–21	1/01/80	RI	18–19	7/01/80
Ind	21			19–20	7/01/81
Iowa	19		SC[l]	18–19	1/01/84
Kan[d]	21		SD[d]	21	
Ky	21		Tenn[m]	19	
La	18		Tex	18–19	9/01/81
Me	20		Utah	21	
Md[g]	18–21	7/01/82	Vt	18	
Mass[h]	20		Va[n]	18–19	7/01/83
Mich	21		Wash	21	
Minn	19		WVa	18–19	7/01/83
Miss	21		Wis	18	
Mo	21		Wyo	19	

[a]Laws apply to all alcoholic beverages except where noted.
[b]10/26/83 from 19 to 21; grandfathered persons born on or before 12/31/64.
[c]12/31/84 from 19 to 21; grandfathered persons 19 before 12/31/84.
[d]18 (3.2% beer).
[e]Grandfathered persons 20 on 1/1/84.
[f]From 19 for beer and wine to 21 for all alcohol.
[g]From 18 for beer and light wine to 21 for all alcohol.
[h]4/16/79 from 18 to 20.
[i]5/24/79 from 18 to 20.
[j]1/2/80 from 18 to 19; grandfathered persons 18 before 1/2/80. 1/1/83 from 19 to 21; grandfathered persons 19 or 20 by 12/31/82.
[k]10/1/83 beer and unfortified wine law raised from 18 to 19.
[l]1/1/84 beer and wine law from 18 to 19.
[m]8/1/84 from 19 to 21; grandfathered persons born before 8/1/65.
[n]7/1/83 from 18 for "on-sale" and 19 for "off-sale" to 19 for all beer.

Methods

All 50 states and the District of Columbia were studied for the years 1979 through 1984. In each state and each year, LDA and injury death rates for persons 15 through 24 years of age were determined. Information on LDA was obtained from the Insurance Institute for Highway Safety (Table 1). Mortality data were collected from the National Center for Health Statistics. Injury deaths (E800–999) were identified, and six categories of death defined on the basis of the *International Classification of Diseases* (9th version; ICD-9) E-code were examined: motor vehicle driver, motorcyclist, pedestrian, unintentional injury excluding motor vehicle, suicide, and homicide (Table 2). The 1980 census provided population data to compute death rates. Census figures were adjusted by extrapolation to arrive at denominator data for the years 1979 and 1981 through 1984 (e.g., persons who were 18 in 1980 were considered 19 in 1981).

For each of the six categories of injury, the data were aggregated into death rates per 100 000 population by calendar year, age, and LDA, thus forming a matrix of 240 rates (6 calendar years × 10 single years of age × 4 LDAs). Calendar year was added to control for temporal trends in injuries. Logistic regression was used for analysis. The following model was fitted to the data:

$$\text{Rate} = a + b_1(\text{CALYR}) + b_2(\text{AGE}) + b_3(\textit{SPILLOVER}) + b_4(\textit{LEGAL}) + b_5(\textit{INITIATION}) + b_6(\textit{EXPERIENCE})$$

where Rate = fatal injury rate, CALYR = calendar year, AGE = age of the fatally injured person, SPILLOVER = LDA − age, if age < LDA (else 0), LEGAL = 1 if drinking legal at age of death or else 0, INITIATION = 1 if LDA = age or else 0, and EXPERIENCE = age − LDA + 1, if age > LDA (else 0). For the variables calendar year and age, a system of indicator variables was constructed to allow for expected nonlinearities in the death rate over time and age.

ACTIVITY 2.7

Collection of data from an experiment

Read the following abstract. We will be using these data for analysis throughout the book. An explanation of each variable is given later in this chapter.

LITERATURE EXAMPLE 2.3

Effects of work equivalent exercise of two different intensities in middle-aged males

D. P. Brown, S. F. Loy, S. J. Hall, M. Heng, G. J. Holland, and S. R. Shaw

Graduate project submitted in partial satisfaction of the requirements for the degree of Master of Arts in Kinesiology and Physical Education, California State University, Northridge, CA 91330

The purpose of this study was to evaluate the effects of two exercise intensities, 65% maximum heart rate (MHR) (LO) vs 85% MHR (HI), 4 days per week for 10 weeks, on aerobic capacity and blood lipid profiles of middle-aged (48.2 ± 1.5 yr) (mean ± SE) sedentary males. Subjects ($n = 23$) were randomly assigned to HI, LO, or non-exercising control. The LO group exercise duration was extended beyond (55.5 + 3.8 minutes by week 10) the HI group's 30 minutes, resulting in work equivalent training (watts/week). Pre- and post-training evaluations included measurement of aerobic capacity, submaximal rate pressure product (RPP), body composition, blood lipids, and dietary assessment. Peak VO2 increased (liters/min^{-1}) ($P < 0.05$) for both HI (2.70 + 0.16 to 3.37 + 0.21) and LO (2.54 + 0.23 to 2.89 + 0.28) with post-training HI different from LO ($P < 0.05$). RPP was lower ($P < 0.05$) in both exercise groups with mean drops of 4,238 and 4,364 units. Control group showed no change in peak VO2 or RPP. All groups showed no change in blood lipid values and body composition. It was concluded that when total work is kept constant, middle-aged males may exercise at a lower intensity and achieve health benefits (RPP changes) similar to high-intensity exercise, although the gains in fitness (peak VO2) may not be substantial.

Name two variables investigated by the authors:

____

Data Used in This Book

Two sets of data are referred to throughout this book: Faculty exercise data (FED) and student questionnaire data (SQD). Lists of the variables in each data set follow.

Faculty Exercise Data Set: List of Variables

(See Activity 2.7 for a description of the experiment.) There are thirty variables in the data set. They are given abbreviated names and can be input into any statistical program as thirty columns, each representing a variable.

Column	Variable name (as it appears on the data sheets in Appendix A) and its units of measurement	Explanation
		Measurements taken before the experiment are indicated in the variable name by the prefix PRE or PR.
C1	AGE (rounded to nearest year)	
C2	PREHR (beats per minute)	Number of ventricular beats per minute. Initial heart rate was recorded using a 3-channel (12-lead) electrocardiograph.
C3	PREMAXHR (beats per minute)	Maximum heart rate. The subjects pedaled a stationary cycle to their maximum capacity and then the heart rate was recorded by the same method as above.
C4	PREVO2 (liters per minute)	Volume of oxygen consumed (VO2) was measured while the participant bicycled at ever-increasing work loads. This is roughly equal to the volume of oxygen utilized in the oxidation of foodstuffs.
C5	PREVE (liters per minute)	Expiratory gas volume per minute
C6	PREWAT	The watt is a measure of power, and it is the rate of doing work. 1 watt is roughly equal to 6 kilograms-meter/minute. (746 watt = 1 horsepower.)

Column	Variable name (as it appears on the data sheets in Appendix A) and its units of measurement	Explanation
C7	PREHR%	Pre heart rate % is the initial heart rate (before exercise) expressed as a percentage of the maximum heart rate given in C3.
C8	PRERQ	Respiratory quotient; the ratio of carbon dioxide produced to the amount of oxygen consumed. RQ indicates what food substrate is the primary source of energy.
C9	PREFAT%	Skin folds from different parts of the body were measured and the percent fat was computed using a formula.
C10	PREWT (lb.)	Weight in pounds
C11	PREHT (inch)	Height in inches
C12	PRERESHR (beats per minute)	Participants rested for 5 minutes and the heart rate was recorded using a 3-channel (12-lead) electrocardiograph.
C13	PRRESSYS (millimeters of mercury)	Systolic blood pressure was measured with a sphygmomanometer after the subject rested for 5 minutes lying down.
C14	PRRESDIA (millimeters of mercury)	Diastolic blood pressure was measured with a sphygmomanometer after the subject rested for 5 minutes lying down.
C15	PRERPP	Submaximal rate pressure product (RPP) is an indication of myocardial oxygen demand. Rate pressure product is calculated as the product of systolic blood pressure and heart rate recorded during the same minute.
		POST or POS prefix indicates measurements taken after the exercise program.
C16	POSTNHR	Corresponds to PREHR
C17	POSTMAXHR	
C18	POSTVO2	
C19	POSTVEE	Corresponds to PREVE
C20	POSTWAT	
C21	POSTHR%	
C22	POSTRQ	

Column	Variable name (as it appears on the data sheets in Appendix A) and its units of measurement	Explanation
C23	POSTFAT	
C24	POSTWT	
C25	POSTHT	This was taken as a routine measurement.
C26	POSTREHR	
C27	POSSYS	
C28	POSDIA	Corresponds to PRRESDIA
C29	POSTRPP	
C30	PARTICIP	Participant code: (1) Non-participant; this group has only pre-experiment data. The rest of the subjects were randomly assigned to groups 2, 3, and 4. (2) Control group (no exercise group). (3) Low-intensity group; exercised at 65 percent of each subject's maximum heart rate (MHR). (Exercise sessions of longer than 30 minutes' duration were adjusted to provide total work equivalent of group 4 for ten weeks.) (4) High-intensity group; exercised at 85 percent of MHR (30 minutes' duration) for nine weeks.

Student Questionnaire Data (SQD) Set: List of Variables

See the questionnaire in Activity 2.4 for a full description of the variables.

Column	Variable name	Explanation
C1	SEX	
C2	AGE	
C3	CLASS	Class status at the university
C4	HEIGHT	
C5	WEIGHT	
C6	PULSE1	Taken for 15 seconds by feeling the carotid artery and multiplying by 4
C7	PULSE2	After running in place for a minute, pulse is obtained by the same method as above.
C8	SMOKE	
C9	ACTIVITY	
C10	STRESS	
C11	FRIENDS	
C12	TIMEOUT	
C13	HEALTH	
C14	TEMP	
	NHPLAN	Values cannot be input into the computer, as it is an unstructured variable.

Self-Assessment

Please assess your competency.

Task:	How well can you do it?					Page
Explain each of the following:	*Poorly*				*Very well*	
1. Variable	1	2	3	4	5	23
2. Case	1	2	3	4	5	28
3. Structured data generation	1	2	3	4	5	24
4. Unstructured data generation	1	2	3	4	5	24
5. Coding a questionnaire	1	2	3	4	5	28
6. Decoding the numbers into phrases	1	2	3	4	5	28
7. Designing a recording sheet for data	1	2	3	4	5	29

Answers for Chapter 2

Answers for questions 1–4, p. 28: 1) 1 = Minimal, 2 = Moderate, 3 = Heavy 2) 1 = Excellent, 2 = Good, 3 = Average, 4 = Below average 3) 1 = True, 2 = False 4) 1 = I never smoked, 2 = I quit smoking, 3 = I smoke, 4 = Other

Answers for questions 1–2, p. 30: 1) Cases 2) Variables

Answers for Activity 2.5, p. 30: 1) Socioeconomic status, body mass index, chronological age, gynecologic age 2) "What is your age?" The other variables were elicited using several questions or examinations.

Answers for Activity 2.6, p. 32: 1) State, LDA, date of change 2) National Center for Health Statistics 3) Insurance Institute for Highway Safety

Answer for Activity 2.7, p. 34: Aerobic capacity, submaximal rate pressure product (RPP), body composition, blood lipids, and dietary assessment are mentioned; however, some of these are names of classes of variables rather than individual variables. The variable VO2 is also mentioned; it probably measures aerobic capacity.

CHAPTER

3

Measurement of Variables

Prelude

"Is this the place where numbers are made?" asked Milo . . . .

"They're not made," [Dodecahedron] replied, as if nothing had happened. "You have to dig for them. Don't you know anything about numbers?"

"Well, I don't think they're very important," snapped Milo, too embarrassed to admit the truth.

"NOT IMPORTANT!" roared the Dodecahedron, turning red with fury. "Could you have tea for two without the two—or three blind mice without the three? Would there be four corners of the earth if there weren't a four? And how would you sail the seven seas without a seven?"

"All I meant was—" began Milo, but the Dodecahedron, overcome with emotion and shouting furiously, carried right on.

"If you had high hopes, how would you know how high they were? And did you know that narrow escapes come in all different widths? Would you travel the whole wide world without ever knowing how wide it was? And how could you do anything at long last," he concluded, waving his arms over his head, "without knowing how long the last was? Why, numbers are the most beautiful and valuable things in the world. Just follow me and I'll show you." He turned on his heel and stalked off into the cave.

— Norton Juster
The Phantom Toll Booth, pp. 176–177

Chapter Outline

Self-Assessment

Please assess your competency.

Tasks:	How well can you do it?					Page
	Poorly				*Very well*	
1. Define and give examples of the following types of variables:						
Nominal	1	2	3	4	5	47
Ordinal	1	2	3	4	5	47
Interval	1	2	3	4	5	47
Ratio	1	2	3	4	5	47
2. Identify the following types of variables in any given questionnaire or table:						
Nominal	1	2	3	4	5	47
Ordinal	1	2	3	4	5	47
Interval	1	2	3	4	5	47
Ratio	1	2	3	4	5	47
3. Convert ratio variables into categorical variables.	1	2	3	4	5	53

Introduction

Everyone needs to measure things from time to time. Sometimes these measurements need to be very precise, sometimes not very precise at all. Scientists and engineers frequently need to make measurements that are extremely precise, and for these tasks they use highly specialized instruments. NASA scientists, for instance, may need to know an astronaut's weight correct to two or three decimal places. For the layperson, however, it is almost always good enough to know someone's weight to the nearest pound. Most of the time, in fact, it's enough just to know that the person is heavy, thin, or of average weight.

Some kinds of characteristics of persons or things cannot be measured numerically, but they can be described. Hair color, for instance, may be given

TABLE 3.1 Characteristics of People and Their Measurements

Characteristics of the Person or Case	Units or Terms	Comment
Weight	Pounds, kilograms, or a descriptive term	Approximate weight or descriptive terms such as heavyset, slender, thin, elephantine, etc., may be used.
Height	Inches, centimeters, or a descriptive term	Approximate height or descriptive terms such as very tall, short, average height, etc., may be used.
Hair color	Brown, white, ash blond, etc.	Descriptive words are usually used.
Appearance	Very attractive, good looking, ugly, etc.	This is a subjective measure and will definitely differ from person to person. Often the descriptive term that is used gives more information about the observer than the observed.
Intelligence	Smart, dumb, good at math, etc.	We don't administer IQ tests to people we meet, but we do often describe them using subjective descriptions of their intellectual attributes. **Write your comments about these measurements below:**
Education	Diploma, degree, year in school, etc.	
Age	Usually in years	
Sex	Male, female	
Kind of clothing	Expensive, casual, etc.	
Personality type	Zestful, reserved, etc.	

as red, blond, auburn, gray, black, or brown. Although intelligence can be measured by giving scores on certain tests, commonly we describe a person's intelligence with phrases like "very clever" or "pretty dull" or by mentioning specific qualities, such as that the person has a good memory. Table 3.1 lists some characteristics that we might use to describe someone we know.

Each one of the characteristics listed in Table 3.1 is called a **variable**. Their values describe the person and are expressed using the units or words given in the second column. We assign numbers to some characteristics and descriptive words to others. The process of assigning numbers or *agreed upon* descriptive words to characteristics is called **measurement**.

All people can be described by referring to the same characteristics; however, the values (words or numbers) assigned to the characteristics will differ from case to case. That is the reason for calling the characteristics variables.

Properties of Numbers

We use numbers frequently as we go about our daily business. Costs, distances, and weights of things are all expressed in numbers. We take it for granted that the numbers will behave in familiar and predictable ways. Their behavior in accordance with well-known rules (for example, $200 is twice as much as $100) is what makes the numbers useful. However, for some uses of numbers the familiar rules do not apply; for other uses, less familiar rules must be used. For example, the numbers that designate addresses on a street rarely start at 1, and in most U.S. cities they do not run consecutively. Generally, odd numbers are on one side of the street and even numbers are on the other. Sometimes numbers are omitted, so the house next door to 113 will almost never be numbered 114 and may not even be numbered 115 or 117. (In some countries, not only do addresses not run consecutively, they seem to have been assigned at random!)

The rules for dealing with street address numbers are different from the rules for dealing with ages. If you are now 21 years old and your sister is 15, she is six years younger than you are: $21 - 15 = 6$. However, 108 Main Street is probably only three houses away from 102, not six houses away, even though $108 - 102 = 6$, and it could be right next door. Sometimes it makes sense to subtract, and sometimes it doesn't.

We need to know what use we are making of numbers, because different uses dictate different rules. We will be misled if we work with addresses as though they are like temperatures or distances. (Salt and sugar look alike too, yet they taste entirely different.)

Numbers can be used to identify, designate, or name—addresses, for instance. Similarly, we designate with numbers when we talk about "the third person in line" or "the second day of the month." When numbers are used in this way, there's often no special connection between the number and whatever has that number. The third person in line can easily become the seventy-ninth person in line if she has to leave for a while to go to the bathroom. Also, if we count from the end of the line instead of the front of the line, person number three could easily become person number seventy-nine. Often assigning numbers is governed by convention: we usually count from the front of the line, we number houses from south to north, etc.

Numbers can also be used to count or to measure. Thus, we say that a certain board is 5 feet 9 inches long or that the month of June has thirty days. When numbers are used in this way, the connection between the thing and the number tends to be more definite, less governed by convention alone.

Of course the length of a board will be a different number if it is measured in meters instead of feet, and June will have a different length if it's measured in hours instead of days. Also, that board could be 5.75 feet long or 5.7483 feet long, depending on how accurately we measure it. Units make a difference and so does degree of accuracy; but once the units and the accuracy have been agreed on, the measurement (the number) is strictly determined by the thing being measured.

The rules that make sense to use vary with the application. For instance, addition makes perfectly good sense in some cases but not in others. If we are laying carpet and add a 5-foot length to a 3-foot length, we will have enough for a room that is 8 feet long. On the other hand, if we mix 5 pints of alcohol with 3 pints of water we will get noticeably less than 8 pints of solution! (Try it if you don't believe it.) Similarly, if you are 18 years old and your dog is 3 years old, then you and your dog are 18 + 3 = 21 years old, but that doesn't mean you will be able to buy a drink at your neighborhood bar if you take your dog with you. It just doesn't make any sense to add ages in this context. Sometimes ordinary arithmetic makes sense, sometimes it doesn't.

Read Table 3.2 and fill in the blanks.

TABLE 3.2 Properties of Numbers in Regular Use

Numerical Expression	Use of Number (to count, to designate, to name, or to identify?)	Possible Operations	Comment or Example
2 people 3 people	To count	Addition (+); for instance, two people may be joined by three more. Subtraction	Addition gives a meaningful result: 2 people + 3 people = [1] ______________.
The board is somewhere between 1 and 2 feet long.	To designate length	Multiplication Division	When we measure we may find that the actual length of the board is 1 foot, 2 feet, 1.2 feet, 1.25 feet, or [2] ______________. It can make sense to multiply using these numbers. For instance, if the board is 1.2 feet long and we need a board twice that long, we multiply by 2: 2.4 is twice the size of 1.2. Similarly, 3.6 is [3] ______________ the size of 1.2.
Joan is first in line, Carlos is second in line, Nguyen is third in line, Satya is fourth in line.	To identify an order relationship between the people	None	The ordering of the people corresponds to the order of the numbers. For example, $1 < 3$, and Joan is in front of Nguyen. The statement $2 < 1$ is false; similarly, Carlos is not in front of Joan.

Table 3.2 Continued

Numerical Expression	Use of Number (to count, to designate, to name, or to identify?)	Possible Operations	Comment or Example
			4 ______________ is less than 4, and similarly, Carlos is in front of Satya.
5°F 10°F 12°F	To measure temperature	Addition Subtraction	Addition and subtraction both make sense. For example, if it is 5°F and the temperature rises 2°, we can add 5 + 2 and say it is 7°F. However, it doesn't make sense to say that when it's 100° it's twice as hot as when it's 50°, because the zero point is not absolute.
Ford = 1 Chevrolet = 2 Toyota = 3 Nissan = 4 Hyundai = 5	To identify	None	In this case the numbers are arbitrarily assigned and have no intrinsic meaning or connection to the cars; no ordering is intended. The numbers could just as easily have been interchanged. The fact that $1 < 2$ should not lead us to conclude in this case that Fords are in any way less than or inferior to Chevrolets.

Variables

Variables represent characteristics of objects or people that can have different values. Numerical variables are said to be measured on a nominal, ordinal, interval, or ratio scale, depending on the particular characteristic being measured.

The **nominal scale** is being used when numbers are used in lieu of names of categories but no ordering or precedence is implied.

The **ordinal scale** is being used when numbers are used to name categories and the categories are ordered, one being ahead of another in some way. The numbers may or may not have intrinsic connections to the categories.

The **interval scale** is being used when the numbers have intrinsic meaning but the zero point is arbitrary. For such a scale, dividing one quantity by another does not make sense. 60° divided by 20° is 3, but this 3 has no meaning; 60° is not three times as warm as 20°. Subtraction does make sense, however. 40°C − 20°C = 20°C and 80°C − 60°C = 20°C. In both cases, the temperature has changed by the same amount. Addition also makes sense.

The **ratio scale** is being used when the numbers have intrinsic meaning and the zero point represents absolute absence of the characteristic being measured. Statements such as that one number is twice the other number make sense.

We now present an example of each scale to illustrate the properties of that scale.

Nominal Scale SEX, coded as 1 for female and 2 for male, is an example of a variable measured on a nominal scale. We can categorize the cases and then count how many are in each category. That's about the only mathematical operation that makes sense, however. The fact that 1 is less than 2, for example, has no meaning correlated to female and male. There is no implication that female precedes male in any way. We could just as easily have assigned 1 to male and 2 to female. It also makes no sense to add 1 + 2, to average a set of 1s and 2s, to subtract 1 from 2, etc.

Ordinal Scale STRESS is a variable that can be measured on an ordinal scale. This variable might be divided into three categories, coded as minimal = 1, moderate = 2, and heavy = 3. Adding these numbers does not make sense (minimal + moderate = heavy?). Intermediate values can make sense in some contexts; for instance, we might average the stress levels for a group of cases and come up with an average stress level of 2.3. This number would indicate that the average stress level was somewhat more than moderate. We can also count the number of cases that occur in each category, i.e., count the number of 1s, the number of 2s, and the number of 3s.

These numbers are ordered in a natural way. Two is bigger than one and moderate stress is more stress than minimal stress. There is a certain arbitrariness, however, in that we might just as easily have coded the categories as minimal = 3, moderate = 2, heavy = 1 or minimal = 0, moderate = 5, heavy = 10.

What matters is that moderate is assigned a number between the numbers assigned to minimal and heavy.

We probably would not want to conclude, however, that the "distance" from minimal to moderate stress is equal to the "distance" from moderate to heavy. Also, it is debatable whether there is a true zero. Probably only dead people are totally stress free.

Interval Scale TEMPERATURE, given in numbers such as 20, 40, or 53.2 degrees centigrade, is a variable that is measured on an interval scale. The values of the variable are the same as the assigned numbers.

All of the operations mentioned so far are possible and all of the properties hold. We can count the number of days, for instance, that a certain temperature was reached. Temperatures are ordered in a natural way; 30° is hotter than 25°, for example. All the numbers between two temperatures are valid as possible temperatures; for instance, 26.0215° is a possible temperature between 25° and 30°. Furthermore, it makes sense to add or subtract temperatures: if the temperature is 18° and it rises 7°, then the temperature is 18° + 7° = 25°. The interval scale is characterized by the fact that the difference between values—the interval between the values—has a physical meaning. Thus, in going from 20° to 30° the temperature has risen by 10°; in going from 50° to 60° the temperature also has risen by 10°. A 10° rise in temperature is a 10° rise regardless of what temperature we started from. The scale has no arbitrary zero, so it is not meaningful to say 40°C is twice as hot as 20°C. Division of one temperature by another does not make sense.

Ratio Scale WEIGHT, given in numbers such as 101, 153.2, or 164 pounds, is a variable measured on a ratio scale. As with the interval scale, the values of the variable are the same as the assigned numbers, and all the previously discussed operations with numbers make sense (counting, ordering, determining intermediate values, or "between-ness," addition, subtraction). Furthermore, we can divide one weight by another, because the zero is not arbitrary. Something that weighs 0 pounds really weighs nothing. It makes sense to say that you weigh twice as much as your little brother if you weigh 110 pounds and he weighs 55 pounds.

ACTIVITY 3.1

Variables measured in a clinic or a laboratory

Suppose someone goes to a physician. His temperature, weight, and other relevant information will be recorded, and he may be given certain laboratory tests. Table 3.3 records the characteristics.

Weight and height are familiar measurements, and each of us can readily measure our own height and weight by using a scale and a tape measure. Some variables, like weight and height, have directly perceivable, concrete attributes. These measurements are **reproducible**, meaning that any person with the same set of instruments can obtain similar values (within the limits of chance variation).

TABLE 3.3 Examples of Variables Measured in a Clinic or a Laboratory

Variable	Measurement	What Scale Is It?
Height	71.5 inches	Ratio scale
Weight	162 pounds	1 ______________ scale
Temperature	101.4°F	2 ______________ scale
Blood pressure	110/80	True zero exists, so the scale is 3 ______________
Cholesterol	184	4 ______________ scale
Serum glucose	80 mg/dl (milligrams per deciliter)	5 ______________ scale
Smoking habit	2 cigarettes per day	6 ______________ scale
Ethnicity	Native American	7 ______________ scale
Sex	Male	8 ______________ scale
Respiratory coefficient	1.24	9 ______________ scale

Temperature measurement is also familiar. Some variables can be detected by the senses but can only be measured indirectly through properties of other substances. Heat, for example, can be measured using the expansion property of mercury. Such results are reproducible. An example of this is body temperature, which is measured in Fahrenheit or centigrade degrees.

Cholesterol, serum glucose, and respiratory coefficient can only be measured by using laboratory procedures. They are not directly accessible to our senses. They too are assigned numbers as a result of the laboratory tests, and the results are reproducible because the laboratory tests are standardized.

Sex is readily determined by observation and may be coded for convenience by assigning a number. The number is chosen arbitrarily. Female may be assigned 1 and male may be assigned 2, as you saw earlier.

We may need to ask some questions before we can decide on ethnicity, and we arbitrarily assign numbers as we did for sex.

ACTIVITY 3.2

Identifying types of variables in the questionnaire

Circle the appropriate type of measurement for each question in the questionnaire:

1. Sex: (1) Female (2) Male (Nominal Ordinal Interval Ratio)
2. Age: _______ (Nominal Ordinal Interval Ratio)
3. Class status: (Nominal Ordinal Interval Ratio)
   (1) Freshman
   (2) Sophomore
   (3) Junior
   (4) Senior
   (5) Graduate
4. Height in inches: _______ (Nominal Ordinal Interval Ratio)
5. Weight in pounds: _______ (Nominal Ordinal Interval Ratio)
6. Pulse (take your pulse for 15 seconds, multiply by 4, and record the number): _______ (Nominal Ordinal Interval Ratio)
7. Run in place for one minute and then take your pulse again: Pulse after running in place for one minute: _______
8. Smoking cigarettes: (Nominal Ordinal Interval Ratio)
   (1) I never smoked
   (2) I quit smoking
   (3) I smoke
   (4) Other
9. Daily physical activity level: (Nominal Ordinal Interval Ratio)
   (1) Minimal
   (2) Moderate
   (3) Vigorous
10. Daily stress level: (Nominal Ordinal Interval Ratio)
    (1) Minimal
    (2) Moderate
    (3) Heavy
11. I have at least three friends or relatives with whom I can discuss the positive and negative events of my day. (Nominal Ordinal Interval Ratio)
    (1) True (2) False
12. I take at least one five-minute "time out" per day to be quiet. (Nominal Ordinal Interval Ratio)
    (1) True (2) False
13. I would rate my overall health as: (Nominal Ordinal Interval Ratio)
    (1) Excellent
    (2) Good
    (3) Average
    (4) Below Average

14. I generally keep the thermostat at my residence set at _______ °F. (Nominal Ordinal Interval Ratio)

15. What is your opinion about national health insurance? ________________________________ ________________________________

Categorical and Measured Variables

Dividing variables into two types, categorical and measured, is useful for determining the statistical techniques that apply to them. Categorical variables (such as sex and ethnicity) are used when we place individuals or things into different categories. Usually it only makes sense to count up how many are in each category (for example, male 27, female 35), although sometimes such variables are on an ordinal scale (for example, physical activity measured as minimal, moderate, or vigorous). Opinions are another example of categorical variables. In this case, between-ness might make sense (for example, in evaluating movies, "terrible" is between "bad" and "abominable").

Categorical variables may be coded with numerals (for example male = 1 and female = 2). It does not make sense to do arithmetic with such coded numbers.

Measured variables such as height and temperature, on the other hand, are measured on either interval or ratio scales. In either case, there is consistent meaning to the interval increase or decrease in such variables.

Quantitative Statements about Categorical Variables: Counts

Categorical variables are used to place individual cases into different categories. We may count the number of persons in each category and compute percentages. We treat counts as though they are ratio data; however, between-ness is not complete, since the values of the count must be whole numbers (positive integers). We can't have 1.5 persons in a category. Counts are **discrete** as opposed to **continuous** variables.

Categorical Variables: Nominal Scale

ACTIVITY 3.3

TABLE 3.4 Sex Composition of Two Statistics Classes at a University

Value	Code	Count	Percent
Female	1	33	64.71
Male	2	18	35.29
Sample size:	$N = 51$		

Fill in the blanks or circle the appropriate words in the following passage.

SEX is a categorical variable measured with the [1] ______________ scale. SEX consists of [2] ______________ values, male and [3] ______________. In this case the values are coded with the numbers [4] _______ and [5] _______. It [6] does/does not make sense to do arithmetic with these two numbers. There [7] are/are no numbers between 1 and 2 in this context that can be values of this variable. The fact that 2 is greater than 1, in this context, [8] does/does not mean that one gender is superior to the other in some way, so SEX [9] is/is not measured on the ordinal scale. It [10] does/does not make sense for 2 to be twice 1 in this context. So it [11] is/is not a ratio scale. The categorical variable SEX places individuals in [12] ______________. We can then count the number in each category. Counts may be treated as measured data, so we can compute [13] ______________.

Categorical Variables: Ordinal Scale

ACTIVITY 3.4

In Table 3.5, physical activity is an ordinal variable with three categories: minimal, moderate, and vigorous.

TABLE 3.5 Daily Activity Level Composition of Two Statistics Classes at a University

Physical Activity	Code	Count	Percent
Minimal	1	6	12.24
Moderate	2	36	73.47
Vigorous	3	7	14.29

Sample size: $N = 49$
Missing values: 2

Fill in the blanks or circle the appropriate words in the following passage.

Physical ACTIVITY is used as a categorical variable and is measured with the [1] ______________ scale. Physical ACTIVITY consists of three values and they are [2] ______________, ______________, ______________. They are coded with three numbers, [3] _______, [4] _______, and [5] _______. It [6] does/does not make sense to do arithmetic with these three numbers. Values between 1 and 2 [7] are/are not possible in this context. The fact that 2 is greater than 1 [8] does/does not make sense in this context. So physical activity

[9]is/is not an ordinal variable. It [10]does/does not have meaning that 2 is twice 1 in this context, so a ratio scale [11]is/is not being used. The physical ACTIVITY variable, when used as a categorical variable, helps us place individuals in [12] ______________ and count them. Counts may be treated as measured data and so we can compute [13] ______________.

Categorical Variables: Categorization of Ratio Data

Listed below, from minimum to maximum, are WEIGHTS of students from the student questionnaire data. WEIGHT intervals representing three categories were formed. They are shown in the left column of Table 3.6. To create Table 3.6, we simply counted the number of WEIGHTS from the listing that fell into each category. In this way, a variable (weight) that would normally be measured on a ratio scale was converted to a categorical variable.

97	98	100	105	107	108	108	108	110	110	110	112	114	115
116	120	120	120	120	120	120	123	125	126	128	130	135	135
140	140	140	145	147	148	150	150	150	155	160	160	175	180
180	182	186	190	190	192	210	220	233					

TABLE 3.6 WEIGHT Composition of Two Statistics Classes at a University

WEIGHT	Wt Code	Count	Percent
91–140 lb.	1	31	60.78
141–190 lb.	2	16	31.37
191–240 lb.	3	4	7.84
Sample size: $N = 51$			

The pattern of variation in the values of a variable is called its **distribution**. We often display distribution in terms of **values** (intervals) and **counts** of their occurrence. The choice of intervals in Table 3.6 is arbitrary. A different choice of intervals will give a somewhat different picture of the weight distribution. This process of creating intervals and counting is called **grouping** the data. When we group data into categories, we lose information.

ACTIVITY 3.5

Fill in the following table. Take information from the ordered list of weights given above.

TABLE 3.7 WEIGHT Composition of Two University Statistics Classes

WEIGHT	Wt Code	[1]Count	Percent
91–116	1	______	______
117–140	2	______	______
141–166	3	______	______
167–190	4	______	______
191–216	5	______	______
217–240	6	______	______

We [2] gain/lose information as we increase the number of intervals.

Examples from the Literature: Categorical Variables, Measured Variables, and Types of Measurement

ACTIVITY 3.6

Sometimes the context determines the classification of variables as nominal, ordinal, interval, or ratio. This will become clearer as we examine examples from the literature. Literature Example 3.1 is from an article entitled "Vaginal Douching among Women of Reproductive Age in the United States: 1988." The first column in Table 1, which accompanies the abstract, lists several variables and their values. The table also lists the variable Race and its values [1] ______________ and [2] ______________. Read the abstract. Then look at the comments and questions in Table 3.8 and enter your answers in the third column.

TABLE 3.8 Variables and the Types of Measurements

Variable	Comment or Question	Circle Your Answer
Place of residence	Values: Central city, Suburb, Nonmetropolitan. What type of variable is it?	[3]Nominal/Ordinal/ Interval/Ratio
Geographic region	Notice the values. Can numbers be arbitrarily assigned to the values?	[4]Yes/No
	Would adding these numbers make any sense?	[5]Yes/No
	What type of variable is it?	[6]Nominal/Ordinal/ Interval/Ratio
Marital status	Is any order implied by the values?	[7]Yes/No
	What type of variable is it?	[8]Nominal/Ordinal/ Interval/Ratio
Poverty level	Suppose the intervals were not equal. What scale of measurement would this be?	[9]Nominal/Ordinal/ Interval/Ratio
	What scale of measurement did the authors use to measure poverty level?	[10]Nominal/Ordinal/ Interval/Ratio
Years of schooling	This variable has been categorized so that we see intervals as values. Suppose the subjects gave the exact years of schooling. What scale of measurement would this variable be measured with in that case?	[11]Nominal/Ordinal/ Interval/Ratio
Age	This variable has been categorized so that we see intervals as values. Suppose the subjects gave their exact age. What scale of measurement would this be?	[12]Nominal/Ordinal/ Interval/Ratio
Race	What are the values?	[13]
	What type of variable is it?	[14]Nominal/Ordinal/ Interval/Ratio

LITERATURE EXAMPLE 3.1

Vaginal Douching among Women of Reproductive Age in the United States: 1988

Sevgi Okten Aral, PhD, William D. Mosher, PhD, and Willard Cates, Jr, MD, MPH

Background. Vaginal douching has been associated with pelvic inflammatory disease (PID) in several epidemiologic studies.

Methods. To determine the extent to which douching is practiced and to describe the population subgroups in which it is most prevalent, we analyzed data from the 1988 National Survey of Family Growth, which is based on a nationally representative sample of 8450 United States women between the ages of 15 and 44 years.

Results. Thirty-seven percent of the sample reported douching; 18% douched at least once a week. The variable most strongly and consistently associated with douching was race: two thirds of Black women, but only one third of White women, reported douching. The practice was least frequent among 15- to 19-year-olds (31%) and most frequent among 20- to 24-year-olds (41%). Douching was more common among women who lived in poverty (50%) than among those who did not (28%). Seventy percent of Black women living in poverty reported douching. Women with less than a high school education were almost four times more likely to report douching as those with 16 or more years of schooling (56% vs 16%). Women with only 1 partner and those with 10 or more partners were less likely to douche than others. Sixteen percent of women who reported douching, compared with 10% of those who did not, also reported a history of PID.

Conclusions. Douching may be a modifiable risk factor for PID, it should be a high priority for future etiologic research. (*Am J Public Health.* 1992;82:210–214)

TABLE 1—Percentage of Women Who Douche Regularly, by Demographic and Ecologic Characteristics

	All Races	White	Black
Age, y			
15–19	31.0*	25.4*	53.5**
20–24	41.1	35.7	63.1
25–29	37.6	32.9	67.6
30–34	36.0	31.5	64.8
35–39	35.1	30.2	70.2
40–44	37.0	33.8	65.8
Total	36.7	32.0	66.5
Marital status			
Currently married	32.7*	30.3	63.5
Never married	38.4*	29.1*	67.8
Formerly married	48.4	44.2	68.4
Years of schooling[a]			
0–11	56.4*	51.5*	75.5
12	45.5*	42.1*	69.5**
13–15	31.4*	26.1*	66.0
16+	16.7	13.6	53.9
Poverty level			
Below poverty	50.1***	40.1	69.7
100–199	42.2**	37.6	68.5
200–399	37.1*	33.5*	65.4
400+	28.2	26.1	60.1
Place of residence			
Central city	41.2*	30.8	66.8
Suburb	33.6*	30.7***	65.2
Nonmetropolitan	39.5	36.2	68.3
Geographic region			
All Regions	36.7	32.0	66.5
West	28.1	26.1	57.9
Northeast	31.4	28.1	62.1
Midwest	32.1*	27.4*	68.2
South	48.0	42.3	68.3

[a]These percentages exclude women < 22 years of age.
*The difference between this percentage and the percentage below is significant at $P < .001$.
**The difference between this percentage and the percentage below is significant at $P < .05$.
***The difference between this percentage and the percentage below is significant at $P < .01$.

TABLE 2—Percentage of Women Who Douche Regularly, by Behavioral Characteristics

	All Races	White	Black
Age at first intercourse, y			
Under 15	45	36	68
15–17	43*	38*	67
18–19	35*	31*	68**
20+	25	22	58
Number of lifetime sex partners			
1	30*	28*	54*
2–9	41*	35	79
10+	36	32	63
Past history of STD			
None	37	32	67
Gonorrhea, chlamydia, or herpes	35	29	59
Genital warts	29	26	61
Past history of PID			
No	35*	31*	65*
Yes	47	41	72

*The difference between this percentage and the percentage below is significant at $P < .001$.
**The difference between this percentage and the percentage below is significant at $P < .05$.

TABLE 3—Prevalence of Specific Douching Practices

	All Women (%)	White %	Black %
Do not douche	63.3	68.0	33.5
Douche regularly	36.7	32.0	66.5
Frequency[a]			
2 times per week	10.5	8.2	24.2
1 time per week	7.7	7.1	11.9
2–3 times per month	10.7	9.2	21.2
≤ 1 time per month	7.7	7.5	9.3
Timing in monthly cycle[a]			
After intercourse	5.5	4.4	9.0
Other times	45.5	50.4	31.4
Both	49.0	45.2	59.6
Timing after intercourse[b]			
Within 30 minutes	36.4	32.3	42.6
After 30 minutes	63.6	67.7	57.4

Note. All White–Black differences in this table are statistically significant at $P < .001$, except for the difference for frequency of douching ≤ 1 time per month, which is statistically significant at $P < .05$.
[a]These figures represent percentages of women who douche regularly.
[b]These figures represent percentages of women who douche after intercourse.

ACTIVITY 3.7

Consider Table 2, "Percentage of Women Who Douche Regularly, by Behavioral Characteristics" and Table 3, "Prevalence of Specific Douching Practices," which are part of Literature Example 3.1. Answer the questions in the next paragraph. Then read Table 3.9 and enter answers to its questions in the third column.

Tables 2 and 3 were derived by counting the number of respondents in each behavioral category and computing percentages. Counts are quantities and are treated as measured variables. So it is appropriate to calculate [1] ____________. The table reports the percentage of blacks and whites manifesting the particular behavior in question. Under the age of 15, [2] more/fewer blacks who douche have intercourse than whites.

TABLE 3.9 Variables and the Types of Measurements

Variable	Comment or Question	Circle Your Answer
Age at first intercourse	Suppose these data had been collected as exact ages. What type of variable would this be?	[3]Nominal/Ordinal/ Interval/Ratio
Number of life-time sex partners	Suppose these data had been collected as exact numbers. What type of variable would this be?	[4]Nominal/Ordinal/ Interval/Ratio
	Can this variable have fraction values (values between 1 and 2)?	[5]Yes/No
	Therefore this variable is	[6]Continuous/Discrete
Past history of STD	What type of variable is this?	[7]Nominal/Ordinal/ Interval/Ratio
Past history of PID	What type of variable is this?	[8]Nominal/Ordinal/ Interval/Ratio
Do not douche	One value of the variable is "Do not douche." What is the other value?	[9]
	What question would elicit the values of the variable?	[10]
	What type of variable is it?	[11]Nominal/Ordinal/ Interval/Ratio
Timing in monthly cycle	What type of variable is it?	[12]Nominal/Ordinal/ Interval/Ratio
Timing after intercourse	Do the values indicate order?	[13]Yes/No
	What type of variable is it?	[14]Nominal/Ordinal/ Interval/Ratio

ACTIVITY 3.8

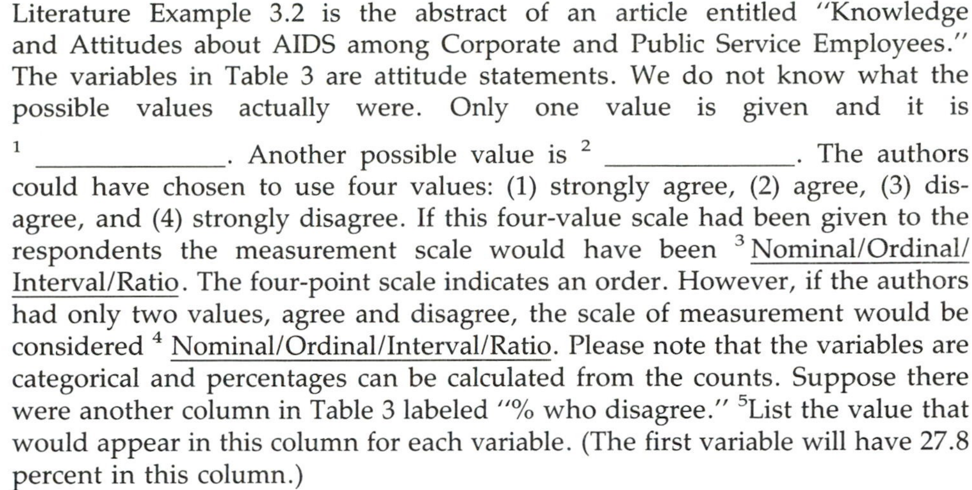

Literature Example 3.2 is the abstract of an article entitled "Knowledge and Attitudes about AIDS among Corporate and Public Service Employees." The variables in Table 3 are attitude statements. We do not know what the possible values actually were. Only one value is given and it is [1] ______________. Another possible value is [2] ______________. The authors could have chosen to use four values: (1) strongly agree, (2) agree, (3) disagree, and (4) strongly disagree. If this four-value scale had been given to the respondents the measurement scale would have been [3] Nominal/Ordinal/Interval/Ratio. The four-point scale indicates an order. However, if the authors had only two values, agree and disagree, the scale of measurement would be considered [4] Nominal/Ordinal/Interval/Ratio. Please note that the variables are categorical and percentages can be calculated from the counts. Suppose there were another column in Table 3 labeled "% who disagree." [5]List the value that would appear in this column for each variable. (The first variable will have 27.8 percent in this column.)

LITERATURE EXAMPLE 3.2

Knowledge and Attitudes about AIDS among Corporate and Public Service Employees

Judith K. Barr, ScD, Joan M. Waring, MA, and Leon J. Warshaw, MD

Background. We examined the relationship between workplace AIDS education efforts and workers' knowledge about HIV transmission and their attitudes toward coworkers with AIDS.

Methods. Questionnaires were mailed to corporate and public service workers at 12 work sites to ascertain the extent of their knowledge about AIDS and their attitudes toward coworkers with AIDS. Each work site had offered an AIDS education program. The average response rate was 40%; 3460 workers returned questionnaires.

Results. Respondents' knowledge was largely consistent with available scientific evidence. However, a substantial minority still believe HIV infection can be transmitted through casual contact. Over 30% endorse the screening of new employees for AIDS, and 23% would fear contagion from an infected coworker. Thirty percent of the respondents expressed skepticism about the veracity of information from government sources and the scientific community. Work site comparisons show that where educational programs are minimal, employees know less about HIV transmission and hold more negative attitudes.

Conclusion. Comprehensive workplace AIDS education programs can reinforce workers' knowledge about HIV transmission, thereby fostering more favorable views toward coworkers with AIDS. (*Am J Public Health.* 1992;82:225–228)

TABLE 3—Workers' Attitudes about People with AIDS (n = 3460)

Attitude Statement	% Who Agree
People with AIDS should be treated at work just like anyone else.	72.2
Employers should have the right to dismiss employees who have AIDS.	8.8
My employer should screen out prospective employees who have AIDS.	31.5
I would be uncomfortable eating lunch with someone who has AIDS.	30.4
I would be afraid of getting AIDS if I worked with someone who has AIDS.	23.4

Activity 3.9

Literature Example 3.3 is from the article "Adolescent Smoking, Weight Changes, and Binge-Purge Behavior: Associations with Secondary Amenorrhea." Table 3 requires some explanation, in particular, the terms **quintile** and **quartile**. These concepts will be discussed at greater length later. Briefly, data can be divided into segments, each containing the same number of units. If the data are divided into five such segments, the segments are called quintiles; if the data are divided into four segments, they are called quartiles. Each quintile contains approximately 20 percent of the data, and each quartile contains approximately 25 percent. In the table with Literature Example 3.3, the 2,156 subjects are placed in quintiles by socioeconomic status (presumably family income). The first quintile contains the 431 girls who come from the families having the lowest family incomes ($431 \approx 2156/5$). The researchers then counted the number of girls in each group who had secondary amenorrhea and divided that number by 4.31 to obtain the rate/100.

What are the four (explanatory) variables in Table 3? [1] _______ [2] _______ [3] _______ [4] _______. You have seen percentages of counts in previous tables. What is reported in the last column of Table 3? [5] ______________. How are rate/100 and percentage different?

[6] ____.

Now answer the questions in the table on the following page.

Variable	Comment or Question	Circle or Fill in Your Answer
Socioeconomic status (SES) quintiles	State the value of Rate/100 Secondary Amenorrhea given in the table for each quintile of the variable SES:	
	7. I (low SES)	7 ________
	8. II	8 ________
	9. III	9 ________
	10. IV	10 ________
	11. V (high SES)	11 ________
	12. What type of measurement scale is being used for the variable SES?	12 Nominal/Ordinal/Interval/Ratio
Body mass index (BMI) quartile	State the value of the rate for each quartile of BMI:	
	13. I (thin girls)	13 ________
	14. II	14 ________
	15. III	15 ________
	16. IV (heavy girls)	16 ________
	17. What type of measurement scale is being used for the variable BMI?	17 Nominal/Ordinal/Interval/Ratio
Chronological age	18. What type of measurement scale was used for this variable?	18 Nominal/Ordinal/Interval/Ratio
Gynecologic age	19. What type of measurement scale was used for this variable?	19 Nominal/Ordinal/Interval/Ratio

LITERATURE EXAMPLE 3.3

Adolescent Smoking, Weight Changes, and Binge-Purge Behavior: Associations with Secondary Amenorrhea

Jim Johnson, PhD, and Agnes H. Whitaker, MD

Background. The association of secondary amenorrhea with extreme forms of substance use, weight control, and exercise in nonrepresentative samples raises questions as to whether adolescents in the general population who engage in these behaviors are at increased risk for secondary amenorrhea. We examined the prevalence and behavioral correlates of secondary amenorrhea in a county-wide high school population of 2544 girls aged 13 to 18.

Methods. A survey questionnaire, which elicited menstrual history as well as weight history, weight control practices, level of exercise, and use of cigarettes, wine, and beer, was administered during school hours; absentees were also surveyed. The completion rate was 91%.

Results. The 1-year prevalence of secondary amenorrhea was 8.5%. Secondary amenorrhea was associated with smoking one or more packs of cigarettes per day (adjusted relative risk [RRa] = 1.96, 1.21–3.10), with multiple binge-eating behaviors in combination with laxative use or self-induced vomiting (RRa = 4.17, 2.54–6.32), and with weight fluctuation due to weight control (RRa = 2.59, 1.33–4.79). There was no association between amenorrhea and alcohol consumption or exercise level.

Conclusions. Estimates of attributable risk are provided and indicate that bulimic behaviors and cigarette smoking may result in a considerable excess of cases of secondary amenorrhea in an adolescent population. (*Am J Public Health.* 1992;82:47–54)

TABLE 3—Rate/100 of Adolescent Females Who Reported Missing Three Consecutive Menstrual Periods in the Past Year, by Background Characteristics (n = 2156)

	Rate/100 Secondary Amenorrhea
Socioeconomic status (SES) quintiles	
I (low SES)	10.4
II	10.2
III	7.5
IV	8.5
V (high SES)	8.5
Body mass index (BMI) quartile	
I (thin girls)	8.9
II	9.3
III	8.1
IV (heavy girls)	7.8
Chronological age	
13	10.8
14	7.8
15	9.4
16	7.9
17	8.1
18	9.8
Gynecologic age	
0	12.5
1	13.5
2	9.7
3	9.6
4	6.1
5	8.5
6	5.3
7	5.4

ACTIVITY 3.10

Literature Example 3.4 is the abstract of an article entitled "Employment Status and Heart Disease Risk Factors in Middle-Aged Women: The Rancho Bernardo Study." What are the three variables in the first column in Table 1? [1] ____________ [2] ____________ [3] ____________. What is the variable specified in the top row of the table? [4] ____________. What are its values? [5] ____________ [6] ____________. What is reported in the rightmost column? [7] ____________. Now answer the questions in the table below.

Variable	Comment or Question	Circle or Fill in Your Answer
Married	What are the values?	[8] ____________ ____________
	What are the percentages of unmarried women who are employed and unemployed?	[9] ______ ______
	What type of variable is this?	[10] Nominal/Ordinal/Interval/Ratio
Hollingshead Index	How was this index determined?	[11]
	If you make a value judgment that says professionals are better than those in other occupations, the Hollingshead Index could be construed as an ordinal scale. We consider it a nominal scale. What is your opinion?	[12] Nominal/Ordinal/Interval/Ratio
Education	What type of scale is used to measure this variable?	[13] Nominal/Ordinal/Interval/Ratio

LITERATURE EXAMPLE 3.4

Employment Status and Heart Disease Risk Factors in Middle-Aged Women: The Rancho Bernardo Study

Donna Kritz-Silverstein, PhD, Deborah L. Wingard, PhD, and Elizabeth Barrett-Connor, MD

Background. In recent years, an increasing number of women have been entering the labor force. It is known that in men, employment is related to heart disease risk, but there are few studies examining this association among women.

Methods. The relation between employment status and heart disease risk factors including lipid and lipoprotein levels, systolic and diastolic blood pressure, fasting and postchallenge plasma glucose and insulin levels, was examined in 242 women aged 40 to 59 years, who were participants in the Rancho Bernardo Heart and Chronic Disease Survey. At the time of a follow-up clinic visit between 1984 and 1987, 46.7% were employed, primarily in managerial positions.

Results. Employed women smoked fewer cigarettes, drank less alcohol, and exercised more than unemployed women, but these differences were not statistically significant. After adjustment for covariates, employed women had significantly lower total cholesterol and fasting plasma glucose levels than unemployed women. Differences on other biological variables, although not statistically significant, also favored the employed women.

Conclusions. Results of this study suggest that middle-aged women employed in managerial positions are healthier than unemployed women. (*Am J Public Health.* 1992;82:215–219)

TABLE 1—Age-Adjusted Comparisons of Employed and Unemployed Women Aged 40–59 on Socioeconomic Variables: Rancho Bernardo, Calif, 1984–1987 (n = 242)

Variable	Unemployed (n = 113)	Employed (n = 129)	χ^2
	Percent	Percent	
Married (% yes)	92.5	79.0	6.86**
Hollingshead Index[a]			
I (professionals, executives)	48.2	37.0	
II (managers, business owners)	23.2	38.6	
III (administrators)	17.0	16.5	
IV–X (secretarial, sales, clerical, other workers)	11.6	7.9	0.19
Education			
Graduate or professional	1.8	8.7	
College graduate	13.4	22.0	
Some college	42.9	33.9	
High school graduate	38.4	33.9	
Some high school or less	3.5	1.5	11.54*

[a]Hollingshead Index to assess social class of household was based on the usual occupation of the head of household.
*$P < .05$, **$P < .01$.

Activity 3.11

Literature Example 3.5 is from an article entitled "Attitudes about Infertility Interventions among Fertile and Infertile Couples." The first column can be thought of as giving values for a variable called "Intervention." Intervention is a [1] Nominal/Ordinal/Interval/Ratio variable. Favorability toward each infertility intervention is scored on a five-point scale. The values in this scale are [2] ______________. (See the first line of the notes underneath the table.) What type of measurement is it? [3] Nominal/Ordinal/Interval/Ratio. The scale is assigned values 1 through 5. Does it make sense to say 1 + 2 = 3 in this context? [4] Yes/No. Could there be values between 1 and 2? [5] Yes/No.

The authors compute mean favorability values of the various interventions. They compare wives and husbands and also infertile and fertile couples on their assessment of favorability values. Then they compute *t*-test scores. (A *t*-test is done to see if the variation in responses that was observed was statistically significant. This test is the subject of Chapter 14.)

Let us look at the mean favorability of infertile couples and fertile couples toward male hormones. The mean for infertile couples = [6] _______; the mean for fertile couples = [7] _______. Both means are between 4 and 5 on the favorability scale. The *t*-test shows that these results are statistically significant. Read the second sentence of the Results section of the abstract. Do the authors think this is a noteworthy result? [8] Yes/No.

Strictly speaking, an ordinal scale is not supposed to lend itself to taking means because it does not really make sense to add different numerical values of the scales. However, practitioners find it useful to take means and test for the differences, so it is done. There are other ways of testing to see if the observed distributions could have happened by chance. These will be explored in Chapter 16.

LITERATURE EXAMPLE 3.5

Attitudes about Infertility Interventions among Fertile and Infertile Couples

L. Jill Halman, RN, PhD, Antonia Abbey, PhD, and Frank M. Andrews, PhD

Background. There has been marked progress in the development of infertility interventions. This paper reports attitudes about 11 interventions for infertility.

Methods. Face-to-face interviews were conducted with each member of 185 infertile and 90 presumed fertile couples in southeastern Michigan.

Results. Seven of these interventions were generally viewed favorably and four were generally viewed negatively, regardless of the couple's fertility status. Infertile couples viewed all interventions, except for adoption, more favorably than did fertile couples. Multidimensional scaling was used to cluster the interventions according to similarity in endorsement. These clusters form a continuum from interventions that allow only one member of the couple to be a biological parent to the most noninvasive techniques. All clusters remain roughly equidistant from adoption, in which neither member of the couple is a biological parent.

Conclusions. Interventions that produce a child who is biologically related to only one member of the couple were viewed most negatively. Members of couples who were receiving fertility treatment made finer discriminations among infertility interventions than did individuals who had not received treatment. (*Am J Public Health.* 1992;82:191–194)

TABLE 1—Favorability toward Infertility Interventions

	Infertile[a]				Fertile			
	Wives	Husbands	Paired *t* test	Couples	Wives	Husbands	Paired *t* test	Couples[b]
Male hormones	4.7	4.3	***	4.5 (342)	4.1	3.8	*	4.0 (166)***
Artificial insemination with husband's sperm	4.6	4.3	***	4.5 (368)	4.1	4.1	NS	4.1 (176)***
Progesterone suppositories[c]	4.5	4.3	NS	4.4 (306)	4.0	3.8	NS	4.0 (153)***
Drugs to stimulate ovulation	4.6	4.1	***	4.4 (364)	3.9	3.7	NS	3.8 (166)***
Tying cervix[c]	4.4	4.0	***	4.3 (276)	3.9	3.7	NS	3.8 (142)***
In vitro fertilization	4.2	4.0	**	4.1 (363)	3.8	4.0	NS	4.0 (176)
Adoption	4.1	4.0	*	4.0 (369)	4.2	4.4	NS	4.3 (179)***
Artificial insemination with husband's and donor's sperm mixed	2.6	2.4	NS	2.5 (355)	2.2	2.4	NS	2.3 (170)*
Artificial insemination with donor's sperm	2.5	2.3	*	2.4 (369)	2.0	2.3	*	2.1 (177)***
Surrogacy (wife's ova, husband's sperm)	2.2	2.0	NS	2.2 (365)	1.9	2.3	***	2.1 (179)
Surrogacy (surrogate's ovum, husband's sperm)	1.9	2.0	NS	2.0 (367)	1.7	2.0	***	1.8 (179)

Notes. Numbers in columns are mean scores; number of cases is in parentheses. The scale ran from 5 (favorable) to 1 (unfavorable). NS = not significant.
[a]Infertile couples are defined as those couples who have seen an infertility specialist.
[b]Mean scores of infertile couples were compared with mean scores of fertile couples using the Student *t* test to determine the significance of differences.
[c]15% or more of the total respondents stated they did not know enough about these interventions to state an opinion.
*$P \leq .1$.
**$P \leq .05$.
***$P \leq .01$.

ACTIVITY 3.12

Conversion of a measured variable into a categorical variable

Listed below are the ordered ages of the students in two statistics classes.

20 20 20 20 20 21 21 21 21 21 21 21 21 21 21 21 21 22
22 22 22 22 23 23 23 23 23 23 23 23 24 24 24 24 24 25
25 25 26 26 28 29 30 30 30 32 33 33 33 38 38

AGE is a measured variable. AGE can also be described as a [1] Nominal/Ordinal/Interval/Ratio variable. We want to convert this measured variable into a categorical variable and count the number of persons in each category. This process is known as grouping the data. Complete the table below.

AGE Interval	Count, or Frequency	Percent (count * 100) ÷ (total number of students)
20–29		
30–39		

Self-Assessment

Please assess your competency.

Tasks:	How well can you do it?					Page
	Poorly				*Very well*	
1. Define and give examples of the following types of variables:						
Nominal	1	2	3	4	5	47
Ordinal	1	2	3	4	5	47
Interval	1	2	3	4	5	47
Ratio	1	2	3	4	5	47
2. Identify the following types of variables in any given questionnaire or table:						
Nominal	1	2	3	4	5	47
Ordinal	1	2	3	4	5	47
Interval	1	2	3	4	5	47
Ratio	1	2	3	4	5	47
3. Convert ratio variables into categorical variables.	1	2	3	4	5	53

Answers for Chapter 3

Answers for Table 3.2: 1) 5 people 2) Any fraction between 1 and 2 3) Three times 4) 2

Answers for Table 3.3: 1) Ratio 2) Interval 3) Ratio 4) Ratio 5) Ratio 6) Ratio 7) Nominal 8) Nominal 9) Ratio

Answers for Activity 3.2: 1) Nominal 2) Ratio 3) Ordinal 4) Ratio 5) Ratio 6) Ratio 7) Ratio 8) Nominal 9) Ordinal 10) Ordinal 11) Nominal 12) Nominal 13) Ordinal 14) Interval 15) The question is unstructured, so no measurement is possible.

Answers for Activity 3.3: 1) Nominal 2) Two 3) Female 4) 1 5) 2 6) Does not 7) Are no 8) Does not 9) Is not 10) Does not 11) Is not 12) Categories 13) Percents

Answers for Activity 3.4: 1) Ordinal 2) Minimal, moderate, vigorous 3) 1 4) 2 5) 3 6) Some researchers do compute means when they wish to compare the activity levels of two groups, so in a very limited way it does make sense. 7) Are; values of ACTIVITY are actually continuous, though we chose three words to measure ACTIVITY level. 8) Does 9) Is 10) Does not 11) Is not 12) Categories 13) Percents

Answers for Activity 3.5: 1)

Wt Code	Count	Percent
1	15	29.41
2	16	31.37
3	9	17.65
4	7	13.73
5	2	3.92
6	2	3.92
N =51		

2) Gain

Answers for Activity 3.6: 1) White 2) Black 3) Nominal 4) Yes 5) No 6) Nominal 7) No 8) Nominal 9) Ordinal 10) Ordinal 11) Ratio 12) Ratio 13) Black, White 14) Nominal

Answers for Activity 3.7: 1) Proportions or Percentages 2) More 3) Ratio 4) Ratio 5) No 6) Discrete 7) Nominal 8) Nominal 9) Douche regularly 10) Do you douche? or an equivalent question 11) Nominal 12) Nominal 13) Yes 14) Ordinal

Answers for Activity 3.8: 1) % who agree 2) % who disagree 3) Ordinal 4) Nominal 5) 27.8%, 91.2%, 68.5%, 69.6%, 76.6%

Answers for Activity 3.9: 1) Socioeconomic status 2) Body mass index 3) Chronological age 4) Gynecologic age 5) Rate/100 of secondary amenorrhea 6) There is no difference; they are exactly the same number. 7) 10.4 8) 10.2 9) 7.5 10) 8.5 11) 8.5 12) Ordinal 13) 8.9 14) 9.3 15) 8.1 16) 7.8 17) Ordinal 18) Ratio 19) Ratio

Answers for Activity 3.10: 1) Marital status 2) Hollingshead Index 3) Education 4) Employment status 5) Employed 6) Unemployed 7) χ^2 is a statistic computed to see whether the values indicate a statistically significant difference. Details can be found in Chapter 16. 8) Yes and no 9) 7.5%, 21% 10) Nominal 11) See footnote in the table; for more details you need to read the article and references. 12) Give your opinion; we gave ours already. 13) Ordinal

Answers for Activity 3.11: 1) Nominal 2) 1 (unfavorable), 2, 3, 4, 5 (favorable) 3) Ordinal 4) No 5) Yes; there can be more than five shades of opinion. 6) 4.5 7) 4.0 8) Yes; they wouldn't have mentioned it if they hadn't thought it was noteworthy.

Answer for Activity 3.12: Ratio

Answers for Table 3.12:

Interval	Count	Percent
20–29	42	82.35
30–39	9	17.65
	$N = 51$	

CHAPTER

4

Sampling from Populations

Prelude

"Yes, it was," agreed Milo, rubbing his head and dusting himself off, "but I think I'll continue to see things as a child. It's not so far to fall."

"A wise decision, at least for the time being," said Alec. "Everyone should have his own point of view."

"Isn't this everyone's Point of View?" asked Tock, looking around curiously.

"Of course not," replied Alec, sitting himself down on nothing. "It's only mine, and you certainly can't always look at things from someone else's Point of View. For instance, from here that looks like a bucket of water," he said, pointing to a bucket of water, "but from an ant's point of view it's a vast ocean, from an elephant's just a cool drink, and to a fish, of course, it's home. So, you see, the way you see things depends a great deal on where you look at them from. . . ."

— Norton Juster
The Phantom Toll Booth, pp. 107–108

Chapter Outline

Self-Assessment

Please assess your competency.

Tasks:	How well can you do it?					Page
	Poorly				*Very well*	
1. Select the following types of samples when a random number table and populations are given:						
Simple random sample	1	2	3	4	5	80
Stratified random sample	1	2	3	4	5	81
Proportional stratified random sample	1	2	3	4	5	83
Multistage sample	1	2	3	4	5	83
2. Assign subjects to experimental treatment levels.	1	2	3	4	5	85
3. Assign subjects as they are recruited into the experiment.	1	2	3	4	5	86

Introduction

We often generalize the traits we observe in a small group to a larger population. If the small group is truly representative of the larger population, our generalizations serve us well. But we always run the risk of being dead wrong when the small group turns out to be significantly different from the larger population. In this chapter we delineate ways of selecting representative samples.

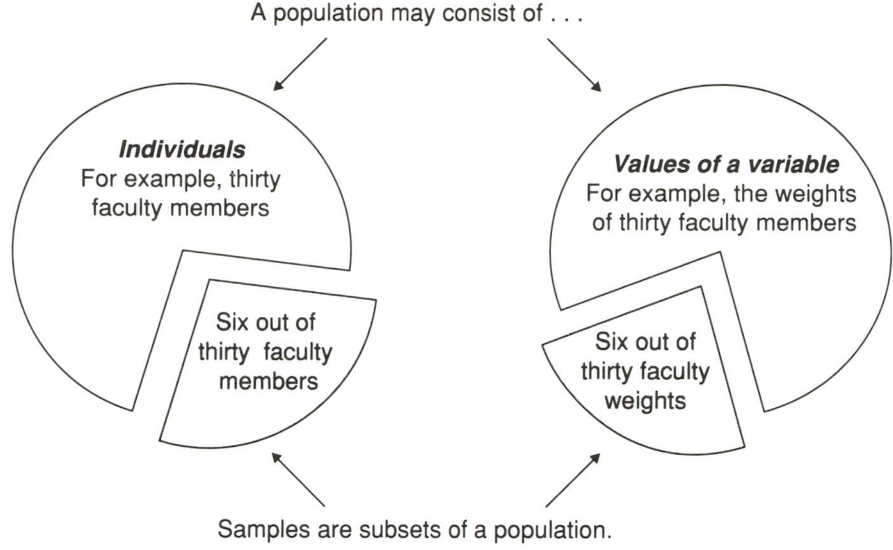
A population may consist of . . .
Individuals
For example, thirty faculty members
Six out of thirty faculty members
Values of a variable
For example, the weights of thirty faculty members
Six out of thirty faculty weights
Samples are subsets of a population.

FIGURE 4.1 Populations and Samples

Samples and Populations

The term **population** is used in two different senses in statistics; it can denote the collection of individuals sharing the characteristics (variables) of interest, or it can denote the collection of values of those variables. Thus, the set of all values of a variable can be called a population. The group of thirty faculty members in the Faculty Exercise Data set is a population in the first sense. The collection of their weights is a population in the second sense. The individual members of a population are called **units**. In this case, the units are each faculty member or each faculty member's weight. A **sample** is a subset of a population, whether the population consists of individuals or values of a variable. Figure 4.1 illustrates these ideas.

ACTIVITY 4.1

Paste each of the 30 cases of Faculty Exercise Data (FED) and the 51 cases of Student Questionnaire Data (SQD) on a playing card or a 3×5 index card. These two decks of cards constitute two populations consisting of units or individuals.

What follows are the weights of all thirty faculty members before exercise (PreWeights, PREWT). Case numbers are given in parentheses. This group of

TABLE 4.1 Sample Drawn by Random Selection from Faculty Exercise Data

	Population	Sample
Individuals	30 cases	6 cases: ______ , ______ , ______ , ______ , ______ , ______
Weights	30 weights	6 weights: ______ , ______ , ______ , ______ , ______ , ______

thirty weights would normally be considered a sample from some larger population (the weights of all male college faculty members in the United States, for instance). For illustrative purposes, however, we will stipulate this set of numbers to be a population.

(01)202	(02)207	(03)160	(04)160	(05)158	(06)225	(07)164	(08)207
(09)180	(10)312	(11)135	(12)146	(13)210	(14)233	(15)255	(16)187
(17)249	(18)210	(19)188	(20)188	(21)170	(22)214	(23)178	(24)162
(25)122	(26)180	(27)312	(28)177	(29)175	(30)163		

Shuffle the FED cards thoroughly and pick six cards. You drew a sample of size six. Sample size is usually represented by the letter n. In this case, $n = 6$. Write the case numbers you drew from the FED deck in Table 4.1. Then write the PREWT that corresponds to each case you selected.

When a deck is well shuffled, drawing cards is one way of obtaining an unbiased selection of cases from the deck. We can also write all the case numbers on pieces of paper, place them in a hat, and draw six cases. These methods are cumbersome, especially when the population is large. A more efficient way to draw unbiased samples is to use a random number table, as elaborated below.

Random Number Tables

"Random" is not synonymous with "haphazard." **Random** has a specific meaning in statistics. When digits ranging from 0 to 9 are arranged in a table in such a way that (1) each one of the digits has an **equal** chance of occurring at each position and (2) each one of the digits has an **independent** chance of occurring (i.e., one digit's occurrence does not influence the occurrence of the next digit), such an arrangement of digits is called a table of random numbers. On the following page is a table of 105 random numbers.

Rows and columns are marked in Table 4.2. The digits were randomly selected. We may read them in groups of one, two, three, four, or more, depending upon the maximum number of digits in the size of the population.

Choosing Samples Using Random Number Tables

If a population has fewer than one hundred members, we can number the population units 01, 02, 03, etc. (Even if the population has ninety-nine members, there will be enough two-digit numbers to label all of its members.)

TABLE 4.2 A Short Table of Random Numbers

	Columns														
	C1	C2	C3	C4	C5	C6	C7	C8	C9	C10	C11	C12	C13	C14	C15
Row 1	1	1	8	9	8	6	6	1	1	8	5	9	7	1	2
Row 2	4	5	1	2	4	9	1	5	8	7	6	7	4	4	8
Row 3	2	1	6	6	3	3	0	5	0	6	6	6	5	0	8
Row 4	8	2	5	2	5	4	5	4	2	2	2	2	1	3	4
Row 5	7	2	3	9	6	9	2	1	8	2	0	6	6	6	8
Row 6	5	2	7	0	9	7	4	6	7	9	1	8	3	6	0
Row 7	6	0	6	8	2	6	8	4	0	3	2	4	0	9	1

Suppose you wanted to choose a sample of size six from FED. Since FED has thirty cases, this procedure is appropriate. You should choose a different starting point in the table of random numbers each time you use it. If you always start from the first number in the table, you will always get the same sequence of numbers. To find a roughly random starting point you can close your eyes and point at a place in the table with your pencil. Suppose it lands on row 3, column 6 (which we will abbreviate R3 C8). Starting there, you read digits in groups of two until you have selected six cases. Discard duplicate numbers and those that exceed 30. If you arrive at the end of the list before you have drawn your whole sample, continue from the beginning of the list. Cases corresponding to the random numbers underlined in the table below will be selected as the sample. The following numbers are discarded: 50, 66, 65, 82, 52, 54, 54, 22 (duplicate), 47, 96, and 92. Therefore, cases included in the sample are ______, ______, ______, ______, ______, ______.

	Columns														
	C1	C2	C3	C4	C5	C6	C7	C8	C9	C10	C11	C12	C13	C14	C15
Row 1	1	1	8	9	8	6	6	1	1	8	5	9	7	1	2
Row 2	4	5	1	2	4	9	1	5	8	7	6	7	4	4	8
Row 3	2	1	6	6	3	3	0	5	0	6	6	6	5	0	8
Row 4	8	2	5	2	5	4	5	4	2	2	2	2	1	3	4
Row 5	7	2	3	9	6	9	2	1	8	2	0	6	6	6	8
Row 6	5	2	7	0	9	7	4	6	7	9	1	8	3	6	0
Row 7	6	0	6	8	2	6	8	4	0	3	2	4	0	9	1

Table 4.3

	Columns														
	C1	C2	C3	C4	C5	C6	C7	C8	C9	C10	C11	C12	C13	C14	C15
Row 1	1	1	8	9	8	6	6	1	1	8	5	9	7	1	2
Row 2	4	5	1	2	4	9	1	5	8	7	6	7	4	4	8
Row 3	2	1	6	6	3	3	0	5	0	6	6	6	5	0	8
Row 4	8	2	5	2	5	4	5	4	2	2	2	2	1	3	4
Row 5	7	2	3	9	6	9	2	1	8	2	0	6	6	6	8
Row 6	5	2	7	0	9	7	4	6	7	9	1	8	3	6	0
Row 7	6	0	6	8	2	6	8	4	0	3	2	4	0	9	1

This procedure can be easily generalized to any size population. If the population has more than one hundred members but less than one thousand, you will need three-digit labels, and you will choose digits from the table in groups of three. Similarly, a population with more than one thousand would require four or more digits in the labels. (One-digit numbers could be used if there were nine or fewer members in the population; however, you would not be likely to need a random sample from such a small population; you would use the entire population.)

Suppose you wanted to select a sample of six cases from a population with 999 members. You would need to select digits in groups of three from the random number table. To do this, you would first close your eyes and point to some spot on the table. Assume that your pencil landed on R2, C8. Start at R2, C8 in Table 4.3, and underline groups of three numbers until you select six three-digit numbers. The first number is underlined; you select the other five numbers.

Suppose we need to select a sample of five cases from a population with 9,999 members. We need to select random digits in groups of four from a random number table. Blindly select a place in the table below and underline the applicable numbers.

	Columns														
	C1	C2	C3	C4	C5	C6	C7	C8	C9	C10	C11	C12	C13	C14	C15
Row 1	1	1	8	9	8	6	6	1	1	8	5	9	7	1	2
Row 2	4	5	1	2	4	9	1	5	8	7	6	7	4	4	8
Row 3	2	1	6	6	3	3	0	5	0	6	6	6	5	0	8
Row 4	8	2	5	2	5	4	5	4	2	2	2	2	1	3	4
Row 5	7	2	3	9	6	9	2	1	8	2	0	6	6	6	8
Row 6	5	2	7	0	9	7	4	6	7	9	1	8	3	6	0
Row 7	6	0	6	8	2	6	8	4	0	3	2	4	0	9	1

TABLE 4.4 One Thousand Random Numbers

7	3	6	7	4	1	8	4	8	8	3	5	9	0	2
9	3	4	2	5	9	6	1	5	6	8	1	7	5	4
0	3	7	2	1	4	6	6	8	1	9	8	2	0	4
6	2	8	3	4	3	9	6	3	6	9	0	2	2	9
4	5	9	2	7	5	7	6	3	9	6	4	0	4	6
2	7	0	9	6	1	9	1	4	1	8	6	4	7	8
3	6	5	4	1	8	0	9	4	8	8	5	5	3	2
9	5	1	5	6	8	7	8	9	6	1	5	6	3	4
5	2	6	0	4	1	7	2	8	8	1	4	9	8	3
7	8	8	3	7	6	5	4	5	9	6	4	2	2	6
6	6	3	4	4	6	9	4	1	1	8	5	3	3	5
9	8	7	5	4	7	7	9	0	5	9	8	3	7	5
1	6	6	3	5	4	1	2	7	0	7	4	6	5	0
3	1	4	6	1	1	2	3	4	4	6	1	2	9	0
3	4	4	3	7	1	3	9	0	4	1	9	7	0	6
1	4	1	6	3	3	7	8	7	8	5	3	4	3	0
8	8	8	6	3	1	3	6	9	5	8	5	6	7	3
4	3	8	8	8	9	9	8	4	3	2	0	3	0	0
8	6	6	6	1	6	6	3	0	7	0	1	5	4	5
2	4	2	5	2	5	3	1	3	0	2	0	3	5	4
6	1	3	9	1	9	2	2	5	0	9	4	1	1	0
9	7	0	3	5	4	2	5	9	9	7	4	8	6	8
3	0	1	1	5	2	2	3	4	7	5	7	5	9	6
5	1	8	4	2	0	1	0	5	0	1	6	8	6	7
8	8	5	8	0	1	8	1	3	3	0	6	0	0	2
1	7	7	1	0	2	9	7	0	2	9	7	1	0	9
2	8	0	3	1	8	1	8	1	6	4	7	7	6	1
1	1	9	8	5	7	8	2	7	1	7	3	9	1	0
9	9	5	3	7	4	3	0	9	1	7	0	3	3	3
4	6	8	6	6	3	4	4	2	9	2	8	0	6	0
1	5	0	1	6	0	1	3	2	5	8	6	4	0	3
9	6	6	9	2	8	6	0	0	4	5	0	3	6	3
8	8	8	9	0	3	9	7	5	8	0	2	7	4	8
1	0	4	5	5	5	8	3	4	1	2	5	6	5	2
6	5	4	6	1	0	8	9	0	0	1	3	4	5	1

TABLE 4.4 Continued

7	4	9	8	0	1	6	4	1	6	0	6	3	3	7
0	1	4	9	9	2	7	0	9	8	0	8	9	8	5
0	3	2	7	6	8	2	3	6	9	1	8	3	0	8
6	5	9	0	0	8	6	6	5	8	6	4	0	0	6
3	0	7	2	2	5	9	5	1	6	5	4	5	9	5
2	2	6	3	3	6	1	0	5	8	3	4	2	0	1
4	3	7	3	7	2	4	3	7	6	9	6	0	6	6
7	8	0	7	9	4	8	5	4	1	9	6	3	8	0
5	6	2	2	0	1	0	8	0	5	5	2	0	0	9
6	1	7	8	7	7	0	9	6	1	1	9	0	3	1
7	5	3	9	7	2	0	3	1	8	5	2	2	1	0
6	6	2	0	4	3	6	0	5	0	7	8	1	3	8
4	1	2	2	2	7	8	1	5	7	6	9	0	1	1
5	6	7	0	3	8	4	8	5	1	4	4	9	1	5
3	7	5	6	9	2	6	7	3	8	2	7	8	3	4
4	4	0	7	4	4	9	2	0	0	7	7	8	4	4
8	3	8	0	4	0	7	0	5	5	0	9	3	1	1
8	3	8	9	8	8	5	2	1	7	1	1	3	0	7
6	8	2	2	5	2	1	7	7	9	1	4	3	3	1
0	6	1	9	3	0	3	2	1	4	9	0	8	4	9
9	3	8	5	9	3	6	0	8	3	9	4	1	2	6
4	4	6	7	3	8	1	9	5	6	1	1	8	1	2
7	5	2	7	8	0	2	5	2	6	5	2	5	1	1
6	3	2	5	0	9	2	8	4	3	8	2	8	7	9
1	3	5	0	0	5	9	9	7	0	7	7	7	4	5
4	8	9	0	6	4	4	7	7	5	4	4	6	2	9
1	1	7	6	1	9	2	2	8	5	8	0	7	2	6
6	7	9	9	9	0	1	0	8	9	3	0	2	3	5
5	3	6	6	9	7	8	6	4	7	4	6	8	2	5
4	9	6	9	1	4	7	4	9	4	5	6	2	3	4
2	6	3	6	0	4	1	9	6	2	2	6	3	9	4
8	8	6	2	4	4	9	3	0	0					

One slight improvement can be made to the technique we have been describing in the rare instance that the population has exactly one hundred, one thousand, ten thousand, one million, . . . members. Use zero to number the first member. Of course, you will need to express zero as 00, 000, 0000, or however many zeros are appropriate. Using this system, two-digit numbers will suffice to label one hundred numbers, three-digit to label one thousand, etc.

Table 4.4 is a longer random number table. Notice that it is laid out in five-row groups. There's no special significance to groups of five; the numbers are grouped that way merely to make the table easier to read. Frequently, random number tables have the columns grouped as well as the rows.

There are many random number tables in print, some of them quite extensive, but current practice is to use computer software to generate random numbers for sampling.

ACTIVITY 4.2

Will we obtain the same cases each time we take a random sample? Obtain nine random samples of size $n = 5$ using Table 4.4 and numbers between 01 and 30.

In Table 4.5, write the weights from the FED list below that correspond to the case numbers you selected. We drew the first sample for you and entered it as Sample 1 on Table 4.5.

(01)202 (02)207 (03)160 (04)160 (05)158 (06)225 (07)164 (08)207
(09)180 (10)312 (11)135 (12)146 (13)210 (14)233 (15)255 (16)187
(17)249 (18)210 (19)188 (20)188 (21)170 (22)214 (23)178 (24)162
(25)122 (26)180 (27)312 (28)177 (29)175 (30)163

TABLE 4.5 Ten Random Samples (n = 5) of Weights from FED

Weights from Selected Cases	Sample									
	1	2	3	4	5	6	7	8	9	10
Value 1	158	____	____	____	____	____	____	____	____	____
Value 2	225	____	____	____	____	____	____	____	____	____
Value 3	170	____	____	____	____	____	____	____	____	____
Value 4	214	____	____	____	____	____	____	____	____	____
Value 5	122	____	____	____	____	____	____	____	____	____

Each time we take a random sample we will obtain a different sample/the same sample. We can take many such random samples. The ten samples we actually obtained are only a few of the many possible samples we might have selected through random sampling from the population. Empirical experience on sampling shows that most of the time we will obtain a sample that gives us useful estimates of the population characteristics. It is possible, however, for a randomly chosen sample to be atypical. For example, our sample could include only the heaviest men in the population, but chances of drawing such a sample are small.

You should always use random number tables, computer programs, or other statistically tested methods for drawing random samples, because methods that you devise yourself may be biased. For example, if you shut your eyes and point to the data set with a pencil, you may unwittingly tend to point toward the lower part of the page. If the data happen to be in order of size, smallest to largest, this would bias your sample toward the larger numbers. The tried and true methods, such as random number tables or random number generators, are not always perfect, but they are far more trustworthy than ad hoc methods that we devise ourselves.

Types of Samples for Surveys

Samples are usually taken because populations are too large to study economically. Data are collected from a sample in the hope that inferences about that sample will also be true of the population from which the sample was drawn. These inferences will only apply to the population from which the sample was randomly drawn, however. If we asked fifty history majors how many hours a week they study, we couldn't generalize the answer to the entire university student body. At best, it would apply only to the population of all history majors at that university.

There are many different types of sampling procedures. The procedures for selecting a simple random sample, a stratified random sample, a proportional stratified random sample, and a multistage sample will be discussed here.

Simple Random Sample

A random sample is a sample that is drawn from the population in such a way that all members of that population have an *equal* and *independent* chance of being selected. Furthermore, if the sample has size n, every set of n individuals from the population has the same chance of being chosen. This is an unbiased way of selecting subjects for a study.

The procedure for selecting a simple random sample using a table of random numbers has three steps:

1. Assign identification numbers to variable values or cases.
2. Select numbers from a random number table.
3. Choose the cases that correspond to the selected random numbers.

ACTIVITY 4.3

Procedure	Example
1. Assign identification numbers (1–30) to variable values or cases.	Listed below are ages of faculty from FED. Assign ID numbers in parentheses. ( )61 ( )52 ( )57 ( )43 ( )48 ( )44 ( )61 ( )58 ( )43 ( )51 ( )57 ( )41 ( )40 ( )62 ( )47 ( )54 ( )40 ( )49 ( )61 ( )41 ( )57 ( )44 ( )52 ( )40 ( )52 ( )41 ( )46 ( )52 ( )40 ( )48
2. Select numbers from a random number table. Select numbers in groups that correspond to the number of digits in the highest ID number.	Select seven ID numbers from this random number table. Row 1: 1 1 8 9 8 6 6 1 1 8 5 9 7 1 2 Row 2: 4 5 1 2 4 9 1 5 8 7 6 7 4 4 8 Row 3: 2 1 6 6 3 3 0 5 0 6 6 6 5 0 8 Row 4: 8 2 5 2 5 4 5 4 2 2 2 2 1 3 4 Row 5: 7 2 3 9 6 9 2 1 8 2 0 6 6 6 8 Row 6: 5 2 7 0 9 7 4 6 7 9 1 8 3 6 0 Row 7: 6 0 6 8 2 6 8 4 0 3
3. Choose the cases or variable values that correspond to the selected random numbers.	Write the ages that correspond to the selected random numbers here. _______, _______, _______, _______, _______, _______, _______

Stratified Random Sample

If a population of interest consists of different, homogeneous strata (each stratum consisting of members sharing a selected characteristic), you must divide the population into these strata or subsets before the random sampling is done if you wish to ensure representation from each of the strata.

For example, the FED data contain four homogeneous strata. To obtain a stratified sample, we first sort the FED cases into four groups on the basis of the PARTICIP variable, which indicates a faculty member's participation status in the study. Sort the FED cards you made earlier into four groups:

1. Non-participant.
2. Control
3. Low-intensity exercise
4. High-intensity exercise

Verify your groups by referring to the following table, and write new ID numbers in the parentheses for the cases in each of the strata. (Assigning new

ID numbers will facilitate use of the random number table. All of the cases in the first stratum and some of the cases in the second stratum have been renumbered for you.) Then select $n = 3$ from random number Table 4.4 for each stratum and circle the cases you selected.

ID Numbers (in parentheses) and Cases

Stratum I Non-Participant	Stratum II Control	Stratum III Low-Intensity Exercise	Stratum IV High-Intensity Exercise
(1)1, (2)2, (3)8, (4)16, (5)19, (6)30	(1)3, (2)10, (3)13, ( )14, ( )18, ( )22, ( )25	( )6, ( )7, ( )9, ( )11, ( )21, ( )24, ( )27, ( )28	( )4, ( )5, ( )12, ( )15, ( )17, ( )20, ( )23, ( )26, ( )29

ACTIVITY 4.4

This activity is designed to illustrate what might happen if we took a simple random sample instead of a stratified random sample. Take a simple random sample $n = 12$ and circle the case numbers in the following table.

FED Cases by Strata

Non-Participants	Control	Low-Intensity Exercise	High-Intensity Exercise
1, 2, 8, 16, 19, 30	3, 10, 13, 14, 18, 22, 25	6, 7, 9, 11, 21, 24, 27, 28	4, 5, 12, 15, 17, 20, 23, 26, 29

Now list below the number of cases selected using the stratified random sample and the simple random sample methods.

	Number of Cases Selected by Each Method	
Participant Status	**Stratified Random Sample**	**Simple Random Sample**
Non-participants	3	
Controls	3	
Low-intensity exercise	3	
High-intensity exercise	3	

Write a sentence giving your observations about both methods. When should you use a stratified random sample?

____

____

Proportional Stratified Random Sample

With a **proportional stratified random sample**, you are interested in having the same proportion of cases in each stratum of the sample as in the population.

ACTIVITY 4.5

Set aside the non-participant cases. The FED population now consists of $30 - 6 = 24$ cases. There are seven controls and seventeen exercise cases. Now there are two strata in the FED population. The proportion of controls in the FED is 0.29 and the proportion of exercise cases is 0.71.

Using Table 4.4, pick eight cases ($n = 8$) to reflect the proportion of cases in each stratum. Pick $0.29 * 8 \approx 2$ of the cases from the control group and pick $(0.71 * 8 \approx 6$ of the cases from the exercise group and circle them in the table below. (Note: The symbol $\approx$ means "approximately equal to.") Begin by assigning ID numbers as in Activity 4.4.

Control Cases	Exercise Cases
3, 10, 13, 14, 18, 22, 25	6, 7, 9, 11, 21, 24, 27, 28, 4, 5, 12, 15, 17, 20, 23, 26, 29

Multistage Sample

Suppose we intend to study the occurrence of violence on university campuses by administering questionnaires to students attending classes. We may do sampling for the study in several stages.

1. Take a simple random sample (for example, $n = 10$) from a list of all the universities.
2. Take simple random samples (for example, $n = 5$) from lists of departments within each university.
3. Take a simple random sample (for example, $n = 2$) from the class lists within each selected department. Now you have one hundred classes you can survey.

Table 4.6 Terminology for Variables in Experiments and Examples from Faculty Exercise Data

Variables	Alternative Names for Variables	Examples	Value	Examples
Explanatory variable	Treatment	Exercise	Different kinds of treatments	Low-intensity exercise, high-intensity exercise, no exercise
	Factor	Exercise	Different levels	Level 1: no exercise, Level 2: low-intensity exercise, Level 3: high-intensity exercise
	Independent variable	Exercise	Different kinds of treatments	
Response variable	Dependent variable	Oxygen consumption	Measured in liters per minute	1.18

Experiments

Terminology: Subjects, Units, and Variables

How does exercise affect a person's oxygen consumption? We will attempt to answer this research question using the FED data. In the faculty exercise experiment, the faculty members who participated in the study are the **subjects**, or units, and the exercise is the **treatment**. Oxygen consumption is the **response variable**, or **dependent variable**, and exercise is the **explanatory variable**, or **independent variable**. The objective of the experiment is to measure the effect of explanatory variables on the response variable(s). Treatments (explanatory variables) are also called **factors** and their different values are called **levels**. Low-intensity exercise and high-intensity exercise are two different levels of the factor exercise. Table 4.6 summarizes these terms.

Activity 4.6

Give two examples of response variables from the FED.

1 ______________, 2 ______________

What are three alternative names for the explanatory variable?

[3] ______________, [4] ______________, [5] ______________

What are the values of factors called? [6] ______________

What is an alternative name for the faculty members who participated in this experiment?

[7] ______________

Assigning Subjects to Treatment Levels

We use a random number table to assign a certain number (determined ahead of time) of the FED cases to each level of treatment. The first seven subjects picked using the random number table will be assigned to the control group. The next eight picked will be assigned to the low-intensity exercise group, and the remaining nine will be assigned to the high-intensity exercise group.

ACTIVITY 4.7

Random assignment of subjects to treatment levels

First write new ID numbers in the parentheses to facilitate use of the random number table:

( )3 ( )4 ( )5 ( )6 ( )7 ( )9 ( )10 ( )11 ( )12 ( )13 ( )14
( )15 ( )17 ( )18 ( )20
( )21 ( )22 ( )23 ( )24 ( )25 ( )26 ( )27 ( )28 ( )29

(The case numbers of the non-participants are missing.) Then complete the following table.

Levels of the factor (treatment) or explanatory variable	
Level 1: No exercise (control group)	Randomly select seven subjects and write their case numbers (not ID numbers) here: ______, ______, ______, ______, ______, ______, ______
Level 2: Low-intensity exercise	Randomly select eight subjects from the remaining seventeen and write their case numbers (not ID numbers) here: ______, ______, ______, ______, ______, ______, ______, ______

Level 3: High-intensity exercise

Write the remaining nine case numbers (not ID numbers) here.

_______, _______, _______, _______, _______,
_______, _______, _______, _______

The case numbers in the table will not coincide with the case assignments in the actual study because each random sample is like/unlike others.

Assigning Subjects as They Are Recruited

Sometimes you have not yet selected the subjects when you first plan an experiment. In this case, you first decide how many subjects are to be assigned to the control and the treatment group. Then you randomly select the required number of subjects for each group and record the numbers. The subjects are then assigned to each group as they are recruited.

Activity 4.8

Random assignment of potential subjects to treatment levels

All subjects who can potentially participate are assigned case numbers. Random numbers are selected to determine which cases will be in each group. As the subjects come in they are assigned case numbers, and then they are assigned to the appropriate group on the basis of their case numbers.

Case numbers of all potential participants:

1 2 3 4 5 6 7 8 9 10 11 12 13 14 15 16 17 18
19 20 21 22 23 24

Levels of the factor (treatment) or explanatory variable

Level 1: No exercise (control group)

Randomly select seven subjects and write their case numbers here:

_______, _______, _______, _______, _______,
_______, _______

Level 2: Low-intensity exercise

Randomly select eight subjects from the remaining seventeen remaining cases and write their case numbers here:

_______, _______, _______, _______, _______,
_______, _______, _______

Level 3: High-intensity exercise	Assign the nine remaining subjects to this group and write their case numbers here:
	_______, _______, _______, _______, _______,
	_______, _______, _______, _______

(These case numbers will not coincide with the case assignments in the actual study because each random sample is different.)

Assigning Factor (Treatment) Levels

Suppose the subjects are already assembled in three comparable groups. Now we can use random numbers to assign factor levels to the different groups. There are fifty-one cases in the Student Questionnaire Data. Let us divide them into three groups: Group I ($n = 17$), Group II ($n = 17$), and Group III ($n = 17$). Use Table 4.4 to assign the subjects to these three groups. Now the treatments are assigned to these groups using the random number table, as Activity 4.9 illustrates.

ACTIVITY 4.9

Random assignment of factor levels (treatments) to the groups

1. Randomly assign 51 cases in the Student Questionnaire Data to one of three groups. Such random assignment into the groups eliminates bias. The groups will be comparable.

   Assign new case numbers to the cases in each group here:

   Group I:

   Group II:

   Group III:

2. Focusing on one group at a time, select a one-digit random number, stipulating that a number between 1 and 3 will suggest assignment of the group under consideration to control (no exercise). A number between 4 and 6 will assign the group

under consideration to the low-intensity exercise treatment. A number between 7 and 9 will assign the group under consideration to the high-intensity exercise treatment.

3. Now consider Group I and draw a random number. The random number is ______. Therefore, we assign Group I to ____________.

4. Now consider Group II and draw a random number. The random number is ______. Therefore, we assign Group II to ____________.

5. Now consider Group III. The unassigned treatment is ____________. Therefore, Group III is assigned to ____________.

Examples from the Literature

Investigators curious about various problems occurring in the society as a whole may glean information through surveys, which may provide a small piece in the larger puzzle. Investigators state the problem in concrete terms and then derive checklists or questionnaires from the problem statement. Often investigators are unable to collect their own data and therefore use data originally collected for other purposes. At best, an investigator is like an artist who paints a landscape in such a way as to help us obtain a new insight.

Samples from Computerized Data Banks

Literature Example 4.1 is from an article entitled "Bicycle-Associated Head Injuries and Deaths in the United States From 1984 Through 1988: How Many Are Preventable?" Answer the following questions about the abstract.

What is the research problem? 1 ____________________________

What is the source of the data? 2 ____________________________

Was a sample taken? 3 Yes/No

What useful purpose does this study serve? 4 ____________________________

LITERATURE EXAMPLE 4.1

Bicycle-Associated Head Injuries and Deaths in the United States From 1984 Through 1988

How Many Are Preventable?

Jeffrey J. Sacks, MD, MPH; Patricia Holmgreen, MS; Suzanne M. Smith, MD; Daniel M. Sosin, MD

Objective.—To estimate the potential benefits from more widespread bicycle safety helmet use.

Design.—Review of death certificates and emergency department injury data for 1984 through 1988. Categorization of deaths and injuries as related to bicycling and head injury. Using relative risks of 3.85 and 6.67 derived from a case-control study and varying helmet usage from 10% to 100%, population attributable risk was calculated to estimate preventable deaths and injuries.

Setting.—Entire United States.

Main Outcome Measures.—Numbers of US residents coded as dying from bicycle-related head injuries, numbers of persons presenting to emergency departments for bicycle-related head injuries, and numbers of attributable bicycle-related deaths and head injuries.

Main Results.—From 1984 through 1988, bicycling accounted for 2985 head injury deaths (62% of all bicycling deaths) and 905 752 head injuries (32% of persons with bicycling injuries treated at an emergency department). Forty-one percent of head injury deaths and 76% of head injuries occurred among children less than 15 years of age. Universal use of helmets by all bicyclists could have prevented as many as 2500 deaths and 757 000 head injuries, ie, one death every day and one head injury every 4 minutes.

Conclusions.—Effective community-based education programs and legislated approaches for increasing bicycle safety helmet usage have been developed and await only the resources and commitment to reduce these unnecessary deaths and injuries.

(*JAMA*. 1991;266:3016-3018)

METHODS

Mortality and Morbidity Ascertainment

We used the National Center for Health Statistics Multiple Cause-of-Death Public Use Data Tapes for 1984 through 1988.[14] We defined a bicycle-associated death (BAD) as any US resident coded as dying as the result of a bicycle crash (E800 through E807, with fourth digit 0.3; E810 through E825, with fourth digit 0.6; E826.1 or E826.9; or E827 through E829, with fourth digit 0.1).[15] A head injury bicycle-associated death (HIBAD) was any BAD that had any of the following HI diagnoses as an associated condition[16]: skull fracture (N800.0 through N801.9 and N803.0 through N804.9); intracranial injury (N850.0 through N854.9); and late effects of either skull fracture (N905.0) or intracranial injury without skull fracture (N907.0).

Samples from Populations with a Known Number of Subjects

Literature Example 4.2 is from the article entitled "How Strictly Do Dialysis Patients Want Their Advance Directives Followed?" Answer the following questions about the abstract.

What is the research problem? 1 ______________________

What is the source of the data? 2 ______________________

What is the sample size? 3 ______________________

What is the sampling procedure? 4 ______________________

What purpose does this study serve? 5 ______________________

LITERATURE EXAMPLE 4.2

How Strictly Do Dialysis Patients Want Their Advance Directives Followed?

Ashwini Sehgal, MD; Alison Galbraith; Margaret Chesney, PhD; Patricia Schoenfeld, MD; Gerald Charles, MD; Bernard Lo, MD

Objective.—The *Cruzan* case and the Patient Self-Determination Act will encourage patients to specify in advance which life-sustaining treatments they would want if they become mentally incompetent. However, strictly following such advance directives may not always be in a patient's best interests. We sought to determine whether patients differ in how strictly they want advance directives followed.

Design.—Interview study.

Setting.—Seven outpatient chronic dialysis centers.

Participants.—One hundred fifty mentally competent dialysis patients.

Intervention.—Using a structured questionnaire, we asked the subjects whether they would want dialysis continued or stopped if they developed advanced Alzheimer's disease. We then asked how much leeway their physician and surrogate should have to override that advance directive if overriding were in their best interests. Subjects granting leeway were also asked what factors should be considered in making decisions for them.

Results.—Subjects varied greatly in how much leeway they would give surrogates to override their advance directives: "no leeway" (39%), "a little leeway" (19%), "a lot of leeway" (11%), and "complete leeway" (31%). Subjects also varied in how much they wanted various factors considered in making decisions, such as pain or suffering, quality of life, possibility of a new treatment, indignity caused by continued treatment, financial impact of treatment on family members, and religious beliefs.

Conclusions.—Strictly following all advance directives may not truly reflect patients' preferences. To improve advance directives, we recommend that physicians explicitly ask patients how strictly they want their advance directives followed and what factors they want considered in making decisions.

(*JAMA*. 1992;267:59-63)

Characteristics of Study Participants (N = 150)

Mean age, y (range)	55 (18-88)
Male, %	58
Ethnicity, %	
Black	48
White	33
Asian	13
Hispanic	3
Education, % completing high school	83
Type of dialysis, % hemodialysis	93
Median time since dialysis started, mo (range)	33 (1-257)
Medical illnesses, %	
Heart disease	28
Diabetes	27
Emphysema or bronchitis	12
Stroke	8
Cancer	7
Hospitalization in past 12 mo, %	67
Living with a partner, %	40
Prior oral advance directive, %	43
Prior written advance directive, %	20
Know of case of withdrawal of dialysis, %	27

METHODS

Subjects

We conducted the study from July to December 1990 at seven chronic dialysis units in the San Francisco, Calif, area. These sites included both free-standing and hospital-based units.

The dialysis unit staff identified all English-speaking outpatients 18 years of age or older who had been receiving chronic dialysis for at least 1 month, who were expected to live at least 6 months, and who were likely to be cooperative. We used a two-step procedure to exclude patients who were cognitively impaired. First, the dialysis unit staff excluded patients who were not considered capable of giving informed consent for medical or surgical procedures. Second, at the beginning of the interview, we described a vignette in which a dialysis patient communicates an advance directive to his wife about wanting dialysis discontinued if he ever gets advanced Alzheimer's disease. The vignette included a standardized description of Alzheimer's disease. We asked subjects to summarize this vignette in their own words. We excluded subjects who could not correctly paraphrase key points of the vignette such as the clinical features of Alzheimer's disease and the decision to stop dialysis. We approached and interviewed eligible subjects until we reached our target sample size.

Samples from Existing Records

Literature Example 4.3 is from the article entitled "No-Smoking Laws in the United States: An Analysis of State and City Actions to Limit Smoking in Public Places and Workplaces." Answer the following questions about the abstract.

What is the research problem? 1 ____________________

What is the source of the data? 2 ____________________

How many cities were cited by both ANR and TFA as having laws? How big a sample was taken to confirm these data? 3 ____________________

____

What is the sampling procedure for the state laws? for cities? for checking? 4 ____________________

What purpose does this study serve? 5 ____________________

LITERATURE EXAMPLE 4.3

No-smoking laws in the United States: An analysis of state and city actions to limit smoking in public places and workplaces

Nancy A. Rigotti, MD, and Chris L. Pashos, PhD

Objective

To assess the prevalence, content, and growth of state and city laws restricting smoking in public places and workplaces in the United States and to identify factors associated with their passage.

Design

A mailed survey of city clerks in US cities with a population of 25 000 or greater (N = 980) and review of existing data sources confirmed the status of smoking restrictions in 902 (92%) of the cities in the sample. State laws were identified by contacting each state's Legislative Reference Bureau (100% response). Content of laws was coded using previously developed categories.

Main outcome measures

Prevalence, comprehensiveness, and cumulative incidence of no-smoking laws in states and in cities with a population of 25 000 or greater.

Results

By July 1989, 44 states and 500 (51%) of the cities in our sample had adopted some smoking restriction, but content varied widely. While 42% of cities limited smoking in government buildings, 27% in public places, 24% in restaurants, and 18% in private workplaces, only 17% of cities and 20% of states had comprehensive laws restricting smoking in all four of these sites. The number of city no-smoking laws increased tenfold from 1980 to 1989. City no-smoking laws were independently associated with population size, geography, state tobacco production, and adult smoking prevalence. Laws were more common in larger cities, Western cities, and states with fewer adult smokers. Laws were less common in tobacco-producing states and in the South.

Conclusions

No-smoking laws are more widespread than previously appreciated, especially at the local level, reflecting a rapid pace of city government action in the 1980s. Nonetheless, comprehensive laws, which are most likely to provide meaningful protection from environmental tobacco smoke exposure, remain uncommon and represent a major gap in smoking control policy. Laws are most needed in smaller and non-Western cities and in states that produce tobacco and have a higher proportion of smokers. (*JAMA*. 1991;266:3162–3167)

Methods: data collection

States

To identify current state laws, we updated our previous compilation[2] by requesting a copy of any state no-smoking law from the director of each state's Legislative Reference Bureau in September 1989. Responses were obtained from 49 to 50 states and the District of Columbia; a telephone call to the nonresponding state (Missouri) confirmed that it had no law.

Cities

Using 1980 census data,[12] we identified all US cities and towns with a population of 25 000 or greater (N = 980); this was our sample. We updated population figures for cities with a population of 50 000 or greater using 1986 data.[13] (Updated figures for cities with a population of 25 000 to 50 000 were not available.) We used two data sources to determine whether a smoking restriction was present: (1) information provided by city clerks in response to a mailed request and (2) existing compendia of laws.[10,11] We requested information from all cities in the sample not cited by both sources as having laws (n = 831). We reasoned that there was little chance that a city cited by two independent sources did not have a law. We confirmed this by obtaining from TFA copies of 100% of ordinances from a random sample of 40 of the 149 cities cited by both ANR and TFA as having laws. From April through June 1989, a mailing to city clerks asked them to indicate whether the city had passed or was considering "any law or regulation that restricts smoking in a public place or workplace" and to mail us a copy of any law. A second mailing was sent to nonresponders. We received responses from 753 of 831 cities (91% response rate). For the analysis, a city was counted as having a smoking restriction if it met one of two criteria: (1) we had a copy of a city or town ordinance or regulation or (2) both compendia reported that the city had a no-smoking ordinance.[10,11] County ordinances and laws in cities with a population of less than 25 000 were excluded.

Samples from Lists

Literature Example 4.4 is from the article entitled "Continuing Education: Do Mandates Matter?" The abstracts we have presented up to this point came from the *Journal of the American Medical Association* or the *American Journal of Public Health*. This abstract is from *Radiologic Technology*. Notice that each journal has its own format and style. Answer the following questions about the abstract.

What is the research problem? 1 ______________________________

What is the source of the data? 2 ______________________________

How large was the sample of hospitals? of radiographers? 3 ______________________________

What is the population of interest? 4 ______________________________

What was the sampling procedure? 5 ______________________________

What purpose does this study serve? 6 ______________________________

LITERATURE EXAMPLE 4.4

Continuing education: Do mandates matter?

Thomas J. Edwards III, EdD, RT(R)

Radiologic Technology, Jan/Feb 1992, Vol. 63, No. 3

This study questions the assumption that radiographers lack the motivation to voluntarily participate in continuing professional education. Results of the author's survey of radiographers in 22 Central Florida hospitals—where state law mandates completion of 12 hours of CPE biannually—showed significant differences in participation due to motivational orientation and selected professional, educational and personal characteristics.

In recent years, radiographers have been challenged by the substantial amount of change associated with the technological advances in medical imaging techniques. According to Houle, a characteristic of a profession is the participation of its members in continuing education activities in order to meet the challenges of change.[1] Citing the importance of continuing education for radiographers to maintain effectiveness, Boissoneau proposed the development of continuing education requirements.[2] Approximately 11 states require radiographers to participate in continuing professional education (CPE) for continuing licensure or certification.

Florida mandates the completion of 12 contact hours of CPE each biennium.[3] Implicit in this mandate is the assumption that radiographers lack the motivation to voluntarily participate in CPE. This study questioned the validity of that assumption by investigating the extent to which the actual and preferred participation of radiographers differed by motivational orientation and selected professional, educational and personal characteristics.

METHODS

The overall design of this study was a causal-comparative investigation using a self-administered questionnaire and personal interviews. The population consisted of 801 radiographers employed in hospitals located in seven counties in Central Florida. Using a table of random numbers, a sample of 22 hospitals was selected from a directory of general and medical hospitals.[15] The subjects consisted of the 401 radiographers employed in these 22 hospitals. Subjects were certified by the State of Florida as general radiographers and registered by the American Registry of Radiologic Technologists. Of the 401 questionnaires distributed, 326 (81.3%) were completed and returned. Thirty-one questionnaires were deleted from the study because of incorrect completion of the questionnaire or lack of participation in CPE. The 295 useable questionnaires represented 73.6% of the sample and 36.8% of the population.

The survey instrument consisted of questions designed to solicit information regarding the following characteristics of the respondents: motivational orientation, educational, professional and personal characteristics, and participation in CPE. The motivational orientations of the respondents were determined through the use of a question consisting of six statements describing reasons for participation in CPE. The statements were developed in consideration of the six motivational clusters associated with the EPS. Rank ordering of the statements provided ordinal level data regarding the motivational orientations of the respondents. Qualitative data were solicited by an interview with the researcher from 25 respondents selected at random.

Samples from Households in Cities

Literature Example 4.5 is from the article entitled "Predictors of Physicians' Smoking Cessation Advice." Answer the following questions about the abstract.

What is the research problem?	1 ____________________
What is the source of the data?	2 ____________________
What was the sample size?	3 ____________________
What was the sampling procedure?	4 ____________________
What purpose does this study serve?	5 ____________________

LITERATURE EXAMPLE 4.5

Predictors of physicians' smoking cessation advice

Erica Frank, MD, MPH; Marilyn A. Winkleby, PhD; David G. Altman, PhD; Beverly Rockhill, MA; Stephen P. Fortmann, MD

Journal of the American Medical Association, 1991, Vol. 266, pp. 3139–3144

Objectives

To determine the percentage of smokers reporting that a physician had ever advised them to smoke less or to stop smoking, and the effect of time, demographics, medical history, and cigarette dependence on the likelihood that respondents would state that a physician had ever advised them to stop smoking.

Design and setting

Data were collected from the Stanford Five-City Project, a communitywide health education intervention program. The two treatment and three control cities were located in northern and central California. As there was no significant difference between treatment and control cities regarding cessation advice, data were pooled for these analyses.

Participants

There were five cross-sectional, population-based Five-City Project surveys (conducted in 1979–1980, 1981–1982, 1983–1984, 1985–1986, and 1989–1990); these surveys randomly sampled households and included all residents aged 12 to 74 years.

Main outcome measures

Improved smoking advice rates over time in all towns was an a priori hypothesis.

Results

Of the 2710 current smokers, 48.8% stated that their physicians had ever advised them to smoke less or stop smoking. Respondents were more likely to have been so advised if they smoked more cigarettes per day, were surveyed later in the decade, had more office visits in the last year, or were older. In 1979–1980, 44.1% of smokers stated that they had ever been advised to smoke less or to quit by a physician, vs 49.8% of smokers in 1989–1990 ($P < .07$). Only 3.6% of 1672 ex-smokers stated that their physicians had helped them quit.

Conclusion

These findings suggest that physicians still need to increase smoking cessation counseling to all patients, particularly adolescents and other young smokers, minorities, and those without cigarette-related disease. (JAMA 1991;266:3139–3144)

Methods

The FCP included two treatment cities (total population, 122 800), two control cities for comparison (total population, 197 500), and one additional control city for monitoring morbidity and mortality rates only. Representative cross-sectional population surveys were conducted from May 1979 to April 1980, May 1981 to July 1982, June 1983 to June 1984, April 1985 to June 1986, and April 1989 to May 1990; these surveys randomly sampled households and included all household residents aged 12 to 74 years. In addition, the original cross-sectional sample was restudied biennially as a cohort. By 1990, five cross-sectional and five cohort surveys had been completed. Cross-sectional data are used in the current study, with the exception of the examination of recall bias, which uses cohort data. Treatment- and control-city data are pooled because there was no substantial difference in reports of receiving smoking cessation advice from physicians between treatment and control cities. In the 1979–1980 survey, 708 smokers were included in our analyses; in 1981–1982, 620; in 1983–1984, 473; in 1985–1986, 453; and in 1989–1990, 456 (a total of 2710 smokers).

Multistage Sample: Sample from National Level

Literature Example 4.6 is from the article entitled "Prevalence of Migraine Headache in the United States." This study attempted to make statements about the population of the United States. It used a complex sampling design, beyond the scope of a beginning statistics text. However, you will frequently come across such sweeping studies, so you should understand what goes into sampling at the national level. Answer the following questions about the abstract.

What is the research problem? 1 ______________________

What is the source of the data? 2 ______________________

How is the panel of households initially selected by NFO? What is the NFO sampling frame skewed toward? Will the conclusion of studies conducted with this NFO sampling frame be applicable to non-whites and/or people from low-income households? 3 ______________________

____

____

____

____

What was the sample size? 4 ______________________

What was the sampling procedure? 5 ______________________

What purpose does this study serve? 6 ______________________

LITERATURE EXAMPLE 4.6

Prevalence of Migraine Headache in the United States

Relation to Age, Income, Race, and Other Sociodemographic Factors

Walter F. Stewart, PhD, MPH; Richard B. Lipton, MD; David D. Celentano, ScD; Michael L. Reed, PhD

Objective.—To describe the magnitude and distribution of the public health problem posed by migraine in the United States by examining migraine prevalence, attack frequency, and attack-related disability by gender, age, race, household income, geographic region, and urban vs rural residence.

Design.—In 1989, a self-administered questionnaire was sent to a sample of 15 000 households. A designated member of each household initially responded to the questionnaire. Each household member with severe headache was asked to respond to detailed questions about symptoms, frequency, and severity of headaches.

Setting.—A sample of households selected from a panel to be representative of the US population in terms of age, gender, household size, and geographic area.

Participants.—After a single mailing, 20 468 subjects (63.4% response rate) between 12 and 80 years of age responded to the survey. Respondents and nonrespondents did not differ by gender, household income, region of the country, or urban vs rural status. Whites and the elderly were more likely to respond. Migraine headache cases were identified on the basis of reported symptoms using established diagnostic criteria.

Results.—17.6% of females and 5.7% of males were found to have one or more migraine headaches per year. The prevalence of migraine varied considerably by age and was highest in both men and women between the ages of 35 to 45 years. Migraine prevalence was strongly associated with household income; prevalence in the lowest income group (<$10 000) was more than 60% higher than in the two highest income groups (≥$30 000). The proportion of migraine sufferers who experienced moderate to severe disability was not related to gender, age, income, urban vs rural residence, or region of the country. In contrast, the frequency of headaches was lower in higher-income groups. Attack frequency was inversely related to disability.

Conclusions.—A projection to the US population suggests that 8.7 million females and 2.6 million males suffer from migraine headache with moderate to severe disability. Of these, 3.4 million females and 1.1 million males experience one or more attacks per month. Females between ages 30 to 49 years from lower-income households are at especially high risk of having migraines and are more likely than other groups to use emergency care services for their acute condition.

(*JAMA.* 1992;267:64-69)

METHODS

Sample

Headache histories were ascertained through a questionnaire mailed to US households. A market research firm, National Family Opinion Inc (NFO), Toledo, Ohio, maintains a panel of 200 000 households nationwide for marketing, opinion, and other types of surveys.

Potential NFO panel households are initially *selected* as a stratified probability sample to be representative of the US population with regard to urban vs rural residence, age of the head of the household, household income, and size. Households are *recruited* by volunteer response to an initial mailing. A follow-up mailing is conducted to obtain a detailed household census and demographic information. Recruited households are randomly assigned to one of 40 blocks of 5000 households each. Every 2 years updated household census and demographic information is obtained. In addition, 30% of each sampling block is replaced. Households that are persistent nonresponders to periodic surveys are removed from the sampling frame. In general, the NFO sampling frame is skewed toward but not limited to upper-income white households.

Survey

In 1989, a self-administered questionnaire was sent to a stratified random sample of 15 000 NFO panel households (three sampling blocks). A designated member from each household responded to the questionnaire by reporting the number of members in the household and the number who suffer from *severe headache*. The study is limited to self-defined severe headache because this constitutes the most significant health problem from the sufferer's point of view. The consequences of this design choice are explored in the "Comment" section.

Each household member with severe headache was asked to complete the questionnaire. Detailed questions were asked about severe headaches, including specific accompanying symptoms and frequency of and disability from severe attacks. A total of 9507 (63.4%) of the 15 000 households responded to a single questionnaire mailing, for a total base population of 23 611 individual household members. Subjects less than 12 years old (n = 3043) were excluded from the analysis because of concerns regarding ability to reliably respond to and interpret questions. We report on responses of the 20 468 subjects between 12 and 80 years of age. One hundred respondents were excluded because gender was not reported.

Multistage Sample: Sample from Selected States and Schools

Literature Example 4.7 is from the article entitled "RJR Nabisco's Cartoon Camel Promotes Camel Cigarettes to Children." This study seeks to make statements about the population of the United States. Answer the following questions about the abstract.

What is the research problem? 1 ____________________

What two groups do the authors wish to compare? 2 ____________________

What is the source of the data for children? for adults? 3 ____________________

How were the states selected? What was the criterion for selecting the schools? What determined the student sample? 4 ____________________

____

____

What was the sample size for children? for adults? 5 ____________________

What is the sampling procedure? 6 ____________________

What purpose does this study serve? 7 ____________________

LITERATURE EXAMPLE 4.7

RJR Nabisco's cartoon camel promotes Camel cigarettes to children

Joseph R. DiFranza, MD; John W. Richards, Jr., MD; Paul M. Paulman, MD; Nancy Wolf-Gillespie, MA; Christopher Fletcher, MD; Robert D. Jaffe, MD; David Murray, PhD

Journal of the American Medical Association, 1991, Vol. 266, pp. 3149–3153

OBJECTIVES

To determine if RJR Nabisco's cartoon-theme advertising is more effective in promoting Camel cigarettes to children or to adults. To determine if children see, remember, and are influenced by cigarette advertising.

DESIGN

Use of four standard marketing measures to compare the effects of Camel's Old Joe cartoon advertising on children and adults.

SUBJECTS

High school students, grades 9 through 12, from five regions of the United States, and adults, aged 21 years and over, from Massachusetts.

OUTCOME MEASURES

Recognition of Camel's Old Joe cartoon character, product and brand name recall, brand preference, appeal of advertising themes.

RESULTS

Children were more likely to report prior exposure to the Old Joe cartoon character (97.7% vs 72.2%; $P < .0001$). Children were better able to identify the type of product being advertised (97.5% vs 67.0%; $P < .0001$) and the Camel cigarette brand name (93.6% vs 57.7%; $P < .0001$). Children also found the Camel cigarette advertisements more appealing ($P < .0001$). Camel's share of the illegal children's cigarette market segment has increased from 0.5% to 32.8%, representing sales estimated at $476 million per year.

CONCLUSION

Old Joe Camel cartoon advertisements are far more successful at marketing Camel cigarettes to children than to adults. This finding is consistent with tobacco industry documents that indicate that a major function of tobacco advertising is to promote and maintain tobacco addiction among children. (*JAMA*. 1991;266:3149–3153)

METHODS: SUBJECTS

Since adolescent brand preferences may vary from one geographic location to another (Saundra MacD. Hunter, PhD, Weihang Bao, PhD, Larry S. Webber, PhD, and Gerald S. Berenson, MD, unpublished data, 1991; D.M., unpublished data, 1991),[14,18,19] we selected children from Georgia, Massachusetts, Nebraska, New Mexico, and Washington, representing five regions. One school in each state was selected based on its administration's willingness to participate. Schools with a smoking prevention program focused on tobacco advertising were excluded.

A target of 60 students in each grade, 9 through 12, from each school was set. In large schools, classes were selected to obtain a sample representative of all levels of academic ability. Students were told that the study concerned advertising and were invited to participate anonymously.

Since adult brand preferences are available from national surveys, adult subjects were recruited only at the Massachusetts site. All drivers, regardless of age, who were renewing their licenses at the Registry of Motor Vehicles on the days of the study during the 1990–1991 school year were asked to participate. Since licenses must be renewed in person, this is a heterogeneous population.

Convenience Sample

Literature Example 4.8 is from the article entitled "Active Enforcement of Cigarette Control Laws in the Prevention of Cigarette Sales to Minors." This study sought to evaluate the effect of cigarette control legislation. The authors measured percentages of stores selling cigarettes to minors and percentage of students who had experimented with cigarettes or were regular smokers. You may find that some of the sampling information is not clear, which is not unusual. You can obtain additional information by examining the entire article and by writing to the authors. Answer the following questions about the abstract.

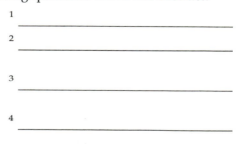

What is the research problem?	1	______________________
What two groups did the authors study?	2	______________________
How were data obtained about stores? about students?	3	______________________
What was the sample size for children? for stores?	4	______________________

What was the sampling procedure? [5] ______________________

How did the researchers determine that an illegal sale had occurred? [6] ______________________

What purpose does this study serve? [7] ______________________

LITERATURE EXAMPLE 4.8

Active enforcement of cigarette control laws in the prevention of cigarette sales to minors

Leonard A. Jason, PhD; Peter Y. Ji; Michael D. Anes; Scott H. Birkhead

Journal of the American Medical Association, 1991, Vol. 266, pp. 3159–3161

Objective

To assess the effect that cigarette legislation would have on reducing merchant sales rates of cigarettes to minors and the effect on adolescent smoking behavior.

Design

Observational survey of merchant selling behaviors and adolescent smoking habits before and after passage of legislation.

Setting

The setting for the merchant survey was Woodridge, Ill (population 25 200), a suburban community of Chicago. The surveys were distributed to adolescents in the local junior high school.

Participants

Convenience sample of both merchants and adolescent students.

Intervention

Passage of community antismoking legislation.

Main outcome measures

Percentage of stores selling cigarettes to minors in Woodridge and percentage of students who had experimented with cigarettes or were regular smokers.

Results

Merchant sales rates in Woodridge decreased from a baseline of 70% before legislation to less than 5% in 1.5 years of compliance checking after legislation. Student surveys showed that the rates of cigarette experimentation and regular use of cigarettes by adolescents were reduced by over 50%.

Conclusion

Cigarette control laws can be effective in significantly reducing the rate of cigarettes sold by merchants and rates of cigarette use by adolescents. Key elements of successful legislation implementation are consistent compliance checking and heightened community awareness of the problems and prevalence of adolescent smoking. (*JAMA*. 1991;266:3159–3161)

Method

This study was initiated as a result of a private citizen's complaint regarding a minor's possession of cigarettes to Officer Bruce Talbot of the Woodridge, IL (population, 25 200), police department. In August 1988, Officer Talbot sent a letter to all cigarette vendors in Woodridge detailing the state law prohibiting cigarette sales to minors. In addition, following a media focus on a DePaul University study of cigarette sales to minors in Chicago (L.A.J. and P.Y.J., unpublished data, 1991), Officer Talbot invited one of the authors (L.A.J.) to assess cigarette sales rates in Woodridge. These measurements were done in August and November 1988 and February 1989. The DePaul University research team also distributed a questionnaire to students at the local junior high school in March 1989 to determine the number of adolescent smokers and their smoking habits. In May 1989, new legislation was passed restricting cigarette sales in Woodridge, and cigarette sales were tracked after the legislation was passed.

During each checking period, minors 12 to 13 years of age (all of whom were rated by independent judges as looking less than 18 years of age) were sent into stores to purchase a pack of cigarettes. Unobtrusively, a Woodridge police official or a DePaul University research assistant observed the transaction. Vending machine sales were also tracked by sending minors to purchase cigarettes from these machines. There was 100% agreement between two independent judges as to whether an illegal sale occurred.

Sales assessments conducted before legislation to determine if Officer Talbot's letter to merchants was effective were made in August and November 1988 and February 1989. In the spring of 1989, Officer Talbot and other Woodridge officials drafted cigarette control legislation modeled after the city's liquor control laws. This was done to treat the sales of tobacco and alcohol, both age-restricted products, in precisely the same manner.

Self-Assessment

Please assess your competency.

Tasks:	How well can you do it?					Page
	Poorly				*Very well*	
1. Select the following types of samples when a random table and populations are given:						
Simple random sample	1	2	3	4	5	78
Stratified random sample	1	2	3	4	5	79
Proportional stratified random sample	1	2	3	4	5	80
Multistage sample	1	2	3	4	5	80
2. Assign subjects to experimental treatment levels.	1	2	3	4	5	81
3. Assign subjects as they are recruited into the experiment.	1	2	3	4	5	82

Answers for Chapter 4

Answer for question on p. 75: 30, 08, 22, 13, 23, 18

Answer for Table 4.3, p. 76: 5 8 7, 6 7 4, 4 8 2, 1 6 6, 3 3 0, 5 0 6

Answer for question on p. 80: A different sample

Answer for Activity 4.4: Stratified sampling ensures that there will be an equal number of cases from each stratum in the population, whereas simple random sampling usually selects a different number of cases from each stratum. Stratified sampling should be used when it is desirable or necessary to have equal numbers in each group.

Answers for Activity 4.6: 1 and 2) Any two of the following: POSTNHR, POSTMAXHR, POSTVO2, POSTVEE, POSTWAT, or other post variables. 3) Factor 4) Independent variable 5) Treatment 6) Levels 7) Subjects or units

Answer for Activity 4.7: Unlike

Answers for Literature Example 4.1: 1) Restate the title as a question: How many head injuries may be prevented if bicyclists wear helmets? 2) Data tapes (This is stated in the first sentence of the Methods section.) 3) No, the researchers used cases collected by the National Center for Health Statistics. This was feasible because all the data are on a computer tape. 4) It provides evidence of the need for education and legislation regarding bicycle helmet use.

Answers for Literature Example 4.2: 1) How strictly do dialysis patients want their physicians to follow their advance directives regarding sustaining their lives should they become mentally incompetent? 2) Dialysis patients in units in San Francisco 3) 150 patients 4) Two steps: Staff excluded patients not considered capable of giving informed consent. The remaining subjects were asked to summarize a vignette. Those who could not do so were eliminated. Eligible subjects were interviewed until a target sample size was achieved. 5) It reinforces the idea that physicians should ask patients how strictly they want their advance directives followed.

Answers for Literature Example 4.3: 1) To assess the prevalence, content, and growth of state and city laws restricting smoking in public places and workplaces 2) State Legislative Reference Bureaus 3) 149 cities; 40 cities 4) Copies of laws were requested from each state. Cities of 25,000 people or larger were selected from census data. A random sample was used to check that certain cities did have laws. 5) The study confirmed that no-smoking laws are still needed in some states.

Answers for Literature Example 4.4: 1) Are radiographers motivated to participate in continuing education? 2) Directory of general and medical hospitals 3) 22 hospitals; all radiographers employed in the 22 hospitals, of whom 295 returned usable questionnaires 4) Radiographers in Florida 5) A table of random numbers was used to select the 22 hospitals. 6) It confirms that radiographers' participation in continuing education varies widely, depending on motivation and professional, educational, and personal characteristics.

Answers for Literature Example 4.5: 1) What percentage of patients say their physicians advised them to smoke less or stop smoking, and what patient characteristics lead physicians to give this advice? 2) Five surveys 3) A total of 2,710 smokers over five years 4) Random sample from households 5) To suggest that physicians need to increase smoking cessation counseling to all patients

Answers for Literature Example 4.6: 1) What is the magnitude and distribution of the public health problem posed by migraine? 2) Self-administered questionnaire mailed using NFO sampling frame. 3) Households selected as a stratified probability sample to be representative of the U.S. population. The NFO sample is skewed toward but not limited to upper-income white households. No, the conclusion will not be applicable to non-whites and low-income households. 4) 20,468 subjects 5) Questionnaire sent to a stratified random sample of 15,000 NFO panel households; household members who reported suffering from severe headache completed another questionnaire. 6) Reveals the extent of the problem and suggests who is likely to use emergency care services

Answers for Literature Example 4.7: 1) Is RJR Nabisco's cartoon theme advertising more effective in promoting Camel cigarettes to children or to adults? Are children influenced by cigarette advertising? 2) High school students and adults 3) Responses from sampled children to four standard marketing measures; responses to the same measures from adults renewing their driver's licenses 4) States were selected to represent five regions; how they were selected is not clear. Schools were selected based on the administration's willingness to participate. Researchers tried to achieve a sample representative of all academic ability; how this was accomplished is not clear. 5) Abstract does not state the sample sizes. 6) 60 students from each grade in each school were chosen. 7) The study reveals that cartoon advertisements are far more successful at marketing cigarettes to children than to adults.

Answers for Literature Example 4.8: 1) Does legislation affect the cigarette selling behavior of merchants and students' smoking habits? 2) Merchants and students 3) Observation of minors sent into stores to purchase cigarettes; survey of students at local junior high 4) The abstract does not specify the size of either sample. 5) Convenience sample 6) Two judges independently observed transactions. 7) It reveals that cigarette control laws can effectively reduce the rate at which cigarettes are sold to and used by children.

CHAPTER

5

Statistical Tools
Tables and Stem and Leaf Plots

Prelude

Up ahead, the road divided into three and, as if in reply to Milo's question, an enormous road sign, pointing in all three directions, stated clearly:

DIGITOPOLIS

5	Miles
1,600	Rods
8,800	Yards
26,400	Feet
316,800	Inches
633,600	Half Inches

AND THEN SOME

"Let's travel by miles," advised the Humbug; "it's shorter."
"Let's travel by half inches," suggested Milo; "it's quicker."

— Norton Juster
The Phantom Toll Booth, p. 171

Chapter Outline

Self-Assessment

Please assess your competency.

Tasks:	**How well can you do it?**					**Page**
When data are given, construct and interpret the following:	*Poorly*				*Very well*	
1. One-way table	1	2	3	4	5	126
2. Two-way table	1	2	3	4	5	126
3. Three-way table	1	2	3	4	5	142
4. Stem and leaf plot	1	2	3	4	5	121

Introduction

Tools enhance our thinking. The microscope magnifies and enables us to see microbes; satellites help us visualize and predict weather patterns; and **statistical tools** help us analyze and display data. Statistics enable us to describe a population.

In this chapter and the next you will practice constructing and interpreting analytical and display tools such as tables, graphs, and numerical summaries, including the range and median. You will use the two data sets we have used in previous chapters (the Student Questionnaire Data, SQD, and the Faculty Exercise Data, FED) for various activities. Examples from the literature are given for each of the tools.

Tables: Patterns of Variation

Our interest in collecting data is to detect a pattern of variation in the values of a variable. Such a pattern is called the **distribution** of that variable. Listed below are the values of the variable SEX in the SQD (Student Questionnaire Data).

1 2 1 1 1 1 2 2 1 1 1 1 1 1 2 2 1 2 1 1 2 1 1 1
2 2 1 2 2 1 2 1 1 1 1 1 1 2 1 2 2 1 1 1 1 2 2 1
2 1 1

From this list of values it's pretty hard to see what's going on with this variable. All we see are a bunch of 1s and 2s (which represent female and male), with each value belonging to a case. How can we make the pattern of variation explicit? Perhaps if we group females and males together the pattern will become clear.

1 1 1 1 1 1 1 1 1 1 1 1 1 1 1 1 1 1 1 1 1 1 1 1
1 1 1 1 1 1 1 1 1 2 2 2 2 2 2 2 2 2 2 2 2 2 2 2
2 2 2

The second listing shows that there are more/fewer females than males.

We can now easily count the number of females and males, but if the sample were larger (e.g., $n = 100$) it would be cumbersome to count the number of occurrences of each value. Let us arrange the SEX data in a table.

TABLE 5.1 Distribution of Values of the Variable SEX Arranged in a Table

SEX	Count
1	33
2	18
	$n = 51$

SEX is a categorical variable with two values, 1 and 2. The pattern is delineated by arranging the numbers into two columns: column 1, values of the variable, and column 2, count for each category. An alternative name for counts is **frequency**. The distribution can be represented or approximated by a table that couples each variable value with its frequency of occurrence. From this **frequency table** we could reconstruct all the values of SEX again if there were any need to do so. Therefore, by forming this distribution table we did not lose any information. This is true for the distribution of any nominal or ordinal (categorical) variable. However, when we categorize a ratio variable by forming a frequency table, we often lose some information. That is, someone who had a frequency table of a ratio variable but not the original (raw) data would be unable to list the exact values of the variable.

Shown below is a list of ages of the students in two statistics classes. Here are the ages of the students in the order that their questionnaires were read and recorded, so-called raw data.

22 25 33 33 21 23 22 23 24 21 21 22 24 32 26 21 23 23
21 21 21 26 25 22 30 29 38 23 22 20 28 33 38 21 23 24
20 20 20 25 21 30 61 20 23 24 21 30 21 23 24

The same ages are listed below, ordered from youngest to oldest.

20 20 20 20 20 21 21 21 21 21 21 21 21 21 21 21 22 22
22 22 22 23 23 23 23 23 23 23 23 24 24 24 24 24 25 25
25 26 26 28 29 30 30 30 32 33 33 33 38 38 61

Table 5.2 presents a frequency table, giving the same ages with the corresponding number of occurrences of each.

TABLE 5.2 Distribution of AGES

AGE	Count, or Frequency
20	5
21	11
22	5
23	8
24	5
25	3
26	2
28	1
29	1
30	3
32	1
33	3
38	2
61	1
	$n = 51$

Table 5.3 presents a different kind of frequency table. The ages are grouped in intervals and the total count for each interval is given.

TABLE 5.3 Categorized AGES (intervals of five years)

AGE Interval	AGE Category	Count
20–24	1	34
25–29	2	7
30–34	3	7
35–39	4	2
40 or older	5	1
		$n = 51$

Finally, Table 5.4 shows the ages grouped differently into intervals.

TABLE 5.4 Categorized AGES (Uneven Intervals)

AGE Interval	AGE Category	Count
20–22	1	16
23–24	2	18
25–29	3	7
30–34	4	7
35–39	5	2
40 or older	6	1
		$n = 51$

It is [1] easy/difficult to discern the pattern in the variation of the ages presented in the first listing. It is easy to notice the following when the ages are ordered, as in the second listing: minimum age is [2] _______; maximum age is [3] _______; most frequently occurring age is [4] _______; least frequent ages are [5] _______. It is easier still to see patterns of variation from Table 5.2: minimum age is [6] _______; maximum age is [7] _______; most frequently occurring age is [8] _______; least frequent ages are [9] _______.

It is easy to see that information has been lost in going from Table 5.2 to Tables 5.3 and 5.4. From Table 5.3, for instance, we can tell that there were seven cases between the ages of 30 and 34, but there is no way to tell that there were three thirty-year-olds and no thirty-one-year-olds. Actually, even in listing the raw data in no order at all (the first listing), we lose some information. We cannot tell, for instance, that the first case was 22 years, 4 months, 8 days, 14 hours, and 26.195 minutes old. Nor do we care. Normally when we are asked to give our age, we give our age as of our last birthday. In fact, continuous data must always be rounded off or otherwise approximated when it is collected.

Professional and technical journals rarely allocate much space to display of data. If only raw data were displayed, the reader would have to redo the analysis that the author has already done in order to discern the pattern. In most journals, data are given in summary form, in tables similar to Tables 5.2, 5.3, and 5.4.

How much detail we need in the display of data depends upon our purpose. At the beginning stages of analysis we often experiment with different ways of displaying the data in the hope of discerning patterns of variation. Categorized ages were presented in Table 5.3. In displaying the data this way, we lost information but gained economy of space. Table 5.3 shows that the age interval with largest frequency is [10] _______. The age interval with the smallest frequency is [11] _______. Just by observing this distribution we are tempted to conclude that there are approximately [12] ______ returning students (those coming back to school after pursuing another career) in this class.

Categorical variables (e.g., SEX, CLASS, HEALTH) and categorized data (e.g., AGE divided into intervals) are used when we group cases into categories. The values of variables may be

1. descriptive phrases, in the case of nominal variables (e.g., female and male);
2. numerals or descriptive phrases that imply order, in the case of ordinal variables (e.g., minimum, moderate, or heavy stress); or
3. intervals, in the case of ratio variables (e.g., age, height).

Cumulative Counts and Percentages

Displays of data are devised to enable us to answer certain kinds of questions with ease. For example, how many students are above the age of 35? How many students are below the age of 34? What percentage of students are above 30 years of age? These questions are best answered by tables of cumulative counts and percentages. Suppose we want to know how the age distributions in the two statistics classes compare. Here is the table for the first class:

AGE Interval	Count	Percent	Cumulative Count	Cumulative Percent
20–24	17	65.4	17	65.4
25–29	5	19.2	22	84.6
30–34	4	15.4	26	100.00

Percents are obtained by dividing counts by sample size (26). The cumulative counts were obtained by adding counts in the following way:

AGE Interval	Count	Cumulative Count
20–24	17	17
25–29	5	17 + 5 = 22
30–34	4	22 + 4 = 26

Fill in the blanks in the corresponding table for Class II:

AGE Interval	Count	Percent	Cumulative Count	Cumulative Percent
20–24	17	68.0	1 ______	2 ______
25–29	2	8.0	3 ______	4 ______
30–34	3	5 ______	6 ______	7 ______
35–39	2	8 ______	9 ______	10 ______
61	1	4.0	11 ______	100.0

Which class has the oldest student? 12 Class I/Class II Which class consists completely of students below age 35? 13 Class I/Class II What percent of students in Class I are younger than age 30? 14 ______ in Class II? 15 ______.

Looking at Entire Distributions: Stem and Leaf Plots

A **stem and leaf plot** is one way of graphically displaying information about a frequency distribution. As with a frequency table, data are ordered from minimum to maximum observation, but in the stem and leaf plot we save some work by not writing down the complete numbers every time they occur. The stems (all but the last digit of the observations) are written only once, followed by all the leaves (the last digits of the observations). For example, the number 61 has 6 as its stem and 1 as its leaf.

For small sets of data a stem and leaf plot is useful and can be readily constructed while the data are being collected. Stem and leaf plots preserve the details of the data while providing graphic information about frequencies.

Following are the grades that a class of students made on a statistics exam:

93 85 80 96 100 80 89 61 98 97 100 73 95 76 90 86 81

Using the units digit as the leaf and the other digits as the stem, we would obtain the following stem and leaf plot:

Stem	Leaves
6	1
7	36
8	050961
9	368750
10	00

From this plot we see immediately that most of the grades are in the 80s and 90s, with fewer grades in the 70s and only one grade in the 60s. We easily see that two people had perfect papers and that the lowest grade was a 61. In fact, all the information that would be contained in a frequency table is contained in this plot, but the stem and leaf plot also gives us a visual picture that we don't get from a frequency table. Thus, a stem and leaf plot is a hybrid representation of a distribution having features in common with a table and features in common with a graph. (In the next chapter we will study the histogram, which is a kind of graph that is very similar to a stem and leaf plot.)

As a second example, we show a stem and leaf plot of the HEIGHT variable from the SQD, and then we show how the plot can be refined, step by step. First we list the heights:

69 66 67 54 65 62 71 73 70 65 62 61 64 62 67 70 61 75
66 63 76 68 62 66 69 70 65 72 72 64 72 62 63 62 64 63
63 72 67 66 73 66 64 63 64 70 67 65 72 64 67

A stem and leaf plot is an excellent tool for quickly producing a rough graph of the distribution.

Each number is divided into a leaf, which is the units digit, and a stem, which is the tens digit. For the maximum height of 76 inches, the stem is [1] _______ and the leaf is [2] _______. Notice that the leaves are listed in the same order the numbers appeared in the original data set. Stem and leaf plots done by hand usually are shown in this way, for there is no particular reason to put the leaves in numerical order. However, if we construct stem and leaf plots using computer software, the software will put the leaves in numerical order. The rest of the plots shown in this book will be given in this form.

Using 5, 6, and 7 as stems, the numbers bunch together and it's hard to see the pattern. The display can be improved by dividing each stem into two parts. The first part will have leaves ranging from 0 to 4 and the second part from 5 to 9.

Stems Leaves

5 4
6 96752521427163826954232433766434754 7
7 13005602222302
$n = 51$

FIGURE 5.1 Stem and Leaf Plot of Heights of Statistics Students (one-part stems)

Stems Leaves

5 4
5
6 1122222233333444444
6 55556666677777899
7 000012222233
7 56

Figure 5.2 Stem and Leaf Plot of Heights of Statistics Students (two-part stems)

Still the heights are bunched together. To get even more detail we can use five-part stems:

part 1, leaves 0 and 1
part 2, leaves 2 and 3
part 3, leaves 4 and 5
part 4, leaves 6 and 7
part 5, leaves 8 and 9

Stems Leaves

5
5
5 4
5
5
6 11
6 22222233333
6 4444445555
6 6666677777
6 899
7 00001
7 2222233
7 5
7 6
7

Note that the stems without leaves on either end are normally omitted.

Figure 5.3 Stem and Leaf Plot of Heights of Statistics Students (five-part stems)

The three largest heights as given in the stem and leaf plot in Figure 5.3 are [3] _______, [4] _______, and [5] _______ inches. There are [6] _______ students 62 or 63 inches tall in these classes. There are [7] _______ parts to the 6 stem. The first part of the 5 stem has leaves 0 and [8] _______. The fifth part of the 6 stem has leaves [9] _______ and [10] _______. We split stems to produce a display that facilitates counting of the leaves.

Stems, Leaves, and Units

Here are some examples showing how to break numbers into stems and leaves:

Number	Stem (all but the last digit)	Leaf (last digit)
25	2	5
127	12	7
1,238	123	8

For numbers with a lot of digits, we may choose those digits that we consider significant digits for the stem and leaf. The reconstruction of the approximate original number is achieved by multiplying the stem and leaf by the appropriate unit value. For example, the number 1,568.7 may be represented by the following stems, leaves, and units:

Stem	Leaf	Unit	Reconstructed Number
1	5	100	1,500
15	6	10	1,560
156	8	1	1,568
1,568	7	.1	1,568.7

The choice of stem is dictated by the precision you want to preserve.

Activity 5.1 will give you practice in constructing a stem and leaf plot for two-digit numbers.

ACTIVITY 5.1

Construction of a stem and leaf plot

We will construct a stem and leaf plot for the same age data that we looked at earlier in this chapter.

The following is a listing of the age data of the statistics classes.

22 25 33 33 21 23 22 23 24 21 21 22 24 32 26 21 23 23
21 21 21 26 25 22 30 29 38 23 22 20 28 33 38 21 23 24
20 20 20 25 21 30 61 20 23 24 21 30 21 23 24

Locate minimum and maximum ages.	[1]Minimum _______ [2]Maximum _______
The last digit (the ones digit) of each number is the leaf; the other digit forms the stem. Write one stem per line, starting with the minimum stem. There are five possible stems: 2, 3, 4, 5, and 6. Note the repetition of each stem. We did this because the 2 stem has forty-one leaves. Dividing each stem into parts makes it easier to count the leaves.	**Stem** 2 Minimum stem 2 3 3 4 4 5 5 [3] _______ Maximum stem
Write leaves of the corresponding stems as you read the data (without paying attention to the numerical order of the leaves).	**Stem Leaves** 2 _______ 2 _______ 3 _______ 3 _______ 4 _______ 4 _______ 5 _______ 5 _______ 6 _______
Rewrite the leaves in numerical order. (This step is sometimes omitted if we are only trying to get a rough idea of the shape of the distribution.)	**Stem Leaves** 2 _______ 2 _______ 3 _______ 3 _______ 4 _______ 4 _______ 5 _______ 5 _______ 6 _______

Now count the leaves and enter the frequencies in the first column. (This step is also optional.)	**Frequency**	**Stem**	**Leaves**
	______	2	__________
	______	2	__________
	______	3	__________
	______	3	__________
	______	4	__________
	______	4	__________
	______	5	__________
	______	5	__________
	______	6	__________

What does a stem and leaf plot tell us?

1. It displays the frequencies in a graphic way. We can see, for instance, which stem has the most cases. For this example, we see that students are bunched between ages 20 and 29. The age interval 20–24 has the largest number of students. There is one person whose age is unusual. That person's age is [5] ______.
2. It points out gaps in the distribution. There were no ages between 39 and 60. Was a mistake in data entry made, or is there a special explanation for the occurrence of so atypical an age as 61? In fact, there was a sixty-one-year-old student in one of the classes, so this entry was not a mistake.
3. We can count the frequencies from the minimum or maximum and find the value of the middle case. We have 51 cases. So the middle case for this distribution is the twenty-sixth case (twenty-five cases on either side of the value). The age of the middle case is [6] ______. The middle value of a distribution is called its **median**.

Activity 5.2

Construction of a stem and leaf plot of the FED ages

Listed below are the ages of the faculty participating in the exercise program.

61 52 57 43 48 44 61 58 43 51 57 41 40 62 47 54 40 49
61 41 57 44 52 40 52 41 46 52 40 48

1. Locate the maximum and minimum ages.

   [1] Minimum ______; Maximum ______

2. The rightmost digit of each number is the leaf; the rest of the digits form the stem. Write one stem per line, starting with the minimum stem. (There are three possible stems: 4, 5, 6.) We

   Minimum stem = 4
   Maximum stem = 6

recommend splitting each stem into five parts.

3. Write leaves of the corresponding stems as you read the data (without paying attention to the numerical order of the leaves). Then rewrite the leaves in numerical order and count to find the frequencies.

Freq.	Stem	Leaves
	4	
	4	
	4	
	4	
	4	
	5	
	5	
	5	
	5	
	5	
	5	
	6	
	6	

Example from the Literature

Literature Example 5.1 is from an article entitled "Canadian Therapists' Priorities for Clinical Research: A Delphi Study." Read the abstract to learn about the study and look at the table. Then answer the following questions.

The stem and leaf display dramatizes which questions received the highest scores.

Which question received the highest score? [1] ________

The authors wanted to identify the ten questions with the highest scores. They could have done this by arranging the scores from highest to lowest with their respective question numbers. Do you think the stem and leaf plot is preferable to this ranking method? [2] Yes/No

Note that this example is different from the examples of stem and leaf plots that we have given previously in that the numbers of the questions are printed above the leaves. Since a stem and leaf plot is mainly a tool for preliminary analysis of data, it is not common to see one in the literature. This was the only example we were able to find.

LITERATURE EXAMPLE 5.1

Canadian Therapists' Priorities for Clinical Research: A Delphi Study

Carole Miles-Tapping
Andrew Dyck
Sandra Brunham
Edith Simpson
Lynne Barber

Clinically relevant physical therapy research questions were developed by a Delphi technique among 55 teaching hospital physical therapists. The Delphi technique used in this study involved three rounds of questionnaires that included characteristics of anonymity, feedback, ranking with statistical scoring, and use of informed respondents. Fifty-eight initial research questions were narrowed to 11 according to their potential benefit for the patient, for the practice of physical therapy, and for decreasing health care costs. A literature review revealed that each of the 11 questions were as yet unanswered. The use of the survey results to guide and plan for clinical research in physical therapy is discussed. [Miles-Tapping C, Dyck A, Brunham S, et al. Canadian therapists' priorities for clinical research: a Delphi study. Phys Ther. 1990;70:448–454.]

The "stem-and-leaves" method was then used to assess the distribution of scores (Table 1). This method was used because it allows quick assessment of the distribution of data and because it retains more data than the more familiar frequency distribution.[12] It was convenient to arrange the scores by tens in order to derive the "stems." For example, the scores 450 to 459 have a stem of 45. The ones digits of the scores form the "leaves." For instance, Question 1 received a score of 457; it was therefore placed alongside the stem 45, over the leaf 7 in the distribution chart (Table 1).

The curve formed by the leaves had three peaks with a trailing lower tail. Our original intent was to select the top 10 questions. However, because the 10th and 11th questions had the same score and because the top 11 questions were contained within a single peak, we selected the top 11 questions for the next round (Table 1).

Table 1. *"Stem-and-Leaves" Method of Showing Distributions of Questions (N = 58) by Round 2 Scores*

Stem[a]	Leaves[b]							
51	11							
	2							
50								
49								
47	37	18						
	0	**2**						
46	26							
	0							
45	4	30	35	14	34	32	1	upper
	0	**0**	**1**	**4**	**5**	**6**	**7**	peak
44	7	28						
	3	**7**						
43	12	42	47	45				
	0	**3**	**4**	**8**				
42	20	44	36	6	17	16		middle
	3	**4**	**6**	**8**	**8**	**9**		peak
41	13	33	31					
	3	**3**	**4**					
40	52	50	51	25				
	0	**1**	**6**	**8**				
39	2	48	41					
	3	**5**	**6**					
38	29	43	3	9				
	0	**2**	**5**	**8**				
37	8	46	49	23	38			
	1	**1**	**1**	**5**	**7**			
36	56							
	5							
35	15	19	57	54	24			lower
	2	**2**	**4**	**5**	**6**			peak
34	5	55	10	39				
	0	**0**	**4**	**5**				
33	21	27						
	0	**5**						
32	40							
	7							
31								
30	22							
	1							
29	58							
	8							
28								
27								lower tail
26								
25								
24	53							
	6							

[a]Scores in tens.

[b]Lightfaced leaves correspond to question numbers in Appendix 1; boldfaced leaves represent ones digits of scores.

Univariate and Multivariate Tables

Frequency tables provide useful summaries of categorical data. Table 5.5 presents different kinds of summary tables:

TABLE 5.5 Introduction to Univariate and Multivariate Tables

Number of Variables	Type of Table	Example
Univariate Tables		
(1) One variable Note: There are 33 females and 18 males in the statistics class.	One-way	SEX — Code — Count, or frequency Female — 1 — 33 Male — 2 — 18 $n = 51$
One variable Note: A variable may be divided into any number of categories; in this case, three.	One-way	ACTIVITY — Count Minimal — 6 Moderate — 36 Vigorous — 7 $n = 49$
(2) Several variables presented together but each tabulated separately	One-way	SEX — Count, or frequency Female — 33 Male — 18 TIMEOUT True — 33 False — 16 Note: Two people did not respond to this question.
Multivariate Tables		
(3) Two variables tabulated to partition cases on two characteristics at once Note: There are 33 females and 16 males. Out of the 33 females, 21 said they took TIMEOUT. Two-way tables allow us to see relationships between two variables. One-way tables will not expose these relationships.	Two-way	Rows: SEX — Columns: TIMEOUT — True — False — ALL Female — 21 — 12 — 33 Male — 12 — 4 — 16 ALL — 33 — 16 — 49
(4) Three variables: Values of one variable are controlled or held constant and the other two variables are tabulated as in a two-way table.	Three-way	Control: DISCUSS = True Rows: SEX — Columns: TIMEOUT — True — False — ALL Female — 20 — 9 — 29 Male — 12 — 4 — 16 ALL — 32 — 13 — 45

TABLE 5.5 Continued

Number of Variables	Type of Table	Example
The first table holds DISCUSS constant at the value "True." Forty-five cases said the statement was true. These cases were tabulated on the SEX and TIMEOUT variables. The second table holds DISCUSS constant at the value "False." Four cases said the statement was false. These four cases were tabulated on the SEX and TIMEOUT variables.		Control: DISCUSS = False Rows: SEX Columns: TIMEOUT True False ALL Female 1 3 4 Male 0 0 0 ALL 1 3 4

When the distribution of one variable is presented in a table, the table is called a [1] univariate/multivariate table. When more than two variables are presented, the table is called a [2] univariate/multivariate table. When frequencies of one variable are presented, the table is a [3] ______________ way table. There are [4] _______ females and 18 [5] ______________ in the statistics class. (2) presents a table with two variables, still it is a [6] _______ variate table because both variables are presented independently. (3) presents a table with variables [7] ______________ and [8] ______________, so we can tell how many females and males take timeout. How many males take timeout? [9] _______ A three-way table holds values of one variable constant; in the first table in (4) the value [10] ______________ of variable [11] ______________ was controlled. This means that this table tabulates cases that answered [12] ______________ to the question about discussing events of the day; the total number that answered "True" is [13] _______. The number of cases in the second table is [14] _______. Both the first and second table (in 4) tabulate SEX by TIMEOUT; the only difference is the response of cases to the variable DISCUSS. Each table is said to control and hold constant one value of the variable [15] ______________.
Only [16] _______ cases do not discuss the events of the day.

Univariate (One-Way) Tables: Examples from the Literature

Literature Example 5.2 is from an article entitled "Use of Nonionic Contrast Media in a High-Risk Outpatient Population." The abstract is accompanied by a one-way table that delineates counts of procedures performed and percentages.

Answer the following questions, which point out some salient features of the one-way table.

The title of a table should briefly describe the table's contents. Frequently titles tell *who* was in the data set, *what* was measured, and *where* and *when* the data were collected. A title with all of these elements will be self-explanatory.

What is the title of Table 1? 1 ________________

Its title is not self-explanatory. What is one possible reason? 2 ________________

What is the variable? 3 ________________

What type of variable is it? 4 Nominal/Ordinal/Interval/Ratio

What are the main values of the variable? 5 ________________

Do the percents that correspond to the main values add up to 100? 6 Yes/No

What is one of the appropriate analyses for a nominal (categorical) variable? 7 ________________

Write a sentence describing the results. 8 ________________

LITERATURE EXAMPLE 5.2

Use of nonionic contrast media in a high-risk outpatient population

James V. Zelch, MD

APPLIED RADIOLOGY, November 1990, Vol. 19, no. 11

Materials and methods

This prospective, open-label, single-center study was designed to permit review of nonionic-contrast-assisted diagnostic procedures in a controlled environment: a newly opened outpatient facility (Cleveland Clinic Florida, Ft. Lauderdale). In the protocol that followed, only iopamidol, a nonionic contrast medium, was used; no patient received more than 50 ml of iopamidol at 300 mgI/ml for intravenous urogram (IVU) or computed tomography (CT) scans; special procedures were performed using digital acquisition and display (no films) with iopamidol 300 or 370; iopamidol M 200 was used for myelographic studies; no patient was denied contrast-assisted intravascular studies because of traditional risk factors (including prior history of "catastrophic reaction" to ionic contrast media), and no sedatives or steroids were administered to any patient prior to diagnostic imaging (including arteriographic studies).

Risk factors were considered to include age over 60 years, previous adverse reaction to ionic contrast media, significant cardiac disease, renal disease, diabetes, history of asthma or drug allergy, and debilitated or malnourished state.

Results: Patient population

Of 1,417 patients who consecutively received nonionic contrast media, 1,175 (82.9%) had at least one risk factor: 924 (65.2%) were 60 years of age or older (figure 1); 241 (17.0%) had significant cardiac disease; 170 (12.0%) had renal disease; 135 (9.5%) had diabetes; 92 (6.5%) had a history of asthma; and 25 (1.8%) had a history of adverse experiences with the use of ionic contrast media (5 patients previously had severe adverse experiences with ionic media). Patient weights ranged from 98 lbs to 328 lbs (mean = 162 lbs.); ages ranged from 16 years to 90 years (mean = 64 years).

Of 1,417 procedures assisted by the nonionic contrast medium, there were 641 CT scans (45.2%), 359 IVUs (25.3%), 250 special angiographic procedures (17.6%), and 167 myelograms (11.8%) (table 1). The proportions of CT scans by body area are shown in table 1. Special procedures included angiography of arch/carotids, renal arteries, visceral arteries, aorta/femoral artery, venogram, and other vessels (table 1).

Procedure Distribution

	N	%
CT scan	641	45
Abdomen/pelvis	366	57
Brain	169	27
Chest	40	6
Other	66	10
IVU	359	25
Special procedures	250	18
Arch/carotids	103	41
Venogram	60	24
Renal	40	16
Visceral	20	8
Aorta/femoral	15	6
Other	12	5
Myelogram	167	12

TABLE 1

Literature Example 5.3 is from the article entitled "How Strictly Do Dialysis Patients Want Their Advance Directives Followed?" The exercises below point out some salient features of the one-way table that accompanies the abstract. Answer the following questions:

Does the title of the table have all the essential elements?	1 Yes/No
What do the two columns show?	2 ______________________
In what form does the table present the nominal variables "Ethnicity" and "Medical illnesses"?	3 ______________________
Who are the participants?	4 ______________________
What type of variable is "Mean Age"?	5 Nominal/Ordinal/Interval/Ratio

LITERATURE EXAMPLE 5.3

How Strictly Do Dialysis Patients Want Their Advance Directives Followed?

Ashwini Sehgal, MD; Alison Galbraith; Margaret Chesney, PhD; Patricia Schoenfeld, MD; Gerald Charles, MD; Bernard Lo, MD

Objective.—The *Cruzan* case and the Patient Self-Determination Act will encourage patients to specify in advance which life-sustaining treatments they would want if they become mentally incompetent. However, strictly following such advance directives may not always be in a patient's best interests. We sought to determine whether patients differ in how strictly they want advance directives followed.

Design.—Interview study.

Setting.—Seven outpatient chronic dialysis centers.

Participants.—One hundred fifty mentally competent dialysis patients.

Intervention.—Using a structured questionnaire, we asked the subjects whether they would want dialysis continued or stopped if they developed advanced Alzheimer's disease. We then asked how much leeway their physician and surrogate should have to override that advance directive if overriding were in their best interests. Subjects granting leeway were also asked what factors should be considered in making decisions for them.

Results.—Subjects varied greatly in how much leeway they would give surrogates to override their advance directives: "no leeway" (39%), "a little leeway" (19%), "a lot of leeway" (11%), and "complete leeway" (31%). Subjects also varied in how much they wanted various factors considered in making decisions, such as pain or suffering, quality of life, possibility of a new treatment, indignity caused by continued treatment, financial impact of treatment on family members, and religious beliefs.

Conclusions.—Strictly following all advance directives may not truly reflect patients' preferences. To improve advance directives, we recommend that physicians explicitly ask patients how strictly they want their advance directives followed and what factors they want considered in making decisions.

(*JAMA*. 1992;267:59-63)

Characteristics of Study Participants (N = 150)

Mean age, y (range)	55 (18-88)
Male, %	58
Ethnicity, %	
Black	48
White	33
Asian	13
Hispanic	3
Education, % completing high school	83
Type of dialysis, % hemodialysis	93
Median time since dialysis started, mo (range)	33 (1-257)
Medical illnesses, %	
Heart disease	28
Diabetes	27
Emphysema or bronchitis	12
Stroke	8
Cancer	7
Hospitalization in past 12 mo, %	67
Living with a partner, %	40
Prior oral advance directive, %	43
Prior written advance directive, %	20
Know of case of withdrawal of dialysis, %	27

METHODS
Subjects

We conducted the study from July to December 1990 at seven chronic dialysis units in the San Francisco, Calif, area. These sites included both free-standing and hospital-based units.

The dialysis unit staff identified all English-speaking outpatients 18 years of age or older who had been receiving chronic dialysis for at least 1 month, who were expected to live at least 6 months, and who were likely to be cooperative. We used a two-step procedure to exclude patients who were cognitively impaired. First, the dialysis unit staff excluded patients who were not considered capable of giving informed consent for medical or surgical procedures. Second, at the beginning of the interview, we described a vignette in which a dialysis patient communicates an advance directive to his wife about wanting dialysis discontinued if he ever gets advanced Alzheimer's disease. The vignette included a standardized description of Alzheimer's disease. We asked subjects to summarize this vignette in their own words. We excluded subjects who could not correctly paraphrase key points of the vignette such as the clinical features of Alzheimer's disease and the decision to stop dialysis. We approached and interviewed eligible subjects until we reached our target sample size.

Literature Example 5.4 is from the article entitled "Predictors of Physicians' Smoking Cessation Advice." The exercise below points out some salient features of the one-way table that accompanies the article. Answer the following questions.

What elements does the table include (who, what, where, when)? — 1 ________________

How does the table define "prevalence"? — 2 ________________

What type of variable is age? occupation? household income? — 3 Nominal/Ordinal/Interval/Ratio

Nominal/Ordinal/Interval/Ratio

Nominal/Ordinal/Interval/Ratio

Which variables have been categorized? — 4 ________________

____

According to this table, males smoke less than females. — 5 True/False

How much of a difference is there between the two groups? — 6 ________

LITERATURE EXAMPLE 5.4

Predictors of physicians' smoking cessation advice

Erica Frank, MD, MPH; Marilyn A. Winkleby, PhD; David G. Altman, PhD; Beverly Rockhill, MA; Stephen P. Fortmann, MD

Journal of the American Medical Association, 1991, Vol. 266, pp. 3139–3144

Objectives

To determine the percentage of smokers reporting that a physician had ever advised them to smoke less or to stop smoking, and the effect of time, demographics, medical history, and cigarette dependence on the likelihood that respondents would state that a physician had ever advised them to stop smoking.

Design and setting

Data were collected from the Stanford Five-City Project, a communitywide health education intervention program. The two treatment and three control cities were located in northern and central California. As there was no significant difference between treatment and control cities regarding cessation advice, data were pooled for these analyses.

Participants

There were five cross-sectional, population-based Five-City Project surveys (conducted in 1979–1980, 1981–1982, 1983–1984, 1985–1986, and 1989–1990); these surveys randomly sampled households and included all residents aged 12 to 74 years.

Main outcome measures

Improved smoking advice rates over time in all towns was an a priori hypothesis.

Results

Of the 2710 current smokers, 48.8% stated that their physicians had ever advised them to smoke less or stop smoking. Respondents were more likely to have been so advised if they smoked more cigarettes per day, were surveyed later in the decade, had more office visits in the last year, or were older. In 1979–1980, 44.1% of smokers stated that they had ever been advised to smoke less or to quit by a physician, vs 49.8% of smokers in 1989–1990 ($P < .07$). Only 3.6% of 1672 ex-smokers stated that their physicians had helped them quit.

Conclusion

These findings suggest that physicians still need to increase smoking cessation counseling to all patients, particularly adolescents and other young smokers, minorities, and those without cigarette-related disease. (*JAMA* 1991;226:3139–3144)

Table 1 Prevalence of Smokers in Study Population*

Characteristics	Smoking Prevalence, %
Survey years (No. of subjects)	
1979–1980 (2487)	30.4
1981–1982 (2330)	27.7
1983–1984 (1846)	27.4
1985–1986 (2360)	20.2
1989–1990 (2405)	20.0
Total (11 428)	25.1
Age, y	
12–17	9.5
18–24	23.0
25–49	28.5
50–74	24.7
Gender	
F	23.9
M	26.4
Ethnic origin	
Non-Hispanic white	25.4
Hispanic	22.1
Education (subjects ≥ 25 y)	
< High school	35.0
High school graduate	32.8
Some college	28.4
≥ College	17.0
Household income ($/y)	
< 10 000	27.1
10 000 to 19 999	30.3
20 000 to 29 999	27.2
30 000 to 39 999	24.2
≥ 40 000	20.5
Occupation	
Blue collar	34.1
White collar	21.4
Homemaker	23.1

*Smoking prevalence is the percentage of the number of subjects that smoke.

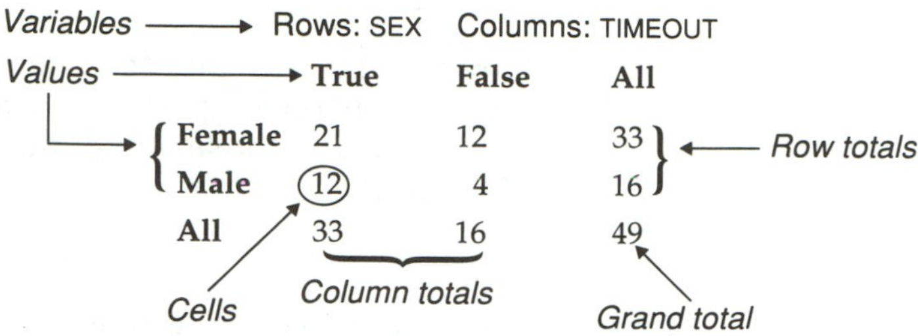

FIGURE 5.4 Sex by Timeout Distribution of Students in Statistics Classes, Spring 1992 at a University

Multivariate Tables: Anatomy of a Two-Way Table

A two-way table has several elements, as shown in Figure 5.4.

ACTIVITY 5.3

Title and other elements of a table

Refer to Figure 5.4 to answer the questions below.

A complete table title includes the following information: who is in the data set; what was being measured; where it was being measured; and when it was measured. In Figure 5.4, who is in the data set? What was being measured? Where was it measured? When was it measured?	1 ____________ 2 ____________ 3 ____________ 4 ____________
What are the variables under consideration?	5 Row variable is ________; column variable is ________.
What are the values of each variable?	6 Row values: ________; column values: ________

What is the number in the cell at the intersection of "True" and "Male"?	[7] ________
What are the row totals in Figure 5.4?	[8] Total number of females: ________; total number of males: ________
What are the column totals?	[9] Total who responded "True": ________; total who responded "False": ________
What is the total number of cases?	[10] ________

Suppose our interest is in finding out if women are more likely to take timeout than men. According to Figure 5.4, twenty-one women took timeout and only twelve men did. But this comparison is misleading because there were more than twice as many females as males in the data set. We cannot make a valid comparison until we standardize the numbers. Percentage **standardizes** the numbers by creating a common denominator (per one hundred).

ACTIVITY 5.4

Two-way tables with counts and percentages

Consider the table below, which shows the same information given in Figure 5.4 but also expresses each count as a percentage.

Rows: SEX **Columns:** TIMEOUT

	True	**False**	**ALL**
Females	21	12	33
Percent	63.64	36.36	100
Males	12	4	16
Percent	75.00	25.00	100
ALL	33	16	49
Percent	67.35	32.65	100

Note how the percents add up to 100 in all three rows. This table shows only row percentages. Out of the 33 females, [1] ________ percent take timeout, in contrast to [2] ________ percent of the males. Compared to females, a [3] higher/lower percentage of males do *not* take timeout. Out of those who take timeout, what percent are females? We cannot answer this question until we calculate column percentages. Since the above table only gives row percentages, it does not provide enough information to answer the question. Out of the 49 people, what percentage take timeout? [4] ________

Rows: SEX	**Columns:** TIMEOUT		
	True	**False**	**ALL**
Females	21	12	33
Percent	63.64	75.00	67.35
Males	12	4	16
Percent	36.36	25.00	32.65
ALL	33	16	49
Percent	100.00	100.00	100.00

Consider the above table representing the same data. Note how the percents add up to 100 in all three columns. This table shows only [5] row/column percentages. Out of those who take timeout, what percent are females? [6] ________. Out of those who do not take timeout, what percent are male? [7] ________. Both percents given for males are small, and this is a reflection of the small number of males in the classes. What percentage of the students are males? [8] ________

Two-Way Tables: Example from the Literature

Literature Example 5.5 is from the article entitled "Breastfeeding and Employment." Read the abstract to learn about the study.

This table has two parts: variables for which means are given and variables for which frequencies are given. Note that all these variables are tabulated against the variable "Plan to work" (values yes and no). For what variables are frequencies given? ("Occupation" and "Mother's prenatal income" could have been tabulated against "Plan to Work" in a pair of two-way tables. For the sake of economy of space they are given in one table. Still, this is considered a two-way table.)

1 ______________________________

What percentage of skilled workers plan to work?

2 _______

What is the mean number of employment hours for those who plan to work? for those who do not plan to work?

3 Plan to work _______; do not plan to work: _______

LITERATURE EXAMPLE 5.5

MARGARET H. KEARNEY, RNC, MS

LINDA CRONENWETT, RN, PhD, FAAN

Breastfeeding and Employment

Breastfeeding problems, outcomes, and satisfaction of married, well-educated first-time mothers who returned to work within six months postpartum were compared to those of mothers with the same characteristics who stayed at home. Mothers who planned to work after giving birth anticipated and experienced shorter durations of breastfeeding than did those who planned to remain at home. Breastfeeding experiences and satisfaction among working mothers differed little from the experiences and satisfaction of their nonworking counterparts; however, employment prior to two months postpartum exerted some negative effects on breastfeeding outcomes.

Accepted: March 1991

JOGGN
Volume 20 Number 6
November/December 1991

Results

Mothers who planned to work did not differ significantly from those planning to stay at home in age, years of education, or expected support for breastfeeding (Table 1). Mothers planning to work worked more hours per week prenatally, had less favorable attitudes toward breastfeeding, and planned to breastfeed for a shorter time than did those not planning to work. (In both groups, mean planned duration of breastfeeding exceeded American Academy of Pediatrics recommendations.)

Table 1.
Prenatal Variables and Plans for Postpartum Employment

	Plan to work				
	Yes (n = *91)*		*No* (n = *29)*		
Variables	***Mean***	***SD***	***Mean***	***SD***	t-***test***
Maternal age	29.0	4.5	28.1	3.5	NS
Years of education	16.1	2.5	15.6	1.9	NS
Employment (hours per week)	27.5	18.0	12.1	15.9	$p < 0.001$
Breastfeeding support	17.0	4.7	17.5	5.5	NS
Breastfeeding attitudes	64.4	5.5	67.1	6.4	$p = 0.03$
Planned duration of breastfeeding (months)	6.8	2.9	8.4	2.8	$p = 0.006$
Occupation and income	***Frequency (%)***		***Frequency (%)***		
Occupation					
Health professional	25		0		
Other professional	26		33		
Skilled worker	24		27		
Manager or retailer	13		10		
Unskilled worker or student	6		10		
Other employment	7		20		
Mother's prenatal income (annual)					
None	10		43		
Under $10,000	15		13		
$10,000 to $20,000	31		30		
More than $20,000	44		13		

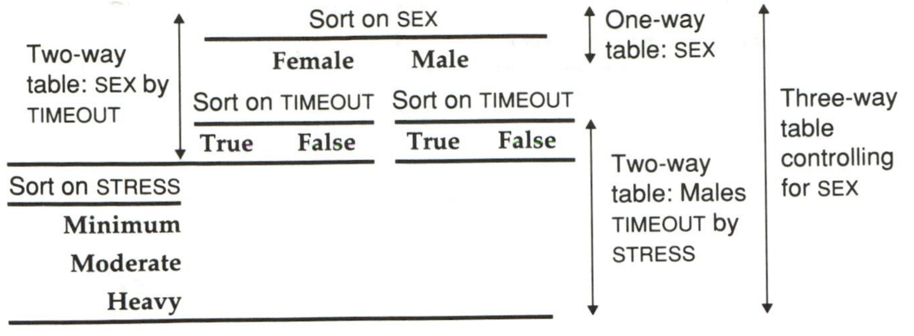

FIGURE 5.5 Overview of Three-Way Table Format

Construction of Three-Way Tables

Consider Figure 5.5. If you sort the SQD cards on the variable SEX, count the cases in the female and male categories, and enter the numbers in a table, you will have created a one-way table. If you take female and male cases separately and sort them on TIMEOUT, you have a two-way table (SEX by TIMEOUT). If you take only females (controlling for SEX) and sort them on TIMEOUT and follow it with a sort on STRESS, you have the female part of a three-way table. Similar sorting on TIMEOUT and STRESS for the male cases will give the male part of the three-way table.

ACTIVITY 5.5

Construction of one-, two-, and three-way tables

Construction of a table showing the distribution of ACTIVITY in the statistics class (one-way table):

1. Sort Student Questionnaire Data (SQD) cards into three piles according to categories (values) of ACTIVITY. Count the cases in each value category. Calculate percentages (number of cases in a category * 100) ÷ (total number of cases).

ACTIVITY	Count	Percent
Minimal (1)		
Moderate (2)		
Vigorous (3)		
Total number of cases n =		
Missing values = 2		

2. Percentages are called **relative frequencies**. By computing percentages we have **standardized** the counts.

The frequency of minimal activity is _______ and its relative frequency is _______. Frequencies when standardized will total to _______.

3. Give the table you created a title. Make sure the title includes all the elements mentioned in Activity 5.3.

3 ____

____

Construction of a two-way table showing the distribution of SEX by ACTIVITY in the statistics class:

4. Sort Student Questionnaire Data (SQD) cards into two piles according to categories (values) of SEX. Count the cases in each SEX category and enter under ALL for each row. Sort "Female" cards into three ACTIVITY values. Count and enter in row 1. Sort "Male" cards into three ACTIVITY values. Count and enter in row 2. Title the table and label the row and column variable names.

Rows:	Columns: Minimal	Moderate	Vigorous	ALL
1 Female				
2 Male				
ALL				

5. Calculate row percentages and add them to the table you just created (number of cases in a category * 100) ÷ (total number of row cases).

	Min. %	Mod. %	Vig. %	ALL
1 Female				
2 Male				
ALL %				

6. Next calculate column percentages (number of cases in a category * 100) ÷ (total number of column cases).

	Min. %	Mod. %	Vig. %	ALL
1 Female				
2 Male				
ALL %				

Construction of a table showing the distribution of SEX by DISCUSS by STRESS in the statistics class (three-way table):

Our interest is in three variables, so we present the frequency distribution of each one. This will tell us if there are any missing values. Note: Though there are three variables in this table, it is not a three-way table.

STRESS	Count	DISCUSS	Count	SEX	Count
1	11	1	46	1	33
2	31	2	5	2	18
3	8	$n = 51$		$n = 51$	
$n = 50$					

One value for STRESS is missing. So we will be processing only 50 cases instead of 51 cases. Our answers will reflect this.

We hold SEX constant (control for SEX). Sort the cards into two piles according to categories (values) of the SEX variable. You have two piles of cards, one for females (33 cases) and one for males (18 cases).

7. Sort the female cards into two piles according to categories (values) of the DISCUSS variable. Construct a one-way table.

Control: SEX = 1
Rows: DISCUSS

DISCUSS	Count
1	
2	
ALL	32

8. Sort each one of the cases into values (True and False) of DISCUSS by categories in the STRESS variable. You will have a two way-table, which is a result of controlling for the SEX variable.

Control: SEX = 1
Rows: DISCUSS Columns: STRESS

	Minimal	Moderate	Vigorous	ALL
True 1				28
False 2				4
ALL	7	19	6	32

9. Construct a two-way table, controlling for SEX = 2.

Control: SEX = 2
Rows: DISCUSS Columns: STRESS

	Minimal	Moderate	Heavy	ALL
True 1				
False 2				
ALL				18

Multivariate Tables: Interpretation of Three-Way Tables

Three-way tables are produced by controlling for one of three variables and then producing a two-way table with the remaining two variables. We can control for each one of the variables; therefore we can produce three sets of tables. By arranging the same data in different ways these tables answer different kinds of questions. For example, the first two tables on the following page control for the variable SEX. You can use them to answer the question: Out of the females who take TIMEOUT, how many are minimally stressed? The next three tables answer the question: Out of those who are minimally stressed, how many females take TIMEOUT? The last two tables answer the question: Out of those who take TIMEOUT, how many females are minimally stressed? All these tables give the same answer. The layout of each table differs and the ease with which that table helps us answer each question differs.

Control: SEX = 1 (Female)
Rows: TIMEOUT Columns: STRESS

	Minimal	Moderate	Heavy	ALL
True	7	16	5	28
False	0	2	2	4
ALL	7	18	7	32

This table cannot answer any questions about males. It controls for SEX = 1 (Female). Out of the 32 females in the sample, how many take TIMEOUT and yet are heavily stressed? [1] ________ Out of the 32 females in the sample, how many do not take TIMEOUT and yet are heavily stressed? [2] ________
Only 12.5 percent of the females do not take TIMEOUT. We do not have a large enough sample to see if TIMEOUT is a crucial factor governing STRESS.

Control: SEX = 2 (Male)
Rows: TIMEOUT Columns: STRESS

	Minimal	Moderate	Heavy	ALL
	1	2	3	
True	3	13	1	17
False	1	0	0	1
ALL	4	13	1	18

Out of the 18 males in the sample, how many take TIMEOUT and yet are heavily stressed? [3] ________
Out of the 18 males in the sample, how many do not take TIMEOUT and yet are heavily stressed? [4] ________
Only 5 percent of the males do not take TIMEOUT. 12.5 percent of the females do not take TIMEOUT. Do males take TIMEOUT more than females? The sample is too small to say for certain.

Control: STRESS = Minimum
Rows: TIMEOUT Columns: SEX

	Male	Female	ALL
	1	2	
True	4	3	7
False	3	0	3
ALL	7	3	10

Out of the 10 persons who are minimally stressed, how many males and how many females take TIMEOUT?
Males: [5] ________
Females: [6] ________

Control: STRESS = Moderate
Rows: TIMEOUT Columns: SEX

	Male	Female	ALL
True	13	8	21
False	5	4	9
ALL	18	12	30

Out of those who are moderately stressed, how many males and how many females do not take TIMEOUT?
Males: [7] ________
Females: [8] ________

Control: STRESS = Heavy
Rows: TIMEOUT Columns: SEX

	Male	Female	ALL
True	3	1	4
False	4	0	4
ALL	7	1	8

Control: TIMEOUT = True
Rows: STRESS Columns: SEX

	Male	Female	ALL
Minimal	4	3	7
Moderate	13	8	21
Heavy	3	1	4
ALL	20	12	32

Control: TIMEOUT = False
Rows: STRESS Columns: SEX

	Male	Female	ALL
Minimal	3	0	3
Moderate	5	4	9
Heavy	4	0	4
ALL	12	4	16

Example from the Literature

Literature Example 5.6 is from the article entitled "Bicycle-Associated Head Injuries and Deaths in the United States From 1984 Through 1988: How many are Preventable?" The exercise below points out some salient features of the three-way table. Answer the following questions:

What essential elements does the title have? 1 ______________________

What do the numbers in the cells represent? 2 ______________________

What are the three variables in this three-way table? 3 ______________________

What might have been the sequence of sorting of variables for this table? 4 ______________________

In the first three columns, what variable was controlled for? 5 ______________________

Head injury death rates are higher for males in all age groups. Why? 6 ______________________

Write the first two lines of the two-way table: Degree of injury by age. 7 ______________________

LITERATURE EXAMPLE 5.6

Bicycle-Associated Head Injuries and Deaths in the United States From 1984 Through 1988

How Many Are Preventable?

Jeffrey J. Sacks, MD, MPH; Patricia Holmgreen, MS; Suzanne M. Smith, MD; Daniel M. Sosin, MD

Objective.—To estimate the potential benefits from more widespread bicycle safety helmet use.

Design.—Review of death certificates and emergency department injury data for 1984 through 1988. Categorization of deaths and injuries as related to bicycling and head injury. Using relative risks of 3.85 and 6.67 derived from a case-control study and varying helmet usage from 10% to 100%, population attributable risk was calculated to estimate preventable deaths and injuries.

Setting.—Entire United States.

Main Outcome Measures.—Numbers of US residents coded as dying from bicycle-related head injuries, numbers of persons presenting to emergency departments for bicycle-related head injuries, and numbers of attributable bicycle-related deaths and head injuries.

Main Results.—From 1984 through 1988, bicycling accounted for 2985 head injury deaths (62% of all bicycling deaths) and 905 752 head injuries (32% of persons with bicycling injuries treated at an emergency department). Forty-one percent of head injury deaths and 76% of head injuries occurred among children less than 15 years of age. Universal use of helmets by all bicyclists could have prevented as many as 2500 deaths and 757 000 head injuries, ie, one death every day and one head injury every 4 minutes.

Conclusions.—Effective community-based education programs and legislated approaches for increasing bicycle safety helmet usage have been developed and await only the resources and commitment to reduce these unnecessary deaths and injuries.

(*JAMA*. 1991;266:3016-3018)

Table 2.—Bicycle-Associated Head Injury Death Rates and Head Injury Rates per Million Population, by Sex and Age Group, United States, 1984 Through 1988

	Head Injury Death Rates			Head Injury Rates		
Age Group, y	**Males**	**Females**	**Total**	**Males**	**Females**	**Total**
0-4	1.1	0.2	0.7	1901	801	1364
5-9	8.2	1.9	5.1	5844	2805	4361
10-14	13.5	3.4	8.6	3258	1173	2243
15-19	9.4	1.2	5.4	1286	358	832
20-24	4.9	0.8	2.8	642	266	454
25-29	3.3	0.4	1.9	442	129	285
30-39	2.3	0.4	1.4	259	104	181
40-49	1.9	0.4	1.1	122	66	94
50-59	1.8	0.3	1.0	73	39	55
60-69	2.0	0.2	1.1	39	23	31
70-79	3.0	0.2	1.4	93	20	50
80+	3.5	0.1	1.2	54	23	33
Total	**4.3**	**0.7**	**2.5**	**1103**	**418**	**751**

METHODS

Mortality and Morbidity Ascertainment

We used the National Center for Health Statistics Multiple Cause-of-Death Public Use Data Tapes for 1984 through 1988.[14] We defined a bicycle-associated death (BAD) as any US resident coded as dying as the result of a bicycle crash (E800 through E807, with fourth digit 0.3; E810 through E825, with fourth digit 0.6; E826.1 or E826.9; or E827 through E829, with fourth digit 0.1).[15] A head injury bicycle-associated death (HIBAD) was any BAD that had any of the following HI diagnoses as an associated condition[16]: skull fracture (N800.0 through N801.9 and N803.0 through N804.9); intracranial injury (N850.0 through N854.9); and late effects of either skull fracture (N905.0) or intracranial injury without skull fracture (N907.0).

Self-Assessment

Please assess your competency.

Tasks:	How well can you do it?					Page
When data are given, construct and interpret the following:	*Poorly*				*Very well*	
1. One-way table	1	2	3	4	5	126
2. Two-way table	1	2	3	4	5	126
3. Three-way table	1	2	3	4	5	142
4. Stem and leaf plot	1	2	3	4	5	121

Answers for Chapter 5

Answer for question on p. 113: More

Answers for questions on pp. 115–116: 1) Difficult 2) 20 3) 61 4) Difficult to determine, but it can be done 5) Again, difficult to determine, but it can be done 6) 20 7) 61 8) 21 9) 28, 29, 32, 61 10) 20–24 11) 40 or older 12) 10; this is just a guess, however, based on age. We do not really know until we ask the question, "Are you a returning student?"

Answers for p. 117: 1) 17 2) 68 3) 19 4) 76 5) 12 6) 22 7) 88 8) 8 9) 21 10) 96 11) 25 12) Class II 13) Class I 14) 84.6 15) 76

Answers for questions on pp. 119–120: 1) 7 2) 6 3) 73 4) 75 5) 76 6) Eleven 7) Five 8) 1 9) 8 10) 9

Answers for Activity 5.1: 1) 20 2) 61 3) 6

4) **Freq.**	**Stem Leaves**
34	2 0000011111111111222223333333344444
7	2 5556689
7	3 0002333
2	3 88
0	4
0	4
0	5
0	5
1	6 1

5) 61 6) 23

Answers for Activity 5.2: 1) Minimum = 40; Maximum = 62;

Frequency	Stem Leaves
7	4 0000111
2	4 33
2	4 44
2	4 67
3	4 889
1	5 1
4	5 2222
1	5 4
3	5 777
1	5 8
3	6 111
1	6 2

Answers for Literature Example 5.1: 1) Question 11 2) This is a matter of opinion. We don't think so, because there seems to be no advantage to the stem and leaf. The graphical aspect gives no useful information that we could discern.

Answers for questions on p. 127: 1) Univariate 2) Multivariate 3) One 4) 33 5) Males 6) Uni- 7) TIMEOUT 8) SEX 9) 12 10) True 11) DISCUSS 12) True 13) 45 14) 4 15) DISCUSS 16) 4

Answers for Literature Example 5.2: 1) Procedure Distribution 2) The table is presented in the article, so readers are seeing it in that context and can get information from the article. 3) Procedures 4) Nominal or categorical 5) CT Scan, IVU, Special Procedures, Myelogram 6) Yes 7) Frequencies and percentages 8) Of the subjects studied, approximately 45 percent received CT scans; 25 percent received IVUs; 18 percent received special procedures; and 12 percent received myelograms.

Answers for Literature Example 5.3: 1) No 2) Left column, variables; right column, percentages, means, or medians 3) As percentages 4) Mentally competent dialysis patients 5) Ratio

Answers for Literature Example 5.4: 1) Who: study population (a phrase that is too vague, in our opinion); what: prevalence of smokers in the population 2) The footnote in the table defines prevalence as the percentage of subjects who smoke. 3) Ratio; nominal; ratio 4) All of the variables have been categorized. 5) False 6) 2.5%

Answers for Activity 5.3: 1) Students in statistics classes 2) The number of students in each of two categories: SEX and TIMEOUT 3) In statistics classes at a university 4) Spring of 1992 5) Row variable is SEX; column variable is TIMEOUT. 6) Row values are Female and Male; column values are True and False. 7) 12 8) Total number of females is 33; total number of males is 16. 9) Total who responded True is 33; total who responded False is 16. 10) 49.

Answers for Activity 5.4: 1) 63.64 2) 75 3) Lower 4) 67.35 5) Column 6) 63.64 7) 25% 8) 32.65%

Answers for Literature Example 5.5: 1) Occupation; mother's prenatal income 2) 24% 3) 27.5; 12.1.

Answers for Activity 5.5:

1)

ACTIVITY	Count	Percent
Minimal (1)	6	12.24
Moderate (2)	36	73.47
Vigorous (3)	7	14.29
	$n = 49$	100%

2) 6; 12.24%; 100% 3) Activity Distribution of Students in Statistics Classes at a University, Spring 1992.

4) Activity Distribution of Students by Sex
Rows: SEX Columns: ACTIVITY

	Minimal	Moderate	Vigorous	ALL
Female	6	24	2	32
Male	0	12	5	17
ALL	6	36	7	49

5)	Min. %	Mod. %	Vig. %	ALL
1 Female	18.75	75.00	6.25	100
2 Male	0	70.59	29.41	100
ALL %	12.24	73.47	14.29	100

6)	Min. %	Mod. %	Vig. %	ALL
1 Female	100	66.67	28.57	65.31
2 Male	0	33.33	71.43	34.69
ALL %	100	100	100	100

7) Control: SEX = 1
Rows: DISCUSS

DISCUSS	Count
True 1	28
False 2	4
ALL	32

8) Control: SEX = 1
Rows: DISCUSS Columns: STRESS

	Minimal	Moderate	Heavy	ALL
True 1	7	17	4	28
False 2	0	2	2	4
ALL	7	19	6	32

9) Control: SEX = 2
Rows: DISCUSS Columns: STRESS

	Minimal	Moderate	Heavy	ALL
True 1	3	13	1	17
False 2	1	0	0	1
ALL	4	13	1	18

Answers for p. 146: 1) 5 2) 2 3) 1 4) None 5) 4 males 6) 3 females 7) 5 males 8) 4 females

Answers for Literature Example 5.6: 1) Who: U.S. Population; What: Rates of head injury and head injury deaths; Where: United States; When: 1984–1988 2) Rates per million population 3) Degree (fatal or not) of injury, Sex, Age 4) Degree of injury, Age groups, Sex 5) Degree of injury 6) It is likely that more males ride bicycles. The denominator for rates is not the total number of people who ride bicycles but one million population. 7) 0–4, 0.7, 1364; 5–9, 5.1, 4361

CHAPTER

6

More Descriptive Tools: Graphs and Summary Statistics

Prelude

" . . . The best way to get from one place to another is to erase everything and begin again. Please make yourself at home."

"Do you always travel that way?" asked Milo as he glanced curiously at the strange circular room, whose sixteen tiny arched windows corresponded exactly to the sixteen points of the compass. Around the entire circumference were numbers from zero to three hundred and sixty, marking the degrees of the circle, and on the floor, walls, tables, chairs, desks, cabinets, and ceiling were labels showing their heights, widths, depths, and distances to and from each other. To one side was a gigantic note pad set on an artist's easel, and from hooks and strings hung a collection of scales, rulers, measures, weights, tapes, and all sorts of other devices for measuring any number of things in every possible way.

— Norton Juster
The Phantom Toll Booth, p. 187

Chapter Outline

Self-Assessment

Please assess your competency.

Tasks:	**How well can you do it?**					**Page**
Construct and/or interpret the following:	*Poorly*				*Very well*	
1. Bar graph	1	2	3	4	5	155
2. Histogram	1	2	3	4	5	155
3. Frequency polygon	1	2	3	4	5	173
4. Scatter plot	1	2	3	4	5	204
5. Time series plot	1	2	3	4	5	195
6. Range	1	2	3	4	5	189
7. Quartile	1	2	3	4	5	190
8. Inter-quartile range	1	2	3	4	5	190
9. Minimum/Maximum	1	2	3	4	5	191
10. Mode	1	2	3	4	5	189
11. Median	1	2	3	4	5	189
12. Five-number summary	1	2	3	4	5	190
13. Box plot	1	2	3	4	5	190

Introduction

Raw data are difficult to make sense of. Unless the data have been organized in some fashion it is difficult to detect patterns in them. One way of organizing data is to put them into a table. Another way is to display them in a graph. Graphs display data in a way that is easy to assimilate. Most of us can see patterns in a graphic display far more easily than we can detect them in a tabular display.

Graphs can be used to present information in an honest fashion, but they can also be used to distort information. The reader must always bear in mind that the goal of the person who has prepared a graph—be that person a scientist or an advertiser—is to persuade you of something.

The Purpose of Graphical Tools

The statistical tools discussed in this chapter serve two purposes: 1) to consolidate and analyze data so we can detect patterns of variation in the variables, and 2) to communicate the information in a form that facilitates the reader's understanding. Figure 6.1 presents various typical elements in a graph. Refer to the figure when answering the questions in Activity 6.1.

Activity 6.1

Elements in a graph

By convention, the vertical and horizontal axes are called the [1] _______ axis and the [2] _______ axis, respectively. In this graph the x axis represents

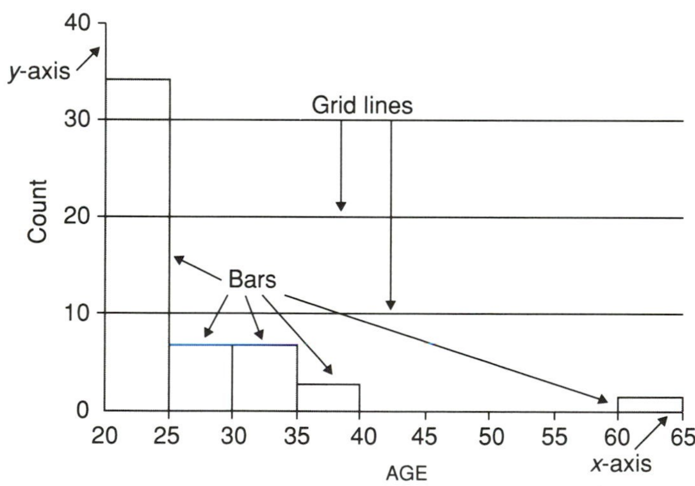

Figure 6.1 Different Elements in the Graph. Age Distribution of Two Statistics Classes at a University, Spring 1992.

[3] _______ and the y axis represents [4] _______. There are [5] _______ bars in the graph and the tallest bar covers the interval [6] _______. The ages are presented in groups of [7] _______. The [8] _______ lines are intended to facilitate reading numbers on the y axis. Similar lines could be placed on the [9] _______ axis. The decision of whether to place grid lines is arbitrary. The title of the graph has the following components: Who? [10] _______ What? [11] _______ When? [12] _______ Where? [13] _______. The most frequently occurring age group is [14] _______; it is called the **mode**. In Figure 6.3 we plotted the same data without grouping the ages. Notice that the pattern of variation of the ages looks different.

Bar Graphs and Histograms

The traditional distinction between a bar graph and a histogram is subtle and some people ignore the distinction, as the two graphs achieve the same purpose. A **histogram** is used for ratio data; the bars in a histogram are drawn touching each other to indicate that the data are continuous. Ratio data are categorized and intervals in categories are governed by the same considerations we used in one-way tables for ratio data. **Bar graphs** are used for nominal and ordinal data; typically, space is left between the bars. Figure 6.2 shows a bar

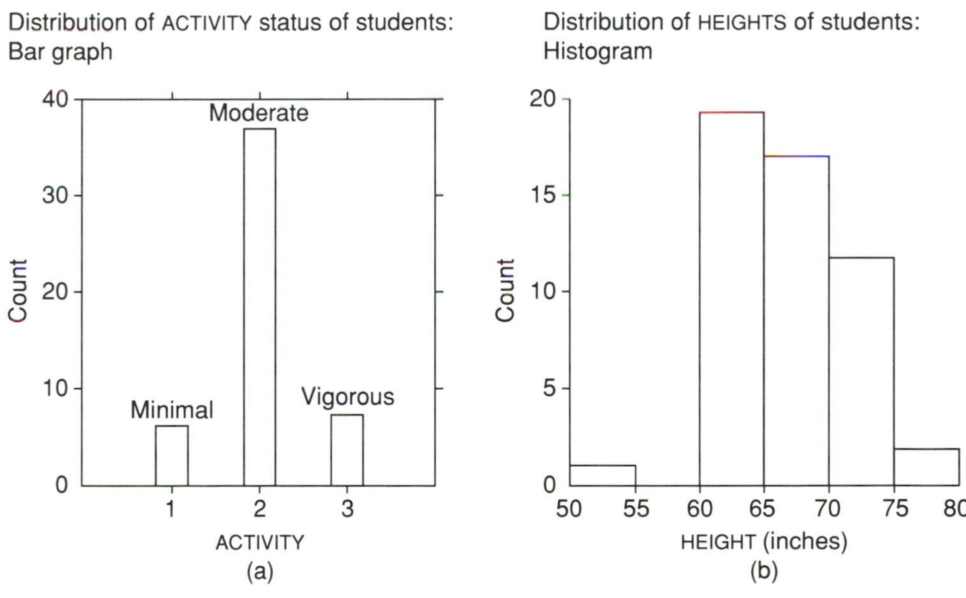

FIGURE 6.2

graph and a histogram. Activity values 1, 2, and 3 represent minimal, moderate, and vigorous. ACTIVITY is a [1] categorical/ratio variable, whereas HEIGHT is a [2] _______ variable. Heights are categorized and have intervals of [3] _______ inches. In both the bar graph and the histogram the y axis represents counts. The bars are separated, in the case of the bar graph; they are [4] separated/contiguous in the case of the histogram. There are no heights in the interval of [5] _______. The mode value for ACTIVITY is [6] _______ and for HEIGHT it is the interval [7] _______.

The histograms in Figures 6.3 and 6.4 present ages of the statistics students without categorization and categorized into ten-year intervals. What is the mode in Figure 6.3? [8] _______ The **range** is the difference between the **maximum** and **minimum** values of the variable. What is the range? [9] _______ Note that this is a bar graph even though it depicts a ratio variable, AGE. Each age is treated as a category.

In Figure 6.4 the age group of 20–29 is the most frequently occurring interval, and it is called the [10] _______. Figures 6.1, 6.3, and 6.4 present the same data but the distributions look different. Three observations are clear from all three graphs: (1) Ages are bunched around the younger end of the spectrum; (2) one person's age does not fit the overall pattern; and (3) the distribution of ages is [11] symmetric/asymmetric.

Consider the following table as it relates to Figure 6.5.

Age Interval	Count
20–24	34
25–29	7
30–34	7
35–39	2
40 or older	1
	$n = 51$

Certain conventions shaped both the table and the graph in Figure 6.5. Both use [12] non-overlapping/overlapping intervals for AGE. The count can be read

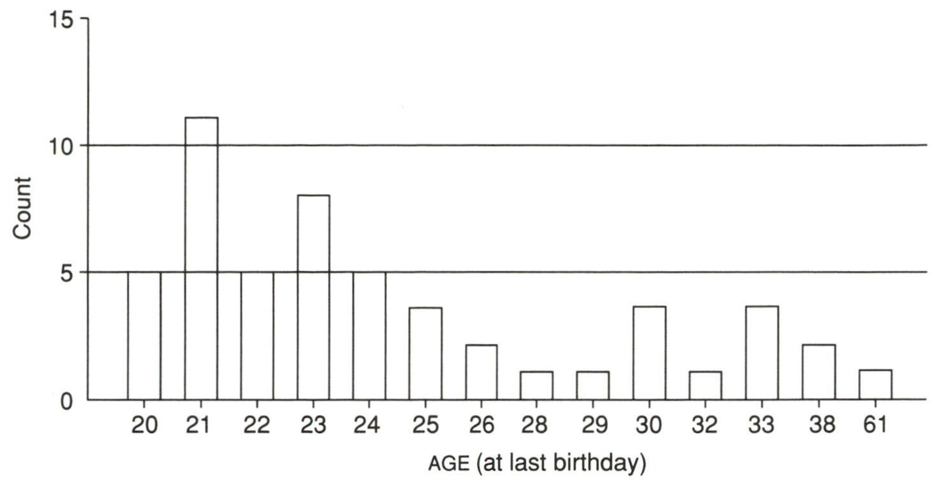

FIGURE 6.3 Age Distribution of Students in Statistics Classes (ages are not grouped)

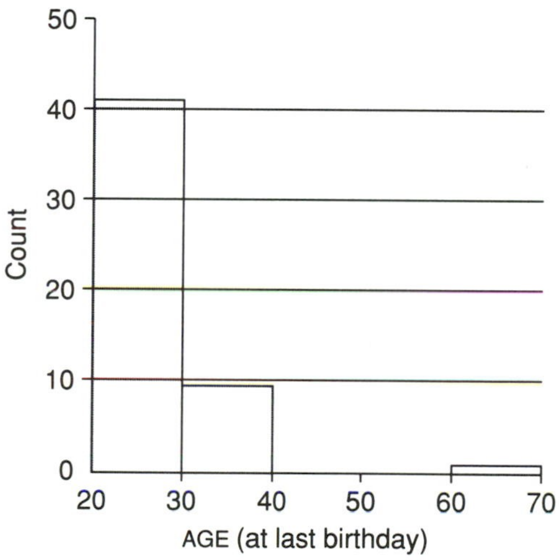

FIGURE 6.4 Ages Grouped in Ten-Year Intervals

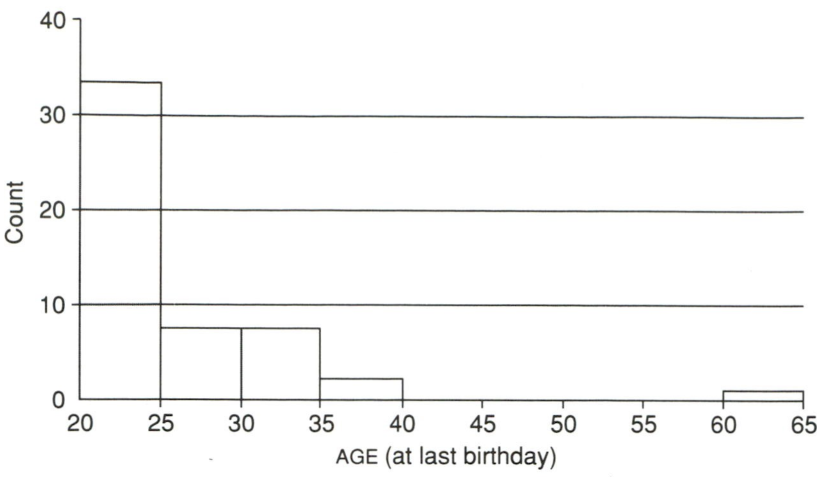

FIGURE 6.5

more readily from a table. The graph uses [1] _______ and [2] _______ axis conventions to show age intervals and [3] _______.

What did we gain by drawing a graph? Sometimes being able to see data in pictorial form dramatizes a phenomenon and helps us remember. The two striking features of the graph are (1) the thirty-four cases in the 20–24 age group and (2) the [4] _______ case in the 60–64 age group; these stand out and are likely to be remembered. Cognitive psychologists postulate that the incorporation of information into schema aids long-term memory. The ability to gain a total impression of the distribution in one glance at the graph mitigates the inconvenience of having to read across to figure out the count. This inconvenience can be eliminated by printing the exact count on or above each bar.

Suppose we change the axes as in Figure 6.6: now AGE is on the [5] _______ axis whereas [6] _______ is on the x axis. It is purely a matter of convention to call the horizontal axis the x axis and the vertical axis the y axis. So we can say that now [7] _______ is on the vertical axis and [8] _______ is on the horizontal axis. The important thing is that we label clearly so we communicate successfully. The labels x and y are arbitrary, though we generally agree to designate the horizontal axis as x and the vertical axis as y.

We displayed the same data in several different formats to show that each of these formats give pretty much the same information. Each format has advantages and disadvantages. Ultimately, each scientist decides which form best communicates the pattern hidden in the raw data. Before choosing a graphical or tabular format you should first study the journals that pertain to your field of study. In each field certain formats tend to dominate. You should also write to the journals in question for their style sheets. These instructions

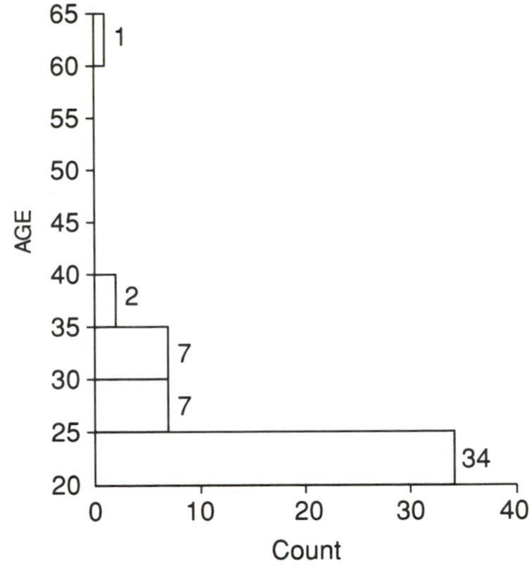

FIGURE 6.6

on preparing manuscripts will make you aware of the specific formats that editors and reviewers prefer.

Examples from the Literature

Literature Example 6.1 is from the article entitled "A Graduate Certification Program." The figure includes the following elements: a title, labels for the x and y axes, and bars.

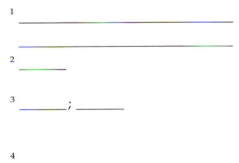

One sentence in the text is depicted in the graph. What is that sentence? [1] ______________________

____

How many variables are depicted in this figure? [2] ________

How many participants hold the most common degree? the least common? [3] ________; ________

Do you find it difficult to read the graph because of the absence of grid lines? [4] ________

LITERATURE EXAMPLE 6.1

A graduate certificate program: Education and training for hazardous waste specialists

R. Powitz; J. McMicking; R. Kummler

Journal of Environmental Health, vol. 52, no. 4

Abstract

Wayne State University has instituted a new concept in graduate education: a graduate certificate in hazardous waste management. The certificate program of 13 semester hours is designed to provide practicing scientists and engineers with the supplementary training in control technology, laws, policy and regulations necessary to manage or regulate corporate hazardous materials programs. A full description of the program and profile of the student body are presented.

A survey of the students in the program was taken in the winter and fall of 1987, 1988 and 1989 for use in future planning. Among the students polled, most held a degree in chemical engineering followed by biology, civil engineering and chemistry (see figure below).

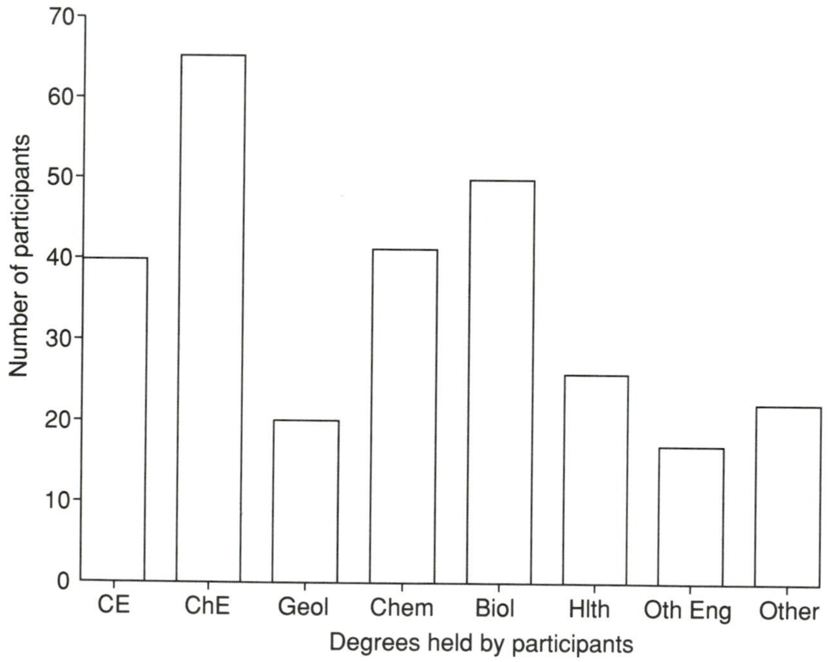

Literature Example 6.2 is the abstract of the article entitled "Measuring Airborne Asbestos." Read the abstract to learn about the study.

Is the distribution symmetric or skewed?	[1] Symmetric/Skewed
What is the length of the fiber that occurs most frequently?	[2] ______
Are these visible under an optical microscope? (Read the text.)	[3] Yes/No
What do you suppose are the implications of this study for the employees who work under conditions where airborne asbestos is present?	[4] ______________________________

LITERATURE EXAMPLE 6.2

Measuring Airborne Asbestos

The significance of small fibers detected under TEM

Bruce E. Lippy, CIH Jeffrey A. Boggs

Abstract

No one in the often rancorous field of measuring airborne asbestos disputes the fact the transmission electron microscope (TEM) is able to see smaller fibers than the optical microscope. The Environmental Protection Agency (EPA), in its guidebooks, points out that TEM routinely can resolve fibers of 0.0025 micrometer diameter and is fully 100 times more sensitive than phase contrast microscopy. The TEM method is, however, a great deal more expensive. Cost versus quality analysis is a dilemma that prompts many questions. How important are the small fibers? Does the ability to detect them really merit the additional expense? Does reporting them dilute data on the longer, reportedly more dangerous fibers? These questions will be answered by evaluating: the differences between phase contrast and TEM in analytical capabilities; the fiber size distribution of asbestos reported in air and in lung tissue; and the medical significance of short fibers.

Journal of Environmental Health, Vol. 52, No. 3

Fiber size distribution in air and lung tissue

With an understanding of the analytical methods, the relationship of the visible to invisible fibers in air samples and in lung tissue can be explored. Of significance is the question of whether an estimate of the invisible portion can be made from the PCM fiber count.

To fully appreciate how totally different the technologies are, note the frequency distribution of fibers on the standard material filters supplied by the National Bureau of Standards for air sample analysis by TEM (Figure 2). This figure accompanies the material and is found in the work of Steel and Small (7). The heaviest distribution of fibers in the samples is around 1 micrometer in length. Almost none of the fibers are above 5 um in length.

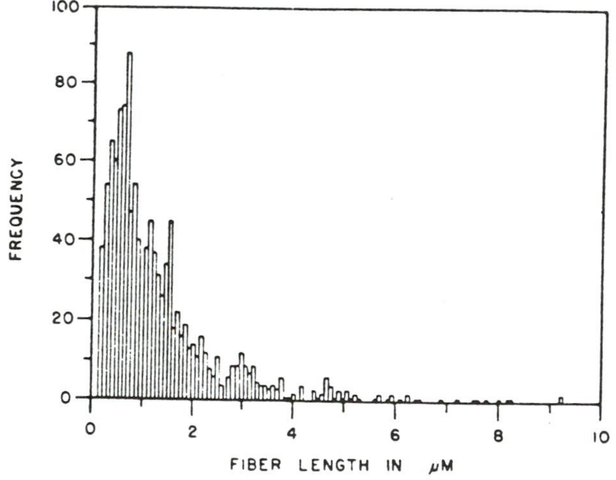

Figure 2

Histogram of fiber lenght for all fibers counted during certification of the SRM filter.

Literature Example 6.3 is from the article entitled "Ventricular Arrhythmias in Patients Undergoing Noncardiac Surgery." Read the abstract to learn about the study.

From the arrows in the figure we can tell which condition the authors contrasted. We extracted the information from Fig. 2 and displayed it in a table. Which format do you find easier to read, the graph or the table?

1 ______________________________

Mention two advantages of giving the figure in addition to the corresponding table.

2 ______________________________

Read the text that describes the figure. In your opinion, which facilitates your understanding of the text more, the figure or the table?

3 ______________________________

LITERATURE EXAMPLE 6.3

Ventricular Arrhythmias in Patients Undergoing Noncardiac Surgery

Brian O'Kelly, MB, MRCPI, FRCPC; Warren S. Browner, MD, MPH; Barry Massie, MD; Julio Tubau, MD; Long Ngo, MS; Dennis T. Mangano, PhD, MD; for the Study of Perioperative Ischemia Research Group

Objective.—To determine the incidence, clinical predictors and prognostic importance of perioperative ventricular arrhythmias.

Design.—Prospective cohort study (Study of Perioperative Ischemia).

Setting.—University-affiliated Department of Veterans Affairs Medical Center, San Francisco, Calif.

Subjects.—A consecutive sample of 230 male patients, with known coronary artery disease (46%) or at high risk of coronary artery disease (54%), undergoing major noncardiac surgical procedures.

Measurements.—We recorded cardiac rhythm throughout the preoperative (mean=21 hours), intraoperative (mean=6 hours), and postoperative (mean=38 hours) periods using continuous ambulatory electrocardiographic monitoring. Adverse cardiac outcomes were noted by physicians blinded to information about arrhythmias.

Main Results.—Frequent or major ventricular arrhythmias (>30 ventricular ectopic beats per hour, ventricular tachycardia) occurred in 44% of our patients: 21% preoperatively, 16% intraoperatively, and 36% postoperatively. Compared with the preoperative baseline, the severity of arrhythmia increased in only 2% of patients intraoperatively but in 10% postoperatively. Preoperative ventricular arrhythmias were more common in smokers (odds ratio [OR], 4.1; 95% confidence interval [CI], 1.2 to 15.0), those with a history of congestive heart failure (OR, 4.1; 95% CI, 1.9 to 9.0), and those with electrocardiographic evidence of myocardial ischemia (OR, 2.2; 95% CI, 1.1 to 4.7). Preoperative arrhythmias were associated with the occurrence of intraoperative and postoperative arrhythmias (OR, 7.3; 95% CI, 3.3 to 16.0, and OR, 6.4; 95% CI, 2.7 to 15.0, respectively). Nonfatal myocardial infarction or cardiac death occurred in nine men; these outcomes were not significantly more frequent in those with prior perioperative arrhythmias, albeit with wide CIs (OR, 1.6; 95% CI, 0.4 to 6.2).

Conclusion.—Almost half of all high-risk patients undergoing noncardiac surgery have frequent ventricular ectopic beats or nonsustained ventricular tachycardia. Our results suggest that these arrhythmias, when they occur without other signs or symptoms of myocardial infarction, may not require aggressive monitoring or treatment during the perioperative period.

(*JAMA*. 1992;268:217-221)

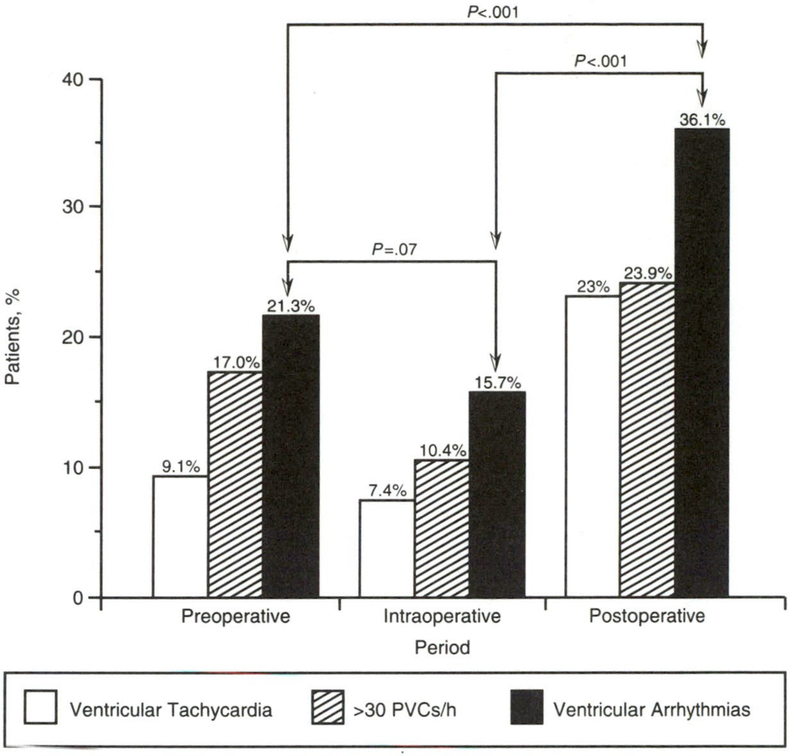

Fig 2.—Proportions of patients with ventricular tachycardia, more than 30 premature ventricular contractions (PVCs) per hour, or either, by time period.

Preoperatively, 39 patients (17.0%) had greater than 30 ventricular ectopic beats per hour, 21 (9.1%) had ventricular tachycardia, and 49 (21.3%) had either arrhythmia. Unadjusted for the length of monitoring, the incidence of ventricular arrhythmias declined to 15.7% during surgery (*P*=.07). During the intraoperative period, a total of 30 episodes of ventricular tachycardia occurred in 17 patients, of which seven (23%) occurred within 30 minutes of tracheal intubation. A greater proportion of patients (36.1%) had ventricular arrhythmias postoperatively (Fig 2), at least in part because of a longer mean monitoring period.

Percent of patients with ventricular tachycardia, more than 30 premature ventricular contractions (PVCs) per hour, or ventricular arrythmias, by time period (Preoperative, Intraoperative, Postoperative)

		Patients %	
	Preoperative	**Intraoperative**	**Postoperative**
Ventricular tachycardia	9.1%	7.4%	23%
> 30 PVs/h	17%	10.4%	23.9%
Ventricular arrhythmia	21.3%	15.7%	36.1%

Literature Example 6.4 is from the article entitled "Quality of Life in the Year before Death." Read the abstract to learn about the study. Conventional practice would have age graphed on the horizontal axis and percentage on the vertical axis. Notice that these authors did it the opposite way.

Different activities are represented by the shaded bars. Which activity is the most frequent across all the age groups?

1 ______________________________

What are the two variables in this graph?

2 ______________________________

LITERATURE EXAMPLE 6.4

The Quality of Life in the Year before Death

Harold R. Lentzner, PhD, Elsie R. Pamuk, PhD, Elaine P. Rhodenhiser, Richard Rothenberg, MD, MPH, and Eve Powell-Griner, PhD

Objectives. Most Americans wish to live a long healthy life, but fear disease and dependency in their last years. Until recently, little has been known about the prevalence of opposite extremes of health in old age, particularly in the period leading up to death.

Methods. We used results from the 1986 National Mortality Followback Survey to estimate proportions of elderly decedents who were "fully functional" or "severely restricted" in the last year of life. Estimates were based on responses from proxies to questions regarding the decedent's functional status, mental awareness, and time spent in institutions.

Results. Approximately 14% of all decedents aged 65 years and older were defined as fully functional in the last year of life; 10% were defined as severely restricted. Proportions varied with the decedent's age and sex, the underlying cause of death, and the presence of other preexisting conditions.

Conclusions. Results from this survey and future surveys can be used to learn more about "successful agers"—their medical histories, their life-styles, and whether their relative number is increasing or decreasing over time. (*Am J Public Health.* 1992; 82:1093–1098)

Statistical Methods

The number and proportion of decedents who were fully functional in their last year and of those who were severely restricted were estimated for three age groups: 65 to 74 years, 75 to 84 years, and 85 years and older. Variances and confidence intervals were computed by using generalized parameters provided by the National Center for Health Statistics.[11]

Results

Characteristics of the Respondents

The relationship of the respondents to the deceased varied according to the characteristics of the decedent. Although a member of the immediate family was most often the proxy respondent, other relatives, friends, and acquaintances were more likely to be proxies for female than for male decedents (24% vs 15%), for Black than for White decedents (25% vs 19%), and for those who died at age 85 years or older (27%) than for those who died at ages 75 to 84 years (20%) or 65 to 74 years (12%).

Components of the Index

The proportion of respondents requiring assistance with ADLs rose with age at death (Figure 1). For all age groups, decedents were least likely to need help eating and most likely to need help bathing.

Among those who died between the ages of 65 and 74, roughly 15% had had trouble knowing where they were, 13% remembering the year, and about 10% recognizing family or good friends (Figure 2). These proportions increased with age; the oldest decedents were more than twice as likely as the youngest to have cognitive limitations.

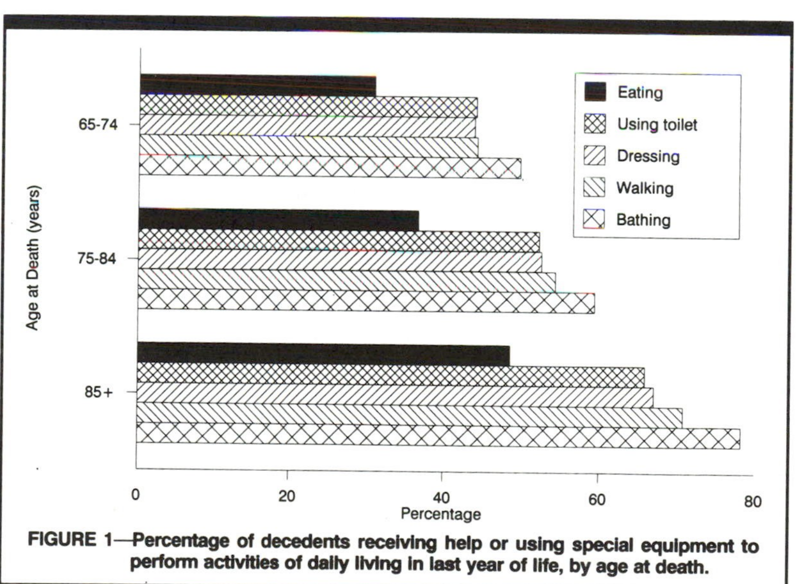

FIGURE 1—Percentage of decedents receiving help or using special equipment to perform activities of daily living in last year of life, by age at death.

Literature Example 6.5 is from the article entitled "HIV Antibody Testing among Adults in the United States: Data from 1988 NHIS." Read the abstract to learn about the study.

What percentage of subjects were aware of the HIV antibody test? [1] ______________________________

What percent of the black adults were aware of the test? [2] ______________________________

Which age group is more aware of the test? [3] ______________________________

How many years of education does the most knowledgeable group have? [4] ______________________________

Among those who said they had little or no knowledge of AIDS, what percent were aware of the test? [5] ______________________________

What percent of those who engaged in high-risk behavior knew about the test? Is this more or less than the percentage of the total population? [6] ______________________________ [7] More/Less

Read the results section to see how the authors described the facts portrayed in the figure.

LITERATURE EXAMPLE 6.5

HIV Antibody Testing among Adults in the United States: Data from 1988 NHIS

ANN M. HARDY, DRPH, AND DEBORAH A. DAWSON, PHD

Data collected from 21,168 adults using the 1988 AIDS supplement to the National Health Interview Survey were examined to determine awareness of and experience with HIV antibody testing in the United States. Three-fourths of adults knew of the blood test for HIV antibodies; awareness was lower among Blacks, Hispanics, older adults, and those less educated. Overall, 17 percent of adults had been tested; of these, 73 percent because of blood donation, 14 percent through other non-voluntary programs (such as military induction), and 16 percent sought testing voluntarily. While a smaller proportion of Black and Hispanic adults had been tested, they were more likely than their White non-Hispanic counterparts to have been tested voluntarily. Persons who reported belonging to groups with high-risk behaviors were also more likely to have been voluntarily tested. Most of those tested voluntarily received their test results, but only one-third also received prevention information. Three percent of adults plan to be tested voluntarily in the next year; about half will seek testing through their doctor or health maintenance organization. (*Am J Public Health* 1990; 80:586–589.)

Results

Overall, 75 percent of adults in the US had heard of the HIV antibody test (Figure 1). The percent who were aware of the test varied by age, race/ethnicity, and education, with lower levels of awareness noted among those over 50 years of age, among Black and Hispanic adults, and among those with less than 12 years of education. People who were more knowledgeable about AIDS and those who reported belonging to one or more of the high-risk behavior groups were more likely to know about the test.

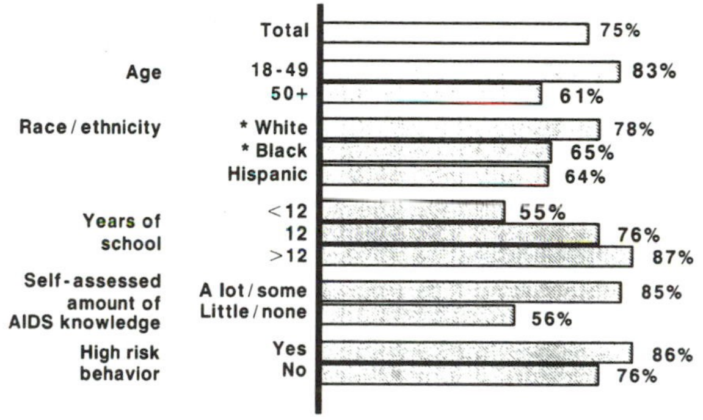

FIGURE 1—Percent of Adults Who are Aware of the HIV Antibody Test, by Demographic Characteristics

Literature Example 6.6 is from the article entitled "The Economic Impact of Injuries: A Major Source of Medical Costs." Read the abstract to learn about the study.

Authors use graphic representation to dramatize the main points they wish to highlight. What is the main point of Figure 1?

1 ______________________________

What variables are represented in Figure 1?

2 ______________________________

What two variables are represented in Figure 2?

3 ______________________________

Which age group has the highest cost for injuries?

4 ______________________________

In your opinion, which figure dramatizes the cost of injuries better? Note that the names of the diagnostic categories are printed in or above the bars in Figure 2 to make the graph clear to the reader. We thought the authors did a particularly good job in the results section of summarizing the information presented in the graphs. Do you agree?

LITERATURE EXAMPLE 6.6

The Economic Impact of Injuries: A Major Source of Medical Costs

LINDA C. HARLAN, PHD, WILLIAM R. HARLAN, MD, AND P. ELLEN PARSONS, PHD

Abstract: Data from the 1980 National Medical Care Utilization and Expenditure Survey were analyzed to place the costs for injuries in the context of all medical costs and to describe the distribution by demographic and diagnostic groups. For the non-institutionalized population, injuries, which include intentional and unintentional, were the second leading cause of direct medical costs, accounting for $16,745 million in medical care expenditures and a major contributor to work loss and disability in the US. For the working-age population (17–64 years) injuries were the leading cost category ($11,341 million) and the third most costly category for persons 65 years of age and over ($3,479 million). The preponderance of costs were attributable to hospital-based care. Direct medical costs were disproportionately greater for males, White and other persons, and for those with household incomes less than $5,000. Injury morbidity also accounts for major indirect costs. Fractures accounted for the highest direct medical costs, greatest per capita charges (based on those with charges), and largest number of restricted activity days. These national estimates document the economic importance of injuries and direct public attention to policy imperatives related to research and prevention. (*Am J Public Health* 1990; 80:453–459.)

Results

Direct Costs

Direct medical costs for injuries comprised the second largest source of expenditures for medical care in the United States during 1980 (Figure 1). The estimated $16.8 billion for injuries represents 12 percent of all direct costs. This amount was exceeded only by an estimated $20 billion in charges for diseases of the circulatory system.[5]

The distribution of medical costs for the four leading condition categories is presented by age groups in Figure 2. Injury and poisoning ranked among the top cost categories for three of the four age groups (<17 years, 17 to 44 years, and >65 years). Injury costs were the greatest ($9 billion) for young adults (17 to 44 years). This diagnostic category was the third largest contributor to medical costs for older Americans (65 years and over) and was exceeded only by costs related to circulatory conditions and neoplasms. When only the working-age population (17–64 years) was considered, full-time or part-time workers had injuries and poisonings as their major source of medical costs. In contrast, persons not employed had greater charges from other diagnostic categories, generally chronic diseases.[6]

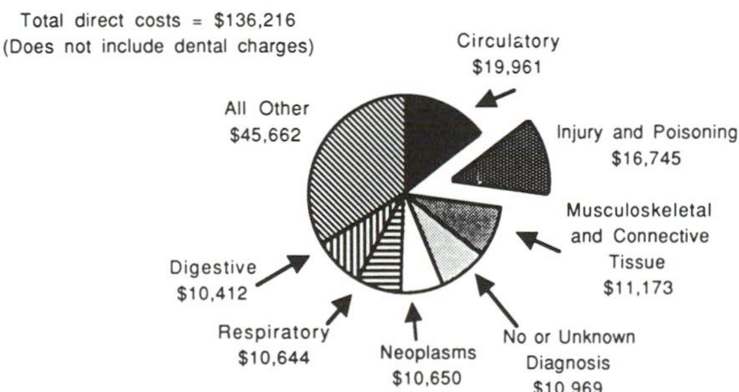

FIGURE 1—Direct Medical Costs for US Civilian Non-institutionalized Population, 1980 (in $ millions)
SOURCE: National Medical Care Utilization and Expenditure Survey

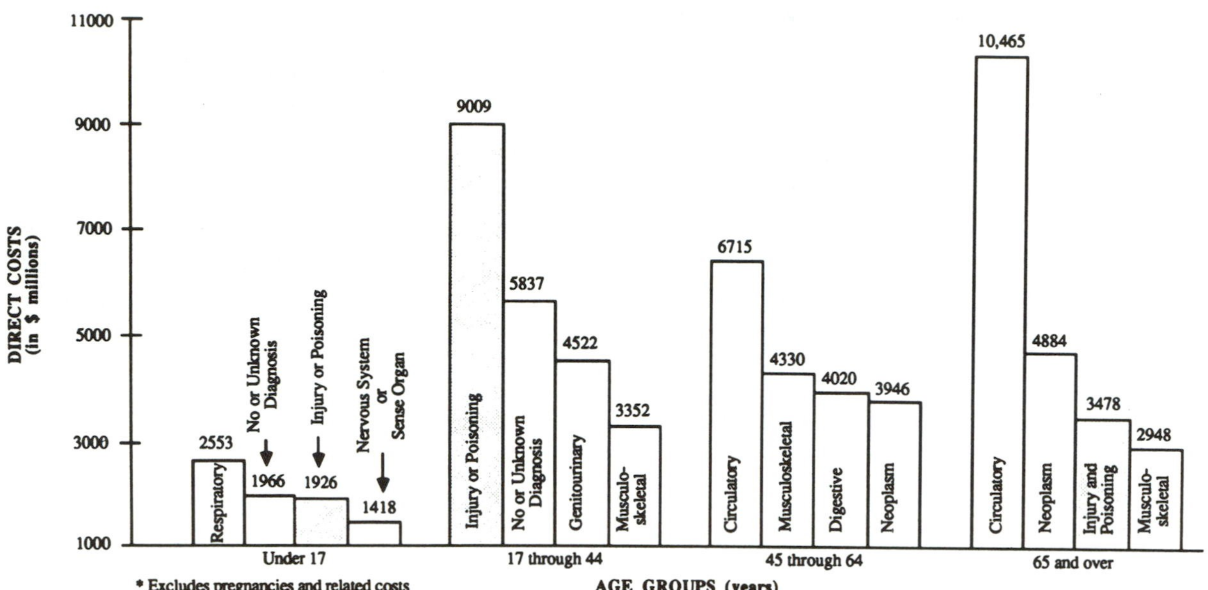

FIGURE 2—Direct Medical Costs for All Persons by Four Leading Diagnostic Categories, Cost, and Age

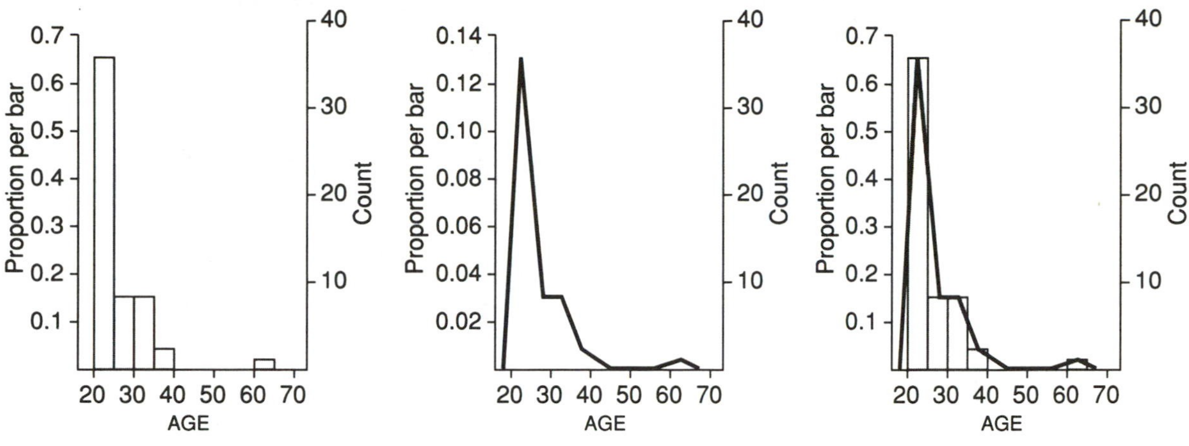

FIGURE 6.7

Frequency Polygons

A **frequency polygon** is another kind of graph that displays the same information as a histogram. To make this kind of graph, the count for each interval is plotted as a dot above the midpoint of that interval. Then the neighboring dots are connected with straight line segments.

Figure 6.7 presents the same distribution of ages in a bar graph and in a frequency polygon. AGE is treated as a categorical variable in both graphs. The mode is [1] the same/different in both graphs. In your opinion, which graph best depicts the distribution?

Literature Example 6.7 is from the article entitled "Use of Nonionic Contrast Media in a High-risk Outpatient Population." Some articles do not include abstracts, so we reproduced the materials and methods section. In which age group were the most procedures performed?

LITERATURE EXAMPLE 6.7

Use of nonionic contrast media in a high-risk outpatient population

James V. Zelch, MD

APPLIED RADIOLOGY, November 1990, Vol. 19, no. 11

Materials and methods

This prospective, open-label, single-center study was designed to permit review of nonionic-contrast-assisted diagnostic procedures in a controlled environment: a newly opened outpatient facility (Cleveland Clinic Florida, Ft. Lauderdale). In the protocol that followed, only iopamidol, a nonionic contrast medium, was used; no patient received more than 50 ml of iopamidol at 300 mgI/ml for intravenous urogram (IVU) or computed tomography (CT) scans; special procedures were performed using digital acquisition and display (no films) with iopamidol 300 or 370; iopamidol M 200 was used for myelographic studies; no patient was denied contrast-assisted intravascular studies because of traditional risk factors (including prior history of "catastrophic reaction" to ionic contrast media), and no sedatives or steroids were administered to any patient prior to diagnostic imaging (including arteriographic studies).

Risk factors were considered to include age over 60 years, previous adverse reaction to ionic contrast media, significant cardiac disease, renal disease, diabetes, history of asthma or drug allergy, and debilitated or malnourished state.

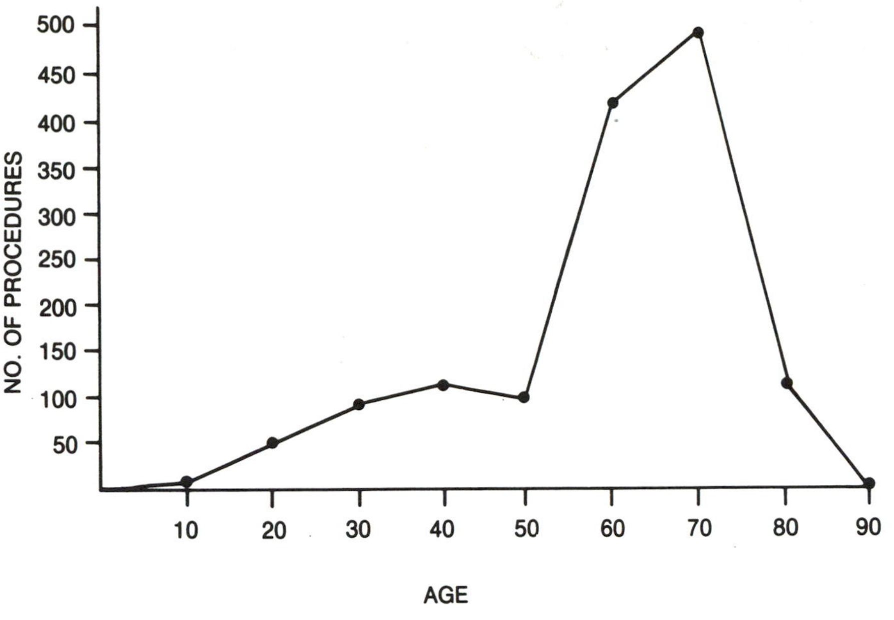

FIGURE 1. Age distribution. Contrast-assisted diagnostic procedures.

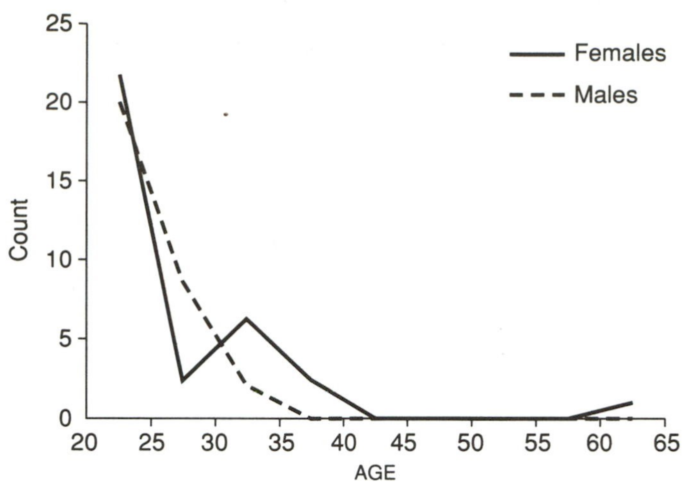

FIGURE 6.8

Overlapping Two Polygons

Suppose we wish to compare the age distribution of males and females in the statistics classes. Refer to Figure 6.8. The ages given on the [1] _______ axis are given in [2] _______ intervals. The points are plotted above the midpoints of the intervals. There is one 61-year-old female. Her age is plotted against the midpoint of the interval between 60 and [3] _______. There are [4] more/fewer females between 30 and 35 than males. Plotting both polygons on the same graph makes it easy to compare the two distributions. We see, for instance, that they are basically the same except that some female students are somewhat older and one is significantly older. These polygons show that these distributions are [5] symmetric/asymmetric.

Example from the Literature

Literature Example 6.8 is from the article entitled "Maternal Recall Error of Child Vaccination Status in a Developing Nation."

How were the two polygons in Figure 1 distinguished from each other? [1] _______________

Do the polygons overlap? [2] _______________

What is the difference between the average doses recalled by mothers and the actual average number? [3] _______________

LITERATURE EXAMPLE 6.8

Maternal Recall Error of Child Vaccination Status in a Developing Nation

Joseph J. Valadez, PhD, SD, and Leisa H. Weld, PhD

In the absence of vaccination card data, Expanded Program on Immunization (EPI) managers sometimes ask mothers for their children's vaccination histories. The magnitude of maternal recall error and its potential impact on public health policy has not been investigated. In this study of 1171 Costa Rican mothers, we compare mothers' recall with vaccination card data for their children younger than 3 years. Analyses of vaccination coverage distributions constructed with recall and vaccination-card data show that recall can be used to estimate population coverage. Although the two data sources are correlated ($r = .71$), the magnitude of their difference can affect the identification of the vaccination status of an individual child. Maternal recall error was greater than two doses 14% of the time. This error is negatively correlated with the number of doses recorded on the vaccination card ($r = -.61$) and is weakly correlated with the child's age ($r = -.35$). Mothers tended to remember accurately the vaccination status of children younger than 6 months, but with older children, the larger the number of doses actually received, the more the mother underestimated the number of doses. No other variables explained recall error. Therefore, reliance on maternal recall could lead to revaccinating children who are already protected, leaving at risk those most vulnerable to vaccine-preventable diseases. (*Am J Public Health.* 1991;82:120–123)

Results

Vaccination Distributions

Distributions of the number of doses from the vaccination cards and from maternal recall are similar (Spearman's $r = .65$; see Figure 1). This suggests that in the absence of vaccination cards, maternal recall can be used as a valid estimate of coverage of the EPI program in Costa Rica. Both sources of data indicate a successful program. A conclusion of policymakers in this circumstance could be to continue the existing EPI program. However, the important public health question remains: Can maternal recall be used to identify an individual child at risk for vaccine-preventable disease?

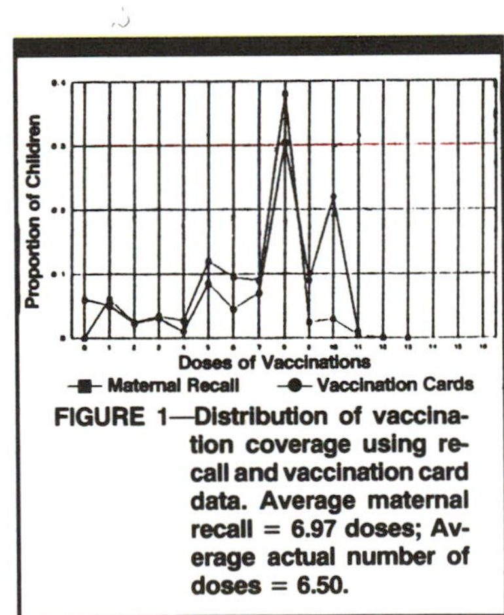

FIGURE 1—Distribution of vaccination coverage using recall and vaccination card data. Average maternal recall = 6.97 doses; Average actual number of doses = 6.50.

Correlation coefficient (r) is mentioned in Literature Example 6.8. The correlation coefficient is a measure of the linear association between two variables. A coefficient of zero indicates no relationship and -1 or $+1$ indicates a perfect linear relationship. Correlation coefficients are discussed in Chapter 15.

Representation of Means with Bar Graphs

Groups may differ in their performance of a task. Such differences can be represented with bar graphs. The arithmetic mean of a sample is the sum of all observations divided by the sample size. Examine Figure 6.9. Is the variation in pulse means (before and after running) noticeably different for sophomores, juniors, seniors, and graduate students? We present a table of the four means and their differences:

Mean PULSE	**Before**	**After**	**Difference**
Sophomore	88	100	12
Junior	77	105	28
Senior	76	96	20
Graduate	65	86	21
ALL	76	98	22

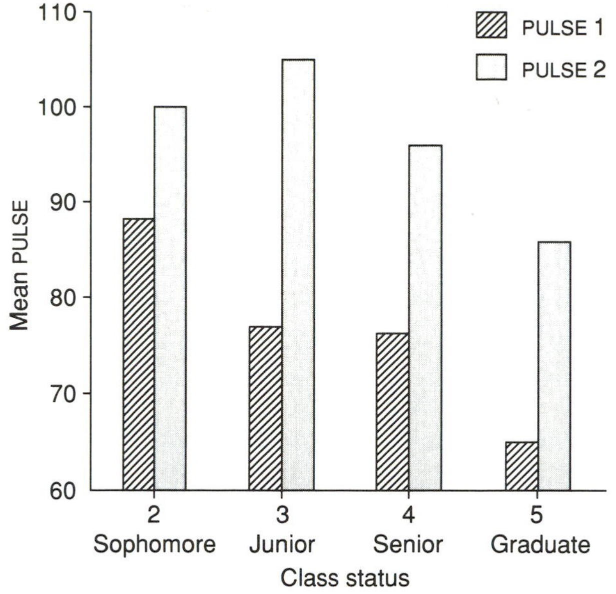

FIGURE 6.9

Which display makes the point about differences in means for each class better—the table or the graph? Variation between classes in before-pulse is interesting, but it might just be an artifact of measurement. What other plots would you like to see to emphasize the differences in classwise pulse reaction to running? Perhaps plotting pulse differences in isolation would be enlightening.

Use the data in the table on page 178 to plot the mean difference between pulse before and after running as a bar graph. Be sure to label the x and y axes.
Note that the bar graph you produce and Figure 6.9 differ from previous bar graphs we have discussed in that for these the vertical axis does not represent frequency. 1

Using the data in the table below, create a bar graph showing ACTIVITY by SEX. Use shading to distinguish activities. Control for SEX and draw activity bars for each sex. You will have two sets of bars, each set consisting of three bars. The numbers on the vertical axis are frequencies again. 2

Rows: SEX **Columns:** ACTIVITY

	Minimal	Moderate	Vigorous	ALL
Female	6	24	2	32
Male	0	12	5	17
ALL	6	36	7	49

Use the same data as above and create a bar graph controlling for ACTIVITY. You will have three sets of bars, each consisting of two bars, one for each sex. 3

Examples from the Literature

Literature Example 6.9 is from the article entitled "Patient 'Dumping' Post-Cobra."

Figure 1 has two variables. What are they? [1] ______________________

What was the total number of patients transferred between June 1 and August 31 of 1986? [2] ______________________

The authors stacked authorized and unauthorized transfers. They also gave us the exact figures for each. If they had depicted these as pairs of adjacent bars instead of stacking them, we could not contrast the totals for each study period as readily. Did admissions decrease in 1987? [3] Yes/Slightly/No

Note how three values of admission status were stacked in Figure 2. Read the authors' description of the figures.

LITERATURE EXAMPLE 6.9

Patient 'Dumping' Post-COBRA

ARTHUR L. KELLERMANN, MD, MPH, AND BELA B. HACKMAN, MD

Abstract: To gauge the impact of the new federal patient transfer provisions following the federal Combined Budget Reconciliation Act of 1985 (COBRA), we monitored all emergency interhospital transfers to a public hospital emergency department in the Memphis, Tennessee area during three identical time periods: June 1 to August 31 of 1986, 1987, and 1988. A high number of transfers in the summer of 1986 diminished only slightly in summer 1987 (following implementation of COBRA). Far greater reductions occurred in summer 1988, when overcrowding forced our hospital to refuse most transfers. In contrast to changes in hospital policy, COBRA alone had little effect in this area. (*Am J Public Health* 1990; 80:864–867.)

Results

During summer 1986, 266 patients were transferred to the Medical Center ED, 243 of which (91 percent) were sent for primarily economic reasons. Somewhat fewer patients (N = 226) were transferred during summer 1987, (when COBRA was in full effect), but the proportion sent for primarily economic reasons did not change. Under the severely overcrowded conditions of summer 1988, total interhospital transfers declined precipitously (N = 85), but the proportion sent for primarily economic reasons remained the same (92 percent).

More than half of patients transferred during the summer of 1986 were unauthorized, including four sent despite refusal by the Medical Center. During summer of 1987 (post COBRA) unauthorized transfers declined by only 18 percent, but 10 patients were sent despite refusal by the Medical Center. During summer 1988, unauthorized transfers declined by fully 61 percent compared to 1986; five transfers arrived despite prior refusal by the Medical Center (Figure 1).

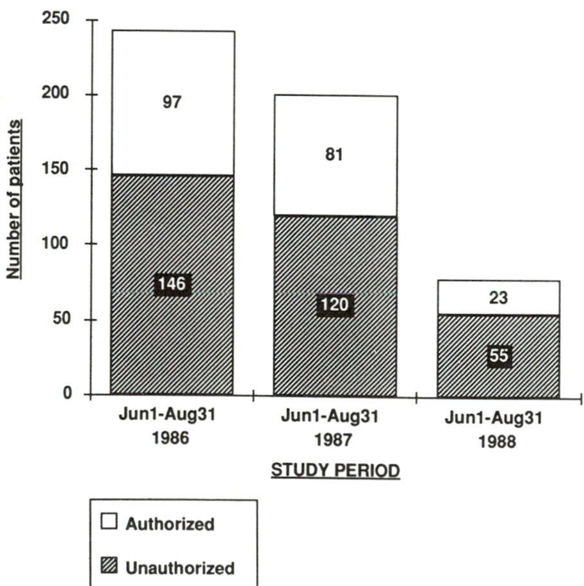

FIGURE 1—Authorized vs Unauthorized Economic Transfers to Memphis Medical Center, Summer 1986, 1987, 1988

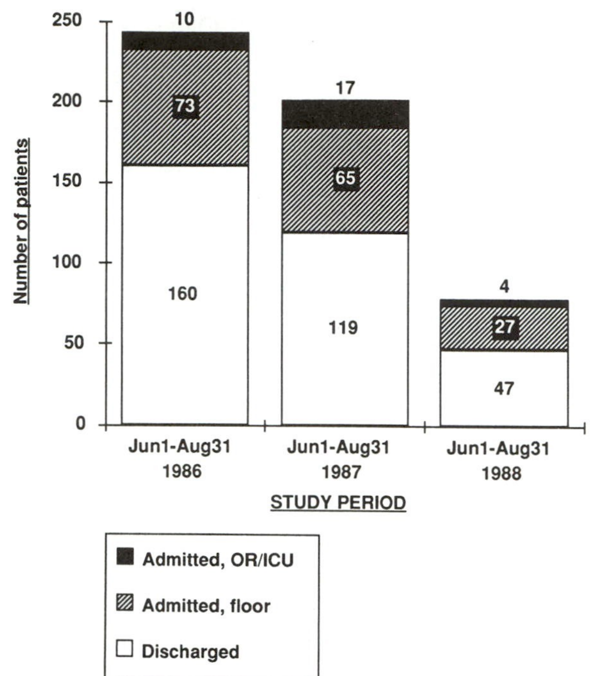

FIGURE 2—Disposition of Economic Transfers Made to Memphis Medical Center Emergency Department, Summer 1986, 1987, 1988

COBRA is intended to stop the inappropriate transfer of unstable patients, particularly when transfer is being accomplished for primarily economic reasons.[2] Based on explicit criteria, 64 of 243 patients (26 percent) transferred to the Medical Center primarily for economic reasons during the summer of 1986 were determined to be unstable. Two of every three had been accepted as "stable for transfer" on the basis of telephone assurances by the sending physician.[3] Unstable economic transfers fell modestly in the summer of 1987 (N = 46) and declined sharply in summer 1988 (N = 13) but the proportion sent without advance authorization did not change. Virtually identical numbers of economic transfers required emergency hospitalization during the summers of 1986 and 1987, and the number requiring emergency surgery and/or intensive care actually increased. Both total admissions and total critical care cases fell sharply during the summer of 1988 (Figure 2). Seven transferred patients died prior to hospital discharge during one of the three study periods: three in summer 1986, three in summer 1987, and one in summer 1988. None of these had been sent without advance authorization by the Medical Center.

Literature Example 6.10 is from the article entitled "The Lowest Birth-Weight Infants and the US Infant Mortality Rate: NCHS 1983 Linked Birth/Infant Death Data."

Consider Figure 1. What are the percentages of deaths among infants born weighing less than 750 grams for blacks and whites?

Blacks: [1] _______

Whites: [2] _______

Consider Figure 2. The stacked bars are not marked with exact percentages, making it difficult to determine those percentages. Estimate the percentages of deaths of infants under 500 grams for blacks and whites.

Blacks: [3] _______

Whites: [4] _______

Read the authors' comments on infant mortality. What do you think is the reason for the differences in the infant mortality figures?

[5] ______________________________

____

LITERATURE EXAMPLE 6.10

The Lowest Birth-Weight Infants and the US Infant Mortality Rate: NCHS 1983 Linked Birth/Infant Death Data

Mary D. Overpeck, MPH, Howard J. Hoffman, MA, and Kate Prager, ScD

The National Center for Health Statistics Linked Birth and Infant Death Data Set, 1983 birth cohort, shows that infants weighing less than 750 g, comprising only 0.3% of all births, account for 25% of deaths in the first year of life and for 41% of deaths in the first week. If interventions had prevented the death of these very small babies, the infant mortality rate would have been 8.3 per 1000 live births instead of 10.9, and the Black/White mortality differential would have been reduced by 25%. (*Am J Public Health.* 1992;82:441–444)

The attributable proportion of Black deaths from births weighing less than 750 g was 33%, compared with 21% for White deaths (Figure 1).

Births weighing less than 750 g resulted in 41% of all early neonatal deaths (Table 3). Of those, 18% were attributable to births weighing less than 500 g. Deaths of the smallest infants contributed to more early neonatal mortality among Blacks than among Whites (Figure 2). Twenty-five percent of Black and 15% of White early neonatal deaths were due to infants weighing <500 grams at birth. Infants weighing less than 750 g contributed 53.2% of Black deaths and 36.4% of White deaths occurring before the first week of life.

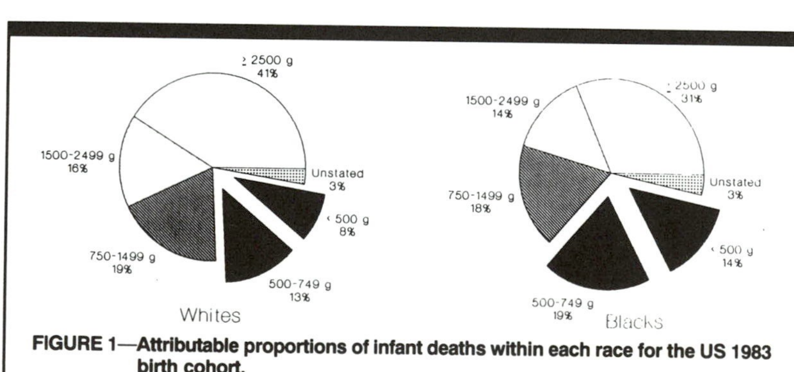

FIGURE 1—Attributable proportions of infant deaths within each race for the US 1983 birth cohort.

Source.—National Center for Health Statistics, National Vital Statistics System (unpublished data).

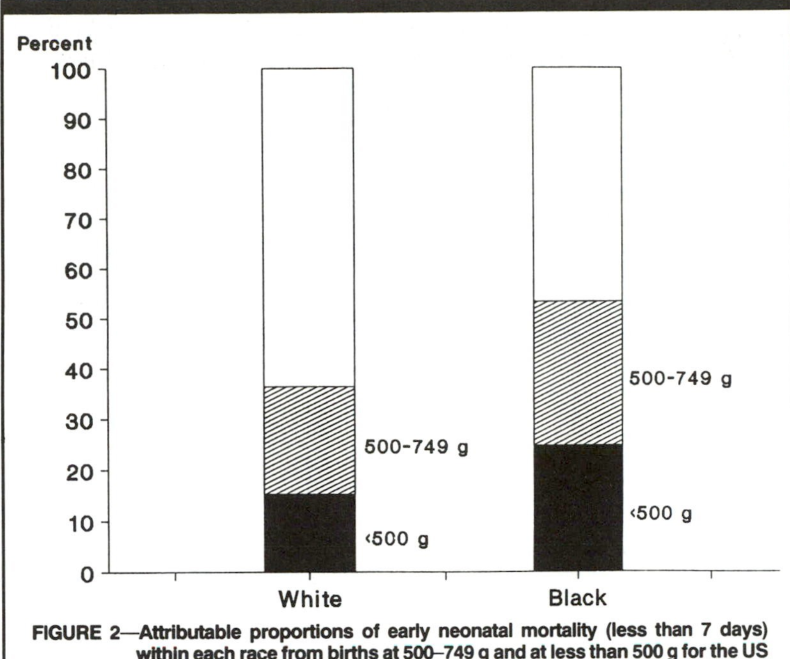

FIGURE 2—Attributable proportions of early neonatal mortality (less than 7 days) within each race from births at 500–749 g and at less than 500 g for the US 1983 birth cohort.

Source.—National Center for Health Statistics, National Vital Statistics System (unpublished data).

Literature Example 6.11 is from the article entitled "The Differential Effect of Traditional Risk Factors on Infant Birthweight among Blacks and Whites in Chicago." Notice the difference between the modes for whites and blacks.

Which race is found predominantly in poorer sections of Chicago? White/Black

Read the results and discussion provided by the authors. "We found that high-risk Whites and Blacks had less divergent LBW rates in the poorest areas than in higher income areas." This finding suggests a reason for differences in infant mortality rates described in Literature Example 6.10.

LITERATURE EXAMPLE 6.11

The Differential Effect of Traditional Risk Factors on Infant Birthweight among Blacks and Whites in Chicago

JAMES W. COLLINS, JR., MD, MPH, AND RICHARD J. DAVID, MD

Abstract: We analyzed 103,072 White and Black births in Chicago from the 1982 and 1983 Illinois vital records, using 1980 median family income of mother's census tract as an ecologic variable. Thirty-one percent of Blacks and 4 percent of Whites resided in census tracts with median family incomes ≤\$10,000/year. Only 2 percent of Black mothers, compared to 16 percent of White mothers, lived in areas where the median family income was greater than \$25,000/year. Among Blacks with incomes ≤\$10,000/year, maternal age, education, and marital status had minimal predictive power on the incidence of low birthweight (LBW) infants. Among high-risk mothers in the poorest areas the proportion of LBW infants in Blacks and Whites was less divergent than in higher income areas. Independent of residential area, low-risk Whites had half the occurrence of LBW infants as Blacks. We conclude that the extremes of residential environments show dramatic racial disparity in prevalence, yet the few low-risk Blacks still do less well than low-risk Whites. Traditional risk factors do not completely explain racial differences in neonatal outcome. (*Am J Public Health* 1990; 80:679–681.)

Results

There were 51,827 Black and 51,245 White births in Chicago during 1982 and 1983. Black neonatal mortality was twice as high as that of Whites (16/1,000 vs 7/1,000) and the LBW proportions were twice as high in Blacks (14 percent vs. 6 percent). Only 2 percent of Black mothers, compared to 16 percent of White mothers, resided in census tracts in which the median family income was greater than \$25,000/year. Conversely, 31 percent of Blacks lived in census tracts in which the average household income was less than \$10,000/year; only 4 percent of White mothers resided in such impoverished neighborhoods (Figure 1).

Discussion

We found that high-risk Whites and Blacks had less divergent LBW rates in the poorest areas than in higher income areas while low-risk Whites had half the occurrence of LBW infants as Blacks regardless of the income of the area in which they lived. In Chicago, the percentages of Black and White mothers at the extremes of residential environment are dramatically different. Contrary to Spurlock, *et al*,[10] who reported that poor and non-poor Black infants in Kentucky had no difference in the incidence of LBW, we found that among all mothers, Black and White, low income is associated with a greater risk of low birthweight.

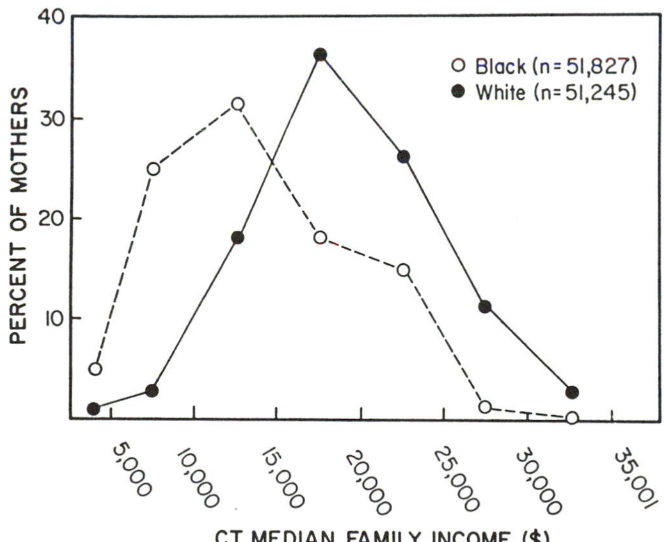

FIGURE 1—Census Tract Median Family Income Distribution of Black and White Mothers

Numerical Descriptions of Distributions

The mean, median, mode, range, and certain percentiles are all used to describe distributions numerically. We will give brief explanations of these terms and relate them to the SQD age data. The cumulative count gives the number of observations to a particular value of the variable. In the stem and leaf plot shown in Figure 6.10, the cumulative count begins at the top, with the lowest age. By looking at the column that lists cumulative count, you can determine how many of the statistics students were a given age or younger. For example, how many students were 24 years old or younger? This is easily seen to be thirty-four from the cumulative count. How many were younger than 33 years of age? To answer this question, we note that forty-one students were younger than 30 years of age, three were 30 years old, none were 31, and one was 32. So $41 + 3 + 0 + 1 = 45$; that is, forty-five students were younger than 33 years old.

Cum. count	Count	Stem Leaves
34	34	2 0000011111111111222223333333344444
41	7	2 5556689
48	7	3 0002333
50	2	3 88
50	0	4
50	0	4
50	0	5
50	0	5
51	1	6 1

Figure 6.10 Stem and Leaf Plot of Ages of Statistics Students $n = 51$

<table>
<tr><th>Definitions and Explanations</th><th>Examples and Questions</th></tr>
<tr><td>The median is the middle observation in a distribution. Approximately 50 percent of the observations will be below the median and approximately 50 percent will be above the median. To find the median, first list the observations in numerical order. (A stem and leaf plot provides such a list.) When n is an odd number, the median is the observation in the $(n + 1)/2$ position. When n is even, the median is the average of the observations in positions $n/2$ and $(n/2) + 1$. That is, it is a number halfway between the two observations, their sum divided by 2.</td><td>Refer to the data in Figure 6.10. $n = 51$, so the sample size is odd. The median is the $(51 + 1)/2 = 26$th observation. What is the median (the 26th observation in the stem and leaf plot)? [1] ________

Suppose the sample size were 32. The median would be halfway between which observations?
$(32/2) = 16$th and [2] ________ observations</td></tr>
<tr><td>The mode is the most frequently occurring observation.</td><td>The mode is [3] ________.</td></tr>
<tr><td>The mean is the sum of all the observations divided by the number of observations.</td><td>Sum of all ages = 1,287
Number of observations, $n = 51$
Mean = [4] ________</td></tr>
<tr><td>The range is the difference between the maximum observation and minimum observation (maximum − minimum). Sometimes the term "range" is used to denote the interval from maximum to minimum.</td><td>Maximum = [5] ________
Minimum = [6] ________
Range = [7] ________
Range in this sense is the interval from [8] ________ to [9] ________.</td></tr>
<tr><td>If p is a number between 0 and 100, the pth percentile is an observation such that p percent of the observations are less than or equal to it. When there is no such observation, the closest one is chosen as the pth percentile.
For example, the fifth observation in the stem and leaf plot (Figure 6.10) is the 10th percentile because 10 percent of 51 is 5.1 and 5 is the closest whole number to 5.1. (Note that some mathematicians define percentiles somewhat differently. Computations based on these other definitions will give slightly different numbers. Percentile is not an exact concept.)</td><td>
<table>
<tr><th>Percentile</th><th>Approximate number of the observation that is that percentile</th><th>Observation</th></tr>
<tr><td>10</td><td>5</td><td>[10] ________</td></tr>
<tr><td>20</td><td>10</td><td>[11] ________</td></tr>
<tr><td>25</td><td>13</td><td>[12] ________</td></tr>
<tr><td>40</td><td>20</td><td>[13] ________</td></tr>
<tr><td>50</td><td>26</td><td>[14] ________</td></tr>
<tr><td>60</td><td>31</td><td>[15] ________</td></tr>
<tr><td>75</td><td>38</td><td>[16] ________</td></tr>
<tr><td>80</td><td>41</td><td>[17] ________</td></tr>
</table>
</td></tr>
</table>

A **quartile** is the 25th percentile (first quartile, denoted Q_1), the 50th percentile (second quartile, denoted Q_2), or the 75th percentile (third quartile, denoted Q_3). The second quartile is the same as the median when there is an odd number of data. For consistency, when there is an even number, the 50th percentile is also taken to be the median.	The first quartile is [18] ________ and the third quartile is [19] ________.
The **inter-quartile range** is the difference between the third quartile (75th percentile) and the first quartile (25th percentile) ($Q_3 - Q_1$).	Inter-quartile range = [20] ________
Quintiles are the 20th, 40th, 60th, and 80th percentiles and are called first, second, third, and fourth quintiles, respectively.	First quintile = [21] ________ Second quintile = [22] ________ Third quintile = [23] ________ Fourth quintile = [24] ________

Box Plots and Five-Number Summaries

The **five-number summary** is a list consisting of the minimum, first quartile, median, third quartile, and maximum. A **box plot** is a graphical way of depicting the five-number summary. From a box plot one can get an immediate idea about the skewness of a distribution. It is also a good graphic device for comparing two or more distributions. Sometimes a box plot also reveals **outliers**, values that do not seem to belong to the distribution because they are unusually large or unusually small. Outliers may be data entry errors or values that need special explanation.

Consider Figure 6.11. An ordinary box plot does not plot outliers in a special way. In an ordinary box plot the whiskers extend to the true maximum value and the true minimum value, regardless of whether these values happen to be outliers. In a modified box plot, data that are more than 1.5 times the inter-quartile range above the upper hinge or below the lower hinge are considered to be outliers and are indicated in some special way (often an asterisk or a solid dot). Data that are more than 3 times the inter-quartile range away from the hinges are extreme outliers and have yet another symbol (we used an open dot).

ACTIVITY 6.2

Reading a modified box plot

Refer to Figure 6.12 and answer the following questions.

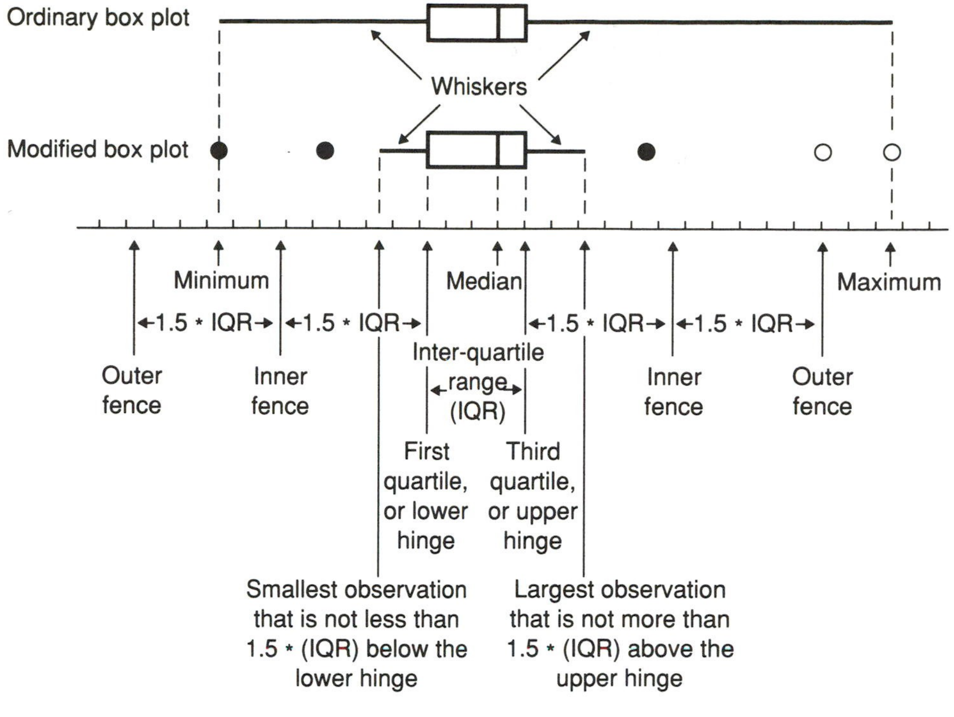

FIGURE 6.11 Anatomy of a Box Plot

The right whisker extends from ages [1] _______ to [2] _______. The median is [3] _______. The minimum is [4] _______; the maximum age that is not an outlier is [5] _______. The first quartile is [6] _______; the lower hinge is [7] _______. The third quartile is [8] _______; the upper hinge is [9] _______. The inter-quartile range is the difference between the upper hinge and the lower hinge, and it is [10] _______. One inner fence is located at 1.5 ∗ IQR to the left of the lower hinge and it is [11] _______; the other is at 1.5 ∗ IQR to the right of the upper hinge, at 33.5. One outer fence is located at 3 ∗ IQR to the left of

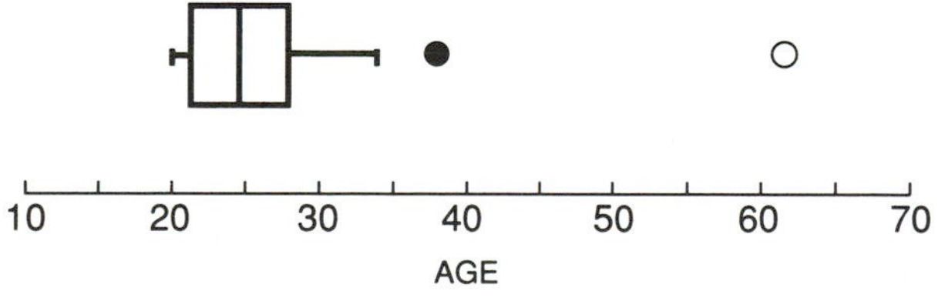

FIGURE 6.12 Box Plot of Ages of Statistics Students

the lower hinge, at 6; the other is at 3 * IQR to the right of the upper hinge, at [12] _______. The value outside the inner fence indicated by an asterisk is [13] _______. The value outside the outer fence (indicated by an open circle) is [14] _______. Box plots, such as the ones in Figures 6.11 and 6.12 reveal, among other things, the five-number summary, namely the [15] _______, [16] _______, [17] _______, [18] _______, and [19] _______. The box plot in Figure 6.12 also indicates values of the variables located outside the inner and outer fences (outliers).

Literature Example 6.12 is from an article entitled "Self-reported Weight and Height."

Box plots give the median to show the center of the data and mark the 25th and 75th percentiles to indicate the variability. This is in contrast to mean and standard deviation (SD) often given in other contexts. (Notice that in addition to the box plots, this article reports the SD of the reporting error in the results section.) Refer to the box plots and notice how easy it is to see how the medians change as the weight status changes.

What is the median for the underweight category? [1] _______
For the severe overweight category? [2] _______

Inter-quartile range is a measure of the amount of variation in the data.

The inter-quartile range for the overweight category is [3] _______.

Note that the author uses a variation on the basic box plot concept that we haven't mentioned. His whiskers end at the 5th and 95th percentiles, instead of at the minimum and maximum. This variation is the author's way of eliminating extreme values, which can mislead a reader about the variability in a distribution.

Based on the 5th and 95th percentiles, what is the range of reporting error in the underweight category? [4] _______
In the severe overweight category? [5] _______

The author makes a point about severely overweight people underreporting their weight, but notice that there is some overreporting too!

Read the part of the text that gives averages (means). Contrast the medians and means.

	Underweight Overreport	Severely Overweight Overreport
Mean	2.3	1.4
Median	[6] _______	[7] _______

LITERATURE EXAMPLE 6.12

Self-reported weight and height

Michael L. Rowland

American Journal of Clinical Nutrition, 1990, vol. 52, pp. 1125–1133

ABSTRACT

The error in self-reported weight and height compared with measured weight and height was evaluated in a nationally representative sample of 11 284 adults aged 20–74 y from the second National Health and Nutrition Examination Survey of 1976–1980. Although weight and height were reported, on the average, with small errors, self-reported weight and height are unreliable in important population subgroups. Errors in self-reporting weight were directly related to a person's overweight status—bias and unreliability in self-report increased directly with the magnitude of overweight. Errors in self-reported weight were greater in overweight females than in overweight males. Race, age, and end-digit preference were ancillary predictors of reporting error in weight. Errors in self-reporting height were related to a person's age—bias and unreliability in self-reporting increased directly with age after age 45 y. Overweight status was also a predictor of reporting error in height. *Am J Clin Nutr* 1990;52:1125–33.

RESULTS: WEIGHT

The distribution of reporting error. The overall tendency was for men to overreport their weight by an average of 0.4 kg, whereas women underreported their weight by an average of 1.0 kg. The variability in reporting error (i.e., self-reported minus measured weight) was similar for men and women: the SD of the reporting error was 3.0 kg for both.

Reporting error varied considerably with weight status and sex. Underweight men overreported their weight by an average of 2.3 kg whereas severely overweight men underreported their weight by an average of 1.4 kg. Underweight women overreported their weight by an average of 0.4 kg whereas severely overweight women underreported their weight by an average of 3.4 kg.

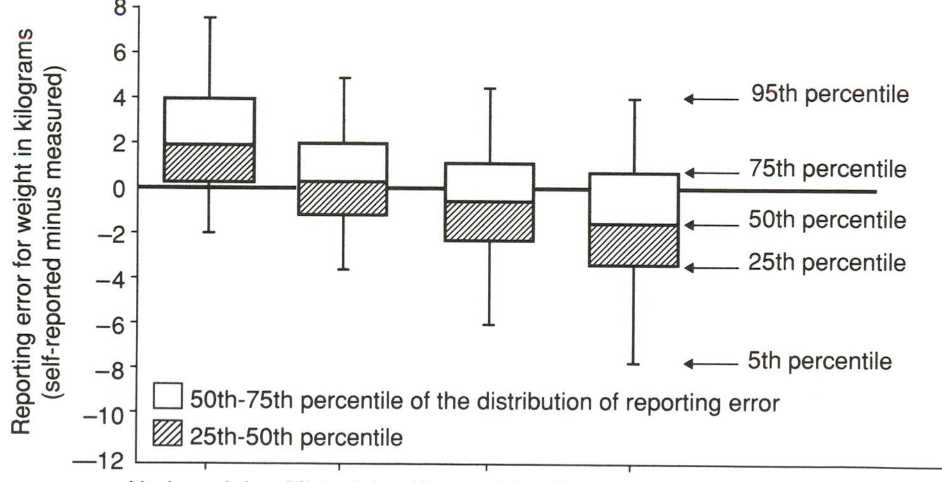

Box Graph of the Distribution of Reporting Error for Weight According to Measured Weight Category for U.S. Men Aged 20–74, 1976–1980

Activity 6.3

Comparing three kinds of graphs

Figure 6.13 presents the ages of students in statistics classes, displayed three different ways. The stem and leaf plot presents the raw data. From it, with a little effort, we can figure the median and the first and third quartiles. The stem and leaf plot also shows that there are no ages in the range [1] _______, as does the box plot. The histogram gives this information as well, but only approximately.

Frequency information can be obtained from the histogram or the stem and leaf plot, but not from the box plot. The mode, as an interval, can be readily obtained from the histogram and from the stem and leaf plot. Since we have chosen the same interval width for both displays, we find from either display that the mode is in the interval [2] _______. From the stem and leaf plot, we can find that the mode is actually [3] _______. The stem and leaf plot and histogram give approximately the same information, but the stem and leaf plot is more detailed.

The box plot shows the five-number summary, information that is not given in the other two displays. None of these figures shows the mean. All three plots give information about the dispersion in terms of maximum and minimum. The box plot also provides an index of dispersion, the interquartile range. [4] _______ of the observations are located between the hinges.

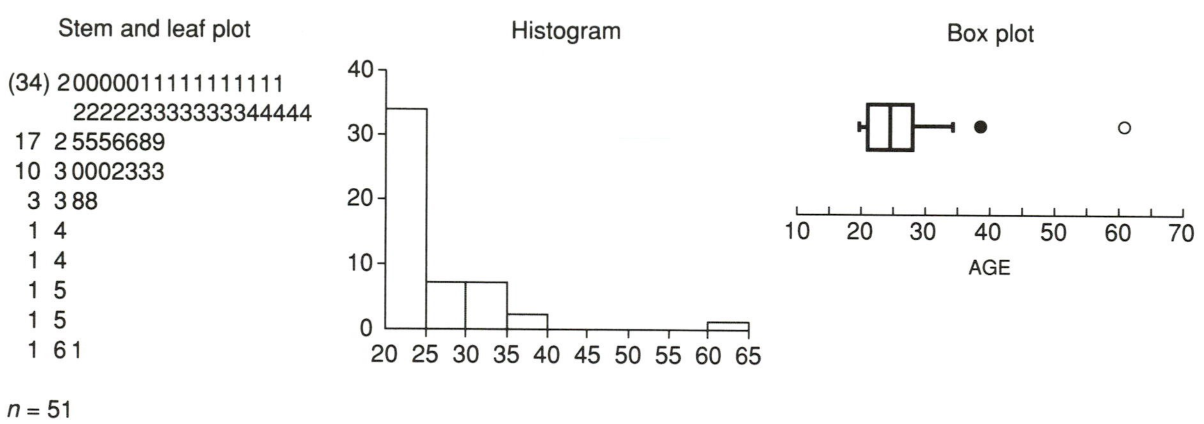

Figure 6.13 Three Displays of Ages of Students in Statistics Classes

Time Series: Plots with Time as One Variable

So far we have constructed bar graphs, histograms, frequency polygons, cumulative frequency polygons, and stem and leaf plots. All of these portray snapshots of what might be a moving scene. How do we portray changes in a variable over a period of time? We simply plot the variable we are interested in against time; that is, we use time as the variable plotted on the x axis. This kind of plot has a special name: **time series**. Thus, a time series is a plot of any variable against time plotted on the x axis.

The following data are from an exercise physiology laboratory. To see how the heart rate varies with time, we simply plot the paired observations, time and heart rate (HR), as a graph; see Figure 6.14. Time is measured in minutes.

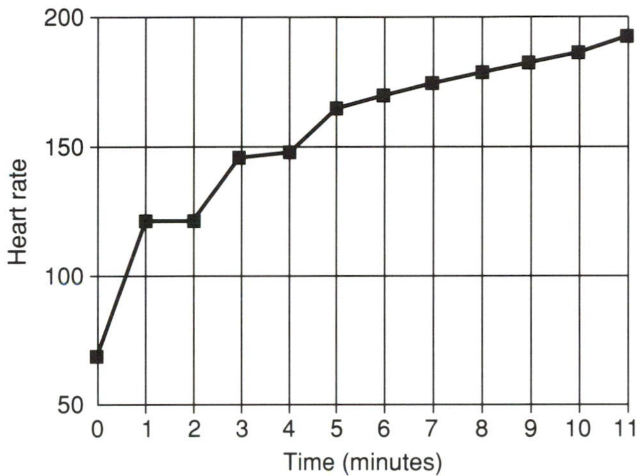

FIGURE 6.14 Heart Rate of an Athlete on a Treadmill

Minutes	Heart Rate
0	70
1	123
2	123
3	148
4	149
5	167
6	172
7	175
8	178
9	182
10	186
11	191

Minutes	Volume of Oxygen
4	2.820
5	3.150
6	3.460
7	3.710
8	3.960
9	4.360
10	4.030
11	4.930

Describe what is happening to the athlete's heart rate with time:

Plot and label the data on volume of oxygen from the right-hand table in Figure 6.15 (we have plotted a few of the points for you). VO2 stands for volume of oxygen consumed.

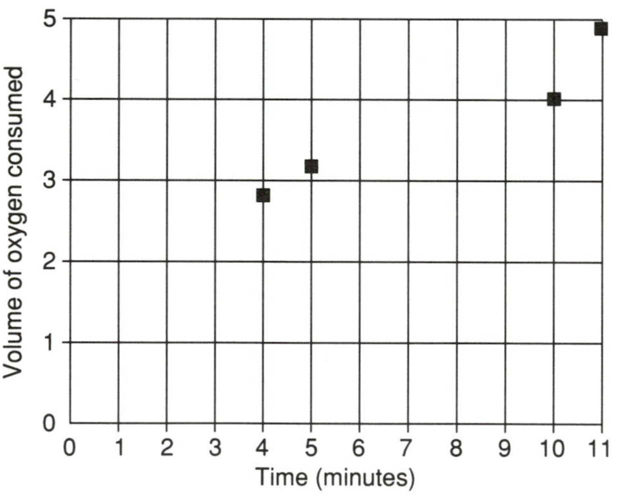

FIGURE 6.15

Describe the behavior of the variable VO2 over time:

____

____

____

The time series you plotted in Figure 6.15 is a special case of a two-variable plot. There is one unique y for each x.

Examples from the Literature

Literature Example 6.13 is from an article entitled "Field Comparison of Several Commercially Available Radon Detectors." Comment on the fluctuations in radon levels with time.

____

____

LITERATURE EXAMPLE 6.13

Field Comparison of Several Commercially Available Radon Detectors

R. William Field, MS, and Burton C. Kross, PhD

To determine the accuracy and precision of commercially available radon detectors in a field setting, 15 detectors from six companies were exposed to radon and compared to a reference radon level. The detectors from companies that had already passed National Radon Measurement Proficiency Program testing had better precision and accuracy than those detectors awaiting proficiency testing. Charcoal adsorption detectors and diffusion barrier charcoal adsorption detectors performed very well, and the latter detectors displayed excellent time averaging ability. Alternatively, charcoal liquid scintillation detectors exhibited acceptable accuracy but poor precision, and bare alpha registration detectors showed both poor accuracy and precision. The mean radon level reported by the bare alpha registration detectors was 68 percent lower than the radon reference level. (*Am J Public Health* 1990; 80:926–930.)

Results

A plot of the hourly measurement results obtained from the CRM is shown in Figure 1. The radon level in the bedroom showed considerable variation, especially during the last two days of the study which was probably related to increased ventilation rates in the basement. The occupant activity log notes that the children living in the home frequently opened the basement door leading to the outside during the evening of March 23 and the afternoon of March 24, 1989.

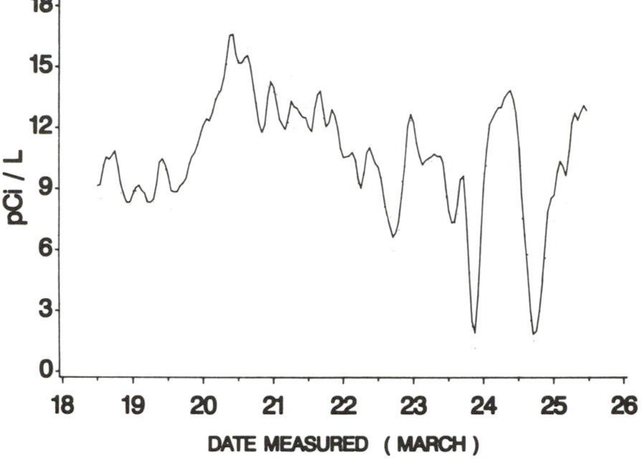

FIGURE 1—Interpolated Hourly Plot of the Radon Measurements Obtained from the Continuous Radon Monitor for the Period 1100 hrs March 18 to 1100 hrs March 25, 1989

Literature Example 6.14 is from the article entitled "Cancer Rates after the Three Mile Island Nuclear Accident and Proximity of Residence to the Plant."

Which polygon shows the highest cancer figures for almost all years? 0–6 km/> 6–12 km/> 12 km

Read the authors' discussion. The authors mention "odds ratio" and "confidence interval." You'll gain an understanding of these terms after we've covered them in Chapters 14 and 16. For now, just try to get the general sense of the study.

LITERATURE EXAMPLE 6.14

Cancer rates after the Three Mile Island nuclear accident and proximity of residence to the plant

Maureen C. Hatch, PhD, Sylvan Wallenstein, PhD, Jan Beyea, PhD, Jeri W. Nieves, MS, and Mervyn Susser, MB, BCh

American Journal of Public Health, 1991, Vol. 81, pp. 719–724

Background

In the light of a possible link between stress and cancer promotion or progression, and of previously reported distress in residents near the Three Mile Island (TMI) nuclear power plant, we attempted to evaluate the impact of the March 1979 accident on community cancer rates.

Methods

Proximity of residence to the plant, which related to distress in previous studies, was taken as a possible indicator of accident stress; the postaccident pattern in cancer rates was examined in 69 "study tracts" within a 10-mile radius of TMI, in relation to residential proximity.

Results

A modest association was found between postaccident cancer rates and proximity (OR = 1.4; 95% CI = 1.3, 1.6). After adjusting for a gradient in cancer risk prior to the accident, the odds ratio contrasting those closest to the plant with those living farther out was 1.2 (95% CI = 1.0, 1.4). A postaccident increase in cancer rates near the Three Mile Island plant was notable in 1982, persisted for another year, and then declined. Radiation emissions, as modeled mathematically, did not account for the observed increase.

Conclusion

Interpretation in terms of accident stress is limited by the lack of an individual measure of stress and by uncertainty about whether stress has a biological effect on cancer in humans. An alternative mechanism for the cancer increase near the plant is through changes in care-seeking and diagnostic practice arising from postaccident concern. (*Am J Public Health*. 1991;81:719–724)

For the "all cancer" grouping, rates in the area close to the plant are higher before the accident as well as afterwards. We corrected for the preaccident pattern by means of the procedure described above; the odds ratio adjusted for risk at baseline was reduced from 1.4 to 1.2 (95% confidence limits = 1.0, 1.4).

In Figure 1 we show incidence rates for all cancers for each year of the study period by proximity to the plant (three distance rings). By 1982, rates in the area closest to TMI were clearly elevated. The numbers of cases among residents of the inner ring rose from a yearly average of less than 50 to a high of 78 (a value that lies outside the 99% CI around 50 (33.7, 71.3), assuming the number of cases to be a Poisson variable). By 1984, the cancer rate had fallen to preaccident levels and by 1985 was lower than the rates in the more distant rings.

Discussion

In an admittedly crude test of an accident-stress hypothesis, we found an increase in cancer following the accident that related to proximity of residence to the Three Mile Island nuclear plant. Cancer rates in those living nearest TMI rose in 1982, three years postaccident, remained raised for another year, and then declined. The relationship with proximity is unlikely to be explained by better case ascertainment for residents of the inner ring since, in the interests of comprehensive case-finding, we reviewed medical records at all hospitals within a 30-mile radius and at regional referral centers.

Residential proximity to the TMI plant was related to cancer rates prior to the accident as well as after, suggesting the presence of risk factors that were not sufficiently controlled by our adjustments for urbanization and social class. As a means of handling the confounding effects of unmeasured factors (e.g., cigarette smoking), we therefore adjusted for baseline risk. With this adjustment, the odds ratio for residential proximity after the accident was reduced to borderline statistical significance. Nonetheless, the postaccident pattern involving a sudden peak of excess cancers in residents near the plant does appear to be distinct from the preaccident pattern which showed cancer rates in the inner ring to be slightly higher throughout the period.

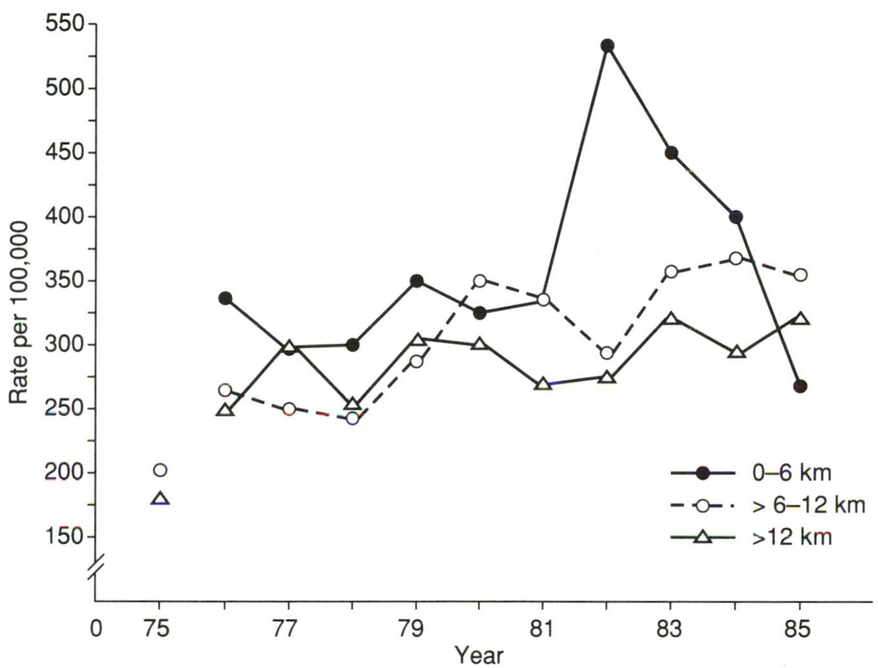

Figure 1. Yearly Age-Adjusted Incidence Rates (100,000) of All Cancers, 1975–1985, for Three Areas Defined by Distance from the TMI Nuclear Plant. Note: In all three areas, figures were low for 1975, when hospital records were initially computerized.

Literature Example 6.15 is from an article entitled "Increase in Condom Sales Following AIDS Education and Publicity, United States." Notice the two scales on the vertical axis of the figure, one for the bar graph and the other for the polygons. The horizontal scale of time is common to both the polygons and the histogram. Combining different modes of representation can be quite useful but can also be misleading.

After the Surgeon General's report, what happened to the number of newspaper abstracts?

1 ______________________________

____

Describe the percentage change in condom sales over time, in your own words.

2 ______________________________

____

LITERATURE EXAMPLE 6.15

Increase in Condom Sales following AIDS Education and Publicity, United States

JOHN S. MORAN, MD, HARLAN R. JANES, THOMAS A. PETERMAN, MD, MSc, AND KATHERINE M. STONE, MD

Abstract: Data from a national probability sample of drug stores show that condom sales rose from 240 million annually in 1986 to 299 million in 1988. The greatest increase occurred in 1987 after the Surgeon General's report on AIDS was released. Sales of latex condoms with spermicide rose 116 percent. Sales of other types of condoms increased less. These data suggest that Americans are using more condoms and probably more effective condoms in response to AIDS education. (Am J Public Health 1990; 80:607–608.)

Results

Drug store condom sales grew slowly from 1984 to 1988 except for a 20 percent increase between 1986 and 1987 (Table 1). Sales of some styles grew more rapidly than others. Between 1986 and 1988, sales of *all* latex condoms increased 25 percent (226.1 to 283.4 million). The biggest percentage growth was in latex condoms with spermicide which increased 116 percent (23.2 to 50.1 million). Latex condoms without spermicide increased 15 percent (202.9 to 233.3 million). Natural membrane condom sales increased 7.8 percent (14.4 to 15.6 million).

Sales increased both in areas with a high incidence of AIDS and in the remaining US between 1986 and 1988 (Figure 1). Sales in the high incidence areas were growing throughout 17 of the 18 two-month periods. In contrast, sales were not growing in the remaining US until the beginning of 1987, and sales stopped increasing in July-August 1988. In both areas, condom sales grew rapidly throughout 1987 and early 1988 following the release of the Surgeon General's report in November 1986. Media attention to condoms also increased. Condoms were rarely mentioned before the report, but were increasingly cited in articles, editorials, and cartoons thereafter, reaching a peak in February 1987, when 182 items appeared in the 19 newspapers indexed by Newspaper Abstracts. (Forty percent of the February items concerned the controversy over whether condom advertisements were appropriate for television.) Throughout the remainder of 1987 and 1988 media attention was greater than in 1986, but slowly diminished.

Discussion

The 20 percent increase in drug store condom sales in the year following the Surgeon General's report suggests that Americans responded to his message. In comparison, cigarette sales fell only 2.4 percent in the year following the Surgeon General's first report on smoking and health.[4]

The increase in sales was probably related to the recommendation that condoms be used to prevent HIV transmission. Sales of latex condoms, which were recommended for AIDS prevention, increased more than sales of natural membrane condoms. In addition, the greatest percent increase was in latex condoms with spermicide which cost more but may provide additional protection. The overall increase in drug store condom sales between 1984 and 1988 was 26 percent.

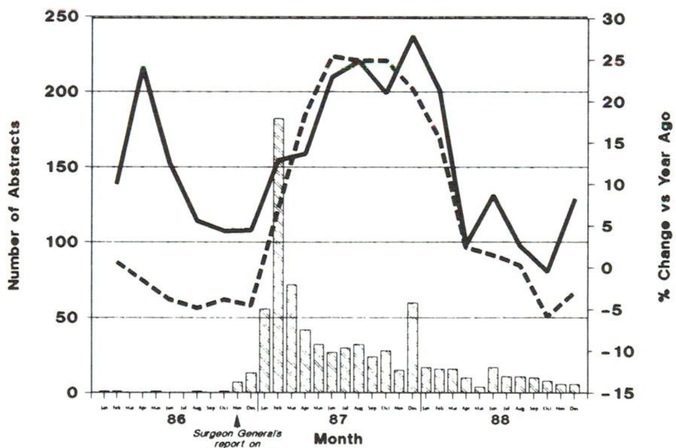

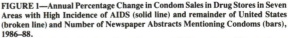

FIGURE 1—Annual Percentage Change in Condom Sales in Drug Stores in Seven Areas with High Incidence of AIDS (solid line) and remainder of United States (broken line) and Number of Newspaper Abstracts Mentioning Condoms (bars), 1986–88.

NOTE: Percent change in sales from the same period one year earlier are shown for two-month periods.

Two-Variable Plots to Study Relationships

If we plot age against weight for a group of people, we may have several weights for each age, because several people the same age may have different weights. Plots of this kind, where two variables are plotted against each other and multiple y values are allowed for the same x, are known as **scatter plots**. Scatter plots can have any two variables plotted against each other. A time series is a special case of a scatter plot in which the variable plotted on the x axis is always time.

Plot the ages and weights listed below on Figure 6.16 to create a scatter plot (we have plotted some of the points for you).

Age	24	20	20	27	47	27	36	36	34	25	20	28	38
Weight	150	113	125	150	148	155	135	210	120	175	116	165	145

Age	24	32	23	27	25	22	23	23	25	28	33	25	40
Weight	110	115	165	130	195	160	123	125	180	155	160	115	145

Age	24	24	24	32	27	30
Weight	145	125	150	110	188	118

Scatter plots are used to answer the question "Are the two variables associated?" For example, are height and weight associated? You've probably noticed that taller people seem to be heavier than shorter people, but exactly how are height and weight related? Could we predict someone's weight if we knew how tall that person was?

In the scatter plot in Figure 6.17, the points seem to indicate a vague relationship between height and weight. Generally, taller people are heavier than shorter people. As height increases by an inch, weight increases by roughly four or five pounds.

Notice there are points that seem to lie outside the "cloud" of points. For

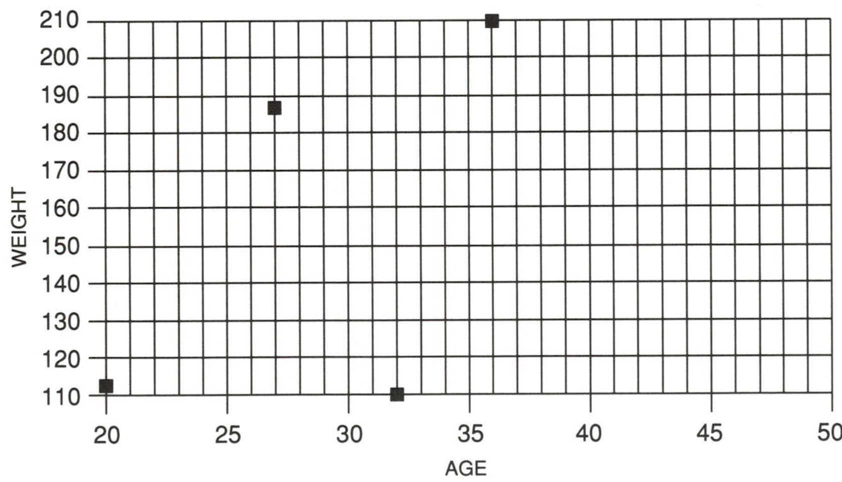

FIGURE 6.16

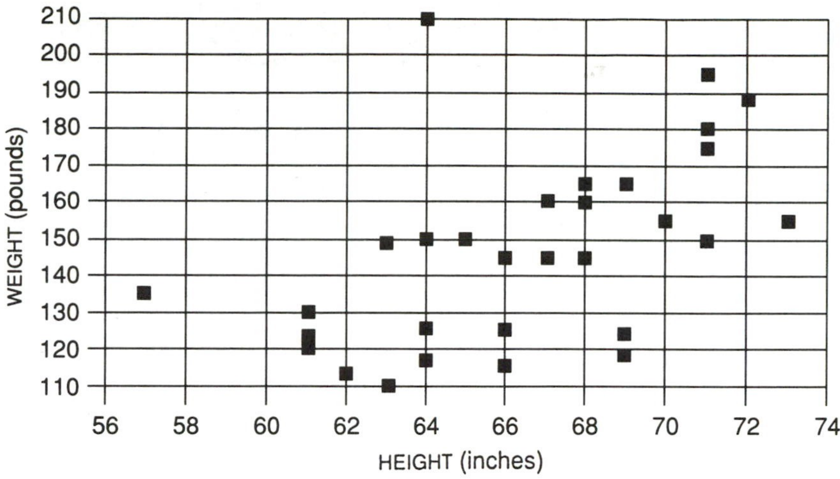

FIGURE 6.17

example, there is a point for a person who weighs 210 pounds but is 64 inches tall. To fit the pattern indicated in the scatter plot, one would expect this person to weigh much less or to be a lot taller. It's possible that this point represents an error in data collection or recording, but it is also possible that there really was a person in the sample with this height and weight. Since we do not have access to the actual people in the data set, we will never know whether it's an error or not.

Examples from the Literature

Literature Example 6.16 is from the article entitled "Relationship Between Resistance to Insulin-Mediated Glucose Uptake, Urinary Uric Acid Clearance, and Plasma Uric Acid Concentration."

Two kinds of figures are shown. The main idea is to portray complex ideas in a pictorial format so that a reader can see the relationships better. How many pairs of variables are compared in Figure 4?

1 ______________________

Try to draw straight lines that best fit the data points in each graph of Figure 3. The scattered points in each of the graphs *do/do not* seem to fall on a straight line.

2 Do/Do not

LITERATURE EXAMPLE 6.16

Relationship Between Resistance to Insulin-Mediated Glucose Uptake, Urinary Uric Acid Clearance, and Plasma Uric Acid Concentration

Francesco Facchini, MD; Y.-D. Ida Chen, PhD; Clarie B. Hollenbeck, PhD; Gerald M. Reaven, MD

Objective.—To define the relationship, if any, between insulin-mediated glucose disposal and serum uric acid.

Design.—Cross-sectional study of healthy volunteers.

Setting.—General Clinical Research Center, Stanford (Calif) University Medical Center.

Participants.—Thirty-six presumably healthy individuals, nondiabetic, without a history of gout.

Measurements.—Obesity (overall and regional), plasma glucose and insulin responses to a 75-g oral glucose load, fasting uric acid concentrations, plasma triglyceride and high-density lipoprotein–cholesterol concentrations, systolic and diastolic blood pressure, insulin-mediated glucose disposal, and urinary uric acid clearance.

Results.—Magnitude of insulin resistance and serum uric acid concentration were significantly related ($r = .69$; $P < .001$), and the relationship persisted when differences in age, sex, overall obesity, and abdominal obesity were taken into account ($r = .57$; $P < .001$). Insulin resistance was also inversely related to urinary uric acid clearance ($r = -.49$; $P < .002$), and, in addition, urinary uric acid clearance was inversely related to serum uric acid concentration ($r = -.61$; $P < .001$).

Conclusions.—Urinary uric acid clearance appears to decrease in proportion to increases in insulin resistance in normal volunteers, leading to an increase in serum uric acid concentration. Thus, it appears that modulation of serum uric concentration by insulin resistance is exerted at the level of the kidney.

(*JAMA*. 1991;266:3008-3011)

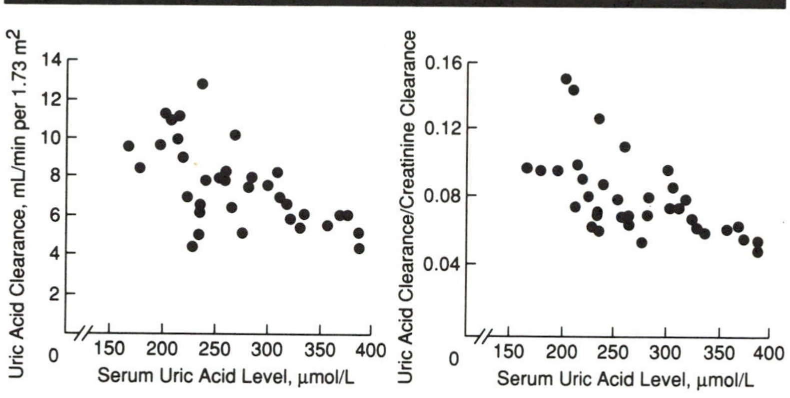

Fig 3.—Relationship between serum uric acid concentration and uric acid clearance ($r = -.61$; $P<.001$) (left panel) and uric acid clearance divided by creatinine clearance (fractional uric acid clearance [$r = -.60$; $P<.001$]) (right panel).

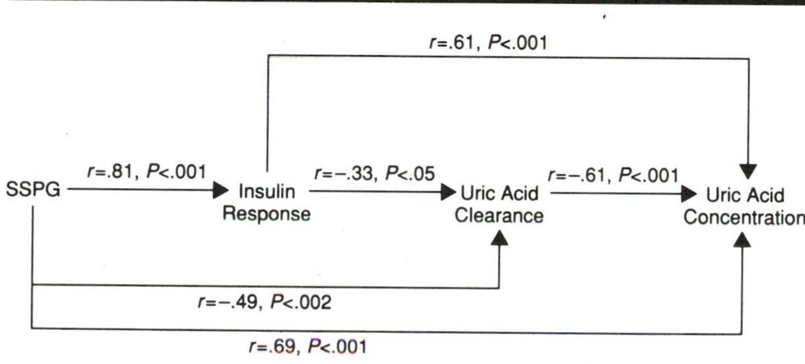

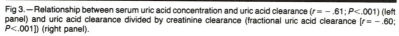

Fig 4.—Summary of the relationships between insulin resistance (steady-state plasma glucose concentration [SSPG]), insulin response, and various measures of uric acid metabolism.

Literature Example 6.17 is from an article entitled "Environmental Tobacco Smoke Exposure during Infancy."

We want to know how infants are exposed to tobacco smoke. Urine cotinine levels indicate degree of exposure. Examine the figure; which category of infant listed on the horizontal axis has the least exposure?

1 ______________________________

Compare breastfed babies' cotinine levels when mother smokes to their levels when others smoke. In which category are babies concentrated above the cutoff line?

2 ______________________________

What observation can you make concerning the data in this scatter plot?

3 ______________________________

The figure is a hybrid graph; it has features of a scatter plot and features of a bar graph. Also note that the vertical axis uses a logarithmic scale. One of the reasons for using this scale is to show the detail at the lower values of cotinine level. Note that the logarithmic scale distorts the enormous increase in cotinine levels. The maximum level goes from approximately 100 in the "no exposure" group to over 600 in the "mother and others" group. Mark the median cotinine levels for each group with a horizontal line. Median levels are given at the bottom of the figure.

LITERATURE EXAMPLE 6.17

Environmental Tobacco Smoke Exposure during Infancy

Barbara A. Chilmonczyk, MD, George J. Knight, PhD, Glenn E. Palomaki, BS, Andrea J. Pulkkinen, MS, Josephine Williams, and James E. Haddow, MD

We collected information about household smoking habits from 518 mothers when they made their first well child visit with a 6 to 8-week old infant. A urine sample was also collected from the infant, the cotinine concentration measured, and the measurement correlated with data provided by the mother. Eight percent of the infant urine cotinine values fell at or above 10 μg/L in the 305 households where no smoking was reported. Corresponding rates were 44 percent in the 96 households where a member other than the mother smoked, 91 percent in the 43 households where only the mother smoked, and 96 percent in the 74 households where both the mother and another household member smoked. In households where the mother smoked, infant urine cotinine levels were lower in the summer, and higher when the infant was breast-fed. A screening question about family smoking habits in conjunction with well child care could effectively define a group of infants exposed to environmental tobacco smoke and thus be at greater risk for respiratory diseases. (*Am J Public Health* 1990; 80:1205–1208.)

Results

Figure 1 contains a scatterplot distribution of urine cotinine levels. Among 305 infants who were not exposed to household tobacco smoke, cotinine levels from 97 (32 percent) fell below 1.0 μg/L, and another 185 (60 percent) fell between 1.0 and 9.9 μg/L. Urine cotinine values of 10 μg/L and above were found in only 23 of the infants (8 percent). Overall, the median urine cotinine level in this group was 1.6 μg/L.

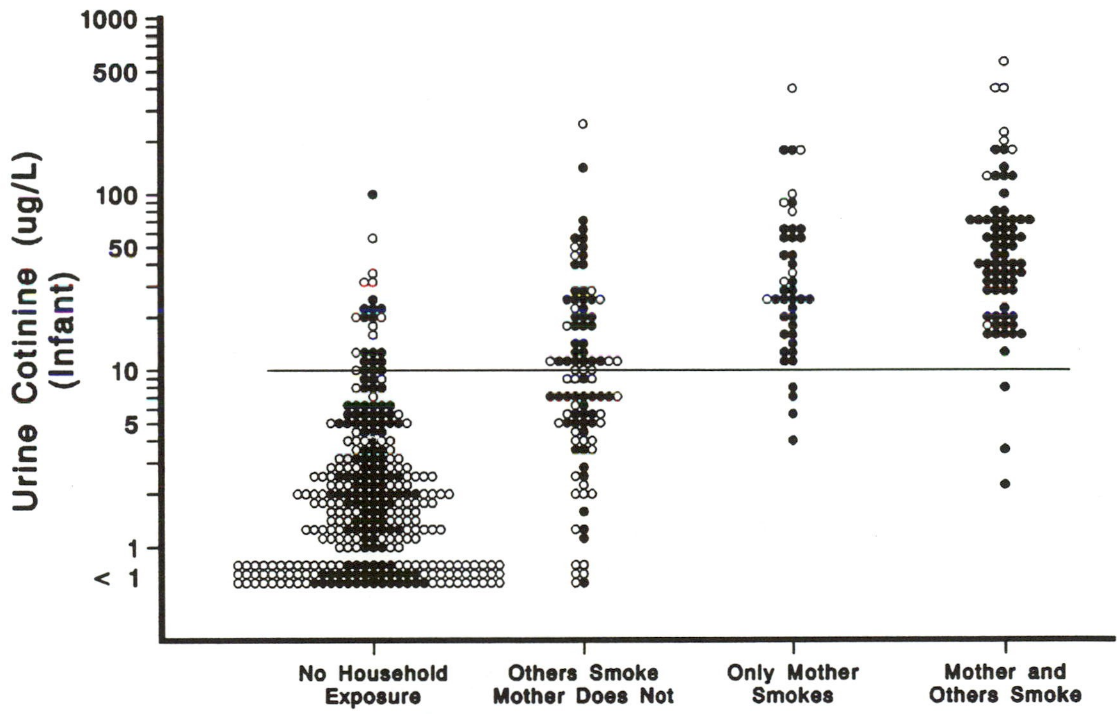

FIGURE 1—Urine Cotinine Levels in Infants, According to Environmental Tobacco Smoke Exposure in the Home.
Cotinine measurements are in spot urine samples. Closed circles indicate that the infant is not breast fed; open circles indicate breast feeding. Cotinine measurements < 1.0 μg/L are considered undetectable, and the horizontal line drawn at 10 μg/L is the cutoff defined from data in the study to indicate significant ETS absorption. Median urine levels in the four groups are as follows: no household exposure—1.6 μg/L; others smoke, mother does not—8.9 μg/L; only mother smokes—28 μg/L; mother and others smoke—43 μg/L.

Activity 6.4

Some aspects of statistical graphs that are worth noting

No matter what type of graph you encounter, you will find it useful to note certain aspects:

1. The most frequently occurring values of the variable(s);
2. How the values are spread;
3. Symmetry or asymmetry of the distribution of the variables;
4. Unusual values that do not quite fit the overall pattern, if any;
5. Whether two continuous variables vary together (relate), and if so, how;
6. How a variable changes over time (if the graph is a time series).

Whenever possible, it is also helpful to compare the distribution shown in a graph with known standard distributions and with distributions of the variable reported in other literature. When two or more distributions are displayed on the same graph, it is useful to compare them.

Note that the above-listed characteristics are worth looking for in any graph, although not all of them will be found in every graph. Consider Figure 6.18, and look for as many of these items as you can find. This will get you into the habit of examining graphs carefully. (See the answers section for descriptions of Figure 6.18.)

The Purpose of Each Type of Table and Graph

We deal with data in three stages. Stage one is collection, stage two is analysis, and stage three is communication. The displays we see in journals are the

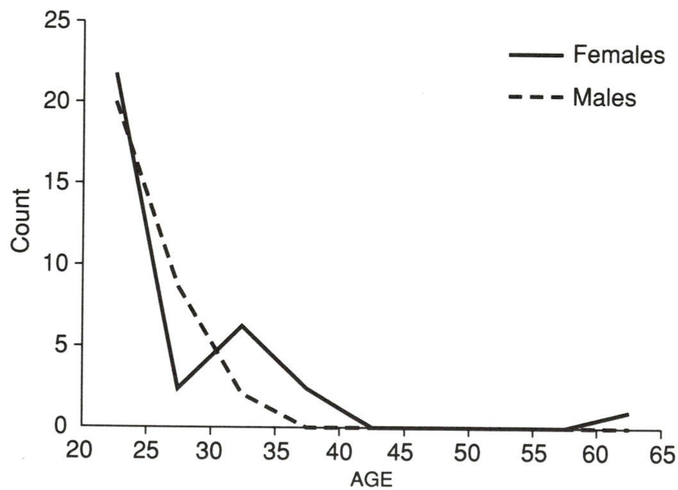

Figure 6.18

final stage; they are intended to communicate with fellow scientists and other interested readers. Each type of display serves a different purpose. Researchers try to choose the one that does the job best. Table 6.1 lists some descriptive statistical tools and indicates specific purposes that they serve.

ACTIVITY 6.5

This activity highlights the main points of Table 6.1. Fill in the blanks.

Discovering [1] _____________ of variation in a variable is one of the interests of an investigator. The statistical tools we use to do this are determined by our [2] _____________. When we have small samples we can display the entire data set by using [3] _____________. The distribution of one variable may be displayed

TABLE 6.1 Selected Statistical Tools for Analysis of Data

Purpose	Statistical Tool
To detect patterns of variation (for small samples	One-way table Stem and leaf plot
To detect the existence of outliers; these may highlight mistakes in data entry or interesting values that need to be explained	Bar graph Histogram Frequency polygon Stem and leaf plot
To reveal the distribution of one nominal or ordinal variable	One-way table Bar graph
To standardize data so two or more distributions from samples of different sizes can be compared	Percents based on counts
To compare distributions of two or more categorical variables	Bar graph Frequency polygon Histogram
To show a relationship between two categorical variables	Two-way table
To show a relationship between three or more categorical variables	Three-way table
To explore relationships between two ratio variables	Scatter plot
To explore how one variable changes with time	Time series plot
To summarize data using a few descriptive numbers	Mean, median, mode, range, minimum, maximum (and other summary statistics that will be defined later)

in a one-way table. We can standardize counts by computing [4] _____________, and this allows comparison of two or more distributions of different sizes. Patterns can be readily seen from [5] _____________. Two or more categorical variables may be compared by using a [6] _____________ graph. The distribution of both categorical and categorized variables can be shown using a [7] _____________ table with counts and percentages. We can discern relationships between two categorical variables by forming [8] _____________ tables. We can discern relationships between three categorical variables by forming [9] _____________ tables. We explore the relationship between two [10] _____________ variables by drawing a scatter plot. When we wish to note how a variable changes with time we draw [11] _____________ plots. Some examples of numerical summaries are [12] _____________, _____________, _____________, _____________, _____________, _____________.

Self-Assessment

Please assess your competency.

Tasks:	**How well can you do it?**					**Page**
Construct and/or interpret the following:	*Poorly*				*Very well*	
1. Bar graph	1	2	3	4	5	155
2. Histogram	1	2	3	4	5	155
3. Frequency polygon	1	2	3	4	5	173
4. Scatter plot	1	2	3	4	5	204
5. Time series plot	1	2	3	4	5	195
6. Range	1	2	3	4	5	189
7. Quartile	1	2	3	4	5	190
8. Inter-quartile range	1	2	3	4	5	190
9. Minimum/Maximum	1	2	3	4	5	191
10. Mode	1	2	3	4	5	189
11. Median	1	2	3	4	5	189
12. Five-number summary	1	2	3	4	5	190
13. Box plot	1	2	3	4	5	190

Answers for Chapter 6

Answers for Activity 6.1: 1) y 2) x 3) Age 4) Count, or frequency 5) Five (unless we also count the "bars" that are zero units high, in which case there are nine 6) 20–25 (Actually some convention must be established to determine which group the "end numbers" get counted in. For instance, this group could include 20 but exclude 25.) 7) Five 8) Grid 9) x 10) Students in two statistics classes 11) Age distribution 12) Spring 1992 13) A university 14) 20–25

Answers for p. 156: 1) Categorical 2) Ratio 3) 5 4) Contiguous 5) 55–60 6) 2, or moderate 7) 60–65 8) 21 9) $61 - 20 = 41$ 10) Mode 11) Asymmetric 12) Non-overlapping

Answers for p. 158: 1) x 2) y 3) Counts 4) One 5) y 6) Count 7) AGE 8) Count

Answers for Literature Example 6.1: 1) "Among the students polled, most held a degree in chemical engineering, followed by biology, civil engineering and chemistry." 2) One 3) Most common, approximately 68; least common, approximately 16 4) We think grid lines would help.

Answers for Literature Example 6.2: 1) Skewed 2) Approximately .7 3) No 4) Since asbestos can cause lung damage in the long run, the study reinforces the need to protect workers from both visible and invisible asbestos particles.

Answers for Literature Example 6.3: 1) We favor the figure in this case. 2) Contrasts can be seen easily; the magnitude of the differences can be seen easily. (You may come up with other answers.) 3) All the information is succinctly summarized in the table, but it is graphically summarized in the figure. The relative sizes are well depicted in the figure.

Answers for Literature Example 6.4: 1) Bathing 2) Age at death, type of help

Answers for Literature Example 6.5: 1) 75% 2) 65% 3) 18–49 4) older than 12 5) 56% 6) 86% 7) more

Answers for Literature Example 6.6: 1) Cost of injury is pulled out of the pie chart to show the importance of cost of injuries. 2) Diagnostic categories and their costs 3) Age and diagnostic categories 4) 17–44

Answer for p. 173: 1) The same

Answer for Literature Example 6.7: The midpoint of the group is age 70. This represents the group of ages from 65 to 75.

Answers for p. 176: 1) x 2) Five-year 3) 65 4) More 5) Asymmetric

Answers for Literature Example 6.8: 1) The legend below the horizontal axis indicates that maternal recall is shown by a square box and vaccination card data are shown by round dots. 2) The polygons overlap for children younger than 6 months but overlap less closely the more doses a child received. 3) .47

Answers for p. 180: 1) Mean difference between pulses before and after running in place:

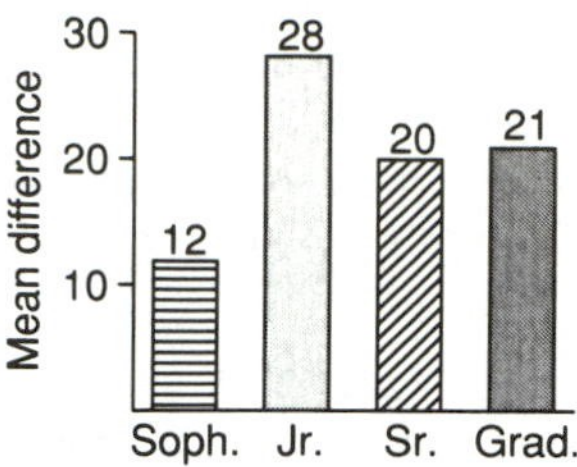

2) Activity by sex:

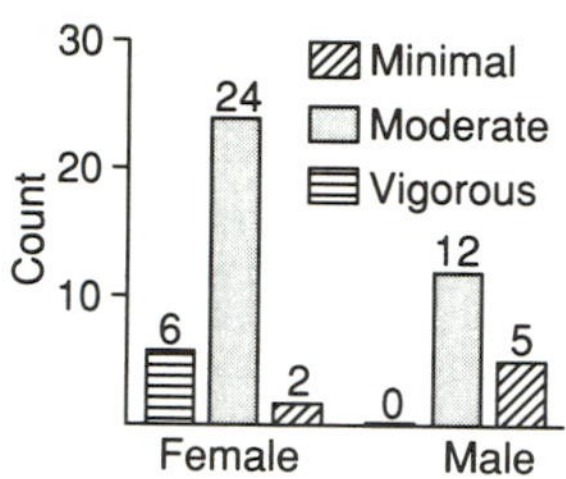

3) Female and male activity status:

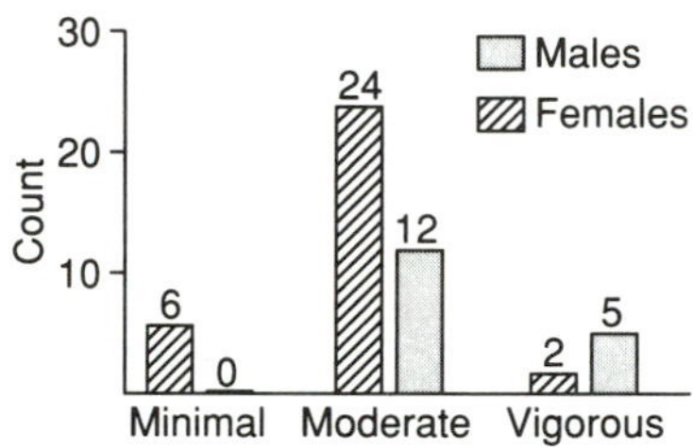

Answers for Literature Example 6.9: 1) Study period and authorization of transfers 2) 243 3) Slightly

Answers for Literature Example 6.10: 1) 33% 2) 21% 3) 25% 4) 15% 5) We don't know, but we'd guess it has to do with racism and poverty. The next literature example treats one aspect of poverty and a factor that could contribute to infant mortality.

Answer for Literature Example 6.11: Black

Answers for pp. 189–190: 1) 23 2) 17th 3) 21 4) 25.24 5) 61 6) 20 7) 41 8) 20 9) 61 10) 20 11) 21 12) 21 13) 22 14) 23 15) 24 16) 26 17) 29 18) 21 19) 26 20) 5 21) 21 22) 22 23) 24 24) 29

Answers for Activity 6.2: 1) 26 2) 33 3) 23 4) 20 5) 33; note that the true maximum age in these data is 61. However, 38 and 61 are both marked as outliers. 6) 21 7) 21 8) 26 9) 26 10) $26 - 21 = 5$ 11) $21 - 7.5 = 13.5$ 12) $26 + 15 = 41$ 13) 38 14) 61 15) Minimum 16) First quartile 17) Median 18) Third quartile 19) Maximum

Answers for Literature Example 6.12: 1) 2 kg 2) −1 kg 3) −2 to 1 (3 kg) 4) −2 to 7.5 5) −7.5 to 4 6) 2 kg 7) Approximately −1 kg

Answers for Activity 6.3: 1) 39 to 60 2) 20 to 25 3) 21 4) 50

Answer for Figure 6.14: Heart rate is increasing rapidly at first, then somewhat more gradually.

Answer for Figure 6.15: The amount of ventilation increases over time, but rate is erratic.

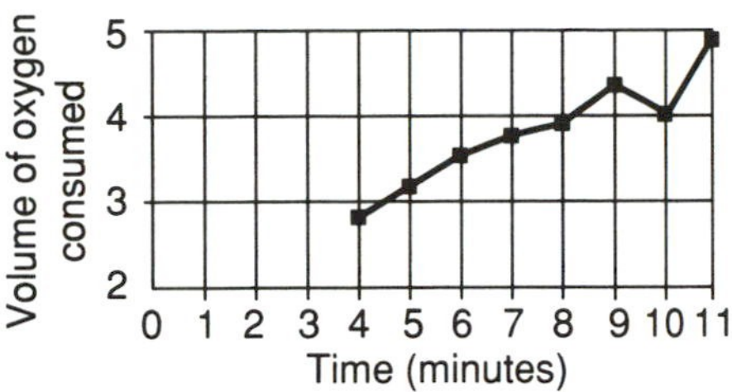

Answer for Literature Example 6.13: Radon levels varied considerably over time; see the results paragraph of the article.

Answer for Literature Example 6.14: The polygon that represents 0–6 km distance from the plant

Answers for Literature Example 6.15: 1) The number of newspaper abstracts increased 2) As the second paragraph in the Results section says, sales increased between 1987 and early 1988.

Answers for Literature Example 6.16: 1) Six. The letter *r* in Figure 4 stands for the correlation coefficient; *P* is probability. These measures facilitate inferences about the relationships between variables. They will be discussed in future chapters. 2) Do not

Answers for Literature Example 6.17: 1) Those who experience no household exposure 2) In the category "Only mother smokes" 3) Urine cotinine levels in infants increase markedly as they are exposed to more tobacco smoke.

Answers for Activity 6.4: We noticed the following about the graph in Figure 6.18: 1) The largest frequencies occur in the youngest ages (20–25) for both males and females. 2) For both sexes, frequencies decrease rapidly with increasing age. 3) There tend to be more females than males age 30 and older. 4) Neither distribution is symmetric. 5) There is one female in the 60–65 age group. 6) The number of students decreases with age. 7) This is not a time-series graph.

Answers for Activity 6.5: 1) Patterns 2) Purposes 3) One-way tables or stem and leaf plots 4) Percentages 5) Bar graphs or histograms 6) Bar 7) One-way 8) Two-way 9) Three-way 10) Ratio 11) Time series 12) Mean, median, mode, range, minimum, maximum

CHAPTER

7

Fun with Formulas

Prelude

"Why, just last month I sent him a very friendly letter, which he never had the courtesy to answer. See for yourself."

He handed Milo a copy of the letter, which read:

4738 1919,

667 394017 5841 62589 85371 14 39588 7190434 203 27689 57131 481206.

5864 98053,

62179875073

— Norton Juster

The Phantom Toll Booth, p. 199

Chapter Outline

Self-Assessment

Please assess your competency.

Tasks:	How well can you do it? *Poorly* — *Very well*					Page
Calculate the following:						
1. Sample mean $\bar{x} = \frac{\Sigma x_i}{n}$	1	2	3	4	5	222
2. Sample variance $s^2 = \frac{\Sigma (x_i - \bar{x})^2}{n - 1}$	1	2	3	4	5	222
3. Sample standard deviation $s = \sqrt{\frac{\Sigma (x_i - \bar{x})^2}{n - 1}}$	1	2	3	4	5	222
4. Standard error of the mean $SE_{\bar{x}} = s_{\bar{x}} = \frac{s}{\sqrt{n}}$	1	2	3	4	5	224
5. Population mean $\mu = \frac{\Sigma x_i}{N}$	1	2	3	4	5	226

6. Population variance 1 2 3 4 5 226

$$\sigma^2 = \frac{\Sigma(x_i - \mu)^2}{N}$$

7. Population standard deviation 1 2 3 4 5 226

$$\sigma = \sqrt{\sigma^2}$$

8. z score 1 2 3 4 5 228

$$z_i = \frac{x_i - \mu}{\sigma}$$

9. Standard deviation and standard error for proportion 1 2 3 4 5 227

$$\sigma_p = \sqrt{\frac{p * (1 - p)}{n}}$$

10. t-statistic 1 2 3 4 5 228

$$t = \frac{(\bar{x} - \mu_{\text{hyp}})}{s/\sqrt{n}}$$

11. χ^2 statistic 1 2 3 4 5 228

$$\chi^2 = \sum \frac{(\mathrm{O}_i - \mathrm{E}_i)^2}{\mathrm{E}_i}$$

12. Correlation coefficient 1 2 3 4 5 229

$$r = \frac{\Sigma\{(x_i - \bar{x})(y_i - \bar{y})\}}{\sqrt{\{\Sigma(x_i - \bar{x})^2\}\{\Sigma(y_i - \bar{y})^2\}}}$$

13. Regression coefficient 1 2 3 4 5 231

$$b = \frac{\Sigma\{(x_i - \bar{x})(y_i - \bar{y})\}}{\Sigma(x_i - \bar{x})^2}$$

Introduction

Formulas are condensed instructions on how to calculate various quantities such as statistics (attributes of samples) and parameters (attributes of populations). In this chapter you will learn to work with most of the formulas that you will be using throughout this course. Other texts present the various formulas when the corresponding concepts are introduced. (For example, the formula for the mean would be given when the concept of mean was introduced.) We have chosen to present all the formulas at the same time, however, because we think it is easier to learn how to use them this way. When you encounter these formulas in later chapters, it will be easier to grasp the concepts, because you will not be bogged down with figuring out how the formulas work. Keep in mind as you work through this chapter that the formulas may not make much sense to you at this point. In later chapters you will learn how the formulas are used; when you have mastered that material you will find that the formulas will make more sense to you.

This chapter is especially designed for the student whose algebra skills are limited. If you find that you already understand some material, you can skim the chapter until you encounter something new to you. We begin with an exercise to test your algebra skills to determine whether you should skip the first part of the chapter.

Compute standard deviation using the following formula and data set:

$$s = \sqrt{\frac{\Sigma\,(x_i - \bar{x})^2}{n - 1}}$$

Data set: 6, 9, 8, 3, 4, 6

s = Sample standard deviation
The x_i are the observations
$\bar{x}$ = mean of the sample
n = sample size

If you already know how to compute s using this formula and you got 2.28 as an answer, please go directly to the section entitled "Why Formulas Can Be Confusing" (page 221). Otherwise, continue to the next section.

Confusion and Clarity

When we understand a certain body of information, that information appears to fit into an organized pattern. Confusion, however, carries with it disorganization, and we may experience unpleasant feelings because of it. More information only tends to increase our confusion rather than leading to understanding.

Note that understanding is not absolute. At different times we attain different levels of understanding, and different people mean different things when they say they understand a concept. For example, a particular medication may be understood in terms of chemical composition by a chemist; purpose and method of administration by a nurse; and effectiveness in curing a disease by a physician.

We will distinguish two levels of understanding for formulas: (1) computational understanding, that is, the ability to use the formulas to obtain correct numerical results; and (2) appropriateness understanding, that is, knowing which formulas are appropriate for particular real-world situations. We will not expect you to have this second level of understanding yet; in this chapter we aim to acquire only computational understanding.

Feelings such as embarrassment, fear, sadness, grief, physical pain, and anger impede our ability to think clearly. Since you might have such feelings about formulas because of past unpleasant experiences with mathematics, we will begin with an exercise to help you gain insight into those feelings so that you can reclaim your power to think clearly.

Assessing Your Own Feelings Toward Formulas

Individuals react to new experiences according to their previous, similar experiences. Pavlov's dog is a prototype. Whether you realize it or not, a part of you reacts in a conditioned way to past experiences. Write six adjectives that describe your feelings about mathematical formulas.

1. ____________________
2. ____________________
3. ____________________
4. ____________________
5. ____________________
6. ____________________

Doing Something About Your Feelings

When we maintain a positive attitude, we see a hurdle as an interesting challenge; our thought processes function well instead of being impeded by the unpleasant feelings that the hurdle might otherwise cause.

Think of an occasion when you encountered a problem that seemed insurmountable. Briefly describe your feelings at that time:

____

____

____

____

____

____

Now suppose that in that situation you had felt absolutely indomitable, thoroughly brilliant, completely powerful. How would you have attacked the problem? (It might help to imagine that you had superhuman powers of some kind.) How would you have felt while doing so?

____

____

____

____

____

____

Dyad Exercise to Explore Feelings and Promote a Positive Attitude

Before starting calculations with formulas, break into dyads and discuss any pleasant subject with your partner (2 minutes each).

If you experience unpleasant feelings (for example, confusion, frustration, or even fear) while you are working on the formulas, break into dyads for one minute each and talk about the pleasant feelings you discussed in the step above. We recommend that you do this any time bad feelings come up (if possible) instead of letting those feelings impede your ability to function well.

Why Formulas Can Be Confusing

Formulas are symbolic representations of procedures (recipes) for calculating. Formulas specify arithmetic operations that need to be performed and the order in which the operations are to be done. There are two possible sources of confusion with regard to a formula:

1. What the *letters* in the formula symbolize may be unclear or unknown;
2. The *operations* indicated by the letters or symbols may not be understood.

In the following pages we will explain both the symbols and the mathematical operations indicated by the symbols in several statistical formulas. Actvity 7.1 introduces the first statistical formulas that we will study and details some conventions used in mathematics. The activity uses two small data sets for ease of computation.

Activity 7.1

Formulas for calculation of sample mean and standard deviation

Explanation	Example
Letters such as x, y, z, t, etc. are used to indicate a quantity that is capable of having different values in a formula. x is often used as the value of an explanatory variable, in contrast to y, which is often used for the response variable. (For an example that uses both x and y, see Activity 7.5.)	Data set of five ages: 20, 25, 30, 32, 28. x may refer to any of the ages.

In the symbol x_i, i is a **subscript** that can take on the values 1, 2, 3, etc., to give x_1, x_2, x_3, etc. Subscripted letters such as these stand for values of the variable. Σ is the uppercase Greek letter sigma. This symbol indicates that we sum the numbers that follow it. Thus, Σx_i means $x_1 + x_2 + x_3$, etc. n represents the size of the sample. The formula

$$\Sigma \frac{x_i}{n}$$

indicates that you are to sum the x_i and divide by n. This is the formula for the **sample mean.**

x_i represents each one of the numbers in the data set of ages. Thus, $x_1 = 20$, $x_2 = 25$, $x_3 =$ [1] ________, $x_4 =$ [2] ________, $x_5 =$ [3] ________.

For the age data,

$$\Sigma x_i = 20 + 25 +$$

[4] ______________ = ______________.

For the age data, $n =$ [5] _______. For our data,

$$(\Sigma x_i)/n = {}^{6}\_\_\_\_\_\_\_.$$

The symbol $\bar{x}$, called "x bar," is the symbol for the mean of a sample. "y bar" is the mean of a sample when we use y to represent the variable. Thus,

$$\bar{x} = \Sigma \frac{x_i}{n}$$

is the formula for calculating the mean of a sample, where n is the sample size. x_i represents successive values of the variable.

Calculate $\bar{x}$ for the following new data: 6, 9, 8, 3, 4, 6.

$\bar{x} =$ [7] _______

s represents **sample standard deviation.** Standard deviation is a measure of how much the individual data tend to deviate from the mean. Because of the many uses of standard deviation, it will become very familiar to you by the end of this book. For now, you need only learn how to compute it.

In the expression s^2, the superscript 2 instructs you to square the value, i.e., to multiply s by itself:

$$s^2 = s * s$$

s^2 is called the **sample variance.** It is also used as a measure of deviation from the mean.

Expressions enclosed in parentheses should be evaluated first. For example, with $(x_i - \bar{x})^2$ we subtract $\bar{x}$ from each x_i and then square the result. When this expression appears in a summation (see below) you do this sequence of operations for each value of the variable x: $(x_1 - \bar{x})^2$, $(x_2 - \bar{x})^2$, etc. These values are called **squared deviations from the mean.**

Earlier you calculated $\bar{x} = 6$ for the second data set. Now calculate each $(x_i - \bar{x})^2$:

$(6 - 6)^2 =$ 8 ______

$(9 - 6)^2 =$ 9 ______

(______)$^2 =$ 10 ______

(______)$^2 =$ 11 ______

(______)$^2 =$ 12 ______

(______)$^2 =$ 13 ______

$$\sum (x_i - \bar{x})^2$$

is the expression for the **sum of the squared deviations from the mean** (SS, or sum of squares, for short). The formula simply instructs us to add up all the squares we computed above.

$\sum (x_i - \bar{x})^2 =$ 14 ______

$$s^2 = \frac{\Sigma (x_i - \bar{x})^2}{n - 1}$$

is the formula for the sum of squared deviations divided by $n - 1$, which gives us the **mean sum of squared deviations** (MS for short). MS is also more commonly called the **variance** of the sample. The variance is a measure of the amount of variation in the values of the variable.

$s^2 = \dfrac{\Sigma (x_i - \bar{x})^2}{n - 1} =$ 15 ______

The expression $n - 1$ is called the number of **degrees of freedom** (df for short). Why do we use $(n - 1)$ instead of n? We use $(n - 1)$ because it yields a better estimate of the population variance. (This is just a strange mathematical fact that you will have to accept on faith!)

df $= n - 1 =$ 16 ______

Above we said that the formula

$$s^2 = \frac{\Sigma (x_i - \bar{x})^2}{n - 1}$$

stands for the variance of a sample.

The square root of the variance, $\sqrt{s^2}$, is s, which we noted earlier is called the **standard deviation** of the

Standard deviation, s, is 17 ______.

sample. It is another measure of the amount of variation in the sample. Taking the square root has the effect of correcting for the squaring operation that was done earlier. Standard deviation is in the same units as the original variable.

Finally, the formula

$$SE_{\bar{x}} = s_{\bar{x}} = \frac{s}{\sqrt{n}}$$

$s_{\bar{x}} =$ [18] _______

the **standard error of the mean**, is a measure of the variation between different means of samples taken from the same population. It is called **SE**$_{\text{mean}}$ for short.

Investigators often need to compute numerical summaries of samples. In Chapter 6 we introduced the summary statistics mean, mode, and median. Each of these statistics gives a different single-number summary of a data set. In describing data sets, one or more of these measures is almost always given. Earlier chapters also introduced range and percentiles, measures of variation of the data. Variance, standard deviation, and standard error of the mean are also measures (or indices) of variation (or **dispersion**). The symbol for sample variance is [1] _______ and the symbol for standard deviation is [2] _______. Both of these are measures of how dispersed the data are. The smaller the variance (and hence the smaller the standard deviation), the more closely packed are the numbers in the data set. If we were to take several random samples from the same population, they all would be likely to have different means. The symbol for the mean is [3] _______. The dispersion of the sample means is measured by the standard error of the mean, denoted [4] _______. MS represents the mean of the [5] _______ of each x_i from $\bar{x}$. This is also called the [6] _______. The square root of the variance is called the [7] _______. Standard error of the mean, or $s_{\bar{x}}$, is symbolized by [8] _______. The symbol that instructs us to sum is [9] _______.

Explanations of the Formulas

Formulas for Samples

Samples are taken from a population in order to learn about the population attributes such as mean and standard deviation.

ACTIVITY 7.2

Computing mean, variance, standard deviation, and standard error for a sample

Suppose the following data represent a sample from a larger population:

5, 4, 3, 7

Calculate the following using these data.

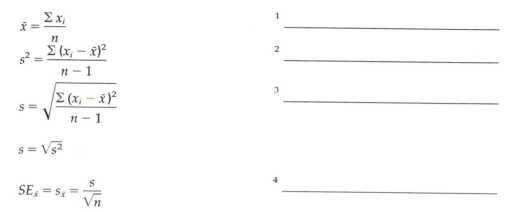

$$\bar{x} = \frac{\Sigma x_i}{n}$$ 1 ____________________

$$s^2 = \frac{\Sigma (x_i - \bar{x})^2}{n - 1}$$ 2 ____________________

$$s = \sqrt{\frac{\Sigma (x_i - \bar{x})^2}{n - 1}}$$ 3 ____________________

$$s = \sqrt{s^2}$$

$$SE_{\bar{x}} = s_{\bar{x}} = \frac{s}{\sqrt{n}}$$ 4 ____________________

Formulas for Populations

Formulas for samples and populations are similar but not identical. The population attributes are called **parameters** and Greek letters are used to denote them. The analogous sample attributes are called **statistics.** (Note that we use the word "statistics" in two senses, to mean both the subject matter of this and similar courses of study and sample attributes such as mean, variance, and standard deviation.) The following table compares the symbols used for samples and populations.

	Sample (Statistics)	Population (Parameters)	
Mean	$\bar{x}$	μ	(read "mu")
Variance	s^2	σ^2	
Standard deviation	s	σ	(read "sigma")
Standard error	$s_{\bar{x}}$	$\sigma_{\bar{x}}$	
Size or number of observations (cases)	n	N	(Note: Greek letter is not used.)

ACTIVITY 7.3

Computing mean, variance, and standard deviation for a population

Assume that an entire population consists of the following data.

10, 11, 12, 14, 15

Calculate the following using these data. Keep in mind that this Activity is artificial, in that population parameters can seldom be computed because the complete data set is almost never available. Usually populations are very large. That is why investigators generally estimate these values by taking samples; the larger the sample, the more accurate the estimate.

$$\mu = \frac{\Sigma x_i}{N} =$$ 1 ______________________

$$\sigma^2 = \frac{\Sigma (x_i - \mu)^2}{N} =$$ 2 ______________________

Notice that the denominator is N and not $N - 1$. (Sometimes mathematics is puzzling!)

$$\sigma = \sqrt{\frac{\Sigma (x_i - \mu)^2}{N}} =$$ 3 ______________________

$$\sigma = \sqrt{\sigma^2} =$$

We wish to find out about the population attributes, or [1] _______, by sampling and calculating statistics such as $\bar{x}$ and s. These symbols stand for [2] _______ and [3] _______. Parameters such as mean and standard deviation are denoted by the symbols [4] _______ and [5] _______, respectively. When calculating sample variance we use $n - 1$ as the denominator, whereas population variance is calculated using [6] _______ as the denominator. Otherwise, the formulas are identical. It is important to be able to discriminate sample from population.

The reason that $n - 1$ is used in the formula for standard deviation (and variance) of a sample is that we want the sample standard deviation to be a good estimate of the standard deviation of the population from which the sample was drawn. For purely mathematical reasons, if n were used in the formula, the resulting numbers would consistently underestimate the population values. We wish to avoid such bias. It can be shown that when $n - 1$ is used, the estimate is unbiased.

In this text we use the FED and SQD sets in various contexts. Both sets are samples, but sometimes we will pretend that the data from one of these sets is a population and take samples from it for pedagogical reasons. When we do this, we will explicitly say so.

Formula for the Standard Deviation and Standard Error of a Proportion

We frequently consider the **proportion**, p, of subjects in a population that have a certain property. This quantity is the number having the property divided by the size of the population, N. For a sample, the definition is analogous: the number in the sample having the property divided by the size of the sample, n. For example, in the Student Questionnaire Data (SQD), the total number of students is 51 and the number of females is 33. Therefore the proportion, p, of females is 33/51 = 0.647.

The formula for computing standard deviation of a proportion is shown below. This formula is only used in the context of a sample, and the size of the sample, n, appears in the formula.

$$\text{Standard deviation of a proportion} = \sigma_p = \sqrt{\frac{p * (1 - p)}{n}}$$

Compute the standard deviation for the proportion of females in the SQD. Assume that a sample of size 10 is being taken and the proportion of females in the population is 0.65.

$$\sigma_p = {}^{1}\ \_\_\_\_\_\_\_$$

Usually an investigator doesn't know what proportion of a population has the property of interest. For this reason, the standard deviation of a proportion must often be estimated using an estimate for p, namely the proportion having the property in the sample. The formula is the same, but in this case it is called the **standard error of the proportion.** Assume that we drew a sample of 10 from the SQD and 6 of the 10 were female. The standard error of the proportion of females would be [2] _______. These formulas measure the deviation of sample proportions from the population proportion.

Formulas for z, t, and Chi-Square

When comparing similar data from different sets with different means and different standard deviations, we can standardize the data by using z **scores.** For example, if you wanted to compare your performance on tests to the average performance of the rest of the class, you could compute the z scores of your test grades. z scores are also used in conjunction with the normal distribution, which you will study in Chapter 11. Additional applications of z scores will be presented in subsequent chapters.

Scientists select a sample from a population and collect data in order to test a hypothesis. A hypothesis is a scientist's precisely formulated educated guess about population parameters. Testing a hypothesis about a mean frequently entails calculation of the **t-statistic,** which is defined below. This statistic will be discussed at greater length in Chapter 14.

Certain relationships among categorical data presented in tables are tested using a statistic called the **chi-square** statistic. χ is the Greek letter chi (pronounced "kye"). Chi-square, χ^2, is defined below. This formula uses **observed counts** (O) and **expected counts** (E) to calculate χ^2. The use of this statistic will be described at greater length in Chapter 16.

ACTIVITY 7.4

Formula	Data	Calculations
$z_i = \dfrac{x_i - \mu}{\sigma}$	Population size = 4: 15, 14, 13, 18 $\mu =$ 1 ______ $\sigma = 1.87$	Calculate z_i for each of the x_i: x_i z_i 15 2 ______ 14 3 ______ 13 4 ______ 18 5 ______
$t = \dfrac{(\bar{x} - \mu_{\text{hyp}})}{\dfrac{s}{\sqrt{n}}}$ or $t = (\bar{x} - \mu_{\text{hyp}})/s_{\bar{x}}$	Sample size = 4 5, 4, 3, 8 $\bar{x} = 5$ $s = 2.16$ $\mu_{\text{hyp}} = 0$	$s_{\bar{x}} =$ 6 ______ $t =$ 7 ______
$\chi^2 = \sum \dfrac{(O_i - E_i)^2}{E_i}$	Observed Expected O_i E_i 21 22.22 12 10.78 12 10.78 4 5.22	$\dfrac{(O_i - E_i)^2}{E_i}$ 8 ______ 9 ______ 10 ______ 11 ______ $\chi^2 = \sum \dfrac{(O_i - E_i)^2}{E_i} =$ 12 ______

The z score standardizes observations (x_i) in units of standard deviation. In other words, it tells us how many standard deviations away any given x_i is from the mean. The t-statistic measures the discrepancy between a calculated mean and a hypothesized mean (μ_{hyp}). The t-statistic in this instance has three symbols in its formula: [1]_______ (sample mean), [2]_______ (hypothesized population mean), and [3]_______ (standard error of the mean). z uses [4]_______ in the denominator instead of $s_{\bar{x}}$ as in the case of t. z deals with the distribution of observations (x_i), whereas t deals with the distribution of sample means ($\bar{x}$). The chi-square statistic is used to measure the relationships of categorical variables, and its symbol is [5]_______. Two important numbers in the formula are [6]_______ count (O) and [7]_______ values (E).

Formulas for Correlation and Regression Coefficients

Is there a relationship between weight and age? Can weight be predicted from age? After they collect data, investigators attempt to unveil relationships by plotting weight versus age (for example) and calculating correlation and regression coefficients. These concepts will be discussed at greater length in Chapter 15.

Since these formulas are more complex than the ones we have previously considered, we present the calculations in stages. We hope this will make computation from the formulas easier to understand.

ACTIVITY 7.5

Computing the correlation coefficient

The formula for calculating the correlation coefficient is as follows:

$$r = \frac{\Sigma\{(x_i - \bar{x})(y_i - \bar{y})\}}{\sqrt{\{\Sigma(x_i - \bar{x})^2\}\{\Sigma(y_i - \bar{y})^2\}}}$$

The main constituents of this equation are (1) the difference between each x and $\bar{x}$ and (2) the difference between each y and $\bar{y}$. We need to calculate this difference for each x and y and put these quantities into the formula. Use the data below to make these calculations:

x values: 18, 15, 11, 8 y values: 3, 2, 1, 2 $\bar{x} =$ [1]_______ and $\bar{y} =$ [2]_______

Complete the following table. We have done the first row to illustrate the computations.

i	x_i	y_i	$(x_i - \bar{x})$	$(y_i - \bar{y})$	$(x_i - \bar{x}) * (y_i - \bar{y})$	$(x_i - \bar{x})^2$	$(y_i - \bar{y})^2$
1	18	3	$18 - 13 = 5$	$3 - 2 = 1$	$5 * 1 = 5$	$5^2 = 25$	$1^2 = 1$
2	15	2					
3	11	1					
4	8	2					

$\Sigma \{(x_i - \bar{x})(y_i - \bar{y})\}$ is found by summing the numbers in the sixth column.

$$\Sigma \{(x_i - \bar{x})(y_i - \bar{y})\} = {}^{3}\_\_\_\_\_\_\_.$$

The sum of the numbers in the seventh column is $\Sigma (x_i - \bar{x})^2 =$ [4]_______ and the sum of the numbers in the last column is $\Sigma (y_i - \bar{y})^2 =$ [5]_______.

The numerator of the formula for r (the correlation coefficient) measures how x and y co-vary. It is simply the sum of the products of the differences. (It is reminiscent of the squared differences in the variance formula for a single variable.) Do you recognize

$$\Sigma (x_i - \bar{x})^2$$

as the numerator of the variance formula? In that formula it was called [6]_______ for short. The corresponding y component of the formula is [7]_______. This too is the sum of squared deviations, the numerator of the [8]_______ formula, but calculated using values of y.

You have already calculated the numerator of the correlation coefficient formula. Now you will calculate the denominator:

$$\{\Sigma (x_i - \bar{x})^2\}\{\Sigma (y_i - \bar{y})^2\} = {}^{9}\_\_\_\_\_\_\_$$

The denominator is found by taking the square root of this number:

$$\sqrt{\{\Sigma (x_i - \bar{x})^2\}\{\Sigma (y_i - \bar{y})^2\}} = {}^{10}\_\_\_\_\_\_\_$$

Now calculate r, the correlation coefficient, by dividing the numerator by the denominator.

$$r = {}^{11}\_\_\_\_\_\_\_$$

The correlation coefficient is a statistic that measures the strength of a certain relationship (called linearity) between two variables (x and y). The symbol for the correlation coefficient of a sample is [1]_______. r is a number between -1 and $+1$. Zero represents no linear correlation and -1 and $+1$ indicate a perfect linear correlation.

The expression for the difference between the mean and each of the observations for x is [2]_______; for y the expression is [3]_______. The expression $\Sigma \{(x_i - \bar{x})(y_i - \bar{y})\}$ measures how x and y vary together (co-vary). The denominator of the correlation coefficient formula requires the computation of the sums of squares of x and y. The expression in the denominator is [4]_______.

The Regression Coefficient

The formula for the regression coefficient is

$$b = \frac{\Sigma \{(x_i - \bar{x})(y_i - \bar{y})\}}{\Sigma (x_i - \bar{x})^2}$$

Compare this formula to the formula for r, the correlation coefficient. The numerators of r and b are [1] the same/different. The denominator for r contains an expression with both x and y, whereas the denominator of b has an expression containing only [2]_______. The denominator of b is the [3] numerator/denominator of the variance formula. The numerator of b is [4]_______ and the denominator is [5]_______. The denominator is also known as the sum of [6]_______ of the deviations of the x_i from their mean. The sum of squares of deviations of the y_i from their mean is used in the denominator of [7] r/b but not in the denominator of [8] r/b.

When a scatter plot of x and y shows a possibility of a linear (straight-line) relationship and the correlation coefficient also indicates such a relationship, we can use our knowledge of x to predict values of y. The regression equation gives us the means to do this. One attribute of a straight line is its slope, a number that measures how tilted the line is. A horizontal line has zero slope; a line that has x and y both increasing together has positive slope; and a line having y decreasing as x increases has negative slope. The formula for computing the slope, b, of the so-called regression line was given above. The next activity leads you step by step through the calculation of b.

ACTIVITY 7.6

Computing the regression coefficient

The formula for the regression coefficient, or slope, is as follows:

$$b = \frac{\Sigma \{(x_i - \bar{x})(y_i - \bar{y})\}}{\Sigma (x_i - \bar{x})^2}$$

Using the data in Activity 7.5 and the values you computed in that activity, you can now compute b.

x values: 18, 15, 11, 8
y values: 3, 2, 1, 2

The numerator of the regression coefficient formula is

$$\Sigma \{(x_i - \bar{x})(y_i - \bar{y})\} = {}^{1}\_\_\_\_\_\_\_.$$

The denominator of the regression coefficient formula is

$$\Sigma (x_i - \bar{x})^2 = {}^{2}\_\_\_\_\_\_\_.$$

$$b = {}^{3}\_\_\_\_\_\_\_$$

Following is another data set:

x values: 1, 3, 5, 7
y values: 3, 6, 9, 10

Calculate $(x_i - \bar{x})$, $(y_i - \bar{y})$, $\{(x_i - \bar{x})(y_i - \bar{y})\}$, and $(x_i - \bar{x})^2$ for these data. Calculate the value of the numerator of the regression coefficient formula for these data: [4]_______. Calculate the value of the denominator of the regression coefficient formula: [5]_______. Now calculate b, the regression coefficient: [6]_______. Finally, calculate r using the data:

$$r = \frac{\Sigma\{(x_i - \bar{x})(y_i - \bar{y})\}}{\sqrt{\{\Sigma(x_i - \bar{x})^2\}\{\Sigma(y_i - \bar{y})^2\}}} = {}^{7}\_\_\_\_\_\_\_$$

Self-Assessment

Tasks:	**How well can you do it?**					**Page**
To calculate the following:	*Poorly*				*Very well*	
1. Sample mean $\bar{x} = \frac{\Sigma x_i}{n}$	1	2	3	4	5	222
2. Sample variance $s^2 = \frac{\Sigma(x_i - \bar{x})^2}{n-1}$	1	2	3	4	5	222
3. Sample standard deviation $s = \sqrt{\frac{\Sigma(x_i - \bar{x})^2}{n-1}}$	1	2	3	4	5	222
4. Standard error of the mean $SE_{\bar{x}} = s_{\bar{x}} = \frac{s}{\sqrt{n}}$	1	2	3	4	5	224
5. Population mean $\mu = \frac{\Sigma x_i}{N}$	1	2	3	4	5	226
6. Population variance $\sigma^2 = \frac{\Sigma(x_i - \mu)^2}{N}$	1	2	3	4	5	226
7. Population standard deviation $\sigma = \sqrt{\sigma^2}$	1	2	3	4	5	226

8. z score 1 2 3 4 5 228

$$z_i = \frac{x_i - \mu}{\sigma}$$

9. Standard deviation and standard error for proportion 1 2 3 4 5 227

$$\sigma_p = \sqrt{\frac{p * (1 - p)}{n}}$$

10. t-statistic 1 2 3 4 5 228

$$t = \frac{(\bar{x} - \mu_{\text{hyp}})}{s/\sqrt{n}}$$

11. χ^2 statistic 1 2 3 4 5 228

$$\chi^2 = \sum \frac{(O_i - E_i)^2}{E_i}$$

12. Correlation coefficient 1 2 3 4 5 229

$$r = \frac{\Sigma\{(x_i - \bar{x})(y_i - \bar{y})\}}{\sqrt{\{\Sigma(x_i - \bar{x})^2\}\{\Sigma(y_i - \bar{y})^2\}}}$$

13. Regression coefficient 1 2 3 4 5 231

$$b = \frac{\Sigma\{(x_i - x)(y_i - y)\}}{\Sigma(x_i - \bar{x})^2}$$

Answers for Chapter 7

Answers for Activity 7.1: 1) 30 2) 32 3) 28 4) . . . 30 + 32 + 28 = 135 5) 5 6) 135/5 = 27 7) 6 8) 0 9) 9 10) (8 − 6), 4 11) (3 − 6), 9 12) (4 − 6), 4 13) (6 − 6), 0 14) 26 15) 26/5 = 5.2 16) 6 − 1=5 17) 2.28 18) 0.93

Answers for p. 224: 1) s^2 2) s 3) $\bar{x}$ 4) $s_{\bar{x}}$ 5) Squared deviation 6) Variance 7) Standard deviation 8) $s/\sqrt{n}$ 9) Σ

Answers for Activity 7.2: 1) 4.75 2) 2.93 3) 1.71 4) 0.86

Answers for Activity 7.3: 1) 12.40 2) 3.44 3) 1.85

Answers for p. 226: 1) Parameters 2) Mean 3) Standard deviation 4) μ 5) σ 6) N

Answers for p. 227: 1) 0.15 2) 0.155

Answers for Activity 7.4: 1) 15 2) 0 3) −0.535 4) −1.07 5) +1.60 6) 1.08 7) 4.63 8) 0.067 9) 0.138 10) 0.138 11) 0.284 12) 0.627

Answers for p. 229: 1) $\bar{x}$ 2) μ_{hyp} 3) $s_{\bar{x}}$ 4) σ 5) χ^2 6) Observed 7) Expected

Answers for Activity 7.5: 1) 13 2) 2

i	x_i	y_i	$(x_i - \bar{x})$	$(y_i - \bar{y})$	$(x_i - \bar{x}) * (y_i - \bar{y})$	$(x_i - \bar{x})^2$	$(y_i - \bar{y})^2$
1	18	3	5	1	5	25	1
2	15	2	2	0	0	4	0
3	11	1	−2	−1	2	4	1
4	8	2	−5	0	0	25	0

3) 7 4) 58 5) 2 6) SS 7) $\Sigma (y_i - \bar{y})^2$ 8) Variance 9) 116 10) 10.8 11) .65

Answers for p. 230: 1) r 2) $(x_i - \bar{x})$ 3) $(y_i - \bar{y})$ 4) $\sqrt{\{\Sigma (x_i - \bar{x})^2\}\{\Sigma (y_i - \bar{y})^2\}}$

Answers for p. 231: 1) The same 2) x 3) Numerator 4) $\Sigma \{(x_i - \bar{x})(y_i - \bar{y})\}$ 5) $\Sigma (x_i - \bar{x})^2$ 6) Squares 7) r 8) b

Answers for Activity 7.6: 1) 7 2) 58 3) 0.12 4) 24 5) 20 6) 1.2 7) 0.98

CHAPTER

8

Measures of Central Tendency and Dispersion

Prelude

"Oh, we're just the average family," he said thoughtfully; "mother, father, and 2.58 children—and, as I explained, I'm the .58."

"It must be rather odd being only part of a person," Milo remarked.

"Not at all," said the child. "Every average family has 2.58 children, so I always have someone to play with. Besides, each family also has an average of 1.3 automobiles, and since I'm the only one who can drive three tenths of a car, I get to use it all the time."

"But averages aren't real," objected Milo; "they're just imaginary."

"That may be so," he agreed, "but they're also very useful at times. For instance, if you didn't have any money at all, but you happened to be with four other people who had ten dollars apiece, then you'd each have an average of eight dollars. Isn't that right?"

"I guess so," said Milo weakly.

"Well, think how much better off you'd be, just because of averages," he explained convincingly. "And think of the poor farmer when it doesn't rain all year: if there wasn't an average yearly rainfall of 37 inches in this part of the country, all his crops would wither and die."

It all sounded terribly confusing to Milo, for he had always had trouble in school with just this subject.

"There are still other advantages," continued the child. "For instance, if one rat were cornered by nine cats, then, on the average, each cat would be 10 percent rat and the rat would be 90 percent cat. If you happened to be a rat, you can see how much nicer it would make things."

"But that can never be," said Milo, jumping to his feet.

"Don't be too sure," said the child patiently, "for one of the nicest things about mathematics, or anything else you might care to learn, is that many of the things which can never be, often are. You see," he went on, "it's very much like your trying to reach Infinity. You know that it's there, but you just don't know where—but just because you can never reach it doesn't mean that it's not worth looking for."

— Norton Juster
The Phantom Toll Booth, pp. 196–197

Chapter Outline

Self-Assessment

Please assess your competency.

Tasks:	How well can you do it?					Page
Decide which of the following is the appropriate measure of central tendency/variation in a given context and calculate it from the data:	*Poorly*				*Very well*	
1. Mean	1	2	3	4	5	237
2. Mode	1	2	3	4	5	238
3. Median	1	2	3	4	5	238
4. Trimmed mean	1	2	3	4	5	238
5. Range	1	2	3	4	5	266
6. Inter-quartile range	1	2	3	4	5	251
7. Variance	1	2	3	4	5	266
8. Standard deviation	1	2	3	4	5	251
9. Proportion	1	2	3	4	5	266
10. Coefficient of variation	1	2	3	4	5	267
11. Variance of proportion	1	2	3	4	5	266
12. Index of dispersion	1	2	3	4	5	279

Introduction

Reality is multifaceted, but it can be analyzed by isolating elements and describing them. We assign numbers in an effort to manipulate aspects of reality symbolically. To interpret the outcomes of our numerical manipulations correctly, we need to remember the ways in which we assigned the numbers in the first place.

In the questionnaire in Chapter 2 we assigned numbers to answers. Each question measures an aspect of each student on one of the four scales (nominal, ordinal, interval, and ratio). Each of the variables has a distribution. In Chapters 5 and 6 we depicted patterns of variation using tables and graphs. In this chapter we will describe these variables by computing numerical summaries (summary statistics) for both categorical and numerical data.

Measures of Central Tendency

Computation of Mean, Median, and Mode for Continuous Data

Mean, median, and mode are called measures of central tendency. You learned how to calculate these measures in Chapters 6 and 7.

If you were asked to give a typical age in your class, which age would you be likely to give—your own? the minimum age? the maximum age? What would you choose as a typical age from the list below of ages of students in two statistics classes?

20 20 20 20 20 21 21 21 21 21 21 21 21 21 21 21 22 22
22 22 22 23 23 23 23 23 23 23 23 24 24 24 24 24 25 25
25 26 26 28 29 30 30 30 32 33 33 33 38 38 61

With a little effort we could come up with the mode, the most frequently occurring age, which is _______, or we could give the median, the age below which 50 percent of the ages lie, or the (arithmetic) mean, which is calculated by using the formula $\bar{x} = (\Sigma x_i)/n$. The median gives the position of one observation. It ignores all other observations. The mean is the result of summing all observations and dividing by the total number of observations. The mean takes into account all the observations, so it is influenced by the extreme values, whereas the median is not.

Influence of Extreme Values on the Mean

Consider the following measures of central tendency for the ages listed above:

Age 61 Included

n	Mode	Mean	Median	Trimmed Mean
51	21	25.235	23.000	24.222

Age 61 Excluded

n	Mode	Mean	Median	Trimmed Mean
50	21	24.520	23.000	24.023

The mode is the most frequently occurring observation.

Is the mode influenced by the extreme value 61? [1] Yes/No Can we influence the mode by including additional observations? [2] Yes/No

The median is the fiftieth percentile.

Is the median influenced by the extreme value? [3] Yes/No

Calculation of the median is based on determining the [4] ________ of observations and finding the middle value. Therefore the magnitudes of most of the values in the distribution do not influence the median. When additional observations are included, does this influence the median? [5] Yes/No

The mean is defined by the formula

$$\bar{x} = (\Sigma x_i)/n$$

Did age 61 influence the calculation of mean? [6] Yes/No

All the values "participate" in the calculation of a mean. Therefore every value has an influence on the mean. However, it is the extreme values that can make the most difference, as age 61 did.

The age data lump the two classes together. For class II (with twenty-five students, including the 61-year-old) we have the following means:

Mean including 61: 26.12

Mean excluding 61: 24.67

The [7] _______ in these means compared to that in the ones on the previous page is more pronounced because of the smaller sample size.

With a trimmed mean, 5 percent of the observations on each end of a distribution are excluded from the calculation of the mean.

The trimmed means of the age data set including the 61-year-old and the same set with that age omitted are almost the same, because 61 is one of the numbers "trimmed" from the set that includes that number. The trimmed mean is the mean of the middle [8] _______ percent of the observations, so it is not influenced by the extreme values on the small- and large-value ends of the distribution. Which observations from the age data set are excluded from the calculation of the trimmed mean? [9] _______ and [10] _______

Computation of Mean for Categorized Data

Sometimes data may be given as a frequency distribution. When data are presented this way, it's easier to calculate the mean, mode, and median. To take the frequencies into account, a slightly different (though logically equivalent) formula is used to compute the mean.

Table 8.1 gives the counts of each of the ages in the data set we listed earlier. By weighting each age (multiplying age by the corresponding frequency), summing the weighted ages, and dividing by n, the total count (or frequency), we obtain the mean,

TABLE 8.1 Computing the Mean for Categorized Age Data from Two Statistics Classes

AGE x	Count, or Weight w	AGE * Count xw
20	5	20 * 5 = 100
21	11	21 * 11 = 231
22	5	22 * 5 = 110
23	8	23 * 8 = 184
24	5	24 * 5 = 120
25	3	25 * 3 = 75
26	2	26 * 2 = 52
28	1	28 * 1 = 28
29	1	29 * 1 = 29
30	3	[1] _______ =
32	1	32 * 1 = 32
33	3	33 * 3 = 99
38	2	38 * 2 = 76
61	1	61 * 1 = 61
$\Sigma w_i = n = 51$		Total = [2] _______ Mean = Total/n

$$\bar{x} = \frac{\Sigma x_i w_i}{\Sigma w_i},$$

where

x_i = the value of each observation, e.g., each age
w_i = the frequency, or count, in this case for each age,
Σw_i = sum of all the frequencies, or total count.

What is the mode? [3] _______ What is the median? [4] _______ What is the mean? [5] _______.

Sometimes data sets are summarized by grouping the data into categories. For example, if we were looking at the age distribution of the people who watch a certain television program, we would probably define age categories such as under 21, 21–30, 31–40, etc. We have grouped the age data from the statistics classes in Table 8.2 as an example. This example is not realistic, because the ages are so concentrated in the 20–24 age group, but we present it as an illustration of how to compute mean, mode, and median from categorized data.

TABLE 8.2 Categorized Ages (Intervals of Five)

Age Intervals	Midpoints x	Count, or Weight w	Midpoint * Weight xw
20–25⁻	22.5	34	22.5 * 34 = 765
25–30⁻	27.5	7	27.5 * 7 = 192.5
30–35⁻	32.5	7	32.5 * 7 =
35–40⁻	37.5	2	[3] ______ = ______
40–45⁻	[1] ____	0	
45–50⁻	[2] ____	0	
50–55⁻	52.5	0	
55–60⁻	57.5	0	
60–65⁻	62.5	1	62.5 * 1 = 62.5
		$n = 51$	Total = [4] ______
			Mean = Total/n

We will not get the same answers as we got using Table 8.1, because we lost information when we categorized the data. For this reason, when accuracy is important, you should always compute summary statistics and parameters from the original data, unless those data are unavailable.

In Table 8.2 the intervals are represented in the form $20\text{–}25^-$, etc. The superscript minus sign indicates that the interval runs up to 25 but does not include 25. Thus, a person whose twenty-fifth birthday was tomorrow would be included in this interval. We could have called this interval 20–24, but then it would have seemed illogical that the midpoint was 22.5.

What is the mean? [5] _______. What is the mode? [6] _______ What is the median? [7] _______

Let us compare the measures we calculated from Table 8.2 above with those obtained when we used the raw data in Table 8.1. The median and mode can be identified by the age interval only. There is a method (interpolation) for approximating them more precisely, but we will not present that method here.

Measure of Central Tendency	From Raw Data	From Categorized Data
Median	23	20–24
Mode	21	20–24
Mean	25.235	25.9

The means are fairly close. In fact, all three measures calculated from the categorized data gave a good approximation to the more exact measures of central tendency from the raw data. Calculations with categorized data are especially useful when a quick approximation to mean, median, or mode is sufficient. They can also be used when the original (raw) data are not available—when all we have are data presented in this categorized format. For maximum accuracy, however, the raw data should always be used, and thanks to computers and calculators, computing from the raw data is not much of a chore.

Computation of Mean for a Dichotomous Variable (Variable with Only Two Values)

A dichotomous variable is one that has only two values, e.g., yes or no. For example, the answers to the questions, "Are you vaccinated against smallpox?" and "What is your gender?" are dichotomous variables. We may arbitrarily assign 1 and 0 to the two values of such a variable. Does the mean for this variable give us any information? Yes, in fact it represents the **proportion** of responses to which we assigned the value 1. In computing the mean for a dichotomous variable with values 1 and 0, we use the same formula we used for the data with frequencies:

$$\bar{x} = \frac{\Sigma x_i w_i}{\Sigma w_i}.$$

Suppose the values represent the answers to the question "What is your gender?" The answers are coded, 1 = female and 0 = male. Study and complete the following table.

Names for the Value of the Dichotomous Variable x	Arbitrary Values Assigned to the Dichotomous Variable	Count, frequency, or weight, w	Value * Weight (xw)
Female	1	33	33
Male	0	18	0
		$n = (\Sigma w_i) =$ [1] ______	Total $(\Sigma x_i w_i) =$ [2] ______
			$\bar{x} = \frac{\Sigma x_i w_i}{\Sigma w_i} =$ [3] ______

When to Use Which Measure of Central Tendency

Mean, mode, and median tell us about the typical central values of the data. Which measure we choose as the typical value depends on the following considerations:

1. If the distribution is symmetrical, mean, median, and mode are identical, so we can choose any one of them as the typical central value.

2. If the distribution is not symmetrical, the mean is heavily influenced by the extreme values, but the median and mode are not. Which measure of central tendency is appropriate depends on the nature of the application. For example, the median would be used to characterize the variable family income if one wanted to minimize the effects of extremely large or extremely small incomes. If one wanted to emphasize these values, then the mean would be used.

Examples from the Literature

Literature Example 8.1 is from the article entitled "Decreased Access to Medical Care for Girls in Punjab, India: Roles of Age, Religion, and Distance."

What is the purpose of the study? 1 ______________________

Who was in the sample and where was the sample taken? 2 ______________________

What measure is used as the index of discrimination? 3 ______________________

Does the male/female ratio change as age increases? 4 ______________________

Which religion has the highest male/female ratio? 5 ______________________

Why do girls have decreased access to medical care? 6 ______________________

Does distance from the hospital have an effect on the ratio? 7 ______________________

Notice the footnote about median and mean ages for boys and girls.

Mean and median are quite different, which indicates that the age distribution is [8] symmetrical/skewed. Whether we consider median or mean, the boys tend to be [9] younger/older than the girls when they are hospitalized.

LITERATURE EXAMPLE 8.1

Decreased Access to Medical Care for Girls in Punjab, India: The Roles of Age, Religion, and Distance

Beverley E. Booth, MD, MPH, and Manorama Verma, MD, DCH, MB, BS

Risk factors that increase the likelihood of discrimination against girls in India have not been well studied. In this study of hospitalized children in Punjab, India, girls were less likely to be in the newborn or infant age groups, to be of the Sikh religion, or to come from far away than were boys. These differences suggest that these factors are significant risk factors for denied access to medical care for girls living in Punjab, India. (*Am J Public Health.* 1992;82:1155–1157)

Results

Of the 2595 patients, 506 (19.5%) were girls, resulting in a M/F ratio of 4.13 (Table 1). The M/F ratio was highest in the newborn age group, in Sikhs, and in those coming from farthest away. The M/F ratio fell rapidly over the first few months of life, continued to decline through early childhood, and then leveled off after 4 years of age (Figure 1).

Discussion

The magnitude of restricted access to hospital care for girls shown by our data is impressive: the M/F ratio of 4.13 suggests that about three out of four girls (75%) who are ill enough to require hospitalization are denied this essential medical care simply because of their sex. Although it might be argued that the higher proportion of boys being admitted to the hospital resulted from a higher proportion of serious illness among boys, this is unlikely, because in Punjab the infant and child mortality rates are higher for girls than for boys.[7]

This study demonstrates that for girls in Punjab, young age, increasing distance from the hospital, and Sikh religion decrease access to hospital care.

TABLE 1—Age, Religion, and Distance from Hospital for Children Admitted to Pediatric Ward, Christian Medical College, Ludhiana, 1987

	Girls[a] (n = 506) No. (%)	Boys[b] (n = 2089) No. (%)	Male/Female Ratio[c]
Age group			
Newborn (<1 month)	84 (16.6)	628 (24.2)	7.48
Infant (1 to 12 months)	158 (31.2)	731 (35.0)	4.63
Child (1 to 14 years)	262 (51.8)	726 (34.8)	2.77
Not known	2 (<0.1)	4 (<0.1)	4.00
Religion			
Sikh	188 (37.2)	1008 (48.3)	5.36
Hindu	305 (60.3)	1046 (50.1)	3.43
Other	13 (2.6)	35 (1.7)	2.69
Distance from hospital			
>80 km	52 (10.3)	403 (19.3)	7.75
60 to 80 km	44 (8.7)	285 (13.6)	6.48
<60 km	111 (21.9)	570 (27.3)	5.14
Ludhiana City	256 (50.6)	702 (33.6)	2.74
Not known/out of state	43 (8.5)	129 (6.2)	3.00

[a]Mean age was 30.1 months (SD = 37.6*); median age was 12 months.
[b]Mean age was 20.9 months (SD = 34.5); median age was 5 months.
[c]The overall male/female ratio was 4.13.
*Compared to boys, $P < .0001$ by Student's t test.

Literature Example 8.2 is from the article entitled "The Differential Effect of Traditional Risk Factors on Infant Birthweight among Blacks and Whites in Chicago."

Extreme values in the distribution affect the mean of a distribution. Is the median similarly affected?	[1] Yes/No
Note the census tract median family incomes of blacks and whites in the figure.	The percent of mothers with lower income is [2] higher/lower for blacks than for whites.
Read the information in Table 1.	The proportion of low birthweight for black and white infants is [3] the same/different for those with median income > $40,000.
Read the second paragraph of the results and check if the authors' assertion coincides with your impression from Table 1.	It [4] does/doesn't coincide.

LITERATURE EXAMPLE 8.2

The Differential Effect of Traditional Risk Factors on Infant Birthweight among Blacks and Whites in Chicago

JAMES W. COLLINS, JR., MD, MPH, AND RICHARD J. DAVID, MD

Abstract: We analyzed 103,072 White and Black births in Chicago from the 1982 and 1983 Illinois vital records, using 1980 median family income of mother's census tract as an ecologic variable. Thirty-one percent of Blacks and 4 percent of Whites resided in census tracts with median family incomes ≤$10,000/year. Only 2 percent of Black mothers, compared to 16 percent of White mothers, lived in areas where the median family income was greater than $25,000/year. Among Blacks with incomes ≤$10,000/year, maternal age, education, and marital status had minimal predictive power on the incidence of low birthweight (LBW) infants. Among high-risk mothers in the poorest areas the proportion of LBW infants in Blacks and Whites was less divergent than in higher income areas. Independent of residential area, low-risk Whites had half the occurrence of LBW infants as Blacks. We conclude that the extremes of residential environments show dramatic racial disparity in prevalence, yet the few low-risk Blacks still do less well than low-risk Whites. Traditional risk factors do not completely explain racial differences in neonatal outcome. (*Am J Public Health* 1990; 80:679–681.)

Results

There were 51,827 Black and 51,245 White births in Chicago during 1982 and 1983. Black neonatal mortality was twice as high as that of Whites (16/1,000 vs 7/1,000) and the LBW proportions were twice as high in Blacks (14 percent vs. 6 percent). Only 2 percent of Black mothers, compared to 16 percent of White mothers, resided in census tracts in which the median family income was greater than $25,000/year. Conversely, 31 percent of Blacks lived in census tracts in which the average household income was less than $10,000/year; only 4 percent of White mothers resided in such impoverished neighborhoods (Figure 1).

The risk of LBW infants among blacks remained essentially twice as high as that of Whites across all maternal income, education, and age groups. Although the higher Black risk appeared to be eliminated at very high incomes, the confidence intervals were broad because of the small number of Blacks (N = 200, or 0.4 percent) residing in census tracts with median family income over $40,000/year (Table 1).

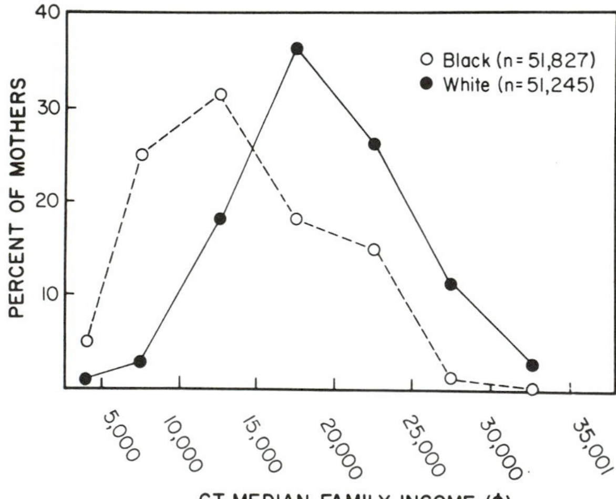

FIGURE 1—Census Tract Median Family Income Distribution of Black and White Mothers

TABLE 1—Proportion of Low Birthweight (<2500 grams) and Relative Risk in Black and White Infants according to Income, Maternal Education, and Age

Variables	% LBW Black	White	Black Relative Risk (95% CI)
Income (per year)			
≤$10,000	15	8	1.92 (1.64, 2.26)
$10,001–$20,000	14	6	2.12 (2.01, 2.24)
$20,001–$30,000	12	6	2.16 (1.99, 2.39)
$30,001–$40,000	10	5	2.26 (1.39, 3.54)
>$40,000	4	4	.98 (.13, 7.29)
Maternal Education (years)			
<12	16	7	2.40 (2.26, 2.55)
12	14	7	2.13 (1.93, 2.42)
13–15	12	5	2.14 (1.93, 2.42)
16	10	5	2.13 (1.76, 2.57)
≥17	9	5	1.87 (1.38, 2.53)
Maternal Age (years)			
≤19	14	8	1.44 (1.37, 1.55)
20–35	14	6	2.41 (2.30, 2.52)
>35	14	8	1.98 (1.64, 2.39)

Literature Example 8.3 is from the article entitled "Environmental Tobacco Smoke Exposure during Infancy."

Infants exposed to tobacco smoke are at greater risk for respiratory diseases. Urine cotinine is an index of smoke absorbed by the infant. What is the urine cotinine level that shows significant environmental tobacco smoke (ETS) absorption? (Read the horizontal line in the figure.)	[1] ________________
Extreme values of a variable influence the arithmetic mean; the median is not influenced by extreme values. In the study described in this article, the median was used to characterize urine cotinine measured in μg/L, perhaps because the authors were trying to present a conservative estimate of central tendency that was not influenced by some extremely large values they observed. What are the median urine cotinine values for the four categories?	No household exposure = [2] _______; others smoke, mother does not = [3] _______; only mother smokes = [4] _______; mother and others smoke = [5] _______
Write a sentence about the effect of tobacco smoking on the ETS absorption by the infant.	[6] ________________
Open circles indicate breastfed infants. Look at the last two categories, in which mothers smoke. Do breastfed infants absorb more ETS?	[7] Yes/No

LITERATURE EXAMPLE 8.3

Environmental Tobacco Smoke Exposure during Infancy

Barbara A. Chilmonczyk, MD, George J. Knight, PhD, Glenn E. Palomaki, BS, Andrea J. Pulkkinen, MS, Josephine Williams, and James E. Haddow, MD

We collected information about household smoking habits from 518 mothers when they made their first well child visit with a 6 to 8-week old infant. A urine sample was also collected from the infant, the cotinine concentration measured, and the measurement correlated with data provided by the mother. Eight percent of the infant urine cotinine values fell at or above 10 μg/L in the 305 households where no smoking was reported. Corresponding rates were 44 percent in the 96 households where a member other than the mother smoked, 91 percent in the 43 households where only the mother smoked, and 96 percent in the 74 households where both the mother and another household member smoked. In households where the mother smoked, infant urine cotinine levels were lower in the summer, and higher when the infant was breast-fed. A screening question about family smoking habits in conjunction with well child care could effectively define a group of infants exposed to environmental tobacco smoke and thus be at greater risk for respiratory diseases. (*Am J Public Health* 1990; 80:1205–1208.)

Results

Figure 1 contains a scatterplot distribution of urine cotinine levels. Among 305 infants who were not exposed to household tobacco smoke, cotinine levels from 97 (32 percent) fell below 1.0 μg/L, and another 185 (60 percent) fell between 1.0 and 9.9 μg/L. Urine cotinine values of 10 μg/L and above were found in only 23 of the infants (8 percent). Overall, the median urine cotinine level in this group was 1.6 μg/L.

Among 96 infants with ETS exposure from other household members only, the median level was 8.9 μg/L; among 43 infants with ETS exposure from the mother only, the median level was 28 μg/L; among 74 infants with ETS exposure from both the mother and other household members, the median level was 43 μg/L. Cotinine values falling at or above 10 μg/L in those categories were: 44, 91, and 96 percent, respectively. The horizontal line drawn at 10 μg/L in Figure 1 represents a reasonable estimate of significant ETS absorption; 94 percent of the infants whose mother smoked had values above this level, as opposed to only 8 percent of the infants in the unexposed category.

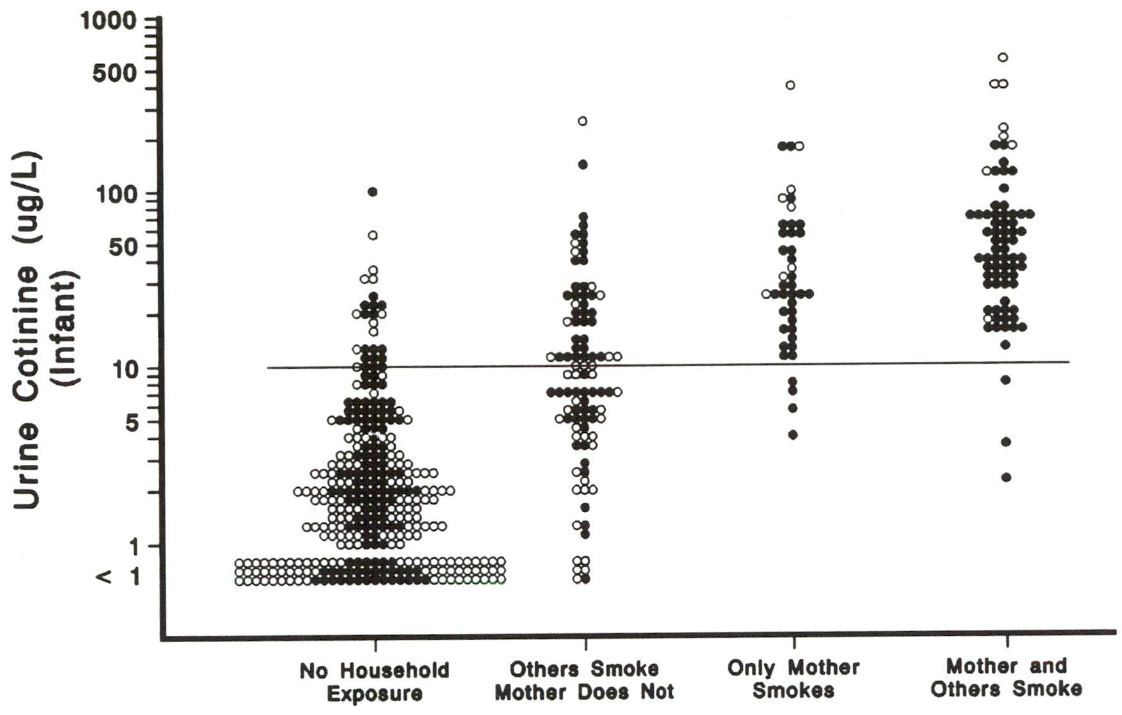

FIGURE 1—Urine Cotinine Levels in Infants, According to Environmental Tobacco Smoke Exposure in the Home.
Cotinine measurements are in spot urine samples. Closed circles indicate that the infant is not breast fed; open circles indicate breast feeding. Cotinine measurements < 1.0 μg/L are considered undetectable, and the horizontal line drawn at 10 μg/L is the cutoff defined from data in the study to indicate significant ETS absorption. Median urine levels in the four groups are as follows: no household exposure—1.6 μg/L; others smoke, mother does not—8.9 μg/L; only mother smokes—28 μg/L; mother and others smoke—43 μg/L.

Cum. Count*	Count	Stems Leaves
34	34	2 0000011111111111222223333333344444
41	7	2 5556689
48	7	3 0002333
50	2	3 88
50	0	4
50	0	4
50	0	5
50	0	5
51	1	6 1

*The count has been cumulated from the frequency of the lowest age, that is, from the top down.

FIGURE 8.1 Stem and Leaf Plot of Ages of the Statistics Students $n = 51$

Measures of Dispersion

Our interest in analyzing data is to learn the pattern of variation of a variable. Measures of central tendency give us a partial view. Measures of dispersion give us additional information about the distribution. Measures of dispersion fall into one of two categories, depending on whether they are measured relative to the mean or the median. The two serve different purposes and are worth exploring, especially at the analysis stage. When we communicate our results we probably will give only one of the measures of dispersion, either percentiles (quartiles, deciles, inter-quartile range, etc.) or standard deviation (or variance).

Measures of Dispersion Centered on the Median

The median is closely related to the concept of percentile. We saw in Chapter 5 that we can calculate percentiles from stem and leaf plots. Here we will review the process briefly, as these concepts are crucial to the understanding of the measures of dispersion centered around the median. First consider Figure 8.1. The sample size is an odd number. The median is the observation at position number $[(n + 1)/2] = 26$. Circle the median (the twenty-sixth) observation in the stem and leaf plot. Median = [1] ________

For p, a number between 0 and 100, the pth percentile is that observation (datum) such that p percent of the observations are less than or equal to that value. When there is no such observation, the closest one is chosen as the pth percentile. Consider the following table.

Percentile	No. of Observation	Observation
10	5	20
20	10	21
25	13	21
40	20	22
50	25	[2] _______
60	30	24
75	38	26
80	41	[3] _______

Note that the numbers in the middle column are rounded to the nearest integer. Note also that 10 percent of 51 ≈ 5.

The inter-quartile range is the difference between observations in the third quartile (seventy-fifth percentile) and the first quartile (twenty-fifth percentile), i.e., Q3 − Q1. The inter-quartile range for the data in Figure 8.1 is [4] _______.

Quintiles are the observations located at the twentieth, fortieth, sixtieth, and eightieth percentiles and are called first, second, third, and fourth quintiles, respectively. For Figure 8.1, the first quintile = 21, the second quintile = 22, the third quintile = [5] _______, and the fourth quintile = 29.

Dispersion is usually measured from the median by giving the five-number summary (see Chapter 6): the minimum, first quartile, median, third quartile, and the maximum. It can be graphically displayed using a box plot, a method that is ideal for comparing the centers (medians) and dispersions of several variables.

Measures of Dispersion Centered on the Mean

Variance and Standard Deviation Neither median nor mode takes into account all values of the variable. The median depends heavily on frequencies, and the mode represents only the most frequently occurring observation. Computation of the mean takes into account all the observations and their frequencies. We discussed the use of percentiles as indicators of dispersion; they give values below which a specified percentage of observations are located. It would seem that another reasonable way to measure dispersion would be to figure the difference between each observation (x_i) and the mean ($\bar{x}$) and average them:

$$\Sigma(x_i - \bar{x})/n.$$

Consider the data set $x_i = 5, 4, 6, 5$.

The mean = [1] _______. If we retain algebraic signs and compute $\Sigma(x_i - \bar{x})/n$, we get [2] _______. This turns out to be useless as a measure of dispersion because the value is always zero.

Suppose we do not retain algebraic signs when we compute, i.e., we average the absolute values.

$\Sigma|(x_i - \bar{x})|/n =$ [3] _______. Throwing away the signs makes this a viable measure of dispersion. It is called the **mean deviation.**

$\Sigma(x_i - \bar{x})^2 =$ [4] _______.

Now calculate $\text{SSq} = \Sigma(x_i - \bar{x})^2$, where SSq stands for sum of squared deviations.

Squaring is another way to make all the signs positive, and it has the useful property of giving greater weight to observations [5] farther from/closer to the mean. This is desirable because it accentuates how bunched the observations are around the mean. When the x_i are closely bunched around the mean, we get a much smaller SSq than when they are farther away. Similarly, if we include outliers in the calculations, we will have a [6] larger/smaller SSq than if we discard the outliers.

SSq itself is not used as a measure of dispersion because it is so dependent on sample size. The measures that are used average SSq over the sample by dividing by n. The first of these, the mean sum of squares, or variance, is given by the formula

$$\text{MS} = s^2 = \Sigma(x_i - \bar{x})^2/n - 1$$

For our small data sets, $s^2 =$ [7] _______. We took the mean of the SSq by dividing by $(n - 1)$, which gives the mean sum of squares (MS), or [8] _______. In practice, for large sample sizes it [9] does/does not make much difference whether we divide by n or $n - 1$.

The square root of MS (or variance) is called standard deviation,

$$s = \sqrt{s^2}$$

The variance of the ages of the students listed at the beginning of this chapter is 47.06. What is the standard deviation? [10] _______. Notice that when we square the deviations of the ages we change the units of the scale. But what does the square of an age mean?

There is no physical meaning for square years. The variance is an index of the pattern of variation of a variable, but its units are not the same as the units of the variable. Taking the square root of the variance brings us back to the original units of the ages (years). Variance has some technical uses in statistical analysis, but standard deviation seems to us to be a more intuitively satisfying measure of dispersion.

One of the uses of standard deviation is in conversion of raw data into standardized z-scores, using the formula

$$z_i = \frac{(x_i - \mu)}{\sigma}$$

Convert ages 20, 25, and 38 into z-scores, given $\sigma = 6.7$ and $\mu = 25.2$. (Here we assume the parameters population mean and population standard deviation are known.)

Age	***z*-score**
20	(20 − 25.2)/6.7 = −.78
25	(25 − 25.2)/6.7 = [11] ______
38	[12] ______

We mentioned that SSq gives more weight to values farther away from the mean because it squares the difference between each value and the mean. Let us see what happens to mean, variance, and standard deviation when we discard the outliers. Consider the information in Table 8.3.

TABLE 8.3 Mean and Variance of Ages of Statistics Students Before and After Discarding the Outlier

	n	Mean	Variance	St Dev
All data values included	51	25.235	47.06	6.866
Outlier discarded	50	24.520	21.48	4.635
Difference		0.715	25.58	

Removing the outlier has a [1] large/small influence on the mean, whereas it has a [2] large/small influence on the variance. This illustrates how the variance weights the values away from the mean.

Examples from the Literature

Literature Example 8.4 is from the article entitled "Does Maternal Tobacco Smoking Modify the Effect of Alcohol on Fetal Growth?"

Birthweight of infants is associated with perinatal infant mortality. How do mothers' smoking and drinking habits influence birthweight?

Mean birthweight of infants born to mothers who neither smoked nor drank = [1] _______

Mean birthweight of infants born to mothers who were heavy smokers (15+ cigarettes per day) and heavy drinkers (120+ grams of alcohol per week) = [2] _______

The difference between these mean birthweights = [3] _______

How does alcohol affect birthweight when the mother is a non-smoker?

The mean birthweight for infants whose mothers did not smoke but did consume 120+ grams of alcohol per week = [4] _______

The difference between this weight and the mean birthweight of infants whose mothers neither smoked nor drank = [5] _______

How does smoking affect birthweight when the mother does not drink?	The mean birthweight of infants whose mothers did not drink but did smoke 15+ cigarettes per day = [6] _______ The difference between this weight and the mean birthweight of infants whose mothers neither smoked nor drank = [7] _______
According to the article, the decrease in birthweight associated with alcohol use is greater among smokers than among non-smokers.	The table shows that the decrease in mean birthweight associated with alcohol use among non-smokers = [8] _______ The decrease in mean birthweight associated with alcohol use among heavy smokers = [9] _______ The difference between these two measures = [10] _______
Which influences infant birthweight more?	[11] Heavy smoking/Heavy alcohol use

LITERATURE EXAMPLE 8.4

Does Maternal Tobacco Smoking Modify the Effect of Alcohol on Fetal Growth?

Jørn Olsen, MD, PhD, Altamiro da Costa Pereira, MD, and Sjurdur F. Olsen, MD

Smoking and drinking habits were registered by a self-administered questionnaire in 36th week of gestation in 11,698 pregnant women, more than 80 percent of all such women in two Danish cities 1984–87. Alcohol consumption of 120 g/week or more was associated with a greater reduction in the average birthweight in the babies of smokers than of non-smokers (about 40 grams for the non-smokers and about 200 grams for the smokers). This is particularly striking considering that the average birthweight for smokers is lower than for non-smokers. A birthweight difference of more than 500 grams was found between babies of mothers who neither smoked nor drank and mothers who smoked and drank heavily. Our data suggest that women's smoking habits should be taken into consideration when giving pregnant women advice about drinking. (*Am J Public Health* 1991; 81:69–73)

Most newborn measures decrease with increasing cigarette and alcohol intake. Among smokers the decrease associated with alcohol is greater than among non-smokers (see Table 3).

TABLE 3—Means of Placental Weight, Birthweight, Birth Length and Head Circumference According to the Mother's Alcohol Consumption during Pregnancy, Stratified by Smoking Status

Smoking Status (Daily)	Alcohol in g/Week	Placental Weight in g	Birthweight in g	Birth Length in cm	Head Circumference in cm
Non-smokers	0	569	3,572	52.6	35.4
	1–	561	3,579	52.6	35.4
	30–	557	3,602	52.8	35.5
	60–	537	3,582	52.6	35.5
	90–	532	3,554	52.8	35.4
	120+	534	3,536	52.4	35.5
Subtotal	Mean	560	3,580	52.6	35.4
	(Stddev)	(132)	(483)	(2.3)	(1.5)
1–4 cigarettes	0	567	3,487	52.2	35.1
	1–	557	3,493	52.3	35.3
	30–	572	3,479	52.0	35.4
	60–	564	3,573	52.6	35.3
	90–	626	3,727	53.4	35.6
	120+	450	3,141	51.3	34.7
Subtotal	Mean	560	3,492	52.2	35.3
	(Stddev)	(135)	(510)	(2.5)	(1.6)
5–9 cigarettes	0	555	3,330	51.6	35.0
	1–	548	3,321	51.5	35.0
	30–	528	3,261	51.5	34.9
	60–	522	3,379	51.9	34.8
	90–	524	3,444	52.1	35.5
	120+	489	3,144	50.6	34.6
Subtotal	Mean	544	3,316	51.5	35.0
	(Stddev)	(121)	(458)	(2.2)	(1.6)
10–14 cigarettes	0	572	3,357	51.5	35.0
	1–	548	3,282	51.4	34.9
	30–	535	3,259	51.4	34.8
	60–	528	3,291	51.4	34.7
	90–	540	3,123	50.5	34.4
	120+	493	3,116	50.9	34.4
Subtotal	Mean	549	3,289	51.4	34.9
	(Stddev)	(132)	(467)	(2.2)	(1.5)
15+ cigarettes	0	546	3,270	51.4	34.9
	1–	561	3,266	51.3	34.8
	30–	531	3,207	51.3	34.6
	60–	524	3,123	50.8	34.6
	90–	539	3,169	50.8	34.8
	120+	472	3,023	50.8	34.0
Subtotal	Mean	548	3,239	51.2	34.7
	(Stddev)	(128)	(459)	(2.3)	(1.5)
Total	Mean	556	3,478	52.2	35.2
	(Stddev)	(132)	(497)	(2.4)	(1.6)

Literature Example 8.5 is from the article entitled "Transdermal Nicotine for Smoking Cessation." The names of various demographic variables that might have influenced the experimental results are given in the first column of the table. The means and standard deviations for these variables for each group are then given. Note that these descriptions pertain to the participants before the treatment. The purpose of this table is to show that all the groups were similar relative to these variables.

1. What are the maximum and minimum mean ages of the groups? _______, _______

2. Compute the standard error of the mean age for all subjects.

$$SE = \frac{s}{\sqrt{n}} = \_\_\_\_\_\_\_$$

   This number gives us the approximate spread in age of participants in samples of size 935 chosen from the same population that this sample was chosen from. Approximately 68 percent of such samples would have a mean age between 42.6 − SE and 42.6 + SE.

3. Approximately how many standard deviations are contained in the age range for the placebo group? (Range/SD) _______

4. Look at the means for the placebo and 21 mg transdermal nicotine groups. What do you notice about the following variables: cigarettes per day; baseline carbon monoxide?

   ____

   ____

   The last column of the table tells us that the differences between the groups are not statistically significant. This means that as far as these demographic variables are concerned these groups are similar.

LITERATURE EXAMPLE 8.5

Transdermal nicotine for smoking cessation

Transdermal Nicotine Study Group

Journal of the American Medical Association, 1991, Vol. 266, pp. 3133–3138

Objective

To evaluate the efficacy of a new transdermal nicotine system for smoking cessation.

Design

Two 6-week, randomized, double-blind, placebo-controlled, parallel group trials were conducted. Successful abstainers from both trials enrolled in a third trial for blinded downtitration from medications (6 weeks) and subsequent off-drug follow-up (12 weeks).

Setting

Nine outpatient clinics specializing in the treatment of smoking cessation.

Patients

Healthy volunteers who smoked one or more packs of cigarettes daily and wanted to participate in a smoking cessation program.

Intervention

Patients were randomly assigned to a transdermal nicotine system delivering nicotine at rates of 21, 14, or 7 mg (in trial 1 only) over 24 hours or to placebo. Group counseling sessions were provided to all participants.

Main outcome measure

Rates of continuous smoking abstinence were determined during 6 weeks of full-dose treatment, a 6-week weaning period (through week 12), and a 3-month follow-up receiving no therapy (through week 24). Abstinence was defined by patient diary reports of no smoking during the designated periods, confirmed by expired-breath carbon monoxide levels of 8 ppm or lower.

Results

The centers enrolled 935 patients. Cessation rates during the last 4 weeks of the two 6-week trials (pooled data) were 61%, 48%, and 27% for 21-

and 14-mg transdermal nicotine and placebo, respectively ($P \leq .001$ for each active treatment vs placebo). Six-month abstinence rates for 21-mg transdermal nicotine and placebo were 26% and 12%, respectively ($P \leq .001$). All transdermal nicotine doses significantly decreased the severity of nicotine withdrawal symptoms and significantly reduced cigarette use by patients who did not stop smoking. Compliance was excellent, and no serious systemic adverse effects were reported.

Conclusions

Transdermal nicotine systems show considerable promise as an aid to smoking cessation. (*JAMA*. 1991;266:3133–3138)

Mean Demographic Results for the Pooled Database (Trials 1 and 2)

	Treatment Group					
	Transdermal Nicotine					
	21 mg/d (*n* = 262)	**14 mg/d (*n* = 275)**	**7 mg/d (*n* = 127)**	**Placebo (*n* = 271)**	**All Subjects (*n* = 935)**	**Treatment Difference***
Women, No. (%)	156 (60)	163 (59)	73 (57)	172 (63)	564 (60)	NS
Age, y						
Mean ± SD	43.1 ± 10.4	42.5 ± 10.6	40.7 ± 9.8	43.2 ± 9.9	42.6 ± 10.3	
Range	22–66	22–68	23–65	21–65	21–68	NS
Weight, kg						
Mean ± SD	73.1 ± 16.3	72.9 ± 16.0	71.3 ± 15.2	71.9 ± 15.7	72.4 ± 15.9	
Range	43–155	43–127	45–131	44–122	43–155	NS
Fagerstrom score						
Mean ± SD	7.2 ± 1.7	7.0 ± 1.7	7.2 ± 1.7	7.1 ± 1.7	7.1 ± 1.7	
Range	2–11	2–11	3–11	2–11	2–11	NS
Time smoking, y						
Mean ± SD	24.9 ± 10.4	24.0 ± 10.4	22.3 ± 9.1	24.2 ± 9.9	24.1 ± 10.1	
Range	1–50	1–50	6–48	2–55	1–55	NS
Cigarettes per day						
Mean ± SD	31.1 ± 10.5	31.0 ± 10.3	29.8 ± 8.4	30.5 ± 10.6	30.7 ± 10.2	
Range	12–90	15–80	20–50	20–80	12–90	NS
No. of previous quit attempts						
Mean ± SD	4.4 ± 4.2	4.1 ± 4.4	4.1 ± 5.4	3.8 ± 3.3	4.1 ± 4.2	
Range	1–30	1–40	1–50	1–20	1–50	NS
Baseline carbon monoxide, ppm						
Mean ± SD	35.0 ± 14.6	33.8 ± 13.3	34.9 ± 12.3	37.0 ± 14.2	35.2 ± 13.8	
Range	10–95	5–74	13–68	5–89	5–95	NS

*NS indicates not significant

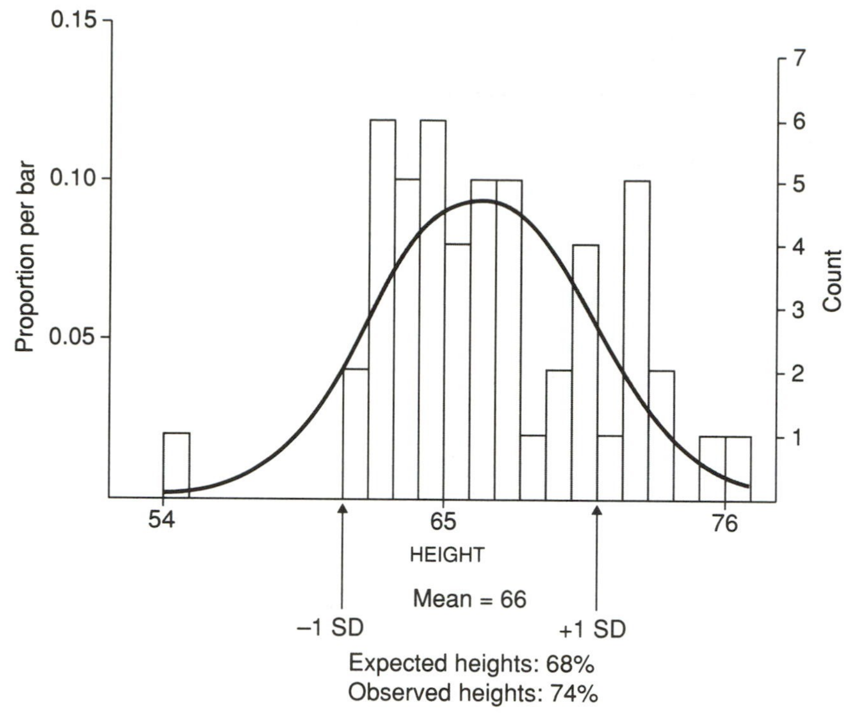

Figure 8.2 Distribution of Heights of Students in Two Statistics Classes

A Useful Property of Standard Deviation

Many large populations have approximately symmetric, bell-shaped distributions with the following characteristics:

1. approximately 68 percent of the observations fall within one standard deviation of the mean;
2. approximately 95 percent of the observations fall within two standard deviations of the mean;
3. almost all the observations fall within three standard deviations of the mean.

Consider Figure 8.2, which represents the distribution of heights of students in two statistics classes. The bars in the histogram represent the actual numbers of cases in each interval. The theoretical distribution (normal curve) superimposed on the histogram uses a mean of 66 inches and a standard deviation approximately 4 inches. The arrows point to heights that are one standard deviation below the mean and one standard deviation above the mean. If our data fit the normal distribution exactly (which they obviously don't), we would expect 68 percent of the data to be between these values.

Figure 8.3 presents the same data as a cumulative distribution. This figure is similar to a histogram, except that the height of the bars is the cumulative count rather than the absolute count of each interval. The theoretical curve

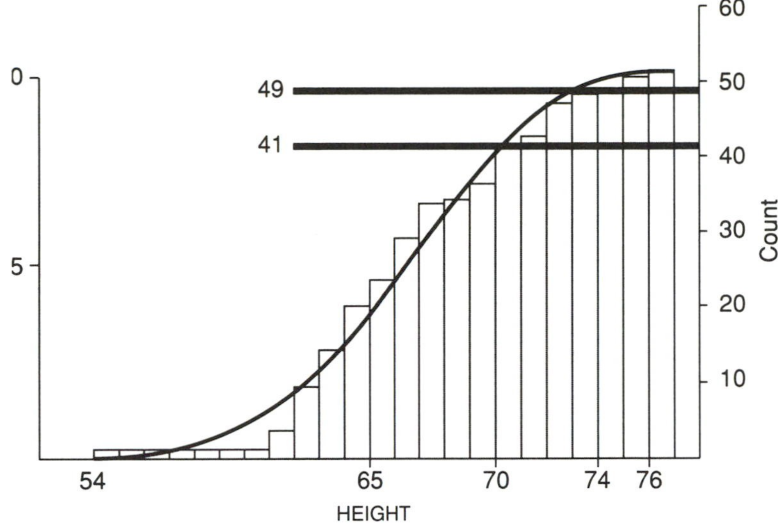

FIGURE 8.3 Cumulative Distribution of Heights of Students in Two Statistics Classes

superimposed on the bars represents the cumulative normal distribution. Use Figures 8.2 and 8.3 to answer the following questions.

	Heights (inches)	Comments
No. of Cases (n)	51	Total number in the sample
Minimum	54.0	Look at Figure 8.2. There are [1] no x_i/some x_i below 54.
Maximum	76.0	[2] _______ of the x_i are ≤ 76. That means that 100 percent of the data are 76 or smaller. As a proportion, it's 51/51 = 1.
Mean	$66.4 \approx 66$	Shade the bars that contain the mean in Figures 8.2 and 8.3.
Variance	18.6	Variance is needed to compute standard deviation.
Standard deviation	$4.3 \approx 4$	Extremely useful statistic

	Heights (inches)	Comments
Mean + 1 * SD	66 + 4 ≈ [3] _______	Shade the bar that represents 70. The count in the cumulative distribution (Figure 8.3) tells how many x_i are below 70: [4] _______
Mean + 2 * SD	66 + 8 ≈ [5] _______	Shade the bar on Figure 8.2 that represents 74. The count in the cumulative distribution (Figure 8.3) tells how many x_i are below 74: [6] _______
Mean + 3 * SD	[7] _______________ ≈ _______	The count in the cumulative distribution (Figure 8.3) tells how many x_i are below 78: [8] _______
Mean ± 1 * SD	66 ± 4 ≈ 62 to 70	We would expect that the theoretical distribution (the smooth curve in Figure 8.2) would have ≈ [9] _______ percent of the x_i between 62 and 70. Our sample actually has [10] _______ percent in this range.
Mean ± 2 * SD	66 ± 8 ≈ [11] _______	We would expect that the theoretical distribution (the smooth curve in Figure 8.2) would have ≈ 95 percent of the x_i between 58 and 74. Our sample actually has 94 percent in that range.
Mean ± 3 * SD	66 ± 12 ≈ [12] _______	[13] _______ percent of the x_i are actually between 54 and 78.

Example from the Literature

Literature Example 8.6 is from the article entitled "Birth Weight and Perinatal Mortality: The Effect of Gestational Age." The mean, $\bar{x}$, and standard deviation, s, of the birth weights of the babies were computed. Using these quantities, the z scores of the birth weights were computed:

$$z = \frac{x - \bar{x}}{s}$$

A z score of -4, for instance, implies that the baby's birth weight was four standard deviations below the mean birth weight. Mortality rates were graphed against the gestational age of the infants for various z scores. Locate the line representing scores of $z = -4$ in Figure 4.

What is the gestational age at the lowest mortality point?	[1] ______________________
Read the authors' interpretation of Figure 4. Are the authors' observations true for the $z = -2$ and $z = -3$ curves?	[2] Yes/No
Comment on the influence of birth weight on infant mortality.	[3] ______________________ ______________________
Notice the use of standard deviation in this figure.	One of the uses of standard deviation is in the computation of the [4] _______ score.

LITERATURE EXAMPLE 8.6

Birth weight and perinatal mortality: The effect of gestational age

Allen J. Wilcox, MD, PhD, and Rolv Skjoerven, PhD

American Journal of Public Health, 1992, Vol. 82, pp. 378–382

BACKGROUND

The strong association between birth weight and perinatal mortality is due both to gestational age and to factors unrelated to gestational age. Conventional analysis obscures these separate contributions to perinatal mortality, and overemphasizes the role of birth weight. An alternative approach is used here to separate gestational age from other factors.

METHODS

Data are from 400 000 singleton births in the Norwegian Medical Birth Registry. The method of Wilcox and Russell is used to distinguish the contributions to perinatal mortality made by gestational age and by relative birth weight at each gestational age.

RESULTS

Gestational age is a powerful predictor of birth weight and perinatal survival. After these effects of gestational age are controlled for, relative birth weight retains a strong association with survival.

CONCLUSIONS

Current public health policies in the United States emphasize the prevention of low birth weight. The present analysis suggests that the prevention of early delivery would benefit babies of all birth weights. (*Am J Public Health*. 1992;82:378–382)

The figure inverts the display to show perinatal mortality risk by gestational age for babies at given *relative* birth weights. Thus, the top curve shows mortality rates for that high-risk group of babies who are four standard deviations below the mean weight for their gestational age. The next-to-bottom curve shows the mortality experience of babies who are at the mean weight for their gestational age. Within each birth weight group there is a strong gradient of mortality risk with gestational age, with the lowest risk occurring at 40 weeks.

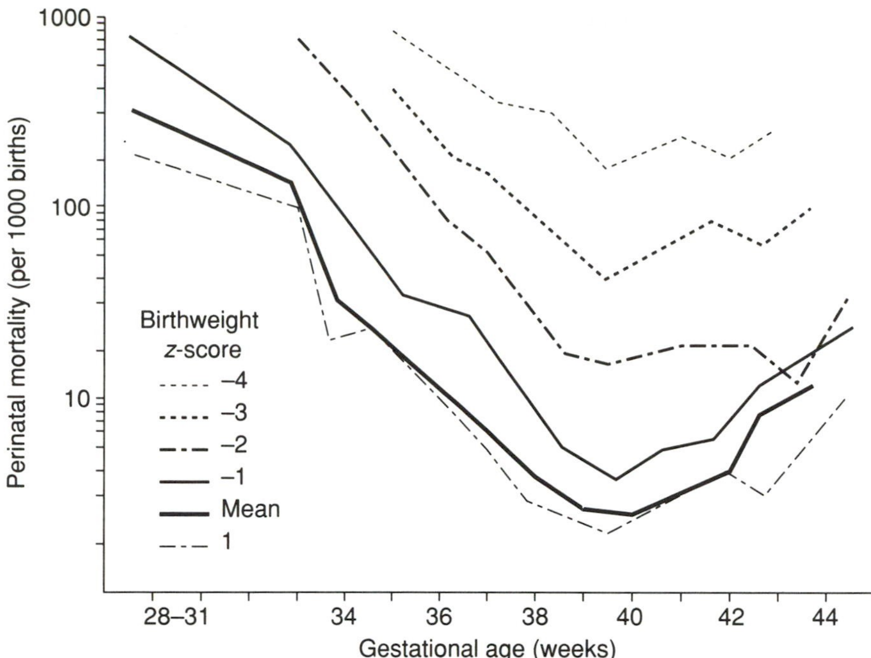

Gestational-Age–Specific Mortality Rates for Babies Stratified by Relative Size (in standard deviations from the Gaussian mean)

Variance and Standard Deviation for a Dichotomous Population (Proportions)

In this section we consider a variable that has only two possible values (such as SEX). Variables of this kind often occur in research in which the experimental procedure either succeeds or fails (for example, when the subject either recovers or does not). For this reason, the two values are often labeled "success" and "failure." When we work with such a variable, frequently we are interested in the proportion of successes in the population. In many cases, the value of the variable that we label "success" is rather arbitrary. For example, for the variable SEX, no value judgment is implied when we choose the value female as "success."

Begin by sorting the Student Questionnaire Data (SQD) cards into two groups, males and females. The proportion of females is [1] _______. This is the proportion when all fifty-one students are considered. If we were to take a sample of twenty from the deck, would we get the same proportion of females? When we did this, we got nine males and eleven females, so our proportion of females was .55. Try it; you will probably get a different proportion.

How is the dispersion of a proportion measured? The variance of a sample proportion is given by the following formula:

$$\text{Proportion variance} = \frac{p(1-p)}{n}$$

where p is the true proportion of individuals in the population having the property, and n is the sample size.

For the proportion of females in samples of size twenty from the SQD, the variance is $(0.65 * 0.35)/20 = .0114$. The standard deviation of the proportion is the square root of the variance, in this case approximately .11. This means that the proportion of females in approximately 68 percent of the samples of size twenty from the SQD would be between $.65 - .11 = .54$ and $.65 + .11 = .76$.

Range as a Measure of Dispersion

The range is the difference between the maximum and minimum, so range is influenced by extreme values of the variable. The range of ages in the Student Questionnaire Data (SQD) is $61 - 20 =$ [2] _______. If the outlier 61 is discarded, the range will be $38 - 20 =$ [3] _______. In small samples, range is a reasonable measure of dispersion. There are tables to estimate population variance from ranges.

Comparison of Different Standard Deviations: The Coefficient of Variation

Suppose we recorded the height data from the student questionnaire both in centimeters and in inches. The size of the numbers would depend on the units of measurement. The heights are the same, though the units differ.

How do we compare different standard deviations? Degree of variability should be an intrinsic property of a data set and should be independent of the

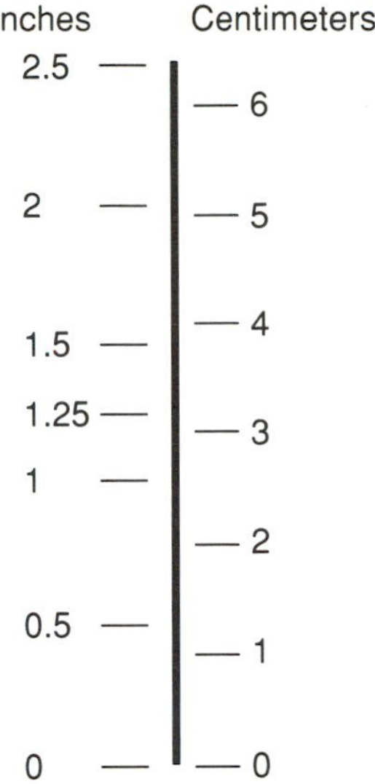

FIGURE 8.4 Comparison of Inches and Centimeters

units in which the variable is measured. The coefficient of variation is a measure that converts standard deviation to a form that is independent of the units of measurement. **Coefficient of variation** is the standard deviation (SD) divided by the mean given as a percent:

$$\text{Coefficient of variation} = 100 * (\text{SD}/\text{mean})$$

Coefficient of Variation =

$$CV = \left\{\frac{s}{\bar{x}}\right\} 100$$

Consider these data:

Height in inches: Mean 66.392
SD 4.313

Height in centimeters: Mean 168.64
SD 10.96

We know that the variation of both must be the same, because these are the heights of the same people. For height in inches, CV = [1] _______;

for height in centimeters, CV = [2] _______.

The coefficient of variation can be used to compare variation in dissimilar variables such as height and weight because it is independent of units.

Is there more variation in weights or in heights? For height in inches, CV = 6.5%. For weight in pounds, CV = [3] _______.

Height in inches: Mean 66.392
SD 4.313
Weight in pounds: Mean 140.45
SD 33.70

The coefficient of variation for weights is [4] smaller/larger than the CV for heights.

Next consider data on pulse rates:
Pulse before running: Mean 75.88
SD 11.30
Pulse after running: Mean 98.00
SD 17.3

Is there more variation in pulse rate after running than before running? For pulse before, CV = [5] _______; for pulse after, CV = [6] _______. Thus, the coefficient of variation for pulse after is [7] smaller/larger than the CV for pulse before.

Examples from the Literature

Literature Example 8.7 is from the article entitled "Performance of the Reflotron in Massachusetts' Model System for Blood Cholesterol Screening Program."

We often assume that measurements of cholesterol levels are accurate. However, the accuracy is dependent upon the skill of the operators and the instruments they use. QC refers to the three levels of quality control material which were analyzed at the beginning of each screening and after every tenth patient sample.

[1] _______ levels of quality control (QC) and [2] _______ Reflotrons were used in this study. The sample sizes in this study range from [3] _______ to [4] _______.

"None of the four analyzers met the 1992 standard for precision, although two met the 1992 standards for accuracy." Precision is measured by the index of variation and accuracy by the mean value as

Table 1 refers only to [5] _______ and not to [6] _______. Compare the CV values shown under Solomon Park High. Which Reflotron would be considered the most precise?
[7] _______

it relates to the mean values from the reference laboratory.

Rank the Reflotrons under Solomon Park Medium according to precision:

Most precise: Machine 2, CV = 2.6

Second most precise: Machine
8 _______, CV = _______

Third most precise: Machine
9 _______, CV = _______

Least precise: Machine 10 _______, CV = _______

Why did the researchers return the Reflotrons and cancel the screenings?

11 ________________________________

____

LITERATURE EXAMPLE 8.7

Performance of the Reflotron in Massachusetts' Model System for Blood Cholesterol Screening Program

Stephen Havas, MD, MPH, MS, Robert Bishop, MS, Lauren Koumjian, MSc, Joel Reisman, MS, and Susan Wozenski, MPH, JD

The precision, accuracy, and durability of the Reflotron were evaluated by the Massachusetts Model Systems for Blood Cholesterol Screening Program. Screenings were conducted in diverse community settings over 16 months. Fingerstick samples from 10 428 individuals were tested. None of the four analyzers met the 1992 standards for precision, although two met the 1992 standards for accuracy. More than 40% of Reflotron values differed from the reference laboratory values by upwards of 5%. More than 16% of individuals were misclassified in terms of their risk category. All four instruments malfunctioned during the project. (*Am J Public Health.* 1992;82:458–461)

Results

For Reflotrons 1, 3, and 4, the weighted average of all values met the current precision goal set by the National Cholesterol Education Program's Laboratory Standardization Panel (LSP) for all three QC levels, but it fell short of the 1992 LSP standards for all levels.[3] For Reflotron 2, the weighted average of the values met the current LSP standards for all QC levels, but the 1992 standards were attained for only two levels (see Table 1).

An unanticipated finding was the malfunctioning of the analyzers. Reflotron 2 had to be returned to the manufacturer when its QC results consistently fell out of range. Subsequently, Reflotron 3 had to be returned for similar reasons. On the last two scheduled days for the project, neither Reflotron 1 nor Reflotron 4 analyzed the QC materials within acceptable ranges, resulting in cancellation of the screenings.

TABLE 1—Precision of the Reflotrons, by Percent

Reflotron	Solomon Park High CV (n)	Solomon Park Medium CV (n)	Solomon Park Low CV (n)
1	3.5 (243)	3.1 (232)	3.5 (218)
2	2.6 (116)	2.6 (84)	3.5 (87)
3	4.4 (24)	3.8 (23)	4.7 (22)
4	4.4 (77)	3.5 (61)	3.7 (51)

Literature Example 8.8 is from the article entitled "Weight Gain Prevention and Smoking Cessation: Cautionary Findings."

For which variables were means and standard deviations given? What were the three groups?

1 ________, 2 ________,
3 ________, 4 ________,
5 ________, 6 ________,
7 ________

Are the mean ages for these three groups different or the same?

8 Different/The same

Consider pretreatment body weights of the three groups. The means are somewhat different. Are the standard deviations just a reflection of the differences in the means? In other words, is the variation similar in each group?

Compute the coefficients of variation for each group and compare them:

CV for innovative condition
9 ________

CV for nonspecific condition
10 ________

CV for standard treatment condition
11 ________

Are the proportions of males and females in each group similar? If they are different, would the sex composition of the groups influence the study?

12 ________________________________

Mean weights are not separated by sex. Is this reflected in the weight differences between the groups? Note that there is a higher percentage of males in the nonspecific condition. What is your opinion about the need to separate data by sex?

13 ________________________________

Compute the difference in mean weights (Post – Pre) for each group.

Innovative condition: 14 _______

Nonspecific condition: 15 _______

Standard treatment condition: 16 _______

Read the conclusions and write a sentence about the importance of computing means and standard deviations.

17 ________________________________

____

LITERATURE EXAMPLE 8.8

Weight Gain Prevention and Smoking Cessation: Cautionary Findings

Sharon M. Hall, PhD, Chrystal D. Tunstall, PhD, Katharine L. Vila, MSW, and Joanne Duffy, PhD

Objectives. Weight gain is a consistent sequela of smoking cessation. A successful intervention might attract smokers who fear weight gain. If the gain causes smoking relapse, such an intervention might reduce smoking relapse risk.

Methods. Using a sample of 158 smokers who completed a 2-week smoking treatment program, we compared an innovative weight gain prevention intervention with both a nonspecific treatment and standard treatment. Subjects were assessed on weight and smoking behavior and followed for 1 year.

Results. A disturbing, unexpected finding was that subjects in both the innovative and nonspecific conditions had a higher risk of smoking relapse than did standard treatment subjects. Some differences were observed between abstinent and smoking subjects in weight gain by treatment condition.

Conclusions. Both active interventions may have been so complicated that they detracted from nonsmoking. Also, caloric restriction may increase the reinforcing value of nicotine, a psychoactive drug, thereby increasing smoking relapse risk. The magnitude of weight gain after smoking cessation may not merit interventions that increase smoking risk. Perhaps attitudinal modifications are the most appropriate. (*Am J Public Health.* 1992;82:799–803)

TABLE 1—Demographic and Descriptive Data by Treatment Condition

Variable	Innovative Condition (n = 53)	Nonspecific Condition (n = 51)	Standard Treatment Condition (n = 54)
Age, y			
Mean	40.68	41.22	39.24
SD	8.71	8.96	8.92
Cigarettes per day			
Mean	25.65	24.91	31.59
SD	12.56	10.80	11.74
Pretreatment body weight, kg			
Mean	72.68	69.23	67.65
SD	17.72	12.38	12.51
Postcessation treatment body weight, kg			
Mean	73.82	69.98	67.97
SD	18.12	10.48	12.37
Gender			
Male	12 (23%)	17 (33%)	14 (26%)
Female	41 (77%)	34 (67%)	40 (74%)
Ethnicity			
Caucasian	48 (91%)	45 (88%)	50 (91%)
Other	5 (9%)	6 (12%)	4 (9%)

Treatment conditions were compared on the same demographic, smoking, and weight history variables used to compare treatment completors and dropouts, as well as on postcessation body weight and postcessation cigarette smoking rates (see Table 1). Only two differences emerged. First, conditions differed on mean number of cigarettes reported smoked per day during the pretreatment week ($F[2,155] = 5.06$, $P < .01$). Subjects in standard treatment smoked significantly more cigarettes than did those in the other two conditions. Second, the three conditions had significantly different variances on pretreatment body weight ($P < .01$). Subjects in the innovative condition had a significantly greater variance in both these measures than did subjects in the nonspecific condition. In tests of the hypotheses, these variables—pretreatment smoking rate and pretreatment body weight—were used as covariates in preliminary analyses if they correlated with the dependent variable, and they were dropped if they did not explain significant proportions of variance or have significant interactions with independent variables. Due to a lack of significant main effects, or interactions, these variables were ultimately dropped from all analyses, except those of abstinence status. No differences due to therapists were found when the main effects of therapist on smoking rate or body weight were examined.

Literature Example 8.9 is from the article entitled "Field Comparison of Several Commercially Available Radon Detectors." The required precision and accuracy for radon detectors has been published by the Environmental Protection Agency.

We want to know whether the detectors tested by the authors meet the standards. To find out, we will need to know the EPA precision standards for radon detectors.

The EPA guideline for seven-day exposure duration detectors =

1 _______

Which detectors did not meet the standard?

2 ________________________________

What was the coefficient of variation for these machines?

3 _______

The Terradex detector has a CV of 4 _______; the EPA precision guidelines require 5 _______ for compliance.

The accuracy of a device is determined by comparing the measurement with that of the reference sample (given in the last column of the table).

Among the two-day exposure detectors, machines from 6 ______________ and 7 ______________ were within the EPA guidelines for both accuracy and precision.

Notice the standard deviations for Rad Elec. detectors in the three time periods (7, 5, and 2 days).

Standard deviations for Rad Elec. detectors for these periods are 8 _______, 9 _______, and 10 _______.

What do the relatively small standard deviations within the three groups indicate?

11 ________________________________
____

LITERATURE EXAMPLE 8.9

Field Comparison of Several Commercially Available Radon Detectors

R. William Field, MS, and Burton C. Kross, PhD

To determine the accuracy and precision of commercially available radon detectors in a field setting, 15 detectors from six companies were exposed to radon and compared to a reference radon level. The detectors from companies that had already passed National Radon Measurement Proficiency Program testing had better precision and accuracy than those detectors awaiting proficiency testing. Charcoal adsorption detectors and diffusion barrier charcoal adsorption detectors performed very well, and the latter detectors displayed excellent time averaging ability. Alternatively, charcoal liquid scintillation detectors exhibited acceptable accuracy but poor precision, and bare alpha registration detectors showed both poor accuracy and precision. The mean radon level reported by the bare alpha registration detectors was 68 percent lower than the radon reference level. (*Am J Public Health* 1990; 80:926–930.)

Results

A plot of the hourly measurement results obtained from the CRM is shown in Figure 1. The radon level in the bedroom showed considerable variation, especially during the last two days of the study which was probably related to increased ventilation rates in the basement. The occupant activity log notes that the children living in the home frequently opened the basement door leading to the outside during the evening of March 23 and the afternoon of March 24, 1989.

The measurement results for each type of detector are listed in Table 1. The measurements for the individual detectors from each company were reduced to the mean, standard deviation, coefficient of variation and MARE (Table 1) as a measure of the detectors' accuracy and precision.

The relatively small standard deviations within the three groups of electret ion chambers exposed for each time period indicate a fairly homogeneous radon concentration in the measurement area.

The MARE for the seven-day exposure detectors were all within the National RMP Programs requirement of ≤ 0.250.[4] The precision for the seven-day exposure duration detectors was below the 10 percent coefficient of variation as per the radon measurement protocol guidelines published by the EPA,[6] except for the radon detectors from the Radon Project which had a coefficient of variation of 25.9 percent.

Terradex's five-day exposure duration bare alpha track registration devices (BARDs) yielded a MARE in excess of the EPA's accuracy guidelines (≤ 0.250 required for accuracy) and a coefficient of variation in excess of the EPA's BARD radon measurement protocol precision guidelines[7] (≤ 20 percent required for precision).

The two-day exposure duration detectors from Ryan Nuclear Labs and Key Technology were both within the EPA guidelines for accuracy[4] and precision.[6]

Figure 2 shows the distribution of individual test results for each company and their relationship to the radon reference level.

The reported radon level for the blanks submitted to the companies were all < 0.5 pCi/L.

TABLE 1—Precision and Accuracy of Radon Detectors

Company	Number of Detectors	Exposure (days)	Radon Level Mean ± S.D. (pCi/L)	*Coefficient of Variation (%)	**Mare	Reference Radon Concentration Mean (pCi/L)
Rad Elec, Inc.	5	7	10.2 ± 0.4	3.9	0.045	10.6
American Radon Services	15	7	10.8 ± 0.8	7.5	0.057	10.6
Air Check, Inc.	15	7	11.4 ± 1.1	9.6	0.098	10.6
The Radon Project	15	7	10.5 ± 2.7	25.9	0.169	10.6
Rad Elec, Inc.	5	5	10.1 ± 1.2	11.8	0.091	10.6
Terradex	15	5	3.4 ± 1.7	50.6	0.679	10.6
Rad Elec, Inc.	5	2	9.6 ± 0.5	5.1	0.052	9.2
Ryan Nuclear Labs	15	2	11.0 ± 0.8	7.7	0.198	9.2
Key Technology	15	2	10.3 ± 0.8	7.5	0.136	9.2
EPA	15	2	11.6 ± 0.5	4.4	0.264	9.2

*The coefficient of variation was calculated using a standard deviation carried out to three decimal places.
**The mean of the absolute values of the relative errors (MARE) must be ≤ 0.250 to pass proficiency. Its calculation is shown below.

$$\text{MARE} = \frac{\sum_{i=1}^{n} \frac{|M_i - T_i|}{T_i}}{n}$$

n = number of detectors exposed
M_i = measured value for detector i
T_i = target value for detector i (reference value in this case)

Literature Example 8.10 is from the article entitled "Effects on Serum Lipids of Adding Instant Oats to Usual American Diets."

Two groups of persons were involved in this study.

One of the groups is called the [1] _______ group; the other is called the [2] _______ group.

Total cholesterol in both groups decreased after eight weeks.

Mean baseline cholesterol for the intervention group = [3] _______; eight-week cholesterol for the intervention group = [4] _______. The difference for the intervention group = [5] _______. Mean cholesterol for the control group = [6] _______; eight-week cholesterol for the control group = [7] _______. The difference for the control group = [8] _______.

Did the blood pressure change?

Mean baseline blood pressure for the intervention group = [9] _______; eight-week blood pressure for the intervention group = [10] _______. The difference for the intervention group = [11] _______. Mean baseline blood pressure for the control group = [12] _______; eight-week blood pressure for the control group = [13] _______. The difference for the control group = [14] _______.

The observed differences need to be assessed in the light of the variations in the measurements that occur because of the instruments used and other errors (which are referred to as errors due to unknown factors).

Coefficients of variation for cholesterol and blood pressure at eight weeks for the intervention group are [15] _______ and [16] _______, respectively.

LITERATURE EXAMPLE 8.10

Effects on Serum Lipids of Adding Instant Oats to Usual American Diets

Linda Van Horn, PhD, RD, Alicia Moag-Stahlberg, MS, RD, Kiang Liu, PhD, Carol Ballew, PhD, Karen Ruth, MS, Richard Hughes, MD, and Jeremiah Stamler, MD

This study was designed as a test of the serum lipid response and dietary adaptation to recommended daily inclusion of instant oats in an otherwise regular diet. Hypercholesterolemic adults were randomly assigned to a control or intervention group. Participants in the intervention group were given packages of instant oats and requested to eat two servings per day (approximately two ounces dry weight), substituting the oats for other carbohydrate foods in order to maintain baseline calorie intake and keep weight stable. Serum lipids were measured in blood collected by venipuncture at baseline, four weeks, and eight weeks. Baseline mean total cholesterol (TC) levels were 6.56 mmol/L and 6.39 mmol/L for intervention and control groups, respectively. After eight weeks, mean serum total cholesterol of the intervention group was lower by −0.40 mmol/L, and mean net difference in TC between the two groups was 0.32 mmol/L (95% CI: 0.09, 0.54). Low-density lipoprotein-cholesterol was similarly reduced with mean net difference of 0.25 mmol/L (95% CI: 0.02, 0.48) between the two groups. Mean soluble fiber intake increased along with slight self-imposed reductions in mean total fat, saturated fat, and dietary cholesterol intake in the intervention group. Neither group changed mean body weight. Daily inclusion of two ounces of oats appeared to facilitate reduction of serum total cholesterol and LDL-C in these hyperlipidemic individuals. (*Am J Public Health* 1991; 81:183–188)

TABLE 2—Group Means and Standard Deviations at Visits 1, 2, and 3 for Serum Lipids, Lipoproteins and Blood Pressure

Variables	Visit 1 Baseline	Visit 2 4 Weeks	Visit 3 8 Weeks
Total Cholesterol			
Group 1			
mmol/L	6.56 ± 0.82	6.28 ± 0.92	6.15 ± 0.86
Group 2			
mmol/L	6.39 ± 0.78	6.32 ± 0.84	6.30 ± 0.82
LDL-C			
Group 1			
mmol/L	4.58 ± 0.81	4.24 ± 0.87	4.16 ± 0.83
Group 2			
mmol/L	4.61 ± 0.73	4.39 ± 0.75	4.44 ± 0.77
HDL-C			
Group 1			
mmol/L	1.29 ± 0.34	1.36 ± 0.28	1.28 ± 0.29
Group 2			
mmol/L	1.26 ± 0.32	1.35 ± 0.26	1.30 ± 0.25
VLDL-C			
Group 1			
mmol/L	0.70 ± 0.50	0.68 ± 0.42	0.71 ± 0.48
Group 2			
mmol/L	0.51 ± 0.23	0.58 ± 0.28	0.56 ± 0.24
Triglycerides			
Group 1			
mmol/L	1.52 ± 1.09	1.48 ± 0.92	1.54 ± 1.04
Group 2			
mmol/L	1.12 ± 0.51	1.26 ± 0.61	1.22 ± 0.53
Blood Pressure (mm Hg)			
Group 1	127.9/80.3 (± 13/10)	121.8/80.8 (± 11/8)	123.5/77.0 (± 14/11)
Group 2	129.1/79.1 (± 20/8)	124.8/80.0 (± 11/8)	124.2/77.2 (± 13/9)

Group 1 = Intervention (oats), n = 42.
Group 2 = Control, n = 38.

Mean group difference in body mass index from visit 1 to visit 3 was not significant (Table 4). Average weight change from visit 1 to visit 3 was less than one pound with a slight decrease in Group 1 and increase in Group 2. Blood pressure changes also were not significant (Table 2).

Index of Dispersion

ACTIVITY and STRESS are two categorical variables from the Student Questionnaire Data (SQD) set. Do the values for ACTIVITY vary less or more than the values for STRESS? Just enumerating the percentages does not seem to answer the question. Consider Table 8.4.

TABLE 8.4 Activity and Stress Status of Students in Two Statistics Classes

ACTIVITY	Count (f_i)	Percent	STRESS	Count (f_i)	Percent
Minimal	6	12.24	Minimal	11	22.00
Moderate	36	73.47	Moderate	31	62.00
Vigorous	7	14.29	Heavy	8	16.00
	$n = 49$			$n = 50$	

From the percentages we can see that the ACTIVITY and STRESS distributions are different, but does one vary more than the other? To decide this we need an index of dispersion similar to the coefficient of variation that we used to compare variation in different variables. The **index of dispersion** is the proportion of dispersion that exists within the observations relative to the maximum dispersion that can possibly exist. This index varies from 0 to 1. The index is computed by using the following formula:

$$\text{Dispersion} = \text{D} = \frac{k(n^2 - \Sigma f_i^2)}{n^2(k-1)}$$

Where

k = number of categories

n = total number of observations

f_i = frequency in each category

We calculated the index of dispersion for ACTIVITY in the left column. Do the same for STRESS in the right column.

Index of Dispersion for ACTIVITY

$k = 3$	n	n^2	f_i	f_i^2
	49	2,401	6	36
			36	1,296
			7	49
				$\Sigma f_i^2 = 1{,}381$

Index of Dispersion for STRESS

$k = 3$	n	n^2	f_i	f_i^2
	1 ______	2 ______	3 ______	4 ______
			5 ______	6 ______
			7 ______	8 ______
				$\Sigma f_i^2 =$ 9 ______

$(n^2 - \Sigma f_i^2) = 2{,}401 - 1{,}381 = 1{,}020$	$(n^2 - \Sigma f_i^2) =$ [10] ______
$k(n^2 - \Sigma f_i^2) = 3(1{,}020) = 3{,}060$	$k(n^2 - \Sigma f_i^2) =$ [11] ______
$(k - 1) = (3 - 1) = 2$	$(k - 1) = (3 - 1) = 2$
$n^2(k - 1) = 2{,}401(2) = 4{,}802$	$n^2(k - 1) =$ [12] ______
$D = \dfrac{k(n^2 - \Sigma f_i^2)}{n^2(k-1)}$	$D = \dfrac{k(n^2 - \Sigma f_i^2)}{n^2(k-1)}$
$= 3{,}060/4{,}802 = 0.637$	$=$ [13] ______
$\approx$ 64 percent of the maximum dispersion	$\approx$ 81 percent of the maximum dispersion

Interpretation of the Index of Dispersion

The index is 0 when all the cases are stacked in one of the cells. The index is 1 when cases are distributed equally. ACTIVITY has about 0.64, or 64 percent, of the maximum dispersion. This becomes more meaningful when we compare it with another sample. Compare it with the index of dispersion for STRESS. STRESS has an index of dispersion of .81, or 81 percent of the maximum dispersion. We can conclude that STRESS is more dispersed than ACTIVITY.

Self-Assessment

Please assess your competency.

Tasks:	**How well can you do it?**					**Page**
Decide which of the following is the appropriate measure of central tendency/variation in a given context and calculate it from the data:	*Poorly*				*Very well*	
1. Mean	1	2	3	4	5	237
2. Mode	1	2	3	4	5	238
3. Median	1	2	3	4	5	238
4. Trimmed mean	1	2	3	4	5	238
5. Range	1	2	3	4	5	266
6. Inter-quartile range	1	2	3	4	5	251
7. Variance	1	2	3	4	5	266
8. Standard deviation	1	2	3	4	5	251

9. Proportion	1	2	3	4	5	266
10. Coefficient of variation	1	2	3	4	5	267
11. Variance of proportion	1	2	3	4	5	266
12. Index of dispersion	1	2	3	4	5	279

Answers for Chapter 8

Answer for p. 237: 21

Answers for pp. 238–239: 1) No 2) Yes 3) No 4) Number 5) Possibly, if we increase the number of observations 6) Yes 7) Difference 8) 90 9) 20 10) 61

Answers for p. 240: 1) $30 * 3 = 90$ 2) 1,287 3) 21 4) 23 5) 25.235

Answers for p. 241: 1) 42.5 2) 47.5 3) $37.5 * 2 = 75$ 4) 1,323.5 5) 25.9 6) 20–25 7) From this table we can't say exactly what the median is; the best we can say is that it is somewhere between 20 and 24.

Answers for p. 242: 1) 51 2) 33 3) 0.65 (or 65 percent) of the total group of 51 were females.

Answers for Literature Example 8.1: 1) To study factors that increase the likelihood of discrimination against girls in India 2) Hospitalized children in Punjab, India 3) Male/female ratio 4) The male/female ratio fell rapidly over the first few months of life and continued to decline until age 4, at which time it leveled off. 5) Sikhs 6) We think that girl babies are less valued than boy babies in this and many other parts of the world. 7) Yes 8) Skewed 9) Younger

Answers for Literature Example 8.2: 1) No 2) Higher 3) The same 4) We think the authors' assertion does coincide with the information in the table.

Answers for Literature Example 8.3: 1) 10 μg/L 2) 1.6 μg/L 3) 8.9 μg/L 4) 28 μg/L 5) 43 μg/L 6) Cotinine levels generally increase as the number of people who smoke in the household with the infant increases. 7) Yes

Answers for pp. 250–251: 1) 23 2) 23 3) 29 4) 5 5) 24

Answers for pp. 251–253: 1) 5 2) 0 3) .5 4) 2 5) Farther from 6) Larger 7) 0.66 8) Variance 9) Does not 10) 6.86 11) −0.03 12) $(38 - 25.2)/6.7 = 1.91$

Answers for Table 8.3: 1) Small 2) Large

Answers for Literature Example 8.4: 1) 3,572 grams 2) 3,023 grams 3) 549 grams 4) 3,536 grams 5) 36 grams 6) 3,270 grams 7) 302 grams 8) 36 grams 9) 247 grams 10) 211 grams 11) Heavy smoking

Answers for Literature Example 8.5: 1) Maximum = 43.1; minimum = 40.7 2) SE $\approx$ .3 3) 4.4 4) The means for those two variables are fairly close in size. Whether they are close enough so that the difference is not statistically significant, however, cannot be determined without some computation. You will learn how to do that computation in Chapter 10.

Answers for pp. 261–262: 1) No x_i 2) All 3) 70 4) 41 5) 74 6) 49 7) $66 + 12 = 78$ 8) 51 9) 68 10) 74 11) 58 to 74 12) 54 to 78 13) 100

Answers for Literature Example 8.6: 1) 40 weeks 2) Yes, although it looks like lowest risk occurs at 43 weeks gestational age in the $z = -2$ group 3) Relative birth weight is strongly associated with infant survival. 4) z

Answers for p. 266: 1) .65 2) 41 3) 18

Answers for pp. 267–268: 1) 6.5% 2) 6.5% 3) ≈ 24% 4) Larger 5) ≈ 15% 6) ≈ 18% 7) Larger

Answers for Literature Example 8.7: 1) Three 2) Four 3) 22 4) 243 5) Precision 6) Accuracy 7) Machine #2 8) 1; 3.1 9) 4; 3.5 10) 3; 3.8 11) The machines did not analyze materials within acceptable ranges.

Answers for Literature Example 8.8: 1) Age 2) Cigarettes per day 3) Pretreatment body weight 4) Postcessation treatment 5) Innovative condition 6) Nonspecific condition 7) Standard treatment condition 8) Different; whether the difference is statistically significant can be determined by using the method given in Chapter 14. 9) 24% 10) 18% 11) 18% 12) The proportions of males and females in each group are different. We think sex composition will influence the study, because the distributions of weights of males and females are generally different. 13) Separating data by sex would make analysis of differences between groups easier. 14) 1.14 15) .75 16) .32 17) Means indicate how much weight was gained and seem to be small when compared to the gains for those in standard treatments. So the authors feel that it may not be worth focusing on the weight as an issue.

Answers for Literature Example 8.9: 1) 10% 2) Those from the Radon Project 3) 25.9 4) 50.6 5) 20% 6) Rayan Nuclear Labs 7) Key Technology 8) 0.4 9) 1.2 10) 0.5 11) A fairly homogeneous radon concentration in the measurement area

Answers for Literature Example 8.10: 1) Intervention 2) Control 3) 6.56 4) 6.15 5) 0.41 6) 6.39 7) 6.30 8) 0.09 9) 127.9 10) 123.5 11) 4.4 12) 129.1 13) 124.2 14) 4.9 15) 14% 16) Systolic, 11%, diastolic, 14%

Answers for pp. 279–280: 1) 50 2) 2,500 3) 11 4) 121 5) 31 6) 961 7) 8 8) 64 9) 1,146 10) 2,500–1,146 = 1,354 11) 3(1,354) = 4,062 12) 2,500(2) = 5,000 13) 4,062 + 5,000 ≈ .81

CHAPTER 9

Research Designs

Prelude

"How do you know all that?" asked Milo.

"Simple," he said proudly. "I'm Alec Bings; I see through things. I can see whatever is inside, behind, around, covered by, or subsequent to anything else. In fact, the only thing I can't see is whatever happens to be right in front of my nose."

"Isn't that a little inconvenient?" asked Milo, whose neck was becoming quite stiff from looking up.

"It is a little," replied Alec, "but it is quite important to know what lies behind things, and the family helps me take care of the rest. My father sees to things, my mother looks after things, my brother sees beyond things, my uncle sees the other side of every question, and my little sister Alice sees under things. . . . Whatever she can't see under, she overlooks."

— Norton Juster
The Phantom Toll Booth, pp. 106–107

Chapter Outline

Self-Assessment

Please assess your competency.

Tasks:	How well can you do it? *Poorly*				*Very well*	Page
1. Given an adequate description of the research context, decide on the appropriate research design or elucidate the conditions under which the following designs should be used:						
Observational designs:						
a. Retrospective						
(i) Case-control study	1	2	3	4	5	305
(ii) Cross-sectional survey	1	2	3	4	5	299
b. Prospective						
(i) Cohort study	1	2	3	4	5	309
Experimental studies:						
RR O_1 X O_2 Expt. Group I RR O_3 O_4 Control Group II RR X O_5 Expt. Group III RR O_6 Control Group IV	1	2	3	4	5	312
XO	1	2	3	4	5	326

Subjects not randomly selected and not assigned randomly to the control and experimental groups; groups are not comparable:

X O_1 O_2	1	2	3	4	5	326

Subjects randomly assigned to experimental and control groups but not selected randomly from the population:

•R X O_1 •R O_2	1	2	3	4	5	326
O_1 X O_2	1	2	3	4	5	327

Subjects not randomly assigned to experiment and control:

O_1 X O_2 O_3 O_4	1	2	3	4	5	327

Subjects randomly assigned to experiment and control but not randomly selected:

•R O_1 X O_2 •R O_3 O_4	1	2	3	4	5	327

Subjects randomly selected and randomly assigned to experiment and control:

RR O_1 X O_2 RR O_3 O_4	1	2	3	4	5	328

Observations on the same group (time series):

O_1 O_2 O_3 O_4	1	2	3	4	5	328

Observations on experimental and control groups not randomly assigned (time series):

O_1 O_2 X O_3 O_4	1	2	3	4	5	328

Observations on experimental and control groups not randomly assigned (time series):

O_1 O_2 X O_3 O_4 O_5 O_6 O_7 O_8	1	2	3	4	5	328

Observations on experimental and control groups; groups randomly assigned (time series):						
•R O_1 O_2 X O_3 O_4 •R O_5 O_6 O_7 O_8	1	2	3	4	5	328
2. Elucidate the strategy for analyzing the data collected using each of the above designs.	1	2	3	4	5	
3. Explain the following indices:						
Relative risk	1	2	3	4	5	309
Odds ratio	1	2	3	4	5	305

Introduction

Researchers are interested in acquiring new knowledge through methodical acquisition of data. Children acquire knowledge through their own instinctual experimentation. They grasp an object, smell, taste, and bang it to decide whether it is worth further attention. A researcher hits upon an idea, formulates a hypothesis, designs an experiment or an observational scheme, collects data, consolidates data, estimates parameters, tests hypotheses, arrives at conclusions, decides whether the subject is worth further study, and if so, repeats the process.

Four kinds of research approaches are common. Figure 9.1 delineates the types. Activity 9.1 gives examples. There are basically two types of investigations: (1) **observational**, in which we simply observe a phenomenon using an appropriate measuring instrument, e.g., a questionnaire to assess knowledge about AIDS, a checklist to extract information from a medical record, or a scale to determine weights of individuals; (2) **experimental**, in which some particular variable is manipulated to determine the effect of the manipulation on a response variable, e.g., feeding different quantities of food (independent, or treatment variable) to animals and determining the weight gain (response variable). Research activity hinges on the type of question an investigator wishes to answer. Different types of questions lead to different research designs.

ACTIVITY 9.1

Investigator's interest and typical research questions

Refer to Figure 9.1 and indicate which type of research approach would provide the desired information. The first item has already been completed.

What are the cigarette smoking habits of college students?	Simple description (Approach #1) 1 ______________________

Description	Comparison of Group Means	Determining Relationships Among Variables: Association	Determining Relationships Among Variables: Prediction	Determining the Effect of Intervention
Simple descriptions (1)	Comparison with values given in the literature (2)	Two categorized variables (5)	The value of one variable from the value of another (8)	Doing experiments (9)
	Comparison of two groups (3)	One categorized variable and one continuous variable (6)		
	Comparison of three or more groups (4)	Continuous data (7)		

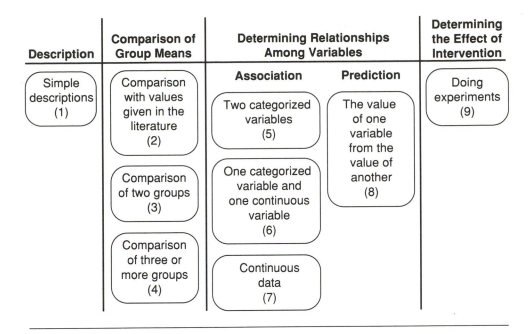

FIGURE 9.1 Tools for Answering Research Questions

Are the smoking habits of students at our college similar to the smoking habits of 18 to 24-year-olds as published in a medical research journal? 2 ______________________

Does the proportion of female smokers differ from the proportion of male smokers? 3 ______________________

Do the proportions of cigarette smokers differ in the various classes, i.e., freshman, sophomore, junior, senior? 4 ______________________

Is gender associated with physical activity level? 5 ______________________

Is activity level (minimal, moderate, vigorous) associated with weight? 6 ______________________

Is there an association between height and weight? 7 ______________________

Can we predict weight from height? 8 ______________________

How much does pulse change after running? 9 ______________________

Vocabulary for Variables Used in Studies

A researcher's interest is in investigating the behavior of selected variables and their relationships. All variables are measured with one of the four scales: nominal, ordinal, interval, or ratio. Variables are classified into two types depending on whether they explain a phenomenon (explanatory variable) or are the outcome (response variable) of a phenomenon. A variable that is labeled one way may be labeled another way in a different context. The following table summarizes terminology associated with variables.

	Name of Variable	
	Variable I	**Variable II**
In many studies we are interested in measuring the subject's response to some action. The variable measured is the **response** variable. The variables that may explain the response are called **explanatory** variables.	Explanatory	Response, or outcome
Variables we hope will explain the response variable are also called **independent** variables. The response variable is called the **dependent** variable because the response is thought to depend on the explanatory variable.	Independent	Dependent
Sometimes researchers intervene and conduct experiments. In that case the response variable is the outcome that is measured. (The response variable is also called the **outcome** variable.) The intervention is called the **treatment** variable. For example, a treatment could be education, and the response variable might be the score on a test. The treatment, or independent, variable is also sometimes called a **factor** or a **stimulus variable**.	Treatment Factor, stimulus	Response, or outcome
The treatment may be administered in different quantities, e.g., education may be delivered in one or two hours. In this case, education is said to have two **quantitative levels**. Treatments may also be qualitatively different, e.g., education by lecture method and by discussion method. These two levels are called **qualitative levels**.	Levels; two types of levels are quantitative and qualitative	Response, or outcome

Alternative names for outcome variables are [1] _______ variable or [2] _______ variable. An alternative name for an independent variable in the context of a survey is [3] _______. Three alternative names for an independent variable in the context of experiments are [4] ______, ______, and ______. An independent variable may have two types of levels, [5] ______ and [6] ______.

ACTIVITY 9.2

Explanatory and response variables in different contexts

Identification of a variable as an explanatory variable or a response variable depends on the context. We will use the student questionnaire data and faculty exercise data to illustrate. Please have at least one card handy from each data set to remind yourself of the variables. (See the explanations of these variables in Chapter 2.) Identify what type each variable is in each context.

Context	Circle the appropriate name
Context I: Our interest is in studying factors that influence the pulse rates of individuals.	PULSE response, factor, dependent, independent, explanatory STRESS response, factor, dependent, independent, explanatory AGE response, factor, dependent, independent, explanatory SMOKE response, factor, dependent, independent, explanatory ACTIVITY response, factor, dependent, independent, explanatory
Context II: Our interest is in studying factors that influence the stress levels of individuals.	SMOKE response, factor, dependent, independent, explanatory ACTIVITY response, factor, dependent, independent, explanatory STRESS response, factor, dependent, independent, explanatory DISCUSS response, factor, dependent, independent, explanatory TIMEOUT response, factor, dependent, independent, explanatory HEALTH response, factor, dependent, independent, explanatory
Context III: Our interest is in studying factors that influence the perceived health of individuals.	SMOKE response, factor, dependent, independent, explanatory ACTIVITY response, factor, dependent, independent, explanatory STRESS response, factor, dependent, independent, explanatory DISCUSS response, factor, dependent, independent, explanatory

(cont.)

Context	Circle the appropriate name
	TIMEOUT response, factor, dependent, independent, explanatory HEALTH response, factor, dependent, independent, explanatory
Context IV: Our interest is in seeing what happens when students run in place for one minute. The intervention is running in place for one minute. This is the experimental part of the study.	PULSE 1 response, factor, dependent, independent, explanatory, treatment PULSE 2 response, factor, dependent, independent, explanatory, treatment Running in place for one minute response, factor, dependent, independent, explanatory, treatment
Context V: Our interest is in seeing what happens to measured variables when faculty members perform low-intensity or high-intensity exercise for ten weeks. Intervention is exercise. There are three levels. What are the levels? ________________ ________________ ________________	HEART RATE response, factor, dependent, independent, explanatory, treatment VO2 response, factor, dependent, independent, explanatory, treatment WEIGHT response, factor, dependent, independent, explanatory, treatment BODY FAT % response, factor, dependent, independent, explanatory, treatment Low-intensity exercise (LI) response, factor, dependent, independent, explanatory, treatment High-intensity exercise (HI) response, factor, dependent, independent, explanatory, treatment Control (C) response, factor, dependent, independent, explanatory, treatment

Why Design Studies?

Suppose that one morning you try an unfamiliar route to get to school. If you get there faster by this route, you accept it as more efficient, and you are likely to use it regularly instead of the old route. In saying that this route is a good route, you have generalized from your (one-time) experience.

Performing an experiment and drawing conclusions from it is similar, but researchers do this in a more structured way so that their generalizations are more dependable and can be used by other people. Research design provides that structure. Research design involves:

1. formulating a clear research question (see Activity 9.1 and Chapter 12);
2. setting up a procedure to select the population of interest;
3. selecting an appropriate sample (subjects for an experiment or study; see Chapter 4);
4. selecting relevant variables and appropriate instruments for measuring those variables;
5. assigning treatments to subjects in a systematic, unbiased fashion (see Chapter 4); and
6. planning for the analysis of data.

This chapter serves as an introduction to selected research designs. Figure 9.2 presents selected types of research designs. Activity 9.3 will help you learn the concepts suggested in Figure 9.2.

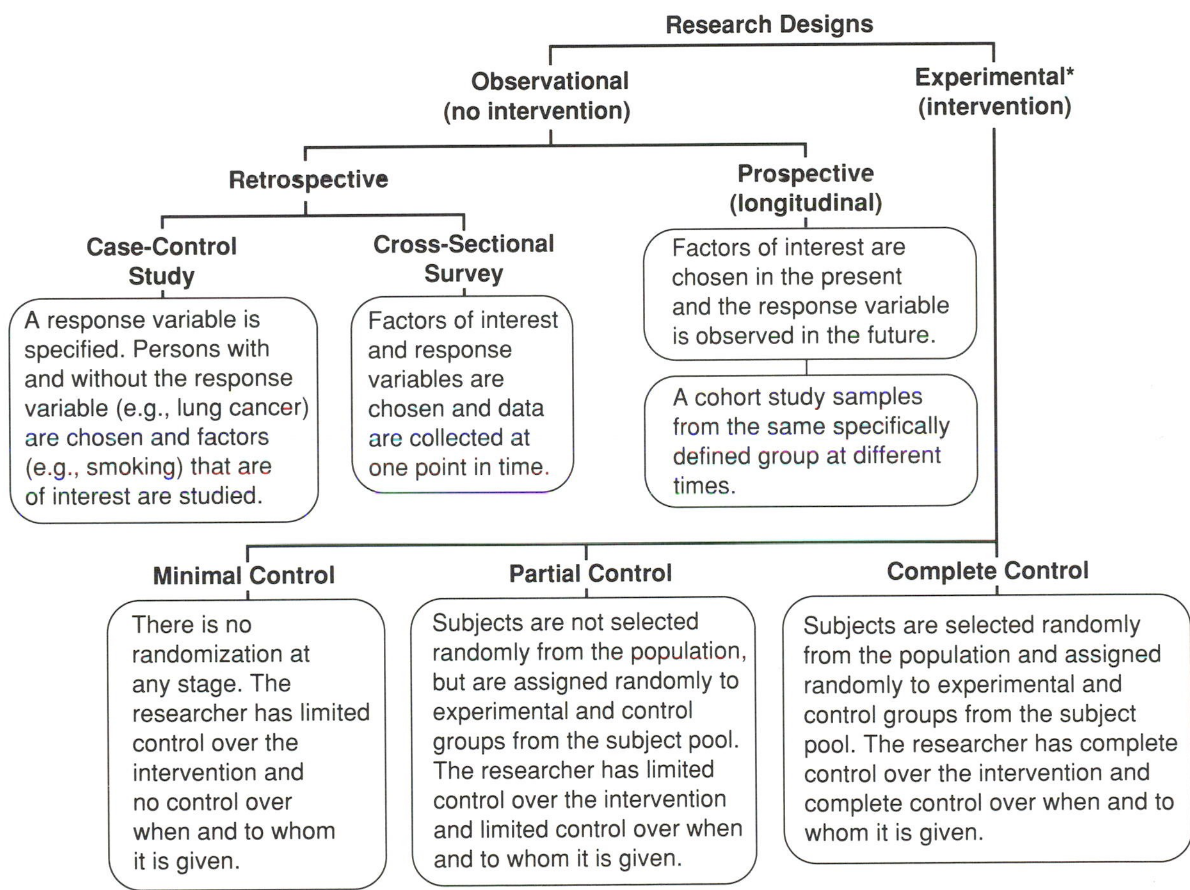

FIGURE 9.2 Research Designs

Activity 9.3

Types of research, their purposes, and sources of data

Observational studies involve no intervention, whereas **experimental studies** involve [1] _______. In observational studies, such as the student questionnaire survey, data on many variables are collected. Many guesses may be checked out by analyzing the data using various tools. We may describe the sample using simple descriptive tools such as tables, graphs, and numerical summaries.

Observational studies are divided into [2] _______ and [3] _______ studies. In **retrospective studies** we decide on the response variable now and [4] look for data in past records/collect data in the future; in **prospective studies** we [5] look for data in past records/collect data in the future. All experiments are [6] prospective/retrospective studies, as they collect data about the response variable in the future. Figure 9.2 mentions two kinds of retrospective studies: the [7] _______ and the [8] _______. A cross-sectional survey collects data on explanatory variables and response variables at [9] _______ _______ in time. A case-control study specifies a response variable and selects survey populations for case and control on the basis of the presence or absence of the [10] _______ variable. Prospective studies collect data over a period of time. When we sample from a specific population with a presumed explanatory variable (e.g., SMOKE) and study it for future occurrence of the [11] _______ variable (e.g., occurrence or non-occurrence of lung cancer) for a specific period of time, we are conducting a **cohort study**. While conducting a cohort study, we include a control group that does not have some values of the presumed explanatory variable. When we select a group of people with a shared characteristic (e.g., born in the same year) and follow them for a specific period of time to note the occurrence of a response variable (e.g., suicides), it too is called a cohort study.

Often a response variable is affected by variables other than the explanatory variables identified by the researcher. In many cases these so-called **extraneous variables** are unknown or unidentified, in which case they are usually called **lurking variables**. For example, spokespersons for the tobacco industry have contended that the positive correlation between cigarette smoking and incidence of lung cancer may really be caused by an unknown factor, a lurking variable, such as a genetic propensity toward both smoking and lung cancer or some life-style factor that tends to cause both.

Random sampling helps to control for extraneous variables because the expectation is that their values will even out in the long run, i.e., the sample will have a full range of values of these variables and they will be fairly uniformly distributed between the various control and experimental groups. The larger a sample is, the more likely it is that this uniformity will be achieved; therefore, large samples are more reliable than small ones. When we discuss specific designs later in this chapter, we will often explicitly state how and to what degree these designs control for extraneous variables.

Measuring phenomena is a critical component of research. The measurements need to be valid and reliable. We will next discuss the concepts of validity and reliability, illustrating with examples from the literature. Later we will describe details of different types of research studies.

Reliability and Validity

A jeweler weighs the same wedding ring five times and gets five different weights. Is it [1] the scale/the jeweler/the ring that is not reliable? Normally we expect the weights of objects to remain constant, yet small variations in temperature and pressure might affect the scale's mechanism to produce different readings at different times. Also, the experimenter may see the readings differently. For instance, the needle may be three-fifths of the way between two marks, but because of the angle of view, the researcher may read it as half-way between. The standard deviation of repeated measurements is a measure of the random error encountered and is an index of how reliable the measurement is. Reliable scales have small random error; unreliable scales have large random error. The extent to which a researcher (or anyone measuring anything) obtains similar numbers on repeated attempts to measure is called **reliability**.

Validity refers to the connection between the measurement and the entity we are attempting to measure. If a jeweler is asked the weight of a man's ring but responds with the weight of a woman's ring, the weight is not valid. This is obvious, but in many situations the lack of validity is quite subtle. A questionnaire, for example, may claim to quantify an individual's anger; however, other established instruments may show that it actually measures resentment. Anger and resentment are not the same, so the questionnaire is not a valid instrument. Reliability and validity are expressions of how trustworthy observations or measurements are.

Example from the Literature

We need to know the accuracy (akin to validity) and precision (akin to reliability) of a measuring device. The following abstract and table are concerned with measuring instruments. Literature Example 9.1 is again from the article entitled "Field Comparison of Several Commercially Available Radon Detectors." Read it and complete Activity 9.4.

ACTIVITY 9.4

Concepts of reliability and validity

What are the reference radon levels? [1] ______________________

What are the lowest and highest mean radon levels recorded by the instruments? [2] ______________________

Can we compare standard deviations directly when the means vary?

3 ________________________________

Accuracy is reflected by how close the mean is to the reference radon level. Because of **random variation** (also called **error**), we do not obtain identical numbers each time we measure something. If we allow ± 0.5 for error (a number arbitrarily chosen for illustrative purposes), which detectors do *not* show acceptable accuracy? (For detectors with a reference level of 10.6, the acceptable accuracy would be between $10.6 - .5 = 10.1$ and $10.6 + .5 = 11.1$; for detectors having a reference level of 9.2, the acceptable accuracy would be between $9.2 - .5 = 8.7$ and $9.2 + .5 = 9.7$.)

4 ________________________________

Precision is reflected by the standard deviation. Since standard deviations are not directly comparable, the authors computed coefficients of variation. If we arbitrarily choose a CV of 10 percent as the cutoff point, which detectors show unacceptable precision?

5 ________________________________

Standard deviation measures the 6 _______ _______ and hence reflects the 7 _______, or precision, of a measure.

The mean when compared with the reference measure gives us an idea of the accuracy (akin to 8 _______) of the measure.

LITERATURE EXAMPLE 9.1

Field Comparison of Several Commercially Available Radon Detectors

R. William Field, MS, and Burton C. Kross, PhD

To determine the accuracy and precision of commercially available radon detectors in a field setting, 15 detectors from six companies were exposed to radon and compared to a reference radon level. The detectors from companies that had already passed National Radon Measurement Proficiency Program testing had better precision and accuracy than those detectors awaiting proficiency testing. Charcoal adsorption detectors and diffusion barrier charcoal adsorption detectors performed very well, and the latter detectors displayed excellent time averaging ability. Alternatively, charcoal liquid scintillation detectors exhibited acceptable accuracy but poor precision, and bare alpha registration detectors showed both poor accuracy and precision. The mean radon level reported by the bare alpha registration detectors was 68 percent lower than the radon reference level. (*Am J Public Health* 1990; 80:926–930.)

TABLE 1—Precision and Accuracy of Radon Detectors

Company	Number of Detectors	Exposure (days)	Radon Level Mean ± S.D. (pCi/L)	*Coefficient of Variation (%)	**Mare	Reference Radon Concentration Mean (pCi/L)
Rad Elec, Inc.	5	7	10.2 ± 0.4	3.9	0.045	10.6
American Radon Services	15	7	10.8 ± 0.8	7.5	0.057	10.6
Air Check, Inc.	15	7	11.4 ± 1.1	9.6	0.098	10.6
The Radon Project	15	7	10.5 ± 2.7	25.9	0.169	10.6
Rad Elec, Inc.	5	5	10.1 ± 1.2	11.8	0.091	10.6
Terradex	15	5	3.4 ± 1.7	50.6	0.679	10.6
Rad Elec, Inc.	5	2	9.6 ± 0.5	5.1	0.052	9.2
Ryan Nuclear Labs	15	2	11.0 ± 0.8	7.7	0.198	9.2
Key Technology	15	2	10.3 ± 0.8	7.5	0.136	9.2
EPA	15	2	11.6 ± 0.5	4.4	0.264	9.2

*The coefficient of variation was calculated using a standard deviation carried out to three decimal places.
**The mean of the absolute values of the relative errors (MARE) must be ≤ 0.250 to pass proficiency. Its calculation is shown below.

$$\text{MARE} = \frac{\sum_{i=1}^{n} \frac{|M_i - T_i|}{T_i}}{n}$$

n = number of detectors exposed
M_i = measured value for detector i
T_i = target value for detector i (reference value in this case)

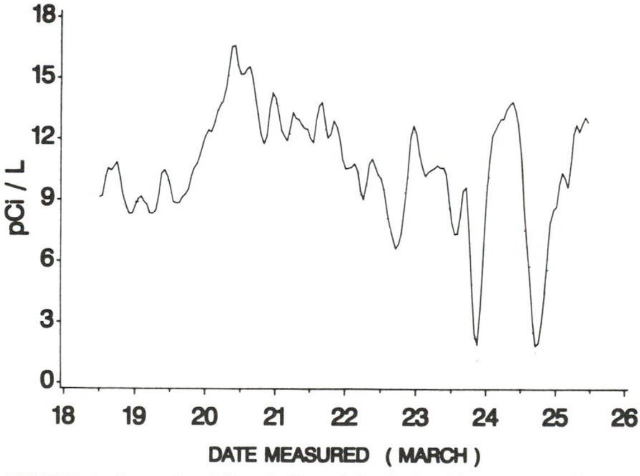

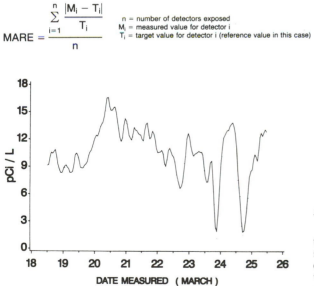

FIGURE 1—Interpolated Hourly Plot of the Radon Measurements Obtained from the Continuous Radon Monitor for the Period 1100 hrs March 18 to 1100 hrs March 25, 1989

Results

A plot of the hourly measurement results obtained from the CRM is shown in Figure 1. The radon level in the bedroom showed considerable variation, especially during the last two days of the study which was probably related to increased ventilation rates in the basement. The occupant activity log notes that the children living in the home frequently opened the basement door leading to the outside during the evening of March 23 and the afternoon of March 24, 1989.

The term validity is used both when referring to instruments (e.g., questionnaires and tests) and when referring to experimental design.

Validity of Instruments

Instruments must exhibit three kinds of validity:

1. Face validity: Does the instrument measure what it says it measures (for example, innate intelligence as opposed to acquired information)?
2. Content validity: Does the instrument include all relevant questions related to the objective of the study? (For example, an instrument designed to study safe sex practices should not limit its questions to the use of condoms; it should cover all known safety measures.)
3. Criterion-related validity: Do measures obtained using this instrument give results comparable to the measures obtained using other direct measures of the traits in question? (For example, a red blood cell count given by an electronic device should compare favorably with a count made by a lab technician using a microscope.)

Validity of Experiments

When we are able to make certain that the treatment is the only factor influencing the outcome of an experiment, the experiment is said to have **internal validity**. **External validity** relates to the generalizability of the results from an experimental sample to a population.

Internal Validity Are the observed changes in the response variable due only to the treatment and not to the influence of other factors? If the answer is yes, the experiment has internal validity. If external variables could have influenced the outcome, the internal validity is jeopardized. Internal validity is ensured by thoughtful designs, which:

1. **Safeguard against interference from factors** other than the experimental intervention. (For example, students did better than expected on a test, and the professor attributed this to his brilliant lecture. Actually, the improved performance was the result of a TV program that most of the students saw the night before the test.)
2. **Choose relevant, measurable response variables**. (For example, if change of behavior is the relevant response variable, measuring knowledge alone would be misleading.)
3. **Choose reliable instruments to measure those variables**;
4. Require that the **intervention is of sufficient potency to ensure some difference**. (For example, if we want to change an entrenched behavior like smoking habits, it is not enough to just mail a pamphlet to smokers. The researcher needs to find a more potent intervention such as a smoking cessation program backed by a support group.)
5. **Assign intervention to groups randomly** so there is no bias.

External Validity External validity answers the question, "Can these results be generalized to other populations?" External validity is ensured by selecting a representative sample from the population of interest. We can generalize only to the population from which we drew the sample.

Example from the Literature

Literature Example 9.2 is from the paper entitled "Effect of Acupuncture-Point Stimulation on Diastolic Blood Pressure in Hypertensive Subjects: A Preliminary Study." Examine it to see how internal and external validity were ensured. Then complete Activity 9.5.

ACTIVITY 9.5

Design considerations that ensure validity

What is the treatment? 1 ____________________

Testing was done in an audiology lab and a pleasant scene was placed on the wall. How does this ensure internal validity? 2 ____________________

What could happen if the investigators wore white coats? 3 ____________________

What would happen if the subjects with hypertension were given the treatment and the others served as controls? 4 ____________________

How did the researchers ensure random assignment of subjects to the treatment and control groups? 5 ____________________

What is the response variable? Is this a standard measure? Is it known to be reliable? 6 ____________________

How was the uniformity of the sample ensured? 7 ____________________

Can these results be generalized to other groups? 8 ____________________

Is the potency of the intervention strong enough to make a difference? 9 ____________________

LITERATURE EXAMPLE 9.2

Effect of Acupuncture-Point Stimulation on Diastolic Blood Pressure in Hypertensive Subjects: A Preliminary Study

Tim Williams
Karen Mueller
Mark W Cornwall

Electrical stimulation of four specific acupuncture points (Liver 3, Stomach 36, Large Intestine 11, and the Groove for Lowering Blood Pressure) was examined in order to determine the effect of this stimulation on diastolic blood pressure in 10 subjects with diastolic hypertension. Subjects were randomly divided into two groups: (1) an Acu-ES group, which received electrical stimulation applied to the four antihypertensive acupuncture points, and (2) a Sham-ES group, which received electrical stimulation applied to non–acupuncture-point areas. A repeated-measures analysis of variance revealed a significant, immediate poststimulation reduction of diastolic blood pressure for the Acu-Es group versus the Sham-ES group. Further studies are needed to determine whether there are other acupuncture points, stimulation characteristics, or modalities that can enhance this treatment effect and whether the treatment effect can last for a clinically significant period of time. [Williams T, Mueller K, Cornwall MW. Effect of acupuncture-point stimulation on diastolic blood pressure in hypertensive subjects: a preliminary study. Phys Ther. 1991;71:523–529.]

Method

Subjects

Subjects were selected on the basis of two inclusion criteria: (1) a resting diastolic BP between 90 and 120 mm Hg and (2) no past or present use of antihypertensive medication.

All subjects were patients under the care of two physicians with a specialty in family practice. In addition, all subjects had been diagnosed with borderline hypertension within the previous 6 months. The diagnosis of borderline hypertension was assigned by these physicians to any patient with a diastolic BP greater than 90 mm Hg during two consecutive office visits. Patients diagnosed with borderline hypertension were counseled by the physicians about dietary modifications and exercise and were instructed to monitor their BP at home. Subsequently, if these patients reported three consecutive at-home diastolic BP readings above 90 mm Hg, their diagnosis was changed from borderline to essential hypertension. All subjects who participated in this study were still considered by these physicians to be borderline hypertensive.

The two physicians contacted 22 subjects who met the study's inclusion criteria, 12 of whom agreed to participate (2 subjects showed normotensive diastolic BPs when they arrived for the first data-collection session and were excluded from the study). Of the 10 subjects who completed the study, 2 were female and 8 were male. Their mean age was 46 years (range=27–72 years). All subjects gave written informed consent.

Test Environment

The testing environment was carefully controlled in order to minimize factors that could influence diastolic BP.[26,27] All testing was performed in an acoustically insulated audiology booth in which the temperature was maintained between 23° and 25°C.

In order to minimize psychological factors affecting diastolic BP, the following steps were taken to ensure a relaxed atmosphere during testing. Subjects were instructed to wear loose and comfortable clothing. The investigators refrained from wearing laboratory coats during all sessions, because the use of such apparel has been associated with increased BP ("white-coat hypertension") in some patients.[28] Except for a standing mercury column sphygmomanometer and a stethoscope, the testing equipment was placed away from the subject's view. Finally, a large poster of a pleasant outdoor scene was placed on the wall facing the subject.

Procedure

Subjects were randomly divided into two groups by the use of a coin toss. Four subjects comprised the Acu-Es group (test group), which received electrical stimulation applied to the four acupuncture points, and six subjects formed the Sham-ES group (control group), which received electrical stimulation applied to non–acupuncture-point areas. The two subjects who were excluded from the study had previously been assigned to the Acu-ES group. Because they were the last two subjects scheduled for data collection, an uneven distribution of subjects resulted.

Research Designs

You try a new route to school one morning at 8 a.m. (experiment). You accept it as an efficient way to reach school (inference). In order to test your inference, you would need to try the new route at 8 a.m. on many different days. After doing this, you obviously could not generalize your results to other times of the day, nor could you generalize to other routes. If you wanted the results to be more general, you would need to take more general samples.

The faculty exercise data (FED) came from volunteer faculty aged 40 or older. Can the conclusions we draw from these data be applied to faculty between 30 and 39? No, since no one in our sample is in that age group. Can we generalize to the rest of the faculty who are in the over-40 age group? No; since all the participants were volunteers, the sample is not representative of the entire faculty. Strictly speaking, conclusions apply only to the participants in the study.

Refer back to Figure 9.2, which divides all studies into observational and experimental studies. The remainder of this chapter will describe each type of observational study and several experimental designs.

Observational Studies

Retrospective Study

With a retrospective study, we look back in time to find the causes of a certain phenomenon or characteristic currently observed in a population. The presence or absence of the characteristic in a subject and perhaps its degree of severity are the values of the response variable. We make guesses about what the presumed explanatory variables (causes) are and prepare a checklist of what is to be noted. For example, STRESS may be one of the response variables of interest, so we decide to ask questions about taking timeout and availability of friends to discuss the day's events. We may use two kinds of studies, each serving a different purpose: 1) the cross-sectional survey, which gives a snapshot of the population of concern on the explanatory and response variables; and 2) the case-control study, in which we collect data about explanatory variables from those with (case) and without (control) the characteristic we observed.

Cross-Sectional Survey A cross-sectional survey is similar to a retrospective study. Whereas a retrospective study collects data about the participants' pasts, a cross-sectional survey collects data about one point in time, the present. Sometimes a questionnaire collects information about both the past and the present; therefore the distinction is often not clear. We like to think of a cross-sectional survey as a kind of retrospective study—one in which the past time period, at least for some of the variables, is very short.

The student questionnaire we gave in Chapter 2 is a cross-sectional survey. It tells about the sample at one particular point in time. That questionnaire, for example, tells us how the students feel about their stress levels. By analyzing the data we may even be able to make some guesses about why they are stressed. The TIMEOUT and DISCUSS variables may be looked upon as explanatory variables. Even if statistical tests show an association between stress and not taking at least five minutes of timeout during the day, we *cannot* assume a causal relationship between TIMEOUT and STRESS. A cross-sectional study helps with the formulation of initial guesses. The AIDS article in Chapter 1 (Literature Example 1.1) is another example of a cross-sectional survey. It describes the knowledge of adolescents about AIDS at one point in time.

After an instrument (e.g., a questionnaire) is designed, tested, and duplicated, it may be administered to an appropriately chosen sample. A cross-sectional survey is an observational study, so there is no intervention. The design may be denoted by the symbol $\mathbf{O_1}$, which stands for one observation. With a cross-sectional survey, observation is done through the administration of a questionnaire.

Analyses depend on the purpose of the investigator. Some investigators might choose to present simple descriptions of the samples (using tables and graphs). Others might perform tests to see if some of the variables are correlated instead or in addition.

Examples from the Literature

Literature Example 9.3 is from the article entitled "Stress Process Among Mothers of Infants: Preliminary Model Testing." Read the article and complete Activity 9.6.

ACTIVITY 9.6

What is the purpose of the study? 1 ______________________

What is the source of the sample? 2 ______________________

What is the instrument for gathering data? 3 ______________________

What is the sample size? How did researchers select the sample? 4 ______________________

What measures did researchers take to ensure an adequate response? 5 ______________________

Why is this a cross-sectional survey? 6 ______________________

LITERATURE EXAMPLE 9.3

Stress process among mothers of infants: preliminary model testing

Lorraine O. Walker

Nursing Research, 1989, vol. 38, no. 1

Maternal employment, cesarean birth, and infant difficultness were used to test the mediating effect of perceived stress and the stress-buffering role of health practices on maternal identity. One hundred seventy-three mothers returned a parenting survey that focused on: stressors, perceived stress, health practices, maternal identity, and a demographic profile. Work status and infant difficultness were related to perceived stress. Neither had direct effects on maternal identity, but were related to it through the mediating effects of perceived stress. While health practices did not show buffering effects between stressors and perceived stress, these did contribute additively to the prediction of stress perception. Also, health practices contributed additively to the prediction of identity. Notable among the health practices predicting identity were self-actualizing expression, nutrition, interpersonal support, and stress management. These findings support a stress process model of parenting in which: (a) effects of stressors on maternal identity are mediated by perception of stress, and (b) health practices contribute positively and directly to maternal identity.

Method

Subjects and Procedures: A random sample of 330 mothers was drawn from birth announcements published in a southwestern newspaper over a 10-month period using a table of random numbers. Mothers were stratified by age of infant with 30 mothers represented for each month of infant age from 2–12 months. Each mother was mailed a questionnaire accompanied by a stamped return envelope. Reminder notices were sent out twice at 2-week intervals to encourage nonresponders to complete and mail the questionnaire. Research procedures were approved by the Institutional Review Board.

Literature Example 9.4 is from the article entitled "No-Smoking Laws in the United States: An Analysis of State and City Actions to Limit Smoking in Public Places and Workplaces." Read the article and complete Activity 9.7.

Activity 9.7

What is the purpose of the study? 1 ______________________

What is the source of the sample? 2 ______________________

What instrument was used for gathering data? 3 ______________________

What is the sample size? How did researchers select the sample? 4 ______________________

____

____

What measures did researchers take to ensure accuracy? 5 ______________________

____

____

Why is this a cross-sectional survey? 6 ______________________

LITERATURE EXAMPLE 9.4

No-smoking laws in the United States: An analysis of state and city actions to limit smoking in public places and workplaces

Nancy A. Rigotti, MD, Chris L. Pashos, PhD

Journal of the American Medical Association, 1991, vol. 266, pp. 3162–3167

Objective

To assess the prevalence, content, and growth of state and city laws restricting smoking in public places and workplaces in the United States and to identify factors associated with their passage.

Design

A mailed survey of city clerks in US cities with a population of 25 000 or greater (N = 980) and review of existing data sources confirmed the status of smoking restrictions in 902 (92%) of the cities in the sample. State laws were identified by contacting each state's Legislative Reference Bureau (100% response). Content of laws was coded using previously developed categories.

Main outcome measures

Prevalence, comprehensiveness, and cumulative incidence of no-smoking laws in states and in cities with a population of 25 000 or greater.

Results

By July 1989, 44 states and 500 (51%) of the cities in our sample had adopted some smoking restriction, but content varied widely. While 42% of cities limited smoking in government buildings, 27% in public places, 24% in restaurants, and 18% in private workplaces, only 17% of cities and 20% of states had comprehensive laws restricting smoking in all four of these sites. The number of city no-smoking laws increased tenfold from 1980 to 1989. City no-smoking laws were independently associated with population size, geography, state tobacco production, and adult smoking prevalence. Laws were more common in larger cities, Western cities, and states with fewer adult smokers. Laws were less common in tobacco-producing states and in the South.

Conclusions

No-smoking laws are more widespread than previously appreciated, especially at the local level, reflecting a rapid pace of city government action in the 1980s. Nonetheless, comprehensive laws, which are most likely to provide meaningful protection from environmental tobacco smoke exposure, remain uncommon and represent a major gap in smoking control policy. Laws are most needed in smaller and non-Western cities and in states that produce tobacco and have a higher proportion of smokers. (*JAMA*, 1991;266:3162–3167)

Methods

Data collection

States To identify current state laws, we updated our previous compilation by requesting a copy of any state no-smoking law from the director of each state's Legislative Reference Bureau in September 1989. Responses were obtained from 49 of 50 states and the District of Columbia; a telephone call to the nonresponding state (Missouri) confirmed that it had no law.
Cities Using 1980 census data, we

identified all US cities and towns with a population of 25 000 or greater (N = 980); this was our sample. We updated population figures for cities with a population of 50 000 or greater using 1986 data. (Updated figures for cities with a population of 25 000 to 50 000 were not available.) We used two data sources to determine whether a smoking restriction was present: (1) information provided by city clerks in response to a mailed request and (2) existing compendia of laws. We requested information from all cities in the sample not cited by both sources as having laws (n = 831). We reasoned that there was little chance that a city cited by two independent sources did not have a law. We confirmed this by obtaining from TFA copies of 100% of ordinances from a random sample of 40 of the 149 cities cited by both ANR and TFA as having laws. From April through June 1989, a mailing to city clerks asked them to indicate whether the city had passed or was considering "any law or regulation that restricts smoking in a public place or workplace" and to mail us a copy of any law. A second mailing was sent to nonresponders. We received responses from 753 of 831 cities (91% response rate). For the analysis, a city was counted as having a smoking restriction if it met one of two criteria: (1) we had a copy of a city or town ordinance or regulation or (2) both compendia reported that the city had a no-smoking ordinance. County ordinances and laws in cities with a population of less than 25 000 were excluded.

Case-Control Study Literature Example 9.5 is the abstract of an article entitled "The Presence and Accessibility of Firearms in the Homes of Adolescent Suicides: A Case-Control Study." In a case-control study, those who have a positive (yes) value for the response variable (i.e., suicide victims) are designated as the cases. Those who have a negative (no) value (i.e., those who never attempted suicide and those who attempted suicide but failed) are selected as the controls. The explanatory variable is presence of a gun in the house. First cases are selected. Next controls are selected from a pool of potential subjects who match the cases on various characteristics that may be identified as alternative explanatory variables. In this example, two control groups were selected from a pool of psychiatric patients by frequency matching individuals in the pool with the subjects on several demographic factors.

The frequencies of the values of the response variable are tabulated for each value of the explanatory variable. In this example, the explanatory variable is the presence or absence of a gun in the house.

Explanatory Variable, or Risk Factor (gun in the house)	**Response Variable (suicide victims)**	
	Positive (cases)	**Negative (controls)**
Present	a	b
Absent	c	d

The crucial part of the analysis of a case-control study is the computation of the odds ratio:

$$\text{Odds ratio (OR)} = \frac{a}{b} \div \frac{c}{d}$$

The formula can be rewritten as ad/bc. The odds ratio is an index of strength of association between the response variable (suicide) and the explanatory variable (the presence of a gun in the house). Risk factor is another name for the [1] explanatory variable/response variable. (a/b) indicates the ratio of cases to controls that [2] have/do not have positive values for the explanatory variable (risk factor); (c/d) is the ratio of cases to controls that [3] have/do not have positive values for the explanatory variable (risk factor). The odds ratio is the [4] _______ of these two ratios. When the explanatory variable is positive, [5] _______ is the ratio of cases to controls. [6] _______ is the ratio of cases to controls when the explanatory variable is negative. The formula

$$\text{Odds ratio} = \frac{a}{c} \div \frac{b}{d}$$

is algebraically equivalent to the formula

$$\text{Odds ratio (OR)} = \frac{a}{b} \div \frac{c}{d}$$

The ratio of cases with the risk factor to those without the risk factor is 7 ______________. The ratio of controls with the risk factor to those without the risk factor is 8 ______________. The odds ratio computed using either formula turns out to be the same.

Study Literature Example 9.5 and complete Activity 9.8.

ACTIVITY 9.8

What is the purpose of the study?	1 ______________
What are the sources of the cases?	2 ______________
What are the sources of the two controls?	3 ______________
What instrument was used for gathering data?	4 ______________
What are the sizes of case and control groups? On what variable did researchers match the controls?	5 ______________
What is the main response variable?	6 ______________
Why is this a case-control study?	7 ______________

LITERATURE EXAMPLE 9.5

The Presence and Accessibility of Firearms in the Homes of Adolescent Suicides

A Case-Control Study

David A. Brent, MD; Joshua A. Perper, MD, LLB, MSc; Christopher J. Allman; Grace M. Moritz, ACSW; Mary E. Wartella, MSW; Janice P. Zelenak, PhD

Objective.—The presence of guns in the home, the type of gun, and the method of storage were all hypothesized to be associated with risk for adolescent suicide.

Design.—Case-control study.

Subjects.—The case group consisted of 47 adolescent suicide victims. The two psychiatric inpatient control groups were 47 suicide attempters and 47 never-suicidal psychiatric controls, frequency-matched to the suicide victims on age, gender, and county of origin.

Setting.—The cases were a consecutive community sample, whereas the inpatients were drawn from a university psychiatric hospital.

Main Outcome Measure.—Odds of the presence of guns in the home of suicide victims (cases) relative to controls.

Results.—Guns were twice as likely to be found in the homes of suicide victims as in the homes of attempters (adjusted odds ratio, 2.1; 95% confidence interval, 1.2 to 3.7) or psychiatric controls (adjusted odds ratio, 2.2; 95% confidence interval, 1.4 to 3.5). Handguns were not associated with suicide to any statistically significantly greater extent than long guns. There was no difference in the methods of storage of firearms among the three groups, so that even guns stored locked, or separate from ammunition, were associated with suicide by firearms.

Conclusions.—The availability of guns in the home, independent of firearms type or method of storage, appears to increase the risk for suicide among adolescents. Physicians should make a clear and firm recommendation that firearms be removed from the homes of adolescents judged to be at suicidal risk.

(*JAMA*. 1991;266:2989-2995)

METHODS AND PROCEDURES

Sample

The suicide completer sample was a consecutive series of suicide victims aged 19 years and younger drawn from 28 Western Pennsylvania counties over a period from July 1986 through February 1988. Of 64 adolescent suicides that occurred over this time, the families of 48 youthful victims agreed to participate (75% compliance rate). Of the 16 suicides not included in the study, the families of six refused outright, and 10 could not be traced. The sample size of completers for the present study was 47, because data on firearms availability were not obtained on one suicide victim. Those victims whose families agreed to participate did not differ from those whose families either refused or could not be contacted with regard to age, sex, race, method of suicide, county of residence, or toxicology. The method of approaching the parents and friends of the suicide victims has been described previously, but this is a new sample distinct from that of previous reports.[3,14] The parents of the suicide victims were contacted by mail approximately 4 months after the death, inviting participation in our study. This letter was followed up by a telephone call by the project coordinator within a week of the receipt of the letter to schedule an interview. Besides parents, siblings and close friends of the victims were also contacted and interviewed.

The suicide attempter sample was drawn from the adolescent inpatient units at Western Psychiatric Institute and Clinic (WPIC), a state-affiliated university hospital that draws from a large geographic area and provides treatment for patients representing the full socioeconomic spectrum. Of 97 attempters approached for participation, 90 agreed to participate (92% compliance). In order to have been considered for this study, the attempter must have been between the ages of 13 and 19 years and have engaged in self-destructive behavior with active suicidal intent that was not purely self-mutilative in nature. Exclusionary criteria were an IQ of less than 70; inability to cooperate with the interview because of delirium or psychosis, and comorbid chronic physical illness, all conditions not found in the above-noted sample of completers.

The never-suicidal psychiatric patient group (referred to in this report as psychiatric controls) was drawn from the same inpatient units at WPIC as noted above, with similar inclusionary and exclusionary criteria, except that the psychiatric controls could not have had current or past clinically significant suicidality (suicidal ideation with a plan or intent, or actual suicidal behavior). Of 116 patients who were approached, 103 consented to participate (89% compliance). These groups of 90 attempters and 103 psychiatric controls consisted of 78% of all patients admitted to WPIC who met study inclusionary and exclusionary criteria and were recruited over the same time period as the completers. The patients recruited for study did not differ with respect to age, race, gender, or social class from those who refused or who, for logistic reasons, were not approached (eg, brief admission, project coordinator unavailable).

The total sample of 90 attempters and 103 psychiatric controls differed from the completers on age, gender, and county of origin. Therefore, 47 attempters and 47 psychiatric controls were selected on these three variables and were frequency matched to the sample of suicide victims, so that there were no statistically significant differences among the three groups on the above-noted demographic factors. Those attempters and psychiatric controls who were chosen for this study did not differ from those who were not selected on firearms variables (physical presence, method of storage, type of gun), family constellation, previous treatment, or psychiatric diagnoses. In fact, analyses comparing the 47 completers with the entire sample of the 90 attempters and 103 psychiatric controls yielded results quite similar to those presented below but required statistical adjustment for several demographic variables. Therefore, in the interest of parsimony, only analyses on the 47 completers and the frequency-matched control samples are presented.

Prospective (Longitudinal) Studies

Cohort Study: Individuals Sampled from the Same Specific Group When we follow a defined population and record the explanatory variables (factors) and response variables over a period of time, we are conducting a cohort study. The explanatory variables are chosen on the basis of the scientist's educated guesses. When there is every reason to believe that an explanatory variable (e.g., smoking) is strongly associated with a given response variable (e.g., lung cancer) we may select two groups with and without the explanatory variable (e.g., groups of smokers and non-smokers) and follow the groups for a specified period of time to note the response (e.g., the occurrence of lung cancer). The number of occurrences of the response variable at any given period of time can be estimated in both groups.

	Response Variable	
Explanatory Variable, or Risk Factor	**Present (cases)**	**Absent (cases that do not manifest the condition in the given period)**
Present	a	b
Absent	c	d

With a prospective study we start with a group of subjects for all of whom the value of the response variable is currently negative (for example, they are all free of lung cancer). These subjects are divided into two groups: one that has the risk factor (explanatory variable positive—smokers) and one that does not (explanatory variable negative—non-smokers). We follow all these subjects for a period of time and observe for how many the response variable changes to positive (how many acquire lung cancer). In contrast, with case-control studies we start with two groups, cases and controls, and look back in time to see how many in each group had positive values for the explanatory variable. In prospective studies it makes sense to ask what proportion of those with the risk factor became cases, $a/(a + b)$, and what proportion of those without the risk factor became cases, $c/(c + d)$. These two proportions are called risks, the value of $a/(a + b)$ expresses the risk of becoming a case when the risk factor is present; the value of $c/(c + d)$ expresses the risk of becoming a case when the risk factor is not present. The ratio of these two proportions is known as relative risk (RR).

$$\text{Relative risk (RR)} = \frac{a}{(a+b)} \div \frac{c}{(c+d)}$$

When a is small compared to $a + b$ and c is small compared to $c + d$ (i.e., there are very few cases of the illness in either group), the odds ratio (OR) is a good estimate of the relative risk.

Example from the Literature

Literature Example 9.6 is from the article entitled "Incidence and Risk Factors for Gout in White Men." Read the abstract and complete Activity 9.9.

ACTIVITY 9.9

What is the purpose of the study? 1 ____________________

What are the sources of the cases? 2 ____________________

Was gout the main focus of the study? 3 ____________________

How were the data gathered? 4 ____________________

____

____

How many cases of gout were detected? 5 ____________________

What is the main response variable? 6 ____________________

Why is this a longitudinal cohort study? 7 ____________________

LITERATURE EXAMPLE 9.6

Incidence and Risk Factors for Gout in White Men

Ronenn Roubenoff, MD, MHS; Michael J. Klag, MD, MPH; Lucy A. Mead, ScM; Kung-Yee Liang, PhD; Alexander J. Seidler, PhD; Marc C. Hochberg, MD, MPH

Objective.—To identify potentially modifiable risk factors for the development of gout.

Design.—Longitudinal cohort study (The Johns Hopkins Precursors Study).

Participants.—Of 1337 eligible medical students, 1271 (95%) received a standardized medical examination and questionnaire during medical school. The participants were predominantly male (91%), white (97%), and young (median age, 22 years) at cohort entry.

Outcome Measure.—The development of gout.

Results.—Sixty cases of gout (47 primary and 13 secondary) were identified among 1216 men; none occurred among 121 women ($P = .01$). The cumulative incidence of all gout was 8.6% among men (95% confidence interval, 5.9% to 11.3%). Body mass index at age 35 years ($P = .01$), excessive weight gain (>1.88 kg/m^2) between cohort entry and age 35 years ($P = .007$), and development of hypertension ($P = .004$) were significant risk factors for all gout in univariate analysis. Multivariate Cox proportional hazards models confirmed the association of body mass index at age 35 years (relative risk [RR]=1.12; $P = .02$), excessive weight gain (RR=2.07; $P = .02$), and hypertension (RR=3.26; $P = .002$) as risk factors for all gout. Hypertension, however, was not a significant risk factor for primary gout.

Conclusions.—Obesity, excessive weight gain in young adulthood, and hypertension are risk factors for the development of gout. Prevention of obesity and hypertension may decrease the incidence of and morbidity from gout; studies of weight reduction in the primary and secondary prevention of gout are indicated.

(*JAMA*. 1991;266:3004-3007)

SUBJECTS AND METHODS

Study Population and Design

The Johns Hopkins Precursors Study was started in 1947 to identify characteristics that would predict development of coronary heart disease and hypertension in medical students. Of 1337 medical students who entered the graduating classes of 1948 to 1964, there were 1271 (95%) who received a standardized medical examination and questionnaire during medical school. The examination included weight, height, blood pressure, and serum cholesterol measurements, and the questionnaire included information about their own and their parents' medical conditions, as well as health and dietary habits. Serum uric acid levels were not measured as gout was not a primary end point in the original design. Since graduation, the cohort has been followed up with yearly questionnaires to detect disease and update risk factor information.

Experimental Studies

Cross-sectional studies collect data on response and explanatory variables simultaneously. These are snapshots of the characteristics of the population at one point in time. Cross-sectional studies may help us detect some significant associations between explanatory and response variables that may later be checked through retrospective studies. The retrospective studies give us confidence in our hunches about the cause of the observed characteristic or phenomenon. To confirm causation we need experimental studies in which we actually treat a group of subjects and assess the response variable.

We can infer that an intervention was effective only if the measurements of the response variable before and after the experimental intervention differ in a significant way. Observation of one individual is not enough. We must measure the effect of intervention on several individuals. Measurements will vary from person to person. In Chapters 14 and 15 we will study statistical methods for determining when the differences in the data from before the intervention and after the intervention are enough larger than the variations within the sample to conclude that the intervention is likely to have caused them. For now, we will present some simple descriptive techniques based on comparisons of graphs and summary statistics that can be used by researchers to compare before and after data. These methods are far from conclusive, but they will reveal gross changes.

We will compare before and after distributions of pulses from the student questionnaire data by using box plots, histograms, and summary statistics. Consider Figure 9.3. The lower and upper hinges of the box plot mark first and third quartiles. Medians of the two pulses do not overlap. The "boxes" of the pulses do not overlap. Some pulses do overlap, however. Running in place seems to raise the pulse.

From the histograms in Figure 9.4 we can see that the mode of the pulses also shifted to the right after running in place. The histogram shows us a second and somewhat different view of what happened to the pulses. For a third view, we can look at the following summary statistics, given by a commonly used computer statistics program, Minitab:

	n	Mean	Median	TrMean	St Dev	SE Mean	Min	Max	Q1	Q3
pulse1	50	76.20	73.00	76.18	11.18	1.58	48.00	104.00	68.00	84.00
pulse2	50	98.00	100.00	98.32	17.13	2.42	54.00	136.00	87.00	113.00

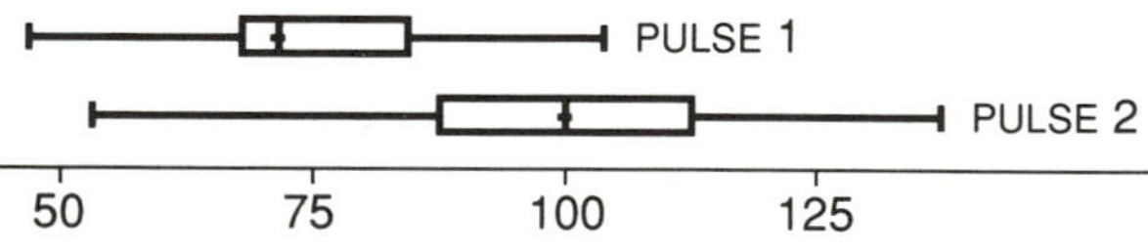

FIGURE 9.3 Box Plots of Student Pulses Before and After Running in Place

We see immediately that the mean of PULSE 2 is greater than the mean of PULSE 1. We know that each time we take a sample from the same population we are likely to obtain a different mean. Is it possible that this is all that is causing the variation in this instance? In other words, can the difference in means be accounted for by sampling variation? An index of such sampling variation (or chance variation) is the standard error of the mean. A commonly used criterion is that if the difference is more than two standard errors of the mean (2 * SE) the difference is probably an effect of the intervention.

A useful technique is to pair before and after pulses for each person and plot the paired differences. In other words, we will examine a new variable, the change in pulse after running (PULSE 2 − PULSE 1) for each subject.

	n	Mean	Median	TrMean	St Dev	SE Mean	Min	Max	Q1	Q3
PULSE 2 −PULSE 1	50	21.80	20.00	21.55	11.59	1.64	0.00	48.00	12.00	30.50

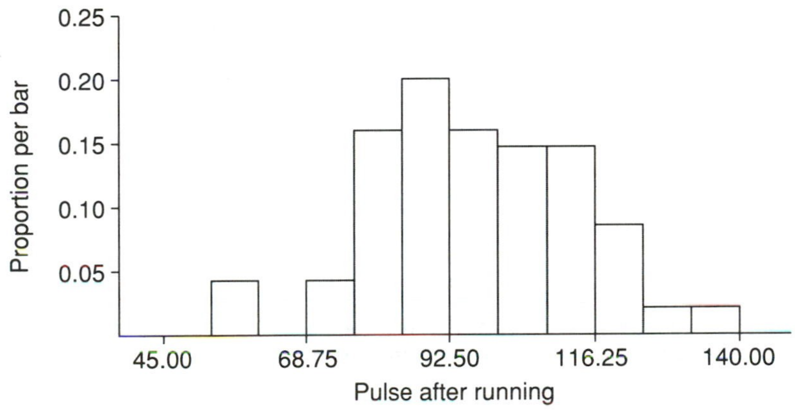

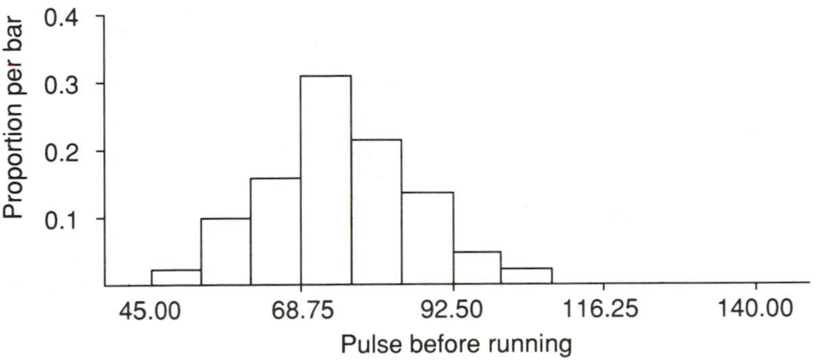

FIGURE 9.4 Histograms of Student Pulses Before and After Running in Place

A difference of zero probably indicates that the intervention had no effect. (Notice that the minimum difference was zero.) But it's also possible that some of the subjects may have cheated by not running. It's also possible that the zero differences represent errors in recording the pulses. We should probably examine the data to see how many such zero differences there are; if there are only one or two we can probably discount them.

Is the difference in pulses something other than zero? The mean of the differences is 21.8, and that seems to suggest that the answer is yes. Again we must ask the same question, namely, could this difference of 21.8 have occurred by chance alone? The decision criterion that we will use at this stage is that the mean should differ from zero by at least two standard errors of the mean. Does it?

Now consider Figure 9.5. It is clear from the histogram that one person's pulse did not get affected. Ultimately, that fact needs to be explained. However, in general we can conclude that it is highly likely that running in place affects pulse.

As another example, consider the faculty exercise study, in which faculty exercised for four months. Did this affect their percent of body fat? Consider Figure 9.6. Median body fat percentage shifted. Contrast this box plot with the previous box plots for student pulses. The boxes representing the middle 50 percent of the observations did not overlap in the case of pulses. Here they did. The effect of the intervention (exercise) is not as clearcut in this case as in the case of pulses. Consider the data below.

	n	Mean	Median	TrMean	St Dev	SE Mean	Min	Max	Q1	Q3
PrFatHi	9	30.79	31.77	30.79	7.08	2.36	21.73	40.67	24.16	37.41
PoFatHi	9	28.86	28.10	28.86	6.09	2.03	18.00	35.52	24.45	35.46
FatDiff	9	−1.94	−0.89	−1.94	4.93	1.64	−12.96	3.39	−4.50	1.60

Note: PrFatHi is the FED variable PREFAT% restricted to the high-intensity exercise group. Similarly, PoFatHi is the variable POSTFAT so restricted. FatDiff is PoFatHi − PrFatHi for each participant.

The body fat differences ranged from −12.96 to 3.39 (the Min and Max). The difference in means is −1.94. Standard error of the mean is 1.64. Again we ask the same question: Is this difference of −1.94 a result of the intervention or is it random fluctuation? Applying the criterion, we see that the mean does not differ from zero by two standard errors of the mean so this difference is probably not caused by the intervention.

What attributes of a sample determine whether the observed difference is significant? It turns out that the most decisive one is the variation in the mean of the differences. This variation is measured by the standard error, and statistical theory tells us that if an intervention made no difference, approxi-

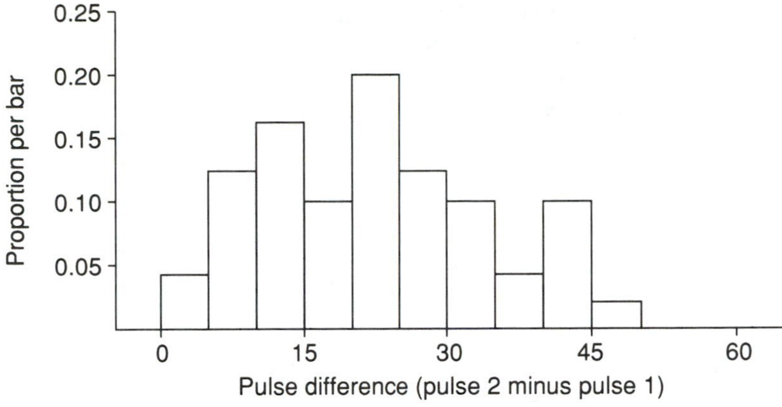

FIGURE 9.5 Histogram of Differences in Pulses Before and After Running in Place

mately 95 percent of the possible samples we could have taken (having the same size as our sample) would have had a mean difference within two standard errors of zero.

Even though a difference may be small, that difference may still be significant if it's large in comparison with the SE. Experiments can be designed to make the standard error of the mean very small. This is desirable, because when there is a significant, though small, difference in the observed variable from before the intervention to after the intervention it is likely to show up as more than two times the standard error. Conversely, if the mean of the differences is less than twice the standard error, it's likely that that difference was caused by random variations in the variables rather than by the intervention. How can SE be made small? Notice the components of the standard error formula:

$$SE_{\bar{x}} = \frac{s}{\sqrt{n}}$$

It has s in the numerator and $\sqrt{n}$ in the denominator. We can make the standard error small by making s small and/or by making n large. The standard deviation

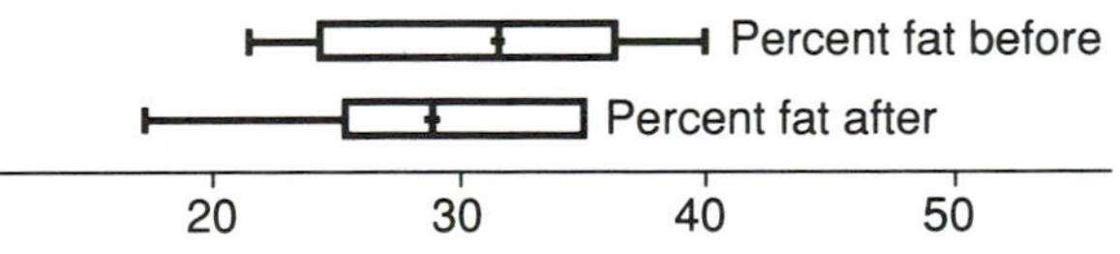

FIGURE 9.6 Percent Body Fat Before and After Four Months of High-Intensity Exercise

of the observations (in this case the differences in some variable before and after an intervention), s, can be made small by selecting the sample from a uniform, or homogeneous, population (a population with small standard deviation). Making n large means selecting a large sample. When n is large, since we divide by $\sqrt{n}$, $SE_{\bar{x}}$ will be small.

By using a powerful enough intervention we can make sure that the mean of the differences is large enough to be detected and not obscured by the standard error. In the faculty exercise research study, exercise itself was not powerful enough to reduce body fat; perhaps it should have been combined with regulation of diet.

Other extraneous variables contribute to the error, or variation (for instance, the effect of white coats on diastolic blood pressure, in Literature Example 9.2). We can control the external variables either by eliminating them, if they are known, or by building them into the design by holding different levels of the variable (e.g., SEX) constant. For example, if researchers suspect that exercise affects the pulse rates of men and women differently, we should compare the before and after means of males and females separately. (Refer to the discussion of three-way tables in Chapter 6 to review the concept of holding a variable constant.) If we randomly assign subjects to treatment and control groups, both groups will be comparable with respect to the unknown external variables, and thus we control for the external factors. The concepts discussed above make up the "MINMAXCON" principle.

MINMAXCON Principle

The following three imperatives should be considered when designing an experiment:

MIN: MINimize variation *within* the sample of subjects (makes s small).
MAX: MAXimize the variation *between* the means of the experimental and control groups, or maximize the mean of the difference between measurements taken before and measurements taken after an intervention.
CON: CONtrol the extraneous variables.

Examples from the Literature

Literature Example 9.7 is from the article entitled "Effects of an Exercise Program on Sick Leave Due to Back Pain." The authors used the MINMAXCON principle in designing the study. Read the abstract and complete Activity 9.10.

ACTIVITY 9.10

One way to minimize the variation within a sample is to select the target group from a homogeneous target population. How did the authors do this?

1 ______________________________

Researchers often select an intervention (treatment or stimulus variable) that is quite different from the subjects' usual practice; this maximizes the difference between the groups. (The subjects' usual practice may be the control.) What intervention did these researchers use?

2 ______________________________

It is best to select subjects randomly from the target population if it is feasible. Then randomly assign subjects to the experimental and control groups so that all variables are randomly distributed in each; in other words, the groups will be comparable on all variables *except* the treatment, stimulus, or independent variable. This will control for extraneous variables. Did these researchers follow this practice?

3 ______________________________

LITERATURE EXAMPLE 9.7

Effect of an exercise program on sick leave due to back pain

Karin M. Kellett, David A. Kellett, Lena A. Nordholm

Physical Therapy, 1991, vol. 71, pp. 283–293

The purposes of this study were to evaluate the effect of a weekly exercise program on short-term sick leave (<50 days) attributable to back pain and to determine whether changes in absenteeism were related to changes in cardiovascular fitness. Subjects were randomly assigned to an exercise group ($n = 58$) and a control group ($n = 53$). Sick leave attributable to back pain was determined in the intervention period of $1\frac{1}{2}$ years and a comparable $1\frac{1}{2}$-year period prior to the study. In the exercise group, the number of episodes of back pain and the number of sick-leave days attributable to back pain in the intervention period decreased by over 50%. Absenteeism attributable to back pain increased in the control group. The decrease in sick leave in the exercise group was not accompanied by any change in cardiovascular fitness. Suggestions for establishing exercise programs are given. [Kellett KM, Kellett DA, Nordholm LA. Effects of an exercise program on sick leave due to back pain. *Phys Ther. 1991;71:283–293.*]

METHOD

SUBJECTS

The participants in this study were employees of Marbodal AB, the biggest employer in Tidaholm, Sweden, and Scandinavia's major producer of kitchen units. The management showed considerable interest in the project and gave their agreement that about 60 employees could take part in an exercise program 1 hour a week for $1\frac{1}{2}$ years during paid working hours. Information about the project was sent to all employees of Marbodal AB. Criteria for inclusion in the study were self-reported current or previous back pain, written commitment to participate in the exercise program during working hours, and a willingness to exercise at least once a week outside working hours for $1\frac{1}{2}$ years. Exclusion criteria were any period of sick leave longer than 50 days, irrespective of cause, during the $1\frac{1}{2}$ years prior to the study, and medical reasons affecting the employee's ability to participate in the exercise program.

Of 850 employees, 143 volunteered to participate in the study. Eighteen volunteers were rejected according to the exclusion criteria. From the remaining 125 volunteers, 58 individuals were randomly selected to form the exercise group, and the remaining 67 individuals formed the control group. Ten individuals were not interested in continuing as part of the control group because they would not be participating in the exercise program, and 4 control group subjects did not participate in the initial cardiovascular fitness test. Thus, the exercise goup consisted of 58 subjects and the control group consisted of 53 subjects.

RESEARCH DESIGN

The study used a prospective, randomized, controlled research design to evaluate the effect of physical exercise intervention on the participants' sick leave attributable to back pain. The number of days of sick leave attributable to back pain and the number of episodes of back pain were recorded during the intervention period (period 2 = November 1, 1986–April 30, 1988) and compared with data recorded during a period prior to intervention (period 1 = November 1, 1984–April 30, 1986). The timing of the intervention and baseline periods was chosen to eliminate the effects of seasonal variation in sick leave. The data relating to sick leave were obtained from the register of the National Social Insurance Board in Tidaholm, Sweden. Participants were guaranteed anonymity. The project was carried out in Tidaholm by personnel from the Primary Health Care Centre and the industrial nurse at the factory of Marbodal AB.

Statistical Significance Versus Medical Significance

Our main interest in experimental studies is to see if a given treatment has a significant effect on the response variable. Historically, the phrase "statistically significant" has been used in a technical sense to mean that it is likely that a change in a variable is due to an experimental intervention or comes about because the subjects are not a part of a certain population (as opposed to being random fluctuation). This technical use of the word "significant" is often confused with the ordinary usage of the word, when it refers to an outcome that is noteworthy to the subject. For example, a reduction in diastolic blood pressure of 5 points may be statistically significant (it is unlikely for it to happen with no intervention) but is probably not medically significant, for it is unlikely to affect the subject's health.

Experimental Designs

Experimental designs differ in the degree of control the investigator is able to impose on them. Randomly choosing subjects from the appropriate target group, randomly assigning subjects to experimental and control groups, and conducting the experiment in a setting in which no external factors could affect the response variable would constitute **complete control**. **Partial control** involves compromise in the random assignment process and/or the setting in which the experiment is conducted. A control group is still required. Designs with no control group would be called designs with **minimal control**. The internal validity of an experiment is dependent upon the degree of control built into the experiment. Depending on the degree of control, designs are classified as true experimental, quasi-experimental, or pre-experimental. (Keep in mind, however, that the literature is far from consistent on the use of these terms.)

We are now ready to describe several experimental research designs. To refresh your memory, consider Figure 9.7, which repeats a portion of Figure 9.2 and shows some characteristics of experimental designs.

Notation

The notation that we will use to describe various designs uses certain symbols. O stands for an observation, and where more than one observation is taken, subscripts distinguish observations from each other. "Observation" is taken to mean a data set collected on a group of subjects. Each row in the symbolic representation of a design represents a different group of subjects. The elements in a row are arranged in the order in which they take place. Thus, $O_1 \quad O_2$ would mean that the second observation was made some time after the first observation, but on the same group of subjects. The letter X signifies an intervention. R stands for randomization, and RS means randomization in choice of subjects and assignment to groups. If the subjects were randomly chosen but not randomly assigned to experimental and control groups, we will denote that situation by R·. If the subjects were randomly assigned to groups but not randomly selected from the population (that is, an existing subset of the population was used as subjects or the subjects were self-selected), we will

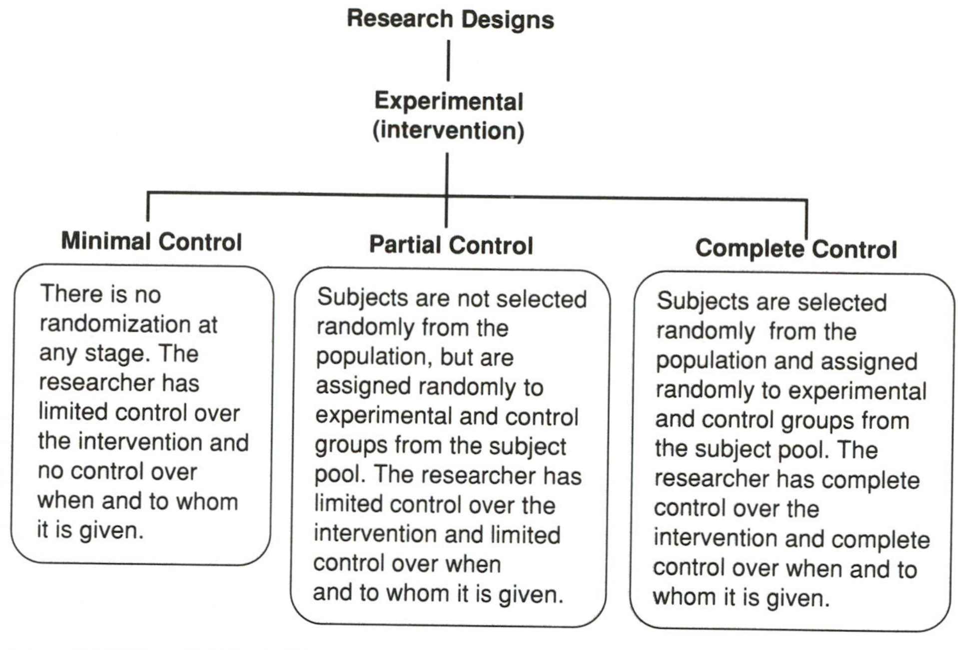

FIGURE 9.7 Experimental Research Designs

denote this situation by ·R. Frequently it is unclear in the literature whether there was one level or two levels of randomization.

Solomon Four-Group Design

The Solomon four-group design uses four groups: two experimental groups and two controls. This elaborate design is seldom used. We will use it to introduce experimental designs because it has all the elements that occur in the other designs we wish to consider. The Solomon four-group design contains several elements to ensure validity of the results.

RR	O_1	X	O_2	Exper. Group I
RR	O_3		O_4	Control Group II
RR		X	O_5	Exper. Group III
RR			O_6	Control Group IV

An experiment that follows the Solomon four-group design has subjects [1]randomly/arbitrarily assigned to [2] ______________ groups. Group I is an experimental group that is tested or observed both before and after the intervention (O_1 and O_2). The other experimental group, Group III, is tested only once, [3] before/after the intervention. Control Group II is tested when Group I is tested; there [4] is/is not an intervention. Group IV [5] is/is not tested before the intervention.

(Mean O_2 − mean O_1)	Gives an estimate of the treatment effect
(Mean O_5 − mean O_6)	Also gives an estimate of the [6] ______________ effect
(Mean O_4 − mean O_3)	Shows what change there was for the [7] ______________ group in the same period of time
(Mean O_4 − mean O_6)	Shows the effect of the pre-test
(Mean O_5 − mean O_2)	Also shows the effect of the [8] ______________

In this design there are two estimates of the treatment effects and two estimates of pre-test effects. (Mean O_2 − mean O_1) − (Mean O_4 − mean O_3) expresses the effect of the treatment minus the change in the response variable in the control, which was presumably due to effects of the pre-test.

Literature Example 9.8 illustrates how this design answers many questions about internal validity. Read the abstract and complete Activity 9.11.

LITERATURE EXAMPLE 9.8

A feasibility study of a values clarification approach in a community drug prevention program

Charles Thomas Kozel

Thesis Submitted for Master of Public Health, Department of Health Science, California State University, Northridge, CA 91330, August, 1974

The purpose of this study was to test the relationship of an individual's valuing process to the following three experimental community educational programs: 1) a Values Clarification approach adapted to drug subject matter, 2) a Values Clarification approach only, and 3) drug content development only. The field experiment was designed to test the relationship of a combined program as well as the separate programs to the changing of an individual's valuing process. Five groups were utilized in the study. Three of the groups had experimental programs presented, whereas the other two groups were control groups (no programs). Analysis was conducted within each separate group between the pre-, post-, and post-post questionnaires by using the *t*-test of significance between the means.

Modified Solomon Four-Group Design

First Week	Expt/Control	Second Week	Fifth Week	Groups	
$O_{1\text{Pre-test}}$	Expt. VCADSM	$O_{1\text{Post-test}}$	$O_{1\text{Post-Post-test}}$	I	$n = 16$
$O_{2\text{Pre-test}}$	Expt. VC	$O_{2\text{Post-test}}$	$O_{2\text{Post-Post-test}}$	II	$n = 10$
$O_{3\text{Pre-test}}$	Expt. DCD	$O_{3\text{Post-test}}$	$O_{3\text{Post-Post-test}}$	III	$n = 15$
$O_{4\text{Pre-test}}$	Control	$O_{4\text{Post-test}}$	$O_{4\text{Post-Post-test}}$	IV	$n = 15$
	Control	$O_{5\text{Post-test}}$	$O_{5\text{Post-Post-test}}$	V	$n = 15$

O represents any set of data collected on subjects. Subscripts are used to denote specific sets of data. Since there were fourteen distinct sets of observations, Kozel chose to double-subscript rather than to use the scheme described in the previous discussion. Thus, all observations on Group I are subscripted with a 1, all observations on Group II are subscripted with a 2, etc.

Treatment Levels
Level 1: VCADSM = Values Clarification Adapted to Drug Subject Matter
Level 2: VC = Values Clarification approach only
Level 3: DCD = Drug Content Development only

ACTIVITY 9.11

Internal Validity Considerations	How Design Addresses Them
Contemporary history: During the experiment something else may happen that will influence the response variable besides the treatment, e.g., subjects may watch a TV program about drugs while participating in the experiment.	If this happens the control and experimental groups are equally affected. So subtracting the control group effect will remove the "contemporary history" effects.
Maturation process: This study lasted five weeks. Five weeks is a long period in a young person's life. The observed effect in Groups I through III may just be due to boys' maturing. What aspect of the design takes care of this concern about maturation?	[1] ________________________________ ________________________________
Pre-testing procedures: The pre-testing procedure itself sensitizes the boys to selectively acquiring information on their own.	Pre-test effects will exist for both control and experimental groups. So pre-test effects will be accounted for. Also, the design gives an estimate of the pre-test effect. What is it? (mean $O_{4\text{Post-test}}$ − mean [2] ________)
Measuring instruments: The proficiency of the person who grades the test may increase and thereby change scores on the post-test. To counteract this possibility a multiple-choice instrument was used.	
Statistical regression: Sometimes the subjects for an experiment are selected because they got either exceptionally good or exceptionally poor scores on a test. A phenomenon called the regression effect causes the high scorers on a test to get lower average scores on the next test and the low scorers to get higher average scores even if there has been no intervention between tests.	Subjects were not chosen for extreme values but just happened to belong to scout troops.

This effect is caused by random factors that affect scores (health, emotional state, and luck, for example). These factors are likely to affect each individual differently on the different tests. So those who were lucky on the first test won't necessarily be lucky on the second test. Samples from a population tend to give a mean close to μ, the mean of the population. We say that scores tend to regress toward the mean. Such changes may be wrongly attributed to the intervention.	
Differential selection of subjects: If the subjects are not randomly assigned to the experimental and control groups, the differences in the tests may be a reflection of the pre-experiment differences rather than the effect of the treatment.	Subjects were randomly assigned to all the groups.
Differential experimental mortality: Subjects who are performing poorly or exceptionally well may drop out of a study. This will affect the means of the response variable.	No dropouts were recorded in this experiment.
Interaction of all the above factors: Even when the pre-test scores of both control and experimental groups are comparable, motivation, intelligence, or some other factor may affect the experimental groups differently. The effects of extraneous variables can be alleviated by random assignment of subjects to experimental and control groups.	Even though the subjects were randomly assigned, the author noted that the first meeting showed clear differences between the groups. So he confined his entire analysis to differences between pre- and post-tests within the groups.

The Solomon four-group design seems to control for many aspects of internal validity, yet the drug prevention education program ran into difficulties because of non-comparability of the groups. The researcher noticed this when he met with the groups to administer tests and give "treatments." Because he needed to go through with the experiment, the researcher made adjustments in the analysis.

Research designs are chosen to serve different purposes and fit different situations. We may sometimes choose designs with lesser control because opportunities present themselves to obtain information that would otherwise not be available. We may want to evaluate a certain procedure when external conditions guarantee that the population has not been exposed to conflicting procedures. The opportunity to do this may be transitory and not allow time to do randomization or pre-testing, so we might be forced to adopt a design with minimal or partial control. All designs have their purposes. It is important, however, to be aware of the concepts of internal and external validity and to fully disclose all limitations in design and their consequences.

Now that you know something about the Solomon four-group design, we will consider the advantages and limitations of some less elaborate research designs.

Research Designs and Analyses of Data

O

Suppose your purpose is to study one group at one point in time; no comparison with any other group is intended. Such a design would be symbolized O_1. (Note that O_1 may involve many response variables or factors for many subjects belonging to one group. It does not signify observation of a single variable.) With such a design you can describe the group on the chosen variables and perhaps compare these variables with reports from the literature. You can describe the distribution of O_1 and compute summary statistics for O_1. For example, from among the faculty exercise data (FED) cards, you might separate the non-participants and analyze BODY FAT % and HEART RATE for the subjects in this group. You could describe these variables with summary statistics such as mean, mode, median, standard deviation, range, and inter-quartile range and display the distribution in tables and graphs. This design is an observational rather than experimental design because there is no intervention.

X O

With this type of design the researcher makes one set of observations of the subjects after conducting an experiment or administering some treatment. Can the researcher claim that the measure of the response variable is due to the treatment? It is difficult to tell, because the subjects were not administered pre-tests, so we do not have a measure of the response variable before the treatment. The amount of control in an experiment that uses this design is minimal.

$$\begin{array}{ll} X & O_1 \\ & O_2 \end{array}$$

With this type of design, subjects are not randomly selected, nor are they assigned randomly to the control and experimental groups. Thus the groups are not comparable. The researcher has partial control (that is, there is slightly more control with this design than with the design described above).

$$\begin{array}{lll} \cdot R & X & O_1 \\ \cdot R & & O_2 \end{array}$$

With this type of design, subjects are randomly assigned to experimental and control groups, but they are not selected randomly from the population. The two groups are comparable because of randomization. However, no data are collected from either group before the intervention, so the researcher has only partial control with this design.

O_1 X O_2

In this design data are collected before and after the intervention, which may give some information, especially if the intervention is very potent. However, the experimenter has no way of determining whether other events that occurred between the pre-test and the post-test contributed to the differences (if any) in the response variable scores. Also, there is no randomization with this design, so it offers only partial control.

Using this design, you would describe response variables corresponding to O_1 and O_2 by comparing statistics such as means. You might display the distributions of pre and post values in tables or graphs. This would allow you to get an intuitive impression of whether the difference between the statistics for O_1 and O_2 was statistically significant. For example, if you were to sort out from the FED cases those who participated in a low-intensity exercise program, compute the mean of the pre and post resting heart rates, and analyze the difference, you would be using this design. If the difference between the means was significant, you would be left with a nagging doubt about whether the intervention actually caused the difference. Even if the difference were not significant this doubt would arise, because there could have been a chance event that canceled out the change caused by the intervention.

O_1 X O_2
O_3 $\quad$ O_4

By comparing a group that does receive the treatment with a group that does not receive it and collecting data twice from both groups, this design addresses the concern expressed above about the effects of events that occur between the pre- and post-tests. However, not randomizing the assignment of subjects to groups can lead to bias. So this design offers only partial control.

As an example of this design and the randomized variations that follow, suppose you were to sort the FED cases into control group and low-intensity exercise group. In this case, $O_2 - O_1$ would indicate the effect of the intervention (exercise) on a particular response variable (for example, heart rate and percent of body fat). If we did not have data on the control group we would never know whether an external event caused the change in the response variable. $O_4 - O_3$ would reveal whether the control group also showed the same difference between pre- and post-intervention. The next question would be whether the difference for the experimental group is larger than that for the control group. This would be investigated by comparing the mean of $O_4 - O_3$ with the mean of $O_2 - O_1$.

•R O_1 X O_2
•R O_3 $\quad$ O_4

With this design subjects are randomly assigned to experimental and control groups but are not randomly selected. This design is similar to the Solomon four-group design but does not include the two groups from which data were collected after the intervention. Thus we cannot estimate any effects caused by the first data collection (pre-test) with this design. The randomization permits partial control.

RR O_1 X O_2
RR O_3 O_4

This design is similar to the one discussed above except that with this design subjects are both randomly selected and randomly assigned to the experimental and control groups. This allows complete control.

Time-Series Designs

O_1 O_2 O_3 O_4

This design calls for the researcher to make a number of separate observations on the same group (a time series). Since there is no intervention, this is actually an observational, not an experimental, study. If the subjects are selected because they are conveniently available, we cannot generalize any results to other groups. How much control do you think the researcher exercises in such a study?

O_1 O_2 X O_3 O_4

This design is the same as the one discussed above, except that there is an intervention. The group serves as its own control in that results before the intervention act as a partial control on the results after the intervention. Clearly the subjects are not assigned randomly to the experimental and control groups. Since there is no randomization, the researcher has minimal control.

O_1 O_2 X O_3 O_4
O_5 O_6 O_7 O_8

With this design the researcher makes several observations over time of both the experimental group and the control group, then administers an intervention of some kind to the experimental group, and then makes several more observations over time of the experimental and control groups. The subjects are neither randomly selected nor randomly assigned.

•R O_1 O_2 X O_3 O_4
•R O_5 O_6 O_7 O_8

This design is like the one discussed above, except that the subjects are randomly assigned to experimental and control groups (though they are not randomly selected). Control in this situation is partial.

$$\begin{array}{llllll} RR & O_1 & O_2 & X & O_3 & O_4 \\ RR & O_5 & O_6 & & O_7 & O_8 \end{array}$$

With this type of time-series design, the subjects are randomly selected and randomly assigned to groups. Control is complete.

Examples from the Literature

Investigators adapt research designs to suit their purposes. Activity 9.12 will shed light on how designs are adapted to answer different research questions. Literature Example 9.9 is from the article entitled "Exercise Effect on Electromyographic Activity of the Vastus Medialis Oblique and Vastus Lateralis Muscles." The experiment used the design X O_1.

ACTIVITY 9.12

Example of X O_1 design

What was the purpose of the experiment? 1 ______________________

What does O_1 refer to? 2 ______________________

How was uniformity of measurements assured? 3 ______________________

____

____

The investigator's interest is in applying the knowledge gained from this experiment to work with patients. In your opinion, can the information they gathered be applied directly in working with patients? 4 ______________________

LITERATURE EXAMPLE 9.9

Exercise Effect on Electromyographic Activity of the Vastus Medialis Oblique and Vastus Lateralis Muscles

William P Hanten
Shane S Schulthies

The purpose of this study was to determine whether the vastus medialis oblique muscle (VMO) had greater electrical activity than the vastus lateralis muscle (VL) when hip adduction and medial (internal) tibial rotation exercises were performed. Electrical activity of the VMO and VL was measured on 25 healthy subjects during maximal-effort isometric contractions of hip adduction and medial tibial rotation. The results showed that the electromyographic activity of the VMO was significantly greater than that of the VL during the hip adduction exercise. Differences noted with medial tibial rotation were not significant. The results suggest that the VMO may be selectively activated by performing hip adduction exercises. Resistive hip adduction exercises, therefore, may be advisable in the treatment of patients with lateral malalignment of the patella with accompanying pain or instability. [Hanten WP, Schulthies SS. Exercise effect on electromyographic activity of the vastus medialis oblique and vastus lateralis muscles. Phys Ther. 1990;70:561–565.]

Procedure

Each subject attended one orientation session and one testing session. The two sessions were between 1 and 7 days apart and consisted of maximal voluntary isometric contractions (MVICs) of hip adduction, medial tibial rotation, and knee extension of the left lower extremity. We chose to test the left lower extremity on all subjects to maintain consistency and to avoid moving the equipment. The isometric knee extension exercise was used to normalize the EMG activity of the VMO and VL collected during hip adduction and medial tibial rotation exercises. The EMG values, therefore, were reported as a percentage of the EMG value obtained during maximal knee extension. This procedure allowed comparisons to be made between the EMG activity of the VMO and the VL.

Literature Example 9.10 is from the article entitled "Effect of Total Contact Cast Immobilization on Subtalar and Talocrural Joint Motion in Patients with Diabetes Mellitus." The experiment it describes uses the design $\begin{matrix} O_1 & X & O_2 \\ O_3 & & O_4 \end{matrix}$. Read the abstract and complete Activity 9.13.

ACTIVITY 9.13

What was the purpose of the experiment? 1 ______________________

What do O_1, O_2, O_3, and O_4 refer to? 2 ______________________

____

____

What was the intervention? 3 ______________________

LITERATURE EXAMPLE 9.10

Effect of Total Contact Cast Immobilization on Subtalar and Talocrural Joint Motion in Patients with Diabetes Mellitus

Jay E Diamond
Michael J Mueller
Anthony Delitto

Background and Purpose. *The purpose of this study was to determine the effect of total contact casting (TCC) on dorsiflexion at the talocrural joint (TCJ) and motion (inversion/eversion) at the subtalar joint (STJ).* ***Subjects.*** *Thirty-seven patients (29 men, 8 women), ranging in age from 32 to 79 years ($\bar{X}$=54, SD=11), with diabetes mellitus and a unilateral plantar ulceration participated in the study.* ***Methods.*** *The subjects were measured with a goniometer for dorsiflexion and STJ range of motion (ROM). The ROMs for each subject's casted and noncasted legs were compared before and after treatment with TCC for neuropathic plantar ulcers by use of a 2×2 repeated-measures analysis of variance design.* ***Results.*** *Mean time of immobilization in TCC (healing time) was 42 days (SD=43, range=8–119). The results indicated (1) ROM was unchanged at the STJ, but dorsiflexion decreased slightly (1°) on both the casted and noncasted sides following the last cast removal, and (2) ROM was less on the ulcerated side prior to casting compared with the nonulcerated side.* ***Conclusion and Discussion.*** *We believe the beneficial effects (healing of wounds) outweigh the minimal detrimental effects (decreased dorsiflexion) of treatment with TCC. [Diamond JE, Mueller MJ, Delitto A. Effect of total contact cast immobilization on subtalar and talocrural joint motion in patients with diabetes mellitus. Phys Ther. 1993;73:310–315.]*

Literature Example 9.11 is from the article entitled "Effect of Feedback on Learning a Vertebral Joint Mobilization Skill." The article describes an experiment that used the design $\begin{matrix} R? & O_1 & X & O_2 \\ R? & O_3 & & O_4 \end{matrix}$. Read the abstract and complete Activity 9.14.

Activity 9.14

What was the purpose of the experiment?

1 ____________________

What do O_1 and O_2 refer to in this instance?

2 ____________________

____

____

What do O_3 and O_4 refer to?

3 ____________________

____

____

What response variable was measured?

4 ____________________

There's a difficulty in this design with respect to randomization. What is it?

5 ____________________

____

____

What degree of control did the researcher have using this design?

6 Minimal/Partial/Complete

LITERATURE EXAMPLE 9.11

Effect of Feedback on Learning a Vertebral Joint Mobilization Skill

Michael Lee
Anne Moseley
Kathryn Refshauge

This study was designed to investigate whether concurrent quantitative feedback of performance could improve the learning of a joint mobilization technique. A group of 110 physical therapy students had been randomly divided into two groups for teaching purposes. All students had previously learned mobilization of peripheral joints and were currently learning spinal mobilization. From one of the groups, 22 students volunteered to comprise a control group, which was taught a spinal mobilization technique in the traditional way. Additional concurrent quantitative feedback of the level of force applied to the patient was given to 31 volunteers from the other group. These students formed the experimental group. A force plate was used for force measurement, and the feedback was given via an oscilloscope. The average force applied by the students' instructors was taken as an "ideal" force. The oscilloscope showed both the applied force and the "ideal" force. Consistency was measured by the variance of the group's performance. Accuracy was assessed by calculating the difference between the applied force and the "ideal" force. Results indicated that this feedback was associated with a significant improvement in accuracy and consistency in the application of the mobilizing force. This improvement was still present at a follow-up test conducted one week later. This result supports a greater use of such feedback in the teaching and practice of joint mobilization techniques, although the need for further research is emphasized. [Lee M, Moseley A, Refshauge K: Effect of feedback on learning a vertebral joint mobilization skill. Phys Ther 70:97–104, 1990]

Literature Example 9.12 is from the article entitled "Reduction of Chronic Posttraumatic Hand Edema: A Comparison of High Voltage Pulsed Current, Intermittent Pneumatic Compression, and Placebo Treatments." The experiment used the design

$$\begin{array}{llll} R & O_1 & X & O_2 \\ R & O_3 & X & O_4 \\ R & O_5 & & O_6 \end{array}$$

. Read the abstract and complete Activity 9.15.

ACTIVITY 9.15

What was the purpose of the experiment?	1 ______________________
What do O_1 and O_2 refer to in this instance?	2 ______________________
What do O_3 and O_4 refer to?	3 ______________________
What do O_5 and O_6 refer to?	4 ______________________
What response variable was measured?	5 ______________________
Which of the following problems of internal validity does this design take care of?	6 ______________________

History

Maturation

Pretests

Instrument

Regression

Selection of subjects

Mortality

Interaction

What degree of control did the researcher have using this design?	7 Minimal/Partial/Complete

LITERATURE EXAMPLE 9.12

Reduction of Chronic Posttraumatic Hand Edema: A Comparison of High Voltage Pulsed Current, Intermittent Pneumatic Compression, and Placebo Treatments

Judy W Griffin
Laurie S Newsome
Susan W Stralka
Phillip E Wright

The purpose of this study was to compare the efficacy of intermittent pneumatic compression (IPC) and high voltage pulsed current (HVPC) in reducing chronic posttraumatic hand edema. Thirty patients with posttraumatic hand edema were randomly assigned to IPC, HVPC, or placebo-HVPC groups (10 patients in each group). Patients received a single application of the respective treatment for 30 minutes. Measurements were made before and after a 10-minute rest period and after the 30-minute treatment. A volumetric method was used to quantify edema reduction. Reduction in hand edema was significant between the IPC and placebo-HVPC groups (p = .01). Differences in edema reduction between the HVPC and placebo-HVPC groups did not reach statistical significance (p = .04), but were considered clinically significant. There was no significant difference between the IPC and HVPC groups. A single 30-minute administration of IPC produced a significant reduction in hand edema. Additional clinical studies are needed to delineate maximally effective treatment protocols for reduction of chronic posttraumatic hand edema. [Griffin JW, Newsome LS, Stralka SW, et al: Reduction of chronic posttraumatic hand edema: A comparison of high voltage pulsed current, intermittent pneumatic compression, and placebo treatments. Phys Ther 70:279–286, 1990]

Literature Example 9.13 is from the article entitled "Progressive Exercise Testing in Closed Head-Injured Subjects: Comparison of Exercise Apparatus in Assessment of a Physical Conditioning Program." Read the abstract and complete Activity 9.16.

ACTIVITY 9.16

A mystery design for you to identify

What was the purpose of the experiment?	1 ______________________
Represent the design symbolically. What do the O and X symbols refer to?	2 ______________________
There is at least one potential validity problem with the design of this experiment. What is it?	3 ______________________

LITERATURE EXAMPLE 9.13

Progressive Exercise Testing in Closed Head-Injured Subjects: Comparison of Exercise Apparatus in Assessment of a Physical Conditioning Program

Marque Hunter
JoAnn Tomberlin
Carol Kirkikis
Samuel T Kuna

Progressive exercise tests were performed on 12 closed head-injured subjects to determine 1) whether results differ when tests are performed on a treadmill, a bicycle ergometer, or mechanical stairs and 2) whether a 3-month general physical conditioning program results in an improvement in exercise performance. The subjects performed progressive exercise tests on each apparatus on entry into a residential transitional rehabilitation program and approximately 3 months later following participation in a physical conditioning program. On both the initial and 3-month exercise tests, maximal oxygen consumption ($\dot{V}O_{2\ max}$) was significantly greater on the treadmill and the mechanical stairs than on the bicycle ergometer. The mean $\dot{V}O_{2\ max}$ was 74% of the predicted value on the initial exercise test and rose to 85% of the predicted value after the 3-month physical conditioning program. Oxygen consumption per kilogram of body weight at a given power output on a given apparatus showed no statistically significant difference between the initial and 3-month tests, indicating no change in exercise efficiency. On the 3-month test, a statistically significant decrease was noted in heart rate at rest and after the 4-minute period of recovery from maximal exercise on any given apparatus. The data obtained in this study indicate that 1) the treadmill and mechanical stairs are more suitable than the bicycle ergometer for assessing maximal exercise performance and 2) improved physical fitness following a physical conditioning program is associated with an improvement in cardiovascular function. The results indicate that progressive exercise tests in closed head-injured subjects can be used to 1) quantify the degree of physical deconditioning, 2) design a physical conditioning program, and 3) assess outcome goals of such a program. [Hunter M, Tomberlin J, Kirkikis C, et al. Progressive exercise testing in closed head-injured subjects: comparison of exercise apparatus in assessment of a physical conditioning program. Phys Ther. 1990;70:363–371]

Conclusion

Experimental design is tricky, and often time and money can be wasted by poor design. Precision (repeatability) and resolution (ability of the experiment to distinguish between small enough differences in response variable) must be taken into account. Furthermore, the effects of known and unknown biases can be reduced by careful design. Too often researchers rely blindly on a common criterion such as the notion that a variation of two standard errors in a response variable is statistically significant. However, such criteria are frequently based on mathematical assumptions, which may not be satisfied in a particular situation. Also, even when the mathematical conditions are satisfied there are many situations in which statistical significance may not correspond to practical significance. The experience of other researchers can be an invaluable asset, and we strongly recommend that you consult with experts at the design stage, especially those with expertise in statistical considerations.

Self-Assessment

Please assess your competency.

Tasks:	How well can you do it? *Poorly*				*Very well*	Page
1. Given an adequate description of the research context, decide on the appropriate research design or elucidate the conditions under which the following designs should be used:						
Observational designs:						
a. Retrospective						
(i) Case-control study	1	2	3	4	5	305
(ii) Cross-sectional survey	1	2	3	4	5	299
b. Prospective						
(i) Cohort study	1	2	3	4	5	309
Experimental studies:						
RR O_1 X O_2 Expt. Group I RR O_3 O_4 Control Group II RR X O_5 Expt. Group III RR O_6 Control Group IV	1	2	3	4	5	312
XO	1	2	3	4	5	326
Subjects not randomly selected and not assigned randomly to the control and experimental groups; groups are not comparable:						
X O_1 O_2	1	2	3	4	5	326

Design						Page
Subjects randomly assigned to experimental and control groups but not selected randomly from the population:						
•R X O_1 •R O_2	1	2	3	4	5	326
O_1 X O_2	1	2	3	4	5	327
Subjects not randomly assigned to experiment and control:						
O_1 X O_2 O_3 O_4	1	2	3	4	5	327
Subjects randomly assigned to experiment and control but not randomly selected:						
•R O_1 X O_2 •R O_3 O_4	1	2	3	4	5	327
Subjects randomly selected and randomly assigned to experiment and control:						
RR O_1 X O_2 RR O_3 O_4	1	2	3	4	5	328
Observations on the same group (time series):						
O_1 O_2 O_3 O_4	1	2	3	4	5	328
Observations on experimental and control groups not randomly assigned (time series):						
O_1 O_2 X O_3 O_4	1	2	3	4	5	328
Observations on experimental and control groups not randomly assigned (time series):						
O_1 O_2 X O_3 O_4 O_5 O_6 O_7 O_8	1	2	3	4	5	328
Observations on experimental and control groups; groups randomly assigned (time series):						
•R O_1 O_2 X O_3 O_4 •R O_5 O_6 O_7 O_8	1	2	3	4	5	328
2. Elucidate the strategy for analyzing the data collected using each of the above designs.	1	2	3	4	5	

3. Explain the following indices:

Relative risk	1	2	3	4	5	309
Odds ratio	1	2	3	4	5	305

Answers for Chapter 9

Answers for Activity 9.1: The answer to each question is the corresponding number in Figure 9.1.

Answers for pp. 288–289: 1) Dependent 2) Response 3) Explanatory variable 4) Treatment; factor; stimulus variable 5) Quantitative level 6) Qualitative level

Answers for Activity 9.2: Context I: PULSE is the response, or dependent variable; all others are factors, or independent (or explanatory) variables. Context II: STRESS is the response, or dependent, variable; all others are factors (or independent or explanatory variables). Context III: HEALTH is the response, or dependent, variable; all others are factors (or independent or explanatory variables). Context IV: PULSE 1 and 2 are the response, or dependent, variables; all others are factors (or independent or explanatory variables or treatments). Context V: The levels of the treatment variable are LI, HI, and C. These are factors (or independent or explanatory or treatment variables); all others are response, or dependent, variables.

Answers for Activity 9.3: 1) Intervention 2) Retrospective 3) Prospective 4) Look for data in past records 5) Collect data in the future 6) Prospective 7) Cross-sectional survey 8) Case-control study 9) One point 10) Response 11) Response

Answer for p. 293: 1) Either the scale or the jeweler

Answers for Activity 9.4: 1) 10.6 and 9.2 pCi/L 2) 3.4 and 11.6 pCi/L 3) No, we need to compute CV. 4) Air Check Inc. and Terradex; Ryan Nuclear Labs, Key Technology, E.P.A. 5) The Radon Project, Rad Elec. Inc., Terradex 6) Random error 7) Error 8) Validity

Answers for Activity 9.5: 1) Acupuncture-point electrical stimulation 2) To minimize psychological factors known to influence blood pressure 3) Previous studies show that blood pressure is influenced by the presence of a white coat. It would be difficult to tell if the observed changes in the response variable were due to treatment or to the white coat. 4) This would introduce bias; the control and experimental groups would not be comparable. 5) Tossed a coin 6) Diastolic blood pressure; it is a reliable standard measure. 7) Patients were screened for hypertension and two who did not fit the criteria were excluded from the sample. 8) 12 out of 22 did not agree to be a part of the study. We wonder if there is something about those 12 that would have changed the outcome if they had been included. We can at least generalize the results to those who are favorably disposed to participating in such a study. 9) In the short run the response variable was affected. Would the treatment have a lasting effect if it were repeated over a period of time? The treatment may not have been complete enough to answer the question.

Answers for Activity 9.6: 1) To test the effects of perceived stress and health practices on maternal identity 2) Birth announcements 3) Questionnaire 4) 330 questionnaires were sent; 173 were returned; sample was selected using a random number table 5) Stamped return envelopes were sent out with questionnaires; two reminders were sent. 6) Researchers only assessed information at one point in time.

Answers for Activity 9.7: 1) To assess the prevalence, content, and growth of laws restricting smoking in public places and workplaces 2) Listing of cities with 25,000 or greater population 3) Mailed request for information 4) 831 cities, of which 753 responded; they tried to obtain information from all cities with population of 25,000 or greater. 5) They obtained a random sample of the cities and sent requests to city clerks. 6) Researchers only assessed information at one point in time.

Answers for pp. 305–306: 1) Explanatory variable 2) Have 3) Do not have 4) Ratio 5) a/b 6) c/d 7) (a/c) 8) (b/d)

Answers for Activity 9.8: 1) To determine how the presence of guns, the type of guns, and methods of gun storage relate to the risk of adolescent suicide 2) Relatives of adolescents who had committed suicide 3) A psychiatric hospital 4) The abstract does not specify; we assume interviews with checklist 5) All three groups have 47 subjects; matched on age, gender, county of origin 6) Odds of the presence of guns in the homes of the suicide victims (cases relative to controls) 7) This is a retrospective study in which cases and controls are used to assess the odds ratio.

Answers for Activity 9.9: 1) To identify risk factors for gout 2) Medical students 3) No; the study originally focused on coronary heart disease and hypertension. 4) Initial health examinations and followup questionnaires 5) 60 6) The development of gout 7) This is a longitudinal cohort study because researchers followed a specific group for a period of time and recorded the number of cases of gout that developed in the group having the risk factor and in the group lacking the risk factor.

Answers for Activity 9.10: 1) The criteria for inclusion in the study were self-reported current or previous back pain and willingness to exercise. 2) One hour of exercise a week for a year and a half for fifty-eight individuals; though the article does not make it explicit, we can assume that this amount of exercise was atypical for the subjects. 3) Subjects were not randomly selected from all the workers. Groups consisted of volunteers. The results would not be applicable to non-volunteers. Subjects were randomly assigned to the treatment and control groups.

Answers for pp. 321–322: 1) Randomly 2) Four 3) After 4) Is not 5) Is not 6) Treatment 7) Control 8) Pre-test

Answers for Activity 9.11: 1) The use of a control group 2) $O_{5\text{Post-test}}$

Answers for Activity 9.12: 1) Comparison of the electrical activity of two muscles 2) O_1 indicates that the electrical activity of two muscles in each subject was measured one time. 3) Electrical activity was measured only in the muscles of the left leg of each subject. 4) In our opinion, clinicians will have to do further work before definitive statements about clinical applications can be made.

Answers for Activity 9.13: 1) To determine the effect of total contact casting (TCC) on dorsiflexion and motion of certain joints 2) All were measures of the range of motion of one leg of each subject; O_1 was on the casted leg before casting; O_2 was on the same leg after casting; O_3 was on the uncasted leg at the same time as O_1; O_4 was on the uncasted leg at the same time as O_2. 3) Total contact casting

Answers for Activity 9.14: 1) To test whether concurrent quantitative feedback improves learning 2) The average force applied by the students before feedback training (O_1) and after feedback training (O_2) 3) The average force applied by the students before usual training (O_3) and after usual training (O_4) 4) The difference between the "ideal" force and force applied by the students 5) Although the two groups were randomly selected, the people in the experimental and control groups were volunteers. It's possible that the difference between the experimental and control groups was influenced by some unknown trait that caused these people to volunteer. The design could have been strengthened by randomly selecting students from one class for the experimental group and students from the other for the control group, or by using the entire classes (since they were already randomly divided). Another difficulty was that the original group of students was not randomly chosen. Therefore the results of the study cannot be generalized to the larger population of all physical therapy students. The population consists of only these particular 110 physical therapy students. 6) Partial

Answers for Activity 9.15: 1) To compare the effectiveness of two treatments in reducing hand edema 2) Before IPC (O_1), after IPC (O_2) 3) Before HVPC (O_3), after HVPC (O_4) 4) Before placebo-HVPC (O_5), after placebo-HVPC (O_6) 5) Quantity of edema 6) All 7) Complete

Answers for Activity 9.16: 1) To test whether results of using three different kinds of exercise machines differ and to see whether a three-month physical conditioning program improves exercise performance 2) From the abstract it is not entirely clear what the design was. It was probably a time-series design of the form O_0? X O_1 X O_2 X O_3 X O_4. It is not clear from the abstract whether there was an observation before the subjects used the first apparatus. Since there were only twelve subjects, it is likely the subjects were not divided into groups and that all the subjects used all the exercise methods. The various Os represent observations taken after each exercise intervention. The Xs represent use of the different exercise devices, and the last X represents the three-month program. 3) Assuming that we are correct about the design, it is possible that residual effects of the first intervention might influence what is observed after the second intervention; residual effects of the first two interventions might influence what is observed after the third; etc. Furthermore, there is no control group, the subjects were not randomly selected, and there were very few of them. Therefore, the observed results apply only to this group of twelve persons; they cannot be generalized to a larger population. Perhaps this was a preliminary experiment done to get some information useful to designing later, more detailed experiments.

CHAPTER

10

Probability

Prelude

"S-S-S-S-S-H-H-H-H-H-H-H," he cautioned, putting his finger up to his lips and drawing Milo closer. "Do you want to ruin everything? You see, to tall men I'm a midget, and to short men I'm a giant; to the skinny ones I'm a fat man, and to the fat ones I'm a thin man. That way I can hold four jobs at once. As you can see, though, I'm neither tall nor short nor fat nor thin. In fact, I'm quite ordinary, but there are so many ordinary men that no one asks their opinion about anything. Now what is your question?"

"Are we lost?" asked Milo once again. . . .

"My, my," the man mumbled. "I know one thing for certain; it's much harder to tell whether you *are* lost than whether you *were* lost, for, on many occasions, where you're going is exactly where you are. On the other hand, you often find that where you've been is not at all where you should have gone, and, since it's much more difficult to find your way back from someplace you've never left, I suggest you go there immediately and then decide."

— Norton Juster
The Phantom Toll Booth, pp. 113–114

Chapter Outline

Self-Assessment

Please assess your competency.

Tasks:	How well can you do it?					Page
	Poorly				*Very well*	
1. Define subjective probability.	1	2	3	4	5	346
2. Given a frequency distribution, calculate probabilities.	1	2	3	4	5	349
3. Given a two-by-two table, calculate expected frequencies.	1	2	3	4	5	356
4. Define density curve.	1	2	3	4	5	351
5. Define probability distribution.	1	2	3	4	5	351
6. Define cumulative probability.	1	2	3	4	5	353
7. When appropriate data are given, calculate probability using the Addition Rule.	1	2	3	4	5	354
8. When appropriate data are given, calculate probability using the Multiplication Rule.	1	2	3	4	5	354

Introduction

People generalize from observed events and guess about unobserved events. We observe smoke and may guess that there is a fire, but do we know for sure? Further observation may strengthen our confidence in the guess. What is your guess about the likelihood of the sun's showing up in the east tomorrow morning? If there are ten red beads and twenty blue beads in a bag and you pick one at random, which color are you more likely to pick? How certain are you that you will get that color?

Activity 10.1

Selecting beads and assigning a subjective certainty level

Assuming a bag contains one hundred red and blue beads (well mixed) in the proportions indicated in each row, and you choose just one bead, which color do you think you will get? Circle that color in the third column of the table below and fill in the proportion of that color, as a fraction, in the fourth column. Also assess your level of certainty in the fifth column.

Red Beads	Blue Beads	Circle the Color You Think You'll Get	Proportion of Beads of the Color You Chose (# of beads of your choice)/(total # of beads)	How Certain Are You? Not sure				Sure
100	0	Red/Blue	________	1	2	3	4	5
99	1	Red/Blue	________	1	2	3	4	5
90	10	Red/Blue	________	1	2	3	4	5
80	20	Red/Blue	________	1	2	3	4	5
70	30	Red/Blue	________	1	2	3	4	5
60	40	Red/Blue	________	1	2	3	4	5
50	50	Red/Blue	________	1	2	3	4	5
40	60	Red/Blue	________	1	2	3	4	5
30	70	Red/Blue	________	1	2	3	4	5
20	80	Red/Blue	________	1	2	3	4	5
10	90	Red/Blue	________	1	2	3	4	5
1	99	Red/Blue	________	1	2	3	4	5
0	100	Red/Blue	________	1	2	3	4	5

When did you switch from red to blue? Perhaps at the seventh row, when the proportions of red and blue beads were 0.5. Now fill in the following frequency distribution table of your certainty levels.

Activity 10.2

Frequency distribution of certainty levels assigned to the bead selection

Count up the number of 1s, 2s, etc., that you chose as your certainty levels in Activity 10.1, and enter those numbers in the third column of the following table.

Certainty Scale	Our Frequency	Your Frequency
1	2	______
2	2	______
3	2	______
4	3	______
5	4	______

As we approach equal numbers of red and blue beads in the bag, the uncertainty increases. At some point we have to say that we are not one hundred percent sure, but we will make a choice anyway, and by so choosing we are willing to be wrong a certain percent of the time. We, the authors, felt quite sure of our choice when the proportion of a given color was 9/10. You may not have felt certain in this case. This was a subjective decision, as are many choices in life. We often can't be absolutely certain, but frequently we must choose anyway.

Probability as Long-Term Relative Frequency

If there were fifty red beads and fifty blue beads, it would be equally likely that you would pick a red or a blue bead. The proportion of each color to the whole is 0.5. Which color you pick is a matter of chance. **Probability** is an estimate of the likelihood of an event's occurring and is expressed as a fraction, a proportion, or a percent. If you were to play this choosing game over and over with fifty red and fifty blue beads in the bag each time, you would expect to choose a red bead 50 percent of the time. If you kept track of the number of red beads actually chosen and divided that number by the number of times you played the game, you would be computing the relative frequency of red beads chosen, and you would expect this number to approach 0.5 the more you played. For this reason, you can think of probability as being an estimate of the long-term relative frequency, or in more mathematical language, the limit of the relative frequency as the number of trials (or the sample size) goes to infinity.

We will be a bit sloppy, mathematically, in this book, because we will not distinguish between relative frequency and probability. Thus, if we chose twenty beads from a container of beads and eight of them turned out to be red, we might call the relative frequency $8/20 = 0.4$ of red beads the probability of getting a red bead in that particular sample.

ACTIVITY 10.3

Coin-tossing experiment

We flipped a coin ten times and recorded our results in the table below. You should also flip ten times and record your results. Then record the frequency distribution of the results in the next table. Note that both heads and tails can't occur on the same flip—the outcomes are said to be **mutually exclusive**. Also, the outcome of one flip does not influence the next flip. The outcome of each flip is **independent** of the other flips.

Results of Coin-Toss Experiment

Flip No.	1	2	3	4	5	6	7	8	9	10
Ours	H	T	H	T	T	T	T	H	T	H
Yours										

Frequency Distribution of the Coin-Toss Experiments

	Observation	Frequency	Fraction	Proportion	Percent
Our Flips	Heads	4	4/10	0.4	40%
	Tails	6	6/10	0.6	60%
Your Flips	Heads				
	Tails				

Are the proportions in our experiment and your experiment the same? We would be somewhat surprised if they were. Your proportions are probably different from the long-term expected frequency of 0.5, as ours were.

Do our experiments contradict the assertion that heads and tails are equally likely events? Would it contradict that assertion if we had gotten ten heads in a row? What if we flipped the coin one hundred times and got one hundred heads in a row? Ten heads in a row is certainly suspicious; one hundred is very suspicious. We would want to examine the coin very closely to see if perhaps it has two heads or is unbalanced. Six heads out of ten, however, is quite common, even given a perfectly balanced coin.

As we mentioned, probability is an estimate of the likelihood of an event's occurring and is expressed as a fraction, a proportion, or a percent. For all practical purposes a probability of 1 represents certainty that an event will occur and 0 represents impossibility. Figure 10.1 illustrates this concept.

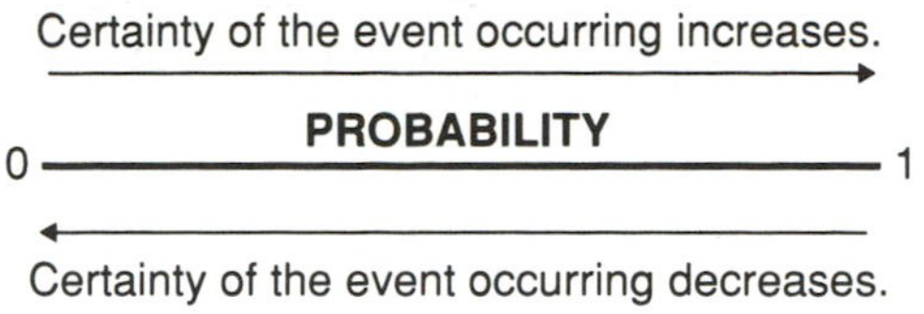
Certainty of the event occurring increases.
PROBABILITY
0
1
Certainty of the event occurring decreases.

FIGURE 10.1 Interpretation of Probability

ACTIVITY 10.4

Probability of picking specific cases from FED

Shuffle the cards with the FED data on them. Pick one card from the deck. What is the probability of your picking case (faculty member) number 10?

1 _______ (Hint: Divide one by the total number of cases.)

ACTIVITY 10.5

The probabilities of selecting participants from each of the groups in FED

Previously you calculated the probability of choosing one particular case from the FED. Now calculate the probabilities of choosing cards from each participant group and enter the probabilities in the table below.

		PROBABILITY	
Participation Status	**Count, or Frequency**	**Proportion (number in the group/total)**	**Percent**
1 = Non-participant (NP)	6	6/30 = .20	$.20 \times 100 =$ 20.00
2 = Control (C)	7	1 _______	2 _______
3 = Low-intensity exercise (LI)	8	3 _______	4 _______
4 = High-intensity exercise (HI)	9	5 _______	6 _______
	$n = 30$		

The above represents what we expect in the long run. However, if we picked ten participants, each time returning the card to the deck and shuffling the deck between choices, what kind of selection distribution would we get?

Activity 10.6 is designed to answer this question empirically; compare the answer with the probabilities obtained in Activity 10.5.

Activity 10.6

Selecting participants from each of the groups in FED

In this activity, you will make an empirical assessment of the probability of choosing a non-participant, a control, or a low- or high-intensity exerciser from the FED data.

Shuffle the FED cards ten times and pick one card after each shuffle, replacing the card before the next draw. For each draw put a tally mark in the appropriate column of the following table.

Shuffle Number	Non-Participant	Control	Low-Intensity	High-Intensity
1	________	____	_______	_______
2	________	____	_______	_______
3	________	____	_______	_______
4	________	____	_______	_______
5	________	____	_______	_______
6	________	____	_______	_______
7	________	____	_______	_______
8	________	____	_______	_______
9	________	____	_______	_______
10	________	____	_______	_______
Total cases = 10	________	____	_______	_______

Activity 10.7

Comparison of the relative frequencies from one experiment and the theoretical probabilities (long-term relative frequencies)

Each time you drew a case in Activity 10.6, that draw was independent of the others. How many cards did you draw from each participant category? Based on this experiment, fill in the following table with the appropriate probabilities for each type of participant.

Participant Category	Proportion of Members of Each Group Actually Chosen (Activity 10.6)	Probability Based on Activity 10.5
Non-participant	______	______
Control	______	______
Low-intensity exercise	______	______
High-intensity exercise	______	______

Based on Activity 10.7 we conclude that the actual relative frequencies are [1] different from/the same as the theoretical ones we arrived at in Activity 10.5.

Now remove non-participants from the deck (case numbers: 1, 2, 8, 16, 19, and 30). This leaves twenty-four cases: seven controls and seventeen low-intensity and high-intensity exercise participants. If we draw one card from the twenty-four, what is the probability that it is a control? [2] ______ What is the probability that it is an exercise case? [3] ______.

Probability Distributions and Density Curves

The **probability distribution** of a variable consists of all the variable values, each paired with a probability; in other words, it is a listing of all possible values and their probabilities. This is also called the **density distribution** and may be presented in the form of a table or graph. The curve that is drawn is called the **density curve** or **probability density curve**.

Area as a Measure of Density

The areas of the bars in a histogram are proportional to the number of cases occurring in the respective intervals. If the total area in all the bars is equal to 1, then the area of each bar is the **relative frequency** of events occurring in that interval. The bars in Figure 10.2 have areas that are proportional to their frequency and hence represent the [4] ______ ______ of the age interval covered by the bar. The total area under the polygon is equal to [5] ______ or [6] ______ percent. This diagram is the density curve for the age data. The probability distribution for the age variable for the theoretical population from which this sample was drawn is approximated by this graph.

You've probably heard of a normal (or bell-shaped) curve. The normal distribution is a bell-shaped theoretical distribution in which 68 percent of the cases are within one standard deviation of the mean, μ. The next chapter will explain the normal distribution further. A great many data sets have distributions that can be approximated by this theoretical distribution. Because so many data sets have approximately normal distributions, it is common to try to fit a

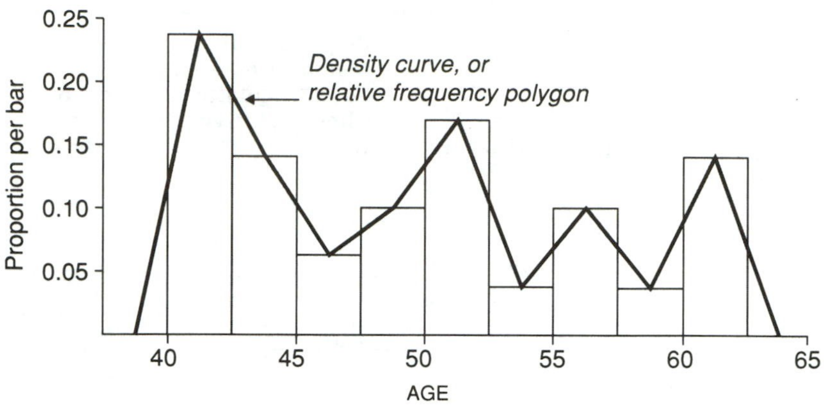

FIGURE 10.2 Histogram of Faculty Ages and a Polygon Connecting the Midpoints of the Bars

normal curve with the same mean and standard deviation to a data set to see if the distribution is approximately normal. Our age data have the following characteristics:

$n = 30$
Mean = 49.4
Standard deviation = 7.4

Figure 10.3 shows a histogram of ages with a normal curve superimposed on it. In this case the normal curve is a terrible fit. The bars in Figure 10.3 have areas that are proportional to their frequency and hence represent the

1 _______ _______ of the age interval covered by the bar. The total area under

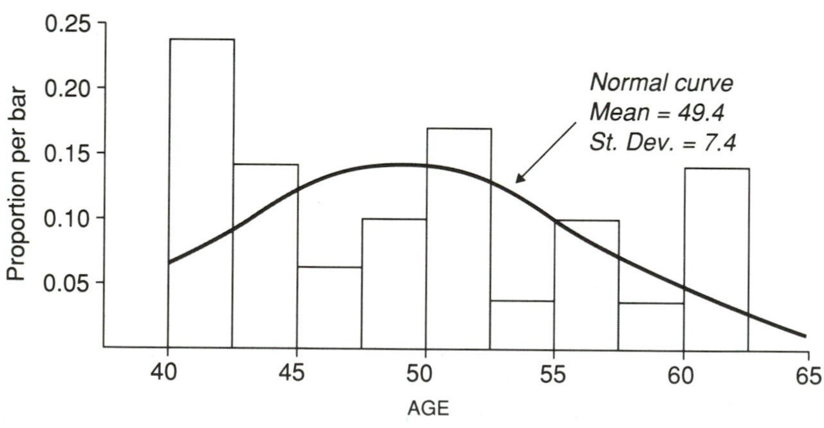

FIGURE 10.3 Histogram of Faculty Ages and a Superimposed Normal Curve

the **normal curve** is equal to [2] _______ or [3] _______ percent. The normal curve represents a probability distribution for a population that is unlikely to be the population from which this sample was drawn.

Calculating the Probabilities of Events

If we draw a card after shuffling the faculty exercise data (FED) cards well, which age are we likely to draw? The probability of drawing a specific age or an age in a specific category estimates the likelihood.

Activity 10.8

Probability and cumulative probability of ages of participants

The table in this activity gives actual ages (and their relative frequencies or probabilities) rather than age intervals, as Figures 10.2 and 10.3 showed. Complete the table.

AGE	Count, or Frequency	Cumulative Frequency	Relative Frequency (probability expressed as a %)	Cumulative Probability
40	4	4	13.33	13.33
41	3	7	10	23.33
43	2	9	6.67	30
44	2	11	6.67	[1] _______
46	1	[2] _______	3.33	40
47	1	13	[3] _______	43.33
48	2	15	6.67	50
49	1	16	3.33	53.33
51	1	17	3.33	56.67
52	4	21	13.33	[4] _______
54	1	22	3.33	73.33
57	3	25	10	83.33
58	1	26	3.33	86.67
61	3	29	10	[5] _______
62	1	30	3.33	100

In Figure 10.4 distributions of the ages of the faculty participants are given in the form of the familiar bar graph. Both actual frequencies and relative

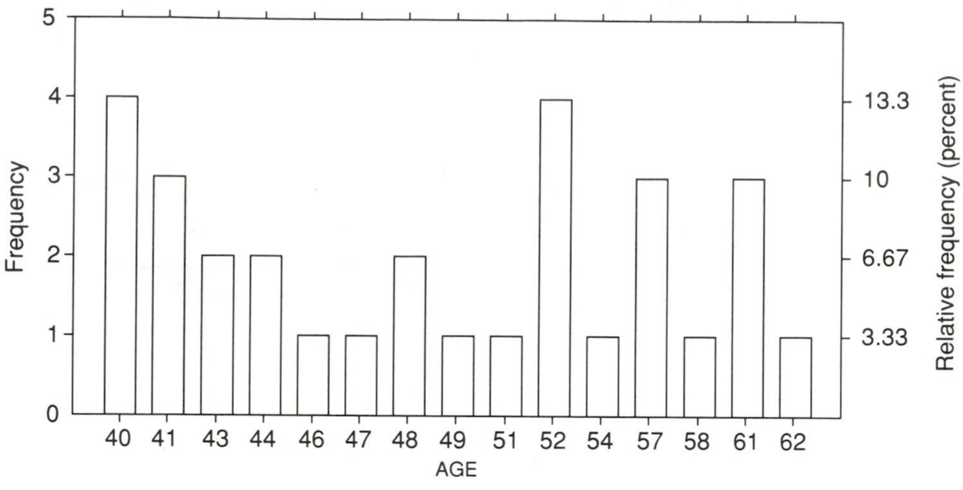

Figure 10.4 Bar Graph of Ages of the FED Participants

frequencies (percentages) are given on the vertical axes. (Note: Missing ages are not represented by bars and have been omitted from the horizontal axis.)

We now give two rules that are useful in calculating the probabilities of events.

Addition Rule

If A and B are mutually exclusive events, then the probability of the event "A or B" can be found by adding the probability of A and the probability of B (probability of A + probability of B). Recall that events are mutually exclusive when they can't both occur at the same time. For example, you can roll a six or a three with a die, but you can't roll both at the same time.

Multiplication Rule

If A and B are independent events, then the probability of the event "A and B" is the product of the probability of A and the probability of B (probability of A * probability of B). Recall that events are independent if the occurrence or non-occurrence of one has no influence on the occurrence or non-occurrence of the other. For example, getting heads on the second flip of a coin is independent of getting heads on the first flip. These rules can be extended to combinations of more than two probabilities and they hold for relative frequencies as well as probabilities. We can read probabilities either from a graph (such as Figure 10.2) or from a table (such as the one in Activity 10.8).

In Activity 10.9, use the Addition Rule to figure the probability of the occurrence of two or more ages in the faculty exercise data set. The various ages are mutually exclusive events. (You can't be two different ages at once.)

ACTIVITY 10.9

Probability of drawing specific ages from faculty data

Imagine drawing one card from the FED data. Fill in the probability of drawing a subject with the given age or in the given age range in the table below.

Age	Probability	Hint
40	1 ____________	Read column 4 or 5 in Activity 10.8.
40 or 41	2 ____________	Add the probabilities in column 4 or read the cumulative probability from column 5 (Activity 10.8). You can read the same information from Figure 10.4.
Less than or equal to 44	3 ____________	
47 to 62	4 ____________	Subtract the cumulative probability of age 46 (in Activity 10.8) from 100.
Between 43 and 57 (including both 43 and 57). This means 43 or 44 or 45 . . . or 57.	5 ____________	Add all the corresponding probabilities.

ACTIVITY 10.10

Computing relative frequencies when two variables are independent

The following two-way table gives the number of male and female participants in each of the activity groups (SQD):

		ACTIVITY		
SEX	**Minimal**	**Moderate**	**Vigorous**	**ALL**
Female	6	24	2	32
Male	0	12	6	18
ALL	6	36	8	50

By dividing a frequency by the sample size, we get the relative frequency for that category. Relative frequencies are much like probabilities. (We mentioned earlier that we would not be careful in the text about drawing a distinction between relative frequencies and probabilities.) Thus, the probability that a member of this sample is female (the relative frequency of females in the sample) is the number of females in the sample divided by the total number in the sample, i.e., 32/50 = 0.64. The probability that a member of the sample was male = [1] ______________. The probability that a member engages in minimal activity = [2] ______________. The probability that a member engages in moderate activity = [3] ______________. The probability that a member engages in vigorous activity = [4] ______________. We are assuming that being female and engaging in vigorous activity are independent events. Therefore, the probability that a member is female and engages in vigorous activity = (probability of being a female member) * (probability of engaging in vigorous activity). Compute this number and put it in the top left space in the following table. Then complete the table.

	Minimal	**Moderate**	**Vigorous**
Female			
Male			

ACTIVITY 10.11

Observed frequencies, probabilities, and expected frequencies

When we multiply the probabilities by the total in the sample we obtain **expected** frequencies. Given that our assumption about independence is correct, these would be the frequencies that we would have expected to have gotten (approximately) in our experimental sample.

Using the probabilities you calculated in Activity 10.9, calculate the expected frequency of each event (n * probability) and fill in the blanks below. Keep in mind that frequencies are counts, and therefore must be whole numbers, so you must round your answers to the nearest whole number.

		ACTIVITY		
SEX	**Minimal**	**Moderate**	**Vigorous**	**ALL**
Female	50 * .08 = 4	[1] _______	[2] _______	32
Male	[3] _______	[4] _______	[5] _______	18
ALL	6	36	8	50

In Activity 10.11 you calculated expected frequencies. If there is no relationship between ACTIVITY and SEX, the observed frequencies will be close to the expected frequencies given in the table in Activity 10.11. If the observed frequencies are different enough, we can conclude that ACTIVITY and SEX are likely to be dependent. In this case, this would probably mean that the groups were not chosen randomly—that somehow sex entered into the way the groups were divided.

Self-Assessment

Tasks:	How well can you do it?					Page
	Poorly				*Very well*	
1. Define subjective probability.	1	2	3	4	5	346
2. Given a frequency distribution, calculate probabilities.	1	2	3	4	5	349
3. Given a two-by-two table, calculate expected frequencies.	1	2	3	4	5	356
4. Define density curve.	1	2	3	4	5	351
5. Define probability distribution.	1	2	3	4	5	351
6. Define cumulative probability.	1	2	3	4	5	353
7. When appropriate data are given, calculate probability using the Addition Rule.	1	2	3	4	5	354
8. When appropriate data are given, calculate probability using the Multiplication Rule.	1	2	3	4	5	354

Answers for Chapter 10

Answer for Activity 10.4: 1/30, or 0.033, or 3.3 percent

Answers for Activity 10.5: 1) 0.2333 2) 23.33% 3) 0.2667 4) 26.67% 5) 0.30 6) 30.0%

Answers for p. 351: 1) Different from (although it's possible that yours were the same; most people will get proportions different from the probabilities) 2) 29.2% 3) 70.8% 4) Relative frequency 5) 1 6) 100

Answers for pp. 352–353: 1) Relative frequency 2) 1 3) 100

Answers for Activity 10.8: 1) 36.67 2) 12 3) 3.33 4) 70 5) 96.7

Answers for Activity 10.9: 1) 13.33 2) 23.33 3) 36.67 4) 60 5) 60

Answers for Activity 10.10: 1) .36 2) .12 3) .72 4) .16

	Minimal	Moderate	Vigorous
Female	.0768	.4608	.1024
Male	.0432	.2592	.0576

Answers for Activity 10.11: 1) 23 2) 5 3) 2 4) 13 5) 3

CHAPTER

11

Reading Theoretical Tables

Prelude

"SILENCE," suggested the king. "Now, young man, what can you do to entertain us? Sing songs? Tell stories? Compose sonnets? Juggle plates? Do tumbling tricks? Which is it?"

"I can't do any of those things," admitted Milo.

"What an ordinary little boy," commented the king. "Why, my cabinet members can do all sorts of things. The duke here can make mountains out of molehills. The minister splits hairs. The count makes hay while the sun shines. The earl leaves no stone unturned. And the undersecretary," he finished ominously, "hangs by a thread. Can't you do anything at all?"

"I can count to a thousand," offered Milo.

"A–A–R–G–H, numbers! Never mention numbers here. Only use them when we absolutely have to," growled Azaz disgustedly.

— Norton Juster
The Phantom Toll Booth, pp. 84–86

Chapter Outline

Self-Assessment

Tasks:	How well can you do it?					Page
	Poorly				*Very well*	
1. Given z scores, read the area from the normal distribution table.	1	2	3	4	5	367
2. Given areas, read z scores from the normal distribution table.	1	2	3	4	5	367
3. Given areas and degrees of freedom, read the t scores from the t table.	1	2	3	4	5	374
4. Given areas and degrees of freedom, read χ^2 from the χ^2 table.	1	2	3	4	5	382
5. Given areas and degrees of freedom, read F from the F table.	1	2	3	4	5	383
6. Read t, z, F, and χ^2 from one table.	1	2	3	4	5	385

Introduction

Tables present large amounts of numerical information efficiently. Information about the normal distribution and statistics such as t and chi-square are presented in tables. Such information could also be acquired by reading graphs or by computing directly from formulas, but these methods are cumbersome. (However, statistical calculators and computer programs may actually make even direct calculation more efficient than looking up values in tables.)

Tables are laid out for ease of reading. However, the user needs to learn the conventions the designers of the table used. The aim of this chapter is to teach you how to read versions of four tables: normal distribution, Student's *t*-distribution, chi-square distribution, and *F*-distribution. Also, we present a table from which you can read all of these statistics.

Some of you are familiar with reading tables and have no apprehensions about reading them. If you feel comfortable reading tables, you can skip Activity 11.1.

Reading Tables Is a Common Experience

You often read information from a calendar, a form of table. Items that vary (*variables*) in a calendar include the day, the week, the month, and the year. From your experience you know that when three of these items are known (given), you can always read the fourth one.

You can add more variables to a calendar, such as morning or afternoon, time of day, or specific appointments. The more information you add, the more complex the table becomes.

ACTIVITY 11.1

Reading information from a calendar

Refer to Tables 11.1 and 11.2 and answer the following questions.

On which day of the week will December 4th fall in 2001? 1 ______________________

What will be the date of the Wednesday of the fourth week in December 2001? 2 ______________________

TABLE 11.1 Calendar for December 2001

			December 2001				
Week	**Sun**	**Mon**	**Tue**	**Wed**	**Thu**	**Fri**	**Sat**
First							1
Second	2	3	4	5	6	7	8
Third	9	10	11	12	13	14	15
Fourth	16	17	18	19	20	21	22
Fifth	23	24	25	26	27	28	29
Sixth	30	31					

TABLE 11.2 Calendar for December 2010

Week	Sun	Mon	December 2010 Tue	Wed	Thu	Fri	Sat
First				1	2	3	4
Second	5	6	7	8	9	10	11
Third	12	13	14	15	16	17	18
Fourth	19	20	21	22	23	24	25
Fifth	26	27	28	29	30	31	

On which day of the week will December 29, 2010 fall? 3 ____________________

In which of these years will December 8th be a Saturday? 4 ____________________

In which week will December 11, 2010 fall? 5 ____________________

Reading Areas from a Normal Curve Marked in Units of Standard Deviation

Many collections of data (populations) are distributed in such a way that they can be described by the following mathematical equation involving the mean, μ, and the standard deviation, σ, of the distribution.

$$y = \frac{1}{(\sqrt{2\pi\sigma})} \exp\left[-\frac{1}{2\sigma^2}(x-\mu)^2\right]$$

The graph of this equation is a bell-shaped curve that is symmetric around the mean and whose spread is governed by the standard deviation. These curves are called **normal** or **Gaussian distributions.** The intimidating equation above is not relevant for our purposes. It was used to compute the values that appear in the tables we will be using, but all you will need to do is learn how to extract the required information from those tables.

Figure 11.1 shows the graph of a normal distribution with $\mu = 0$ and $\sigma = 1$. You may think of the total area under the curve as equal to 1 (it's this proportion of the total area) or 100 percent (it's this percentage of the total area). We tend to use percentages more, so we chose 100 percent as the total area, and we represented the areas of sections of this graph as percentages of the total. The x-axis is marked in units of standard deviation with the mean,

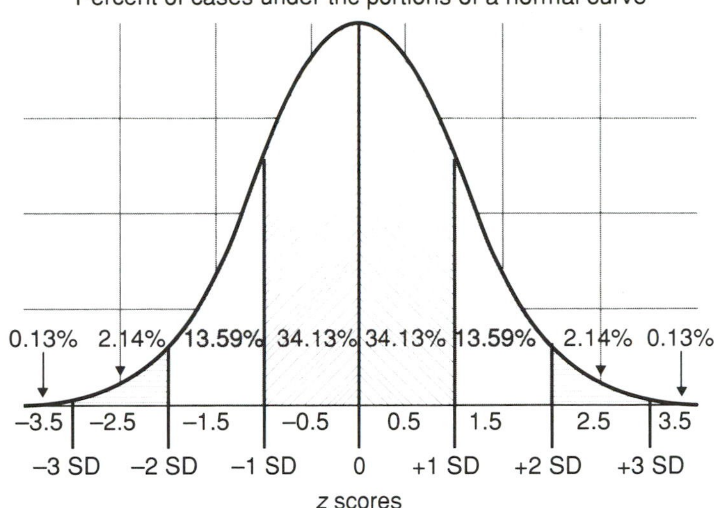

FIGURE 11.1 The Standard Normal Distribution

0, in the center. The y-axis is not marked at all because it is not relevant for reading areas.

The normal distribution with $\mu = 0$ and $\sigma = 1$ is called the **standard** normal distribution. All normal distributions are shaped like the standard normal distribution; they are all bell-shaped and symmetric around the mean. Essentially, the only difference between different normal distributions is that the numbers on the x-axis change. (Note that distributions of other shapes are not called "abnormal" distributions. So many data sets tend to fit the normal distribution that at one time distributions that did not were indeed considered to be "abnormal," but we now know that many perfectly fine distributions don't fit the normal model.) One characteristic of the normal distribution is that 68 percent of the area is within ± 1 standard deviation of the mean and 95 percent is within ± 2 standard deviations.

Figure 11.1 shows a bell-shaped theoretical curve (normal distribution). The x-axis is marked in standard deviation units (z-scores). Percentages of area between portions of the graph are marked for increments of one standard deviation. Use the curve in Figure 11.1 to answer the following questions.

What is the area under the curve between z-scores -3 and -2? 1 ______________

Between $+3$ and $+2$? 2 ______________

Between -2 and -1? 3 ______________

Between $+1$ and $+2$? 4 ______________

Between -1 and 0? 5 ______________

From Figure 11.1 you can also easily calculate the cumulative area from left of the curve. The area for z-scores < −2 SD is 2.14 + 0.13 = 2.27, which may be rounded to 2 percent. What is the area for z-scores < −1? 6 ______________

For z-scores < 0? 7 ______________

For z-scores < +1? 8 ______________

For z-scores < +2? 9 ______________

From Figure 11.1 you can also calculate the middle area contained between two z-scores (For example, ±z). What is the area for ±1? 10 ______________

For ±2? 11 ______________

Graphs such as the one in Figure 11.1 are constructed so that areas under the curve correspond to probabilities. For example, for the normal curve, the area under the curve between the vertical lines at $z = -1$ and $z = +1$ is equal to the probability that a normally distributed variable will have a z-score between −1 and +1. If this concept seems terribly complicated to you, don't despair; as you use the normal distribution, the concept will become more clear. For now, you can concentrate on learning to read areas from the graph and from the tables.

When we speak of areas under the curve, we are really talking about probabilities. We like to talk about area because it is concrete, easy to visualize, whereas probability is abstract, to us, and we find it difficult to visualize. We will generally give the areas as percentages, i.e., percentages of the total area, to emphasize the connection to probabilities (relative frequencies).

ACTIVITY 11.2

Reading areas from a graph given standard deviations

Refer to Figure 11.1. What area lies to the left of the following standard deviations?

−2 SD 0.13% + 2.14% = 1 ______________

−1 SD 2 ______________

+2 SD 3 ______________

+1 SD 4 ______________

0 5 ______________

What area lies to the right of the following standard deviations?

−1 SD	6 ______
−2 SD	7 ______
+1 SD	8 ______

ACTIVITY 11.3

Reading areas from a table of cumulative percentages against standard deviations

Standard Deviations	Rounded Cumulative Percentages, or Areas
−2 SD	2%
−1 SD	16%
0 SD	50%
+1 SD	84%
+2 SD	98%

Standard Deviation	Rounded Cumulative Percentage
0	1 ______
−1	2 ______
+2	3 ______
−2	4 ______
+1	5 ______
6 ______	2%
7 ______	84%
8 ______	98%
9 ______	16%
10 ______	50%

The Normal Distribution

The following are descriptive statistics for selected variables from the faculty exercise data (FED). Each variable has a different mean and standard deviation. We substituted mean and standard deviation into the normal distribution formula and instructed a computer to draw each bell-shaped curve on the same axis system. These are shown in Figure 11.2.

	PREHR	PREVO2	PREWAT	PREWT
n of cases	30	30	30	30
Minimum	67.0	21.4	144.0	122.0
Maximum	122.0	48.0	326.0	312.0
Mean	86.7	31.1	221.9	194.3
Standard dev	12.6	6.4	39.7	44.7

The curves appear to have different shapes, but that is because we maintained the same units on the x-axis for all four curves. You can see that each is bell-shaped, however. Graphing all four with the same x-axis is somewhat misleading because the units for each variable are different. PREWT is measured in pounds, for example, whereas PREWAT is measured in watts.

The four distributions are all normal distributions and hence 68 percent of the cases are within one standard deviation of the mean and 95 percent of the cases are within two standard deviations.

It would be impossible to make a table for every normal distribution, because there are infinitely many of them. However, transforming the data values

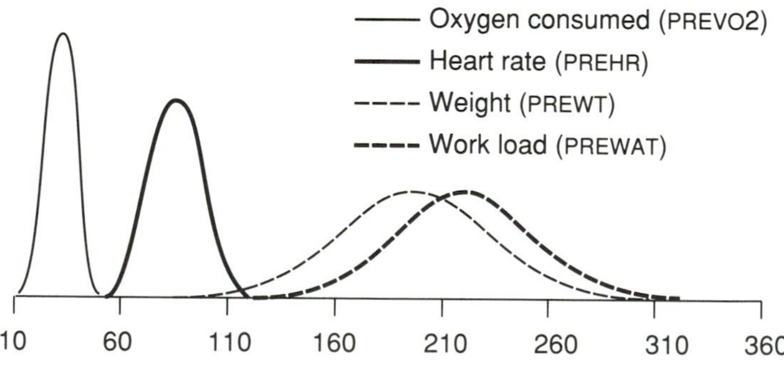

FIGURE 11.2 Normal Curves for Four Variables from the Faculty Exercise Data

to z-scores (using the following formula) allows us to use the standard normal distribution to compute values for any normal distribution.

$$z\text{-score formula: } z = \frac{(x - \mu)}{\sigma}$$

Any given x can be converted to its z-score using this formula. Then you can look up z in the normal distribution table and find the percent of cases whose z-scores are less than this z. The percent of cases less than x will be the same. The same table can be used to find z (and hence x) when the area (probability) is known.

Cumulative Normal Distribution

Table 11.3 gives values of the cumulative normal distribution. This table is similar to the table in Activity 11.3 except that it has many more intermediate z-values. You use this table just as you used the table in Activity 11.3. We find it helpful to include a graph of the normal distribution curve with the normal table so that when we are using it we know exactly which region we are finding the area of. (Note that Table 11.3 also appears on page 688 as Table B-1 in Appendix B.)

ACTIVITY 11.4

Reading areas when z-scores are given

Table 11.3 has three columns. These columns are labeled z, [1] _______, and [2] _______. z refers to the z-score as defined above, and the area is the cumulative area under the curve up to the specified z (the area in the left-hand tail of the curve). 1 − area is the area under the curve to the right of the specified z (the area in the right-hand tail). After converting data values to their z-scores we can read cumulative values for any normal distribution from Table 11.3. For example, the area in the left-hand tail when $x = 19$, $\mu = 31$, and $\sigma = 6$ is [3] _______. This number is the probability that $x < 19$. If the area in the right-hand tail is known to be 0.066807, the corresponding z-score would be [4] _______, and assuming the same mean and standard deviation as before, x is [5] _______.

By this time you are quite familiar with the area values corresponding to the z-scores −2, −1, 0, 1, and 2. Refer to Table 11.3 and note the cumulative areas for −2.5, −1.5, −0.5, 0.5, 1.5, and 2.5 in the blanks given in Figure 11.3.

ACTIVITY 11.5

Reading z-scores when areas are given

Write the z-score corresponding to the areas given in Figure 11.4 on the arrow shafts.

TABLE 11.3 Cumulative Standard Normal Distribution, or z

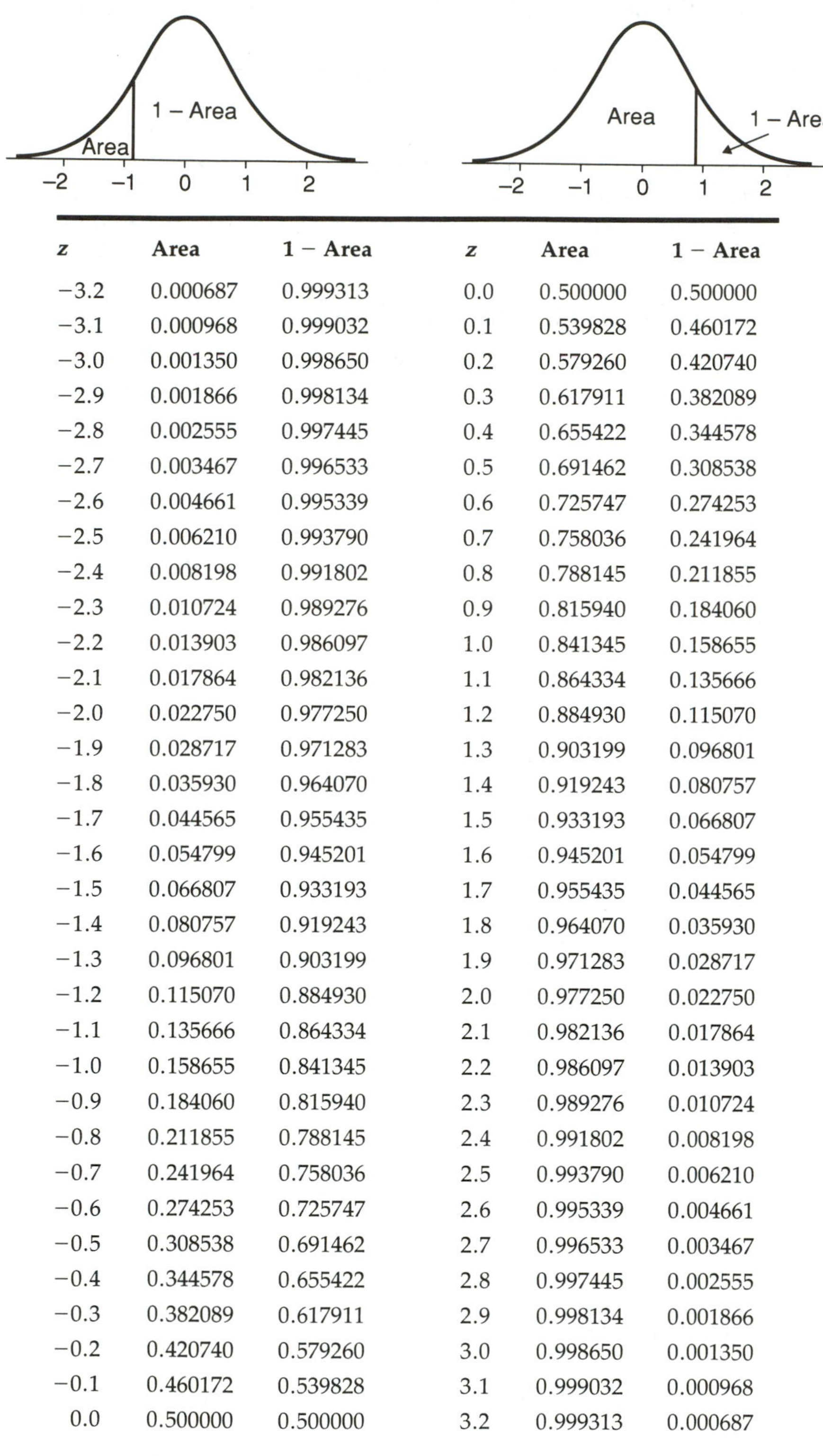

z	Area	1 − Area	z	Area	1 − Area
−3.2	0.000687	0.999313	0.0	0.500000	0.500000
−3.1	0.000968	0.999032	0.1	0.539828	0.460172
−3.0	0.001350	0.998650	0.2	0.579260	0.420740
−2.9	0.001866	0.998134	0.3	0.617911	0.382089
−2.8	0.002555	0.997445	0.4	0.655422	0.344578
−2.7	0.003467	0.996533	0.5	0.691462	0.308538
−2.6	0.004661	0.995339	0.6	0.725747	0.274253
−2.5	0.006210	0.993790	0.7	0.758036	0.241964
−2.4	0.008198	0.991802	0.8	0.788145	0.211855
−2.3	0.010724	0.989276	0.9	0.815940	0.184060
−2.2	0.013903	0.986097	1.0	0.841345	0.158655
−2.1	0.017864	0.982136	1.1	0.864334	0.135666
−2.0	0.022750	0.977250	1.2	0.884930	0.115070
−1.9	0.028717	0.971283	1.3	0.903199	0.096801
−1.8	0.035930	0.964070	1.4	0.919243	0.080757
−1.7	0.044565	0.955435	1.5	0.933193	0.066807
−1.6	0.054799	0.945201	1.6	0.945201	0.054799
−1.5	0.066807	0.933193	1.7	0.955435	0.044565
−1.4	0.080757	0.919243	1.8	0.964070	0.035930
−1.3	0.096801	0.903199	1.9	0.971283	0.028717
−1.2	0.115070	0.884930	2.0	0.977250	0.022750
−1.1	0.135666	0.864334	2.1	0.982136	0.017864
−1.0	0.158655	0.841345	2.2	0.986097	0.013903
−0.9	0.184060	0.815940	2.3	0.989276	0.010724
−0.8	0.211855	0.788145	2.4	0.991802	0.008198
−0.7	0.241964	0.758036	2.5	0.993790	0.006210
−0.6	0.274253	0.725747	2.6	0.995339	0.004661
−0.5	0.308538	0.691462	2.7	0.996533	0.003467
−0.4	0.344578	0.655422	2.8	0.997445	0.002555
−0.3	0.382089	0.617911	2.9	0.998134	0.001866
−0.2	0.420740	0.579260	3.0	0.998650	0.001350
−0.1	0.460172	0.539828	3.1	0.999032	0.000968
0.0	0.500000	0.500000	3.2	0.999313	0.000687

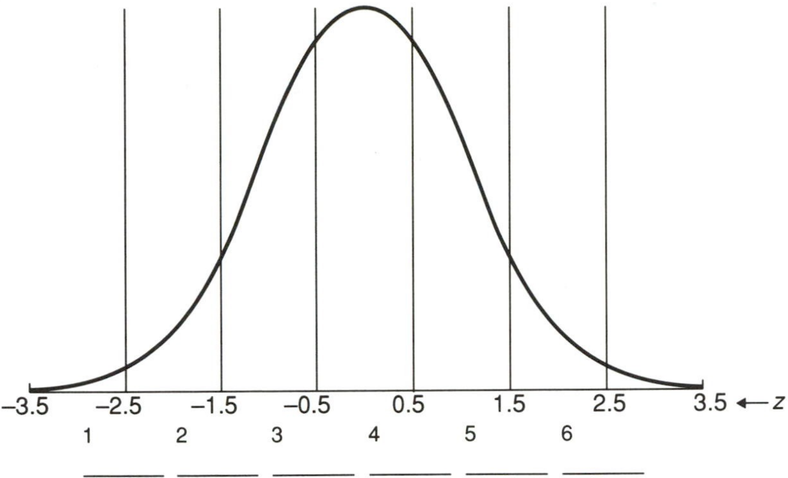

FIGURE 11.3 Normal Distribution Curve

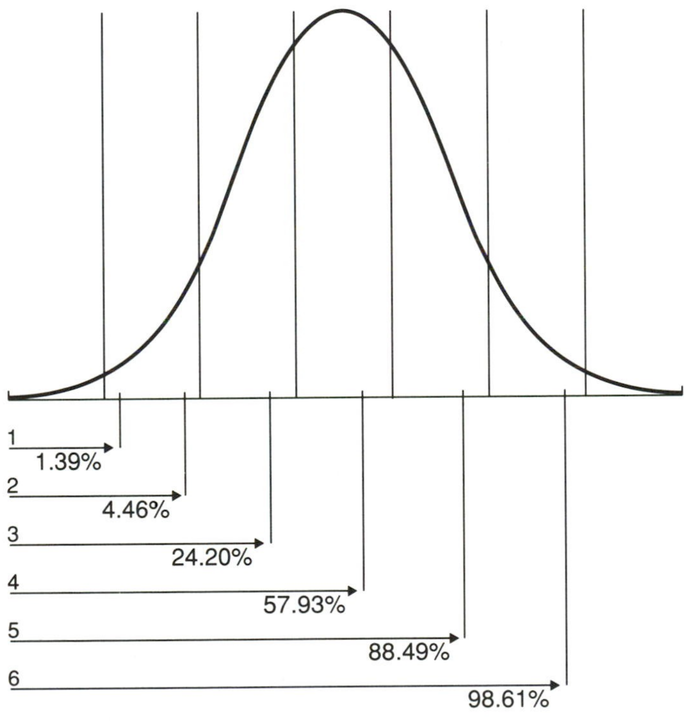

FIGURE 11.4 Cumulative Areas and the Corresponding z-Scores

ACTIVITY 11.6

Calculation of x when μ, σ, and z are given

Calculate the value of x, given the values below.

Mean, μ, and SD, σ	z	Values of $x = \mu \pm z\sigma$
$\mu = 31$, $\sigma = 6$	-1.5	1 ________
$\mu = 87$, $\sigma = 13$	$+1$	2 ________
$\mu = 222$, $\sigma = 40$	-1.0	3 ________
$\mu = 194$, $\sigma = 45$	$+2$	4 ________

Normal Distribution: Central Area

Table 11.4 is another form of the normal distribution table. It has the following headings: 1 _______, 2 _______, 3 _______, and 4 _______. If you add values for the left area (area in the left-hand tail), the central area, and the right area (area in the right-hand tail), you get 5 _______. (Note that Table 11.4 also appears on page 690 as Table B-2 in Appendix B.)

ACTIVITY 11.7

Reading central area percentages when z is given

Four z-intervals are indicated by arrows in Figure 11.5. Read the appropriate central area value from Table 11.4 and write it on the arrow indicating each interval. Round off your answers to three decimal places.

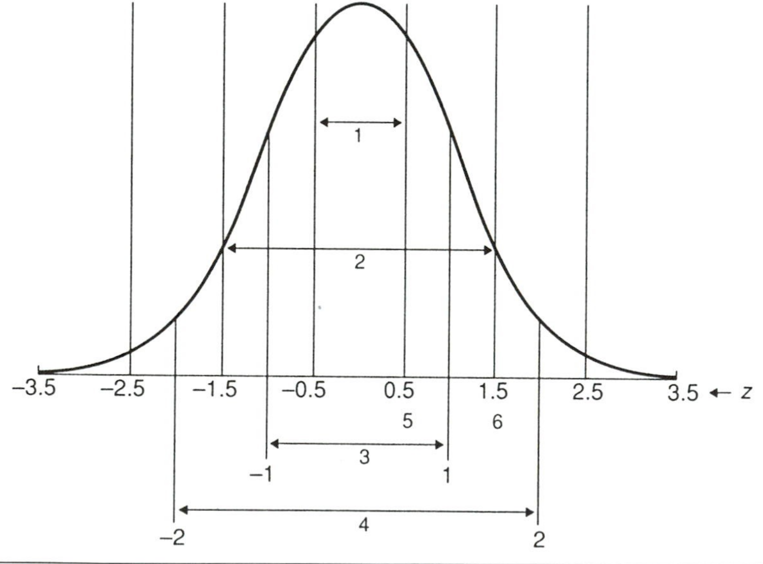

FIGURE 11.5 Central Area Percentages Under the Curve When z-Score Is Given

TABLE 11.4 Normal Distribution: Central Area (incremented *z*-scores)

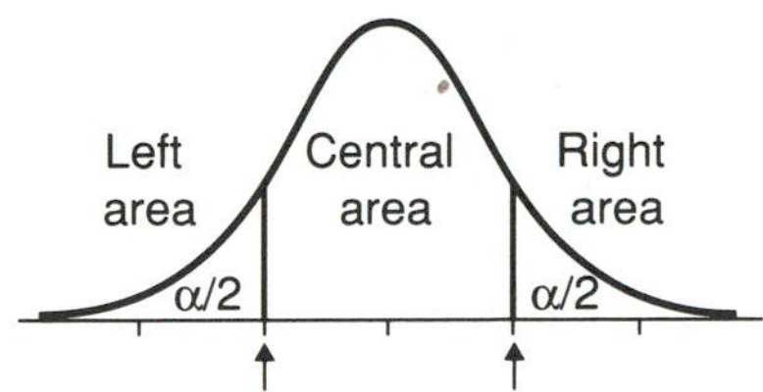

$\pm z$	Left Area	Central Area	Right Area
±3.2	0.000687	0.998626	0.000687
±3.1	0.000968	0.998065	0.000968
±3.0	0.001350	0.997300	0.001350
±2.9	0.001866	0.996268	0.001866
±2.8	0.002555	0.994890	0.002555
±2.7	0.003467	0.993066	0.003467
±2.6	0.004661	0.990678	0.004661
±2.5	0.006210	0.987581	0.006210
±2.4	0.008198	0.983605	0.008198
±2.3	0.010724	0.978552	0.010724
±2.2	0.013903	0.972193	0.013903
±2.1	0.017864	0.964271	0.017864
±2.0	0.022750	0.954500	0.022750
±1.9	0.028717	0.942567	0.028717
±1.8	0.035930	0.928139	0.035930
±1.7	0.044565	0.910869	0.044565
±1.6	0.054799	0.890401	0.054799
±1.5	0.066807	0.866386	0.066807
±1.4	0.080757	0.838487	0.080757
±1.3	0.096801	0.806399	0.096801
±1.2	0.115070	0.769861	0.115070
±1.1	0.135666	0.728668	0.135666
±1.0	0.158655	0.682690	0.158655
±0.9	0.184060	0.631880	0.184060
±0.8	0.211855	0.576289	0.211855
±0.7	0.241964	0.516073	0.241964
±0.6	0.274253	0.451494	0.274253
±0.5	0.308538	0.382925	0.308538
±0.4	0.344578	0.310843	0.344578
±0.3	0.382089	0.235823	0.382089
±0.2	0.420740	0.158519	0.420740
±0.1	0.460172	0.079656	0.460172
0.0	0.500000	0.000000	0.500000

Activity 11.8

Reading z-scores, central area, or tail area

Notice the graph that appears with Table 11.4. Because of the symmetry of the curve, the area in the left-hand tail is the same as the area in the right-hand tail. Use Table 11.4 to fill in the blanks below (two-decimal-place accuracy is sufficient).

Central area	Tail area	*z*-score
45.15%	1 ______	2 ______
7.97%	3 ______	4 ______
99.86%	0.07%	5 ______
6 ______	11.51%	7 ______
94.26%	8 ______	9 ______
91.09%	10 ______	11 ______
12 ______	5.48%	13 ______
99.07%	14 ______	15 ______
16 ______	8.08%	17 ______

Table 11.5 is similar to Table 11.4, but instead of incrementing the *z*-scores by a fixed amount, this table increments the central areas. Table 11.5 can be used to find the *z*-scores when certain standard areas are given. Fill in the *z*-scores for the areas given below.

Central area	*z*-score
90%	1 ______
95%	2 ______
99%	3 ______
80%	4 ______
50%	5 ______
60%	6 ______

TABLE 11.5 Normal Distribution with Incremented Central Areas (areas given as percents)

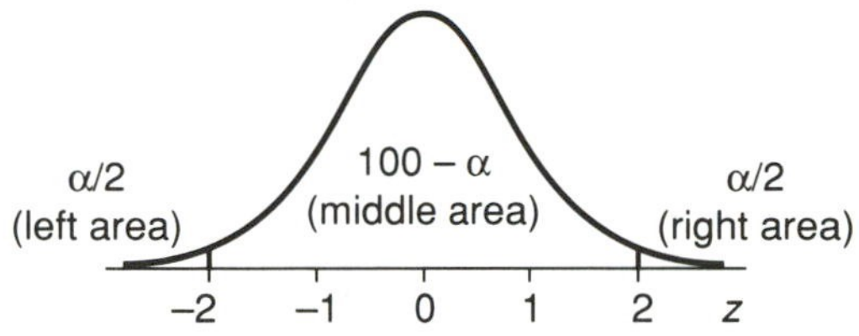

Area at the Left Side under the Curve (%)	$-z$	Middle Area under the Curve (%)	$\pm z$	Area at the Right Side under the Curve (%)	$+z$
25	−0.674	50	±0.674	25	0.674
20	−0.842	60	±0.842	20	0.842
15	−1.036	70	±1.036	15	1.036
10	−1.282	80	±1.282	10	1.282
5	−1.645	90	±1.645	5	1.645
2.5	−1.960	95	±1.960	2.5	1.960
0.5	−2.576	99	±2.576	0.5	2.576

Uses of the Normal Distribution

Normal distributions have many applications, some of which will be presented in later chapters. Some of the major uses include:

1. Testing hypotheses. To do this a researcher computes z from the experimental data and looks up the corresponding areas in a normal distribution table. This process will be discussed at greater length in Chapter 14.
2. Estimating the number of cases below or above a given observation of a normally distributed variable, or between two given observations.
3. Using a sample mean to estimate the size of a population mean. Even when the population is not normally distributed, sample means are normally distributed, provided the sample size is not too small.

Student's t-Distribution

Student's t-distribution was derived by W. S. Gossett, an Irish mathematician who worked for a Dublin brewery and was probably the first statistician em-

Table 11.6 Values of t for Various Cumulative Areas, df = 25 and ∞

df	$t_{.70}$	$t_{.80}$	$t_{.90}$	$t_{.95}$	$t_{.975}$	$t_{.99}$	$t_{.995}$
25	.531	.856	1.316	1.708	2.060	2.485	2.787
∞ (normal distribution)	.524	.842	1.282	1.645	1.960	2.236	2.576
	$-t_{.30}$	$-t_{.20}$	$-t_{.10}$	$-t_{.05}$	$-t_{.025}$	$-t_{.01}$	$-t_{.005}$

ployed by a major corporation other than an insurance company. Because his employers did not want their competition to know that they employed a statistician, Gossett published his mathematical discoveries under a pseudonym. He published the derivation of the t-distribution in 1908 under the name of A. Student, and to this day, the distribution is known as Student's t-distribution.

Like the normal distributions, the t-distributions are bell-shaped curves with the total area under the curve equal to 1, or 100 percent. Unlike the normal distributions, the cumulative area under the t-distribution curve depends on sample size, n, as well as on mean and standard deviation. For certain technical reasons, these curves are classified by a variable called **degrees of freedom,** denoted df, where df $= n - 1$. There is a different t-distribution for each value of the variable df.

t-distributions are generally heavier in the tails than are normal distributions. As df (or n) increases, the t-distribution begins to look more like a normal distribution, so for df $= \infty$ the t-distribution values are identical to the normal distribution values. In practice, t-values are usually rounded off to one decimal place. Therefore, for all practical purposes the t-distribution is approximated by the normal distribution above df = 30. t-distributions are mainly used for two purposes: comparison of means and comparison of slopes of regression lines.

Often the tail area of a t-distribution is denoted by α. The value of the t-statistic corresponding to a cumulative area α is labeled t_α. Table 11.6 gives the values of t_α for various commonly used values of α. The corresponding values of z for a normal distribution are also given in the row labeled df $= \infty$. Notice that the columns are labeled with positive values of t at the top and with negative values of t at the bottom. Thus, the value of t for $\alpha = .80$ is .856; the value of t for $\alpha = .20$ is $-.856$.

Activity 11.9

Reading t-values when cumulative areas are given

Figure 11.6 shows various cumulative areas of a t-distribution. Refer to Table 11.6 and write the corresponding value of t on each arrow. Table 11.6 gives the

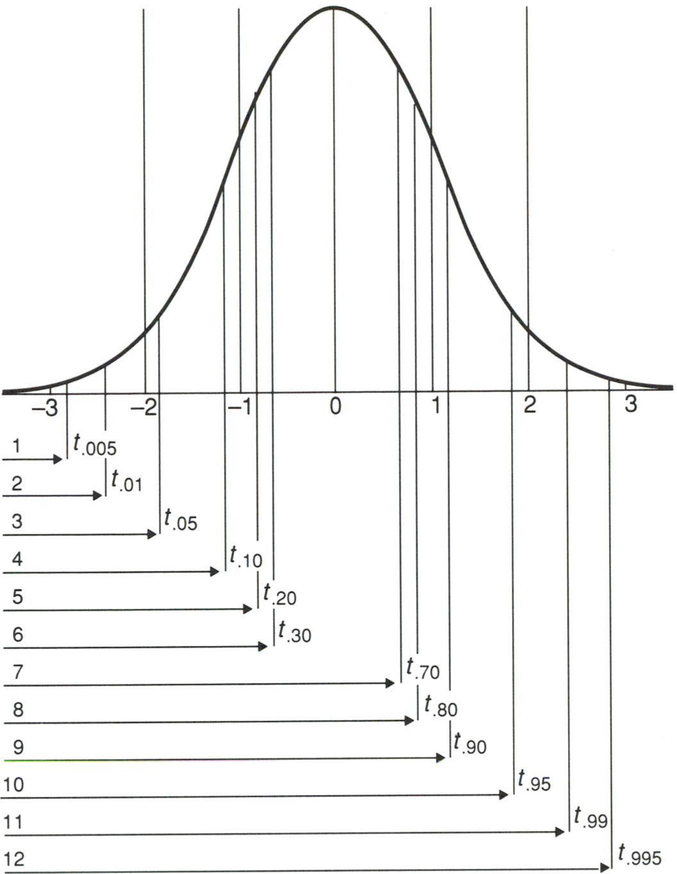

FIGURE 11.6 *t*-Distribution (df = 25)

t-distributions for df of 25 and ∞. When rounded to one decimal place, the values for the two distributions are [1] very similar/quite dissimilar.

In Tables 11.7 and 11.8, each row represents a *t*-distribution for a different df. The values of *t* are given only for $\alpha = .05$ and $\alpha = .01$. Table 11.7 gives the values of *t* for areas of .05 and .01 in the right-hand tail and for the same areas in the left-hand tail. Table 11.8 is similar, but is designed to be used for reading *t*-values either when the central area is known (as is typical when doing calculations of confidence intervals; see Chapter 14) or when the area in question is split equally between the two tails (as is typical when testing a two-sided hypothesis; see Chapter 14). For now, your task is only to learn how to read the table. When you get to Chapter 14 you will be able to concentrate on learning how the numbers you get from the tables are used. (Note that Tables 11.7 and 11.8 also appear on pages 691 and 692 as Tables B-3 and B-4 in Appendix B.)

TABLE 11.7 Critical Values of the *t*-Distribution for One-Tailed Tests

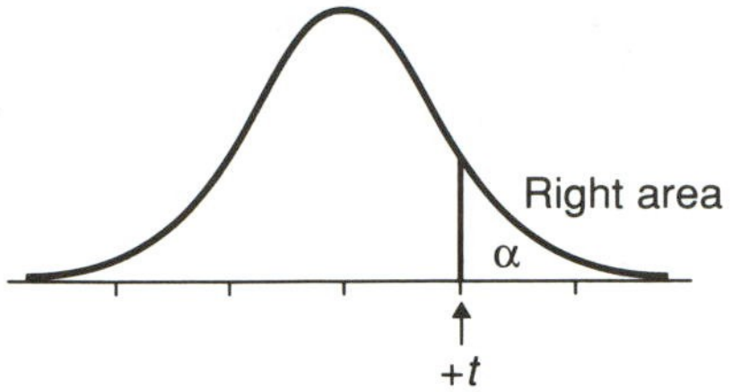

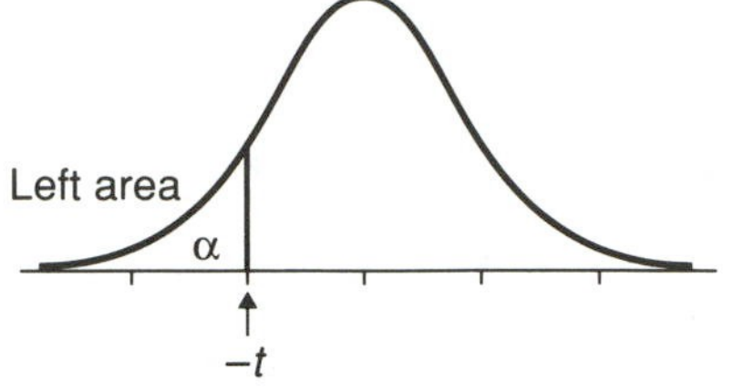

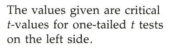

The values given are critical *t*-values for one-tailed *t* tests on the right side.

The values given are critical *t*-values for one-tailed *t* tests on the left side.

df	$\alpha = .05$	$\alpha = .01$	df	$\alpha = .05$	$\alpha = .01$
5	2.015	3.365	5	−2.015	−3.365
6	1.943	3.143	6	−1.943	−3.143
7	1.895	2.993	7	−1.895	−2.993
8	1.860	2.896	8	−1.860	−2.896
9	1.833	2.821	9	−1.833	−2.821
10	1.812	2.764	10	−1.812	−2.764
11	1.796	2.718	11	−1.796	−2.718
12	1.782	2.681	12	−1.782	−2.681
13	1.771	2.650	13	−1.771	−2.650
14	1.761	2.624	14	−1.761	−2.624
15	1.753	2.602	15	−1.753	−2.602
16	1.746	2.583	16	−1.746	−2.583
17	1.740	2.567	17	−1.740	−2.567
18	1.734	2.552	18	−1.734	−2.552
19	1.729	2.539	19	−1.729	−2.539
20	1.725	2.528	20	−1.725	−2.528
21	1.721	2.518	21	−1.721	−2.518
22	1.717	2.508	22	−1.717	−2.508
23	1.714	2.500	23	−1.714	−2.500
24	1.711	2.492	24	−1.711	−2.492
25	1.708	2.485	25	−1.708	−2.485
26	1.706	2.479	26	−1.706	−2.479
27	1.703	2.473	27	−1.703	−2.473
28	1.701	2.467	28	−1.701	−2.467
29	1.699	2.462	29	−1.699	−2.462
30	1.697	2.457	30	−1.697	−2.457
40	1.684	2.423	40	−1.684	−2.423
60	1.671	2.390	60	−1.671	−2.390
120	1.658	2.358	120	−1.658	−2.358
∞	1.645	2.326	∞	−1.645	−2.326

At df = ∞, the *t*-distribution and the standard normal distribution are exactly the same.

At df = ∞, the *t*-distribution and the standard normal distribution are exactly the same.

TABLE 11.8 Critical Values of the *t*-Distribution for Two-Tailed Tests and Confidence Intervals

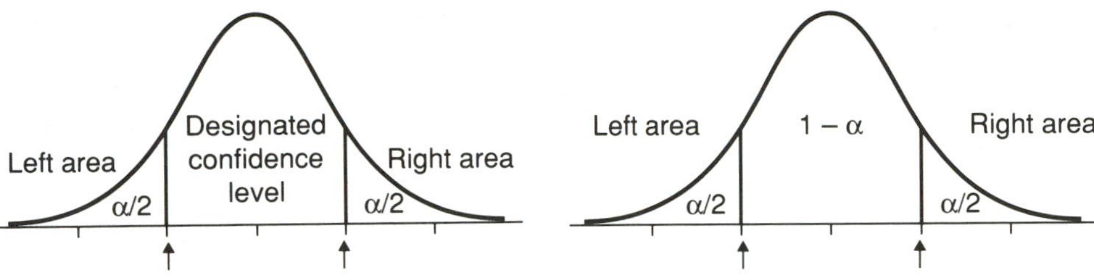

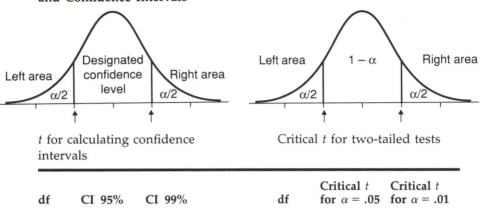

t for calculating confidence intervals

Critical *t* for two-tailed tests

df	CI 95%	CI 99%
5	±2.571	±4.032
6	±2.447	±3.707
7	±2.365	±3.499
8	±2.306	±3.355
9	±2.262	±3.250
10	±2.228	±3.169
11	±2.201	±3.106
12	±2.179	±3.055
13	±2.160	±3.012
14	±2.145	±2.977
15	±2.131	±2.947
16	±2.120	±2.921
17	±2.110	±2.898
18	±2.101	±2.878
19	±2.093	±2.861
20	±2.086	±2.845
21	±2.080	±2.831
22	±2.074	±2.819
23	±2.069	±2.807
24	±2.064	±2.797
25	±2.060	±2.787
26	±2.056	±2.779
27	±2.052	±2.771
28	±2.048	±2.763
29	±2.045	±2.756
30	±2.042	±2.750
40	±2.021	±2.704
60	±2.000	±2.660
120	±1.980	±2.617
∞	±1.960	±2.576

At df = ∞, *t*-distribution and standard normal distribution are exactly the same.

df	Critical *t* for $\alpha = .05$	Critical *t* for $\alpha = .01$
5	±2.571	±4.032
6	±2.447	±3.707
7	±2.365	±3.499
8	±2.306	±3.355
9	±2.262	±3.250
10	±2.228	±3.169
11	±2.201	±3.106
12	±2.179	±3.055
13	±2.160	±3.012
14	±2.145	±2.977
15	±2.131	±2.947
16	±2.120	±2.921
17	±2.110	±2.898
18	±2.101	±2.878
19	±2.093	±2.861
20	±2.086	±2.845
21	±2.080	±2.831
22	±2.074	±2.819
23	±2.069	±2.807
24	±2.064	±2.797
25	±2.060	±2.787
26	±2.056	±2.779
27	±2.052	±2.771
28	±2.048	±2.763
29	±2.045	±2.756
30	±2.042	±2.750
40	±2.021	±2.704
60	±2.000	±2.660
120	±1.980	±2.617
∞	±1.960	±2.576

At df = ∞, *t*-distribution and standard normal distribution are exactly the same.

Activity 11.10

Reading values of t when one tail is given

In the following table, df and α are given. L and R indicate left or right tail, respectively, for α. Use Table 11.7 to fill in the values of t in the last column.

df	α, L/R	t
16	.05, R	1 __________
8	.01, R	2 __________
20	.01, L	3 __________
30	.05, R	4 __________
7	.05, L	5 __________
60	.01, L	6 __________

Figure 11.7 relates the central area under a t-distribution curve to the tail areas, which are usually expressed as subscripts on t. When the central area is known, we compute α by subtracting the central area from 1 and dividing the result by 2 (because the remaining area is split between the two tails). We could then find t from Table 11.7. However, similar tables are specifically designed for this purpose. Table 11.8 is an example of such a table. The left-hand and right-hand parts of the table are identical except that the headings are designed for different purposes. A CI of 95 percent means that the central area is 95 percent.

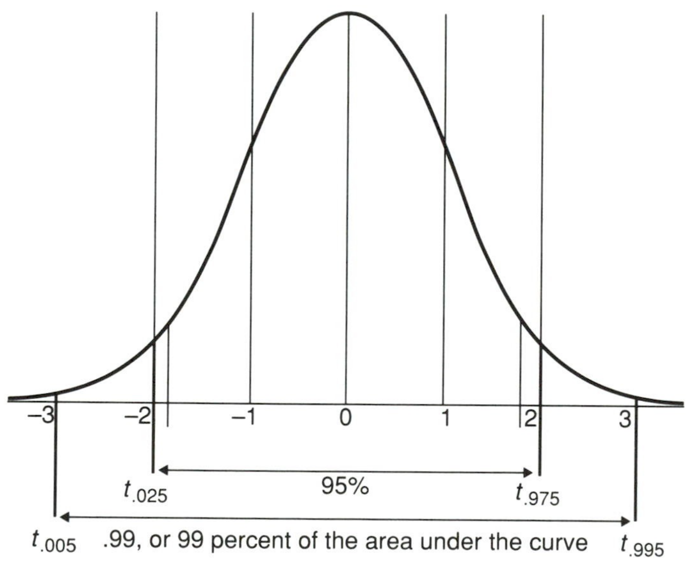

Figure 11.7 Connection Between Central Areas and Tail Areas

ACTIVITY 11.11

Finding t-statistics or central areas

Use Table 11.8 to fill in the missing values in the following table.

df	Central area	*t*
10	95%	[1] ± __________
15	99%	[2] ± __________
20	95%	[3] ± __________
25	[4] __________	± 2.787
6	99%	[5] ± __________
5	95%	[6] ± __________
15	[7] __________	± 2.131
∞	99%	[8] ± __________
∞	95%	[9] ± __________

χ^2 (Chi-Square) Distributions

χ^2 distributions, like *t*-distributions, are a family of distributions indexed by a variable called degrees of freedom. Values of χ^2 are given in Table 11.9. (Note that Table 11.9 also appears on page 693 as Table B-5 in Appendix B.) χ^2 and its respective degrees of freedom are calculated in different ways using different sets of formulas, depending on the application.

χ^2 is used to test whether the frequency distribution of a variable measured on a categorical scale fits a certain known theoretical distribution. For this application,

$$\chi^2 = \sum \frac{(O_i - E_i)^2}{E_i}$$

where O_i stands for the observed frequencies of the variable for each category and E_i stands for the expected values for that variable predicted by the theoretical distribution. In this case, the value of df $= k - 1$, where k is the number of categories of the variable.

A variation of this application for χ^2 is testing whether two categorical variables are independent. In this case the frequencies are easiest to work with if presented in a two-way table, each cell of which contains both the observed and expected frequency for that cell. Expected frequencies are computed as described in Chapter 10. χ^2 is computed using the above formula, summing over all cells in the table. The only difference is that df $= (r - 1) * (c - 1)$, where r is the number of rows in the table (i.e., the number of categories of the row variable) and c is the number of columns (i.e., the number of categories of the column variable).

TABLE 11.9 Critical χ^2 for Chosen Tail Areas or Percentiles

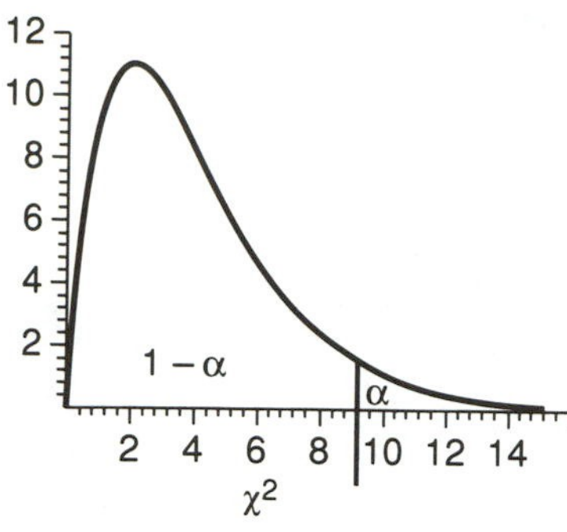

df	$\alpha = .10$, or P_{90}	$\alpha = .05$, or P_{95}	$\alpha = .01$, or P_{99}
1	2.71	3.84	6.63
2	4.61	5.99	9.21
3	6.25	7.81	11.34
4	7.78	9.49	13.28
5	9.24	11.07	15.09
6	10.64	12.59	16.81
7	12.02	14.07	18.48
8	13.36	15.51	20.09
9	14.68	16.92	21.67
10	15.99	18.31	23.21
11	17.28	19.68	24.73
12	18.55	21.03	26.22
13	19.81	22.36	27.69
14	21.06	23.68	29.14
15	22.31	25.00	30.58
16	23.54	26.30	32.00
18	25.99	28.87	34.81
20	28.41	31.41	37.57
24	33.20	36.42	42.98
30	40.26	43.77	50.89
40	51.81	55.76	63.69
60	74.40	79.08	88.38
120	140.23	146.57	158.95

The following table has three columns and two rows. So for this table, df = (2 − 1) ∗ (3 − 1) = 2.

	Stress				
	Minimal	**Moderate**	**Heavy**	**Total**	
Male	7	18	7	32	←Observed
	7.04	19.84	5.12		←Expected
Female	4	13	1	18	←Observed
	3.96	11.16	2.88		←Expected

χ^2 is also used to compare a sample variance with a population variance. For this application, the formula used to calculate χ^2 is

$$\chi^2/\text{df} = \frac{s^2}{\sigma^2}.$$

In this formula s^2 is the sample variance and σ^2 is the population variance. Here df = $n - 1$, where n is the sample size. Notice that this χ^2 is divided by df. The values of χ^2 in Table 11.9 are not divided by df.

All t-distributions are symmetric with mean 0 and standard deviation 1, whereas χ^2 distributions are generally not symmetric and χ^2 only takes positive values. When df ≥ 100, the χ^2 distribution is roughly symmetrical and can be approximated by the normal distribution with mean = df and variance = 2df.

In Figure 11.8, write the values of χ^2 from Table 11.9 that correspond to each of the given α values on the line to the right of the α value.

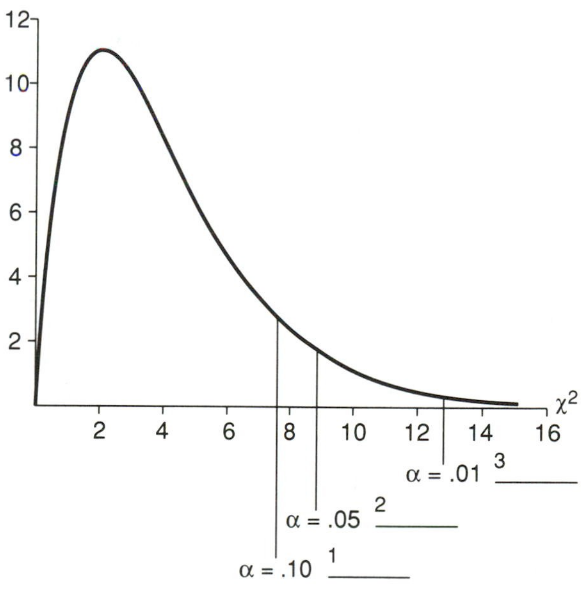

Figure 11.8 χ^2 Distribution with df = 4

ACTIVITY 11.12

Reading percentiles or χ^2

Use Table 11.9 to fill in the missing values below.

df	χ^2	Percentile
120	1 ______	99
1	2 ______	95
3	3 ______	99
4	4 ______	90
1	6.63	5 ______
2	4.61	6 ______
5	7 ______	99
6	12.59	8 ______
120	146.57	9 ______

F-Distribution

F is the ratio

$$s_1^2/s_2^2$$

and it is primarily used in analysis of variance, a procedure used to test hypotheses regarding equality of three or more means (see Chapter 15). In contrast, *t* is used for testing hypotheses regarding equality of two means.

F-distributions, like *t* and χ^2, are a family of curves computed from mathematical formulas, one for each combination of two separate degrees of freedom. In the formula

$$F = s_1^2/s_2^2$$

s_1 is the variance for one of the samples and s_2 is the variance of another. If the hypothesis that $\sigma_1^2 = \sigma_2^2$ is true, both s_1^2 and s_2^2 estimate the same variance, and therefore differ only because of the sampling error. Usually each variance has different degrees of freedom. Frequently *F* is denoted in the following way:

$F_{5,6}$ ← Denominator df

↑ Numerator df

(Note: The 5 and 6 are examples only; these numbers can be anything from 1 to infinity. Also, by convention, the numerator df is written before the denominator df.)

Tables for the *F*-distribution are used as tables for *t* and χ^2 are used. In hypothesis testing, when the value arrived at experimentally is less than the value in the table, the so-called null hypothesis is accepted, whereas when the value is greater than the value in the table, the so-called alternative hypothesis is accepted. Hypothesis testing using these distributions will be discussed at greater length in Chapter 14. For now, your only task is to understand how to find values in the tables.

Look at Table 11.10. (Note that Table 11.10 also appears on page 694 as Table B-6 in Appendix B.) Note that for any combination of degrees of freedom for numerator and denominator, the 95th percentile *F* statistic ($F_{.95}$) is given. (For *t*, our tables gave 95th and 99th percentiles. For the normal distribution, we were able to read a great many different percentiles.) Tables for other percentiles of the *F* statistic can be found in some libraries. These percentiles can also be computed using most statistical computer software.

ACTIVITY 11.13

Reading $F_{.95}$ values when df of numerator and denominator are given

Use Table 11.10 to fill in the missing values below.

Numerator df, denominator df	***F***
10, 3	1 ____________
5, 4	2 ____________
2, 5	3 ____________
5, 11	4 ____________
8, 21	5 ____________
4, 16	6 ____________
15, 30	7 ____________
20, 20	8 ____________

Relationship of z, t, F, and χ^2

z, *t*, *F*, and χ^2 are related. Each has a distribution that indicates the percentages of the occurrence of the statistics we compute. *z*, *t*, and χ^2 can be viewed as special cases of the *F*-distribution. The deciding factor is the degrees of freedom for numerator and denominator.

Figure 11.9 shows the relationships among the four fundamental sampling distributions. At the top of the diagram is the general *F*-distribution, indexed

TABLE 11-10 *F* Distribution, Upper 5 Percent Values for Various Degrees of Freedom

df for Denominator	df for Numerator																		
	1	2	3	4	5	6	7	8	9	10	12	15	20	24	30	40	60	120	∞
1	161	200	216	225	230	234	237	239	241	242	244	246	248	249	250	251	252	253	254
2	18.5	19.0	19.2	19.2	19.3	19.3	19.4	19.4	19.4	19.4	19.4	19.4	19.4	19.5	19.5	19.5	19.5	19.5	19.5
3	10.1	9.55	9.28	9.12	9.01	8.94	8.89	8.85	8.81	8.79	8.74	8.70	8.66	8.64	8.62	8.59	8.57	8.55	8.53
4	7.71	6.94	6.59	6.39	6.26	6.16	6.09	6.04	6.00	5.96	5.91	5.86	5.80	5.77	5.75	5.72	5.69	5.66	5.63
5	6.61	5.79	5.41	5.19	5.05	4.95	4.88	4.82	4.77	4.74	4.68	4.62	4.56	4.53	4.50	4.46	4.43	4.40	4.37
6	5.99	5.14	4.76	4.53	4.39	4.28	4.21	4.15	4.10	4.06	4.00	3.94	3.87	3.84	3.81	3.77	3.74	3.70	3.67
7	5.59	4.74	4.35	4.12	3.97	3.87	3.79	3.73	3.68	3.64	3.57	3.51	3.44	3.41	3.38	3.34	3.30	3.27	3.23
8	5.32	4.46	4.07	3.84	3.69	3.58	3.50	3.44	3.39	3.35	3.28	3.22	3.15	3.12	3.08	3.04	3.01	2.97	2.93
9	5.12	4.26	3.86	3.63	3.48	3.37	3.29	3.23	3.18	3.14	3.07	3.01	2.94	2.90	2.86	2.83	2.79	2.75	2.71
10	4.96	4.10	3.71	3.48	3.33	3.22	3.14	3.07	3.02	2.98	2.91	2.85	2.77	2.74	2.70	2.66	2.62	2.58	2.54
11	4.84	3.98	3.59	3.36	3.20	3.09	3.01	2.95	2.90	2.85	2.79	2.72	2.65	2.61	2.57	2.53	2.49	2.45	2.40
12	4.75	3.89	3.49	3.26	3.11	3.00	2.91	2.85	2.80	2.75	2.69	2.62	2.54	2.51	2.47	2.43	2.38	2.34	2.30
13	4.67	3.81	3.41	3.18	3.03	2.92	2.83	2.77	2.71	2.67	2.60	2.53	2.46	2.42	2.38	2.34	2.30	2.25	2.21
14	5.60	3.74	3.34	3.11	2.96	2.85	2.76	2.70	2.65	2.60	2.53	2.46	2.39	2.35	2.31	2.27	2.22	2.18	2.13
15	4.54	3.68	3.29	3.06	2.90	2.79	2.71	2.64	2.59	2.54	2.48	2.40	2.33	2.29	2.25	2.20	2.16	2.11	2.07
16	4.49	3.63	3.24	3.01	2.85	2.74	2.66	2.59	2.54	2.49	2.42	2.35	2.28	2.24	2.19	2.15	2.11	2.06	2.01
17	4.45	3.59	3.20	2.96	2.81	2.70	2.61	2.55	2.49	2.45	2.38	2.31	2.23	2.19	2.15	2.10	2.06	2.01	1.96
18	4.41	3.55	3.16	2.93	2.77	2.66	2.58	2.51	2.46	2.41	2.34	2.27	2.19	2.15	2.11	2.06	2.02	1.97	1.92
19	4.38	3.52	3.13	2.90	2.74	2.63	2.54	2.48	2.42	2.38	2.31	2.23	2.16	2.11	2.07	2.03	1.98	1.93	1.88
20	4.35	3.49	3.10	2.87	2.71	2.60	2.51	2.45	2.39	2.35	2.28	2.20	2.12	2.08	2.04	1.99	1.95	1.90	1.84
21	4.32	3.47	3.07	2.84	2.68	2.57	2.49	2.42	2.37	2.32	2.25	2.18	2.10	2.05	2.01	1.96	1.92	1.87	1.81
22	4.30	3.44	3.05	2.82	2.66	2.55	2.46	2.40	2.34	2.30	2.23	2.15	2.07	2.03	1.98	1.94	1.89	1.84	1.78
23	4.28	3.42	3.03	2.80	2.64	2.53	2.44	2.37	2.32	2.27	2.20	2.13	2.05	2.01	1.96	1.91	1.86	1.81	1.76
24	4.26	3.40	3.01	2.78	2.62	2.51	2.42	2.36	2.30	2.25	2.18	2.11	2.03	1.98	1.94	1.89	1.84	1.79	1.73
25	4.24	3.39	2.99	2.76	2.60	2.49	2.40	2.34	2.28	2.24	2.16	2.09	2.01	1.96	1.92	1.87	1.82	1.77	1.71
30	4.17	3.32	2.92	2.69	2.53	2.42	2.33	2.27	2.21	2.16	2.09	2.01	1.93	1.89	1.84	1.79	1.74	1.68	1.62
40	4.08	3.23	2.84	2.61	2.45	2.34	2.25	2.18	2.12	2.08	2.00	1.92	1.84	1.79	1.74	1.69	1.64	1.58	1.51
60	4.00	3.15	2.76	2.53	2.37	2.25	2.17	2.10	2.04	1.99	1.92	1.84	1.75	1.70	1.65	1.59	1.53	1.47	1.39
120	3.92	3.07	2.68	2.45	2.29	2.18	2.09	2.02	1.96	1.91	1.83	1.75	1.66	1.61	1.55	1.50	1.43	1.35	1.25
∞	3.84	3.00	2.60	2.37	2.21	2.10	2.01	1.94	1.88	1.83	1.75	1.67	1.57	1.52	1.46	1.39	1.32	1.22	1.00

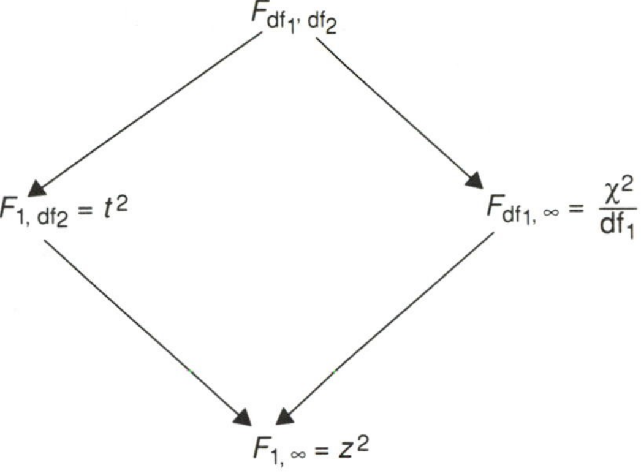

FIGURE 11.9 Relationships Among the Four Fundamental Sampling Distributions (based on normal parental populations)

by df of numerator and df of denominator. If $df_1 = 1$, then the value of F is the same as t^2 for df_2 degrees of freedom; if $df_2 = \infty$, the value of F is χ^2/df_1; and if $df_1 = 1$ and $df_2 = \infty$, the value of F is z^2.

Reading z, t, F, and χ^2 Values from One Table

Consider Table 11.11. Notice that top row gives df of the [1] ____________ and the first column gives df of the [2] ____________. The second column gives $t_{.95}$ for various degrees of freedom in the denominator. The next column (df = 1) gives t^2. We read F as we did before, by using numerator and denominator df. We read χ^2 from the last row using the numerator df. z is given as $t_\infty = 1.96$. The next entry to the right for df = 1 gives z^2. (Note that Table 11.11 also appears on page 695 as Table B-7 in Appendix B.)

ACTIVITY 11.14

Reading z, t, F, and χ^2 from one table

Use Table 11.11 to find the missing numbers in the last column of the following table.

Numerator df	Denominator df	z, t, F, or χ^2
1	∞	$z = 1.96$
1	6	$t =$ [1] ____________
1	11	$t =$ [2] ____________
4	11	$F =$ [3] ____________

(cont.)

Numerator df	Denominator df	z, t, F, or χ^2
8	25	$F =$ [4] ________
∞	3	$\chi^2 =$ [5] ________
∞	8	$\chi^2 =$ [6] ________
5	100	$F =$ [7] ________
1	16	$t =$ [8] ________

TABLE 11.11 *z, t, F,* **and** χ^2 **Distribution Values of Variance Ratio** *F* **Exceeded in 5 Percent of Random Samples**

df of Denominator	t	df of Numerator 1	2	3	4	5	6	7	8
6	2.45	5.99	5.14	4.76	4.53	4.39	4.28	4.21	4.15
7	2.365	5.59	4.74	4.35	4.12	3.97	3.87	3.79	3.73
8	2.31	5.32	4.46	4.07	3.84	3.69	3.58	3.50	3.44
9	2.26	5.12	4.26	3.86	3.63	3.48	3.37	3.29	3.23
10	2.23	4.96	4.10	3.71	3.48	3.33	3.22	3.14	3.07
11	2.20	4.84	3.98	3.59	3.36	3.20	3.09	3.01	2.95
12	2.18	4.75	3.88	3.49	3.26	3.11	3.00	2.91	2.85
13	2.16	4.67	3.80	3.41	3.18	3.02	2.92	2.83	2.77
14	2.145	4.60	3.74	3.34	3.11	2.96	2.85	2.76	2.70
15	2.13	4.54	3.68	3.29	3.06	2.90	2.79	2.71	2.64
16	2.12	4.49	3.63	3.24	3.01	2.85	2.74	2.66	2.59
17	2.11	4.45	3.59	3.20	2.96	2.81	2.70	2.61	2.55
18	2.10	4.41	3.55	3.16	2.93	2.77	2.66	2.58	2.51
19	2.09	4.38	3.52	3.13	2.90	2.74	2.63	2.54	2.48
20	2.086	4.35	3.49	3.10	2.87	2.71	2.60	2.51	2.45
25	2.06	4.24	3.38	2.99	2.76	2.60	2.49	2.40	2.34
30	2.04	4.17	3.32	2.92	2.69	2.53	2.42	2.34	2.27
40	2.02	4.08	3.23	2.84	2.61	2.45	2.34	2.25	2.18
60	2.00	4.00	3.15	2.76	2.52	2.37	2.25	2.17	2.10
100	1.98	3.94	3.09	2.70	2.46	2.30	2.19	2.10	2.03
∞	1.96	3.84	3.00	2.605	2.37	2.21	2.10	2.01	1.94
	χ^2	3.84	5.99	7.815	9.49	11.07	12.59	14.07	15.51

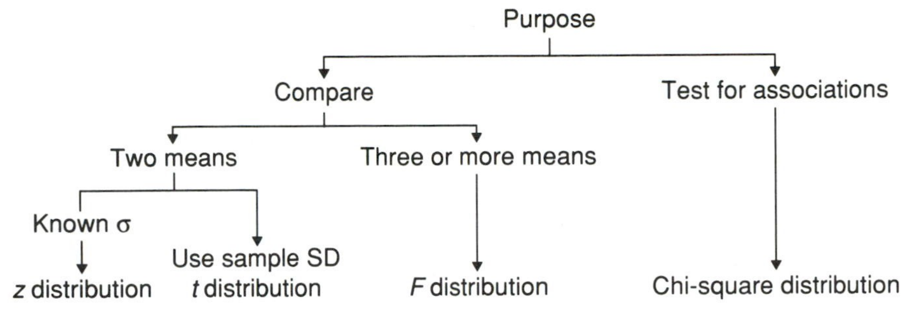

FIGURE 11.10

When to Use Each Distribution

Figure 11.10 shows when to use each of the distributions discussed in this chapter. When your purpose is to compare two means and the standard deviation of the population is known, use the [1] _______ distribution; if the standard deviation is not known, use the [2] _______ distribution. If you want to compare three or more means, use the [3] _______ distribution. If your intention is to test for associations, use the [4] _______ distribution.

Self-Assessment

Tasks:	How well can you do it? Poorly				Very well	Page
1. Given z scores, read the area from the normal distribution table.	1	2	3	4	5	367
2. Given areas, read the z scores from the normal distribution table.	1	2	3	4	5	367
3. Given areas and degrees of freedom, read the t scores from the t table.	1	2	3	4	5	374
4. Given areas and degrees of freedom, read χ^2 from the χ^2 table.	1	2	3	4	5	382
5. Given areas and degrees of freedoms, read F from the F table.	1	2	3	4	5	383
6. Read t, z, F, and χ^2 from one table.	1	2	3	4	5	385

Answers for Chapter 11

Answers for Activity 11.1: 1) Tuesday 2) 19th 3) Wednesday 4) 2001 5) Second

Answers for Figure 11.1: 1) 2.14% 2) 2.14% 3) 13.59% 4) 13.59% 5) 34.13% 6) 15.86% 7) 50% 8) 84.13% 9) 97.72% 10) 68.26% 11) 95.44

Answers for Activity 11.2: 1) 2% 2) 16% 3) 98% 4) 84% 5) 50% 6) (100% − 16%) = 84% 7) 98% 8) 16%

Answers for Activity 11.3: 1) 50% 2) 16% 3) 98% 4) 2% 5) 84% 6) −2 7) +1 8) +2 9) −1 10) 0

Answers for Activity 11.4: 1) Area 2) 1 − Area 3) 0.022750 4) 1.5 5) 31 + (1.5 * 6) = 40

Answers for Figure 11.3: 1) .6% 2) 6.7% 3) 30.85% 4) 69.15% 5) 93.3% 6) 99.4%

Answers for Activity 11.5: 1) −2.2 2) −1.7 3) −.7 4) .2 5) 1.2 6) 2.2

Answers for Activity 11.6: 1) (31 − 1.5 * 6) = 22 2) 100 3) 182 4) 284

Answers for p. 370: 1) $\pm z$ 2) Left Area 3) Central Area 4) Right Area 5) 1 (the total area under the curve)

Answers for Activity 11.7: 1) z-interval = ± 0.5, central area value = .383 2) ±1.5, .866 3) ±1.0, .683 4) ±2.0, .955

Answers for Activity 11.8: 1) 27.43% 2) ±0.6 3) 46.02% 4) ±0.1 5) ±3.2 6) 76.99% 7) ±1.2 8) 2.87% 9) ±1.9 10) 4.46% 11) ±1.7 12) 89.04% 13) ±1.6 14) 0.47% 15) ±2.6 16) 83.85% 17) ±1.4

Answers for p. 372: 1) ±1.645 2) ±1.960 3) ±2.576 4) ±1.282 5) ±0.674 6) ±0.842

Answer for Activity 11.9: Very similar

Answers for Figure 11.6: 1) $t_{.005} = -2.787$ 2) −2.485 3) −1.708 4) −1.316 5) −0.856 6) −0.531 7) 0.531 8) 0.856 9) 1.316 10) 1.708 11) 2.485 12) 2.787

Answers for Activity 11.10: 1) 1.746 2) 2.896 3) −2.528 4) 1.697 5) −1.895 6) −3.232

Answers for Activity 11.11: 1) 2.228 2) 2.947 3) 2.086 4) 99% 5) 3.707 6) 2.571 7) 95% 8) 2.576 9) 1.960

Answers for Figure 11.8: 1) 7.78 2) 9.49 3) 13.28

Answers for Activity 11.12: 1) 158.95 2) 3.84 3) 11.34 4) 7.78 5) 99.0 6) 90 7) 15.09 8) 95 9) 95

Answers for Activity 11.13: 1) 8.79 2) 6.26 3) 5.79 4) 3.20 5) 2.42 6) 3.01 7) 2.01 8) 2.12

Answers for p. 385: 1) Numerator 2) Denominator

Answers for Activity 11.14: 1) 2.45 2) 2.20 3) 3.36 4) 2.34 5) 7.815 6) 15.51 7) 2.30 8) 2.12

Answers for Figure 11.10: 1) z 2) t 3) F 4) χ^2

CHAPTER

12

Formulating Hypotheses

Prelude

"I'm not very good at problems," admitted Milo.

"What a shame," sighed the Dodecahedron. "They're so very useful. Why, did you know that if a beaver two feet long with a tail a foot and a half long can build a dam twelve feet high and six feet wide in two days, all you would need to build Boulder Dam is a beaver sixty-eight feet long with a fifty-one-foot tail?"

"Where would you find a beaver that big?" grumbled the Humbug as his pencil point snapped.

"I'm sure I don't know," he replied, "but if you did, you'd certainly know what to do with him."

"That's absurd," objected Milo, whose head was spinning from all the numbers and questions.

"That may be true," he acknowledged, "but it's completely accurate, and as long as the answer is right, who cares if the question is wrong? If you want sense, you'll have to make it yourself." . . .

"But if all the roads arrive at the same place at the same time, then aren't they all the right way?" asked Milo.

"Certainly not!" he shouted, glaring from his most upset face. "They're all the *wrong* way. Just because you have a choice, it doesn't mean that any of them *has* to be right."

— Norton Juster
The Phantom Toll Booth, pp. 174–176

Chapter Outline

Self-Assessment

Tasks:	How well can you do it?					Page
	Poorly				*Very well*	
1. Given the research context and a description of variables, formulate research hypotheses.	1	2	3	4	5	401
2. Formulate hypotheses regarding means.	1	2	3	4	5	401
3. Formulate hypotheses regarding correlation.	1	2	3	4	5	402
4. Formulate hypotheses regarding prediction.	1	2	3	4	5	402
5. Generate an example illustrating all the steps that culminate in a statistical hypothesis.	1	2	3	4	5	395

Research Questions

Research questions grow out of our own research experience or from a study of the relevant literature. From these questions we formulate hypotheses to test. Four kinds of research questions are common. In Figure 12.1 these questions are grouped into nine categories based on the tools used to answer them. Below you will practice writing questions in each of the categories shown in Figure 12.1.

Suppose that a new illness has been observed in a certain population. As researchers we do not know much about this illness. Our first task might be to conduct a cross-sectional survey and describe the population on selected variables such as age, gender, weight, and stress level. For this we would use research questions that can be answered using simple [1] ______________. We

Description	Comparison of Group Means	Determining Relationships Among Variables		Determining the Effect of Intervention
		Association	**Prediction**	
Simple descriptions (1)	Comparison with values given in the literature (2)	Two categorized variables (5)	The value of one variable from the value of another (8)	Doing experiments (9)
	Comparison of two groups (3)	One categorized variable and one continuous variable (6)		
	Comparison of three or more groups (4)	Continuous data (7)		

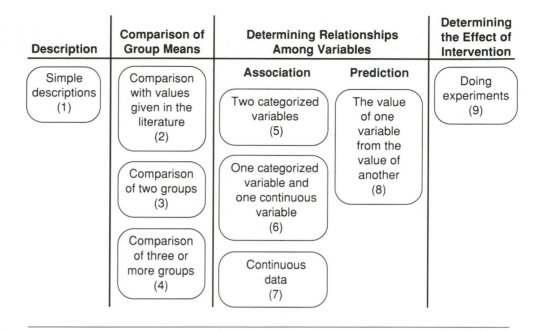

FIGURE 12.1 Types of Research Questions

might be trying to discover how the afflicted group within the population differs from other (unafflicted) groups on each one of these variables. To do this we could treat the afflicted group as a sample and compare it with [2] ______________. Another possibility might be to ask the question, "Are the afflicted persons older than the healthy ones in our population?" This research question would involve comparison of [3] ______________.

Suppose our data on stress level fell into three categories, light, moderate, and heavy, and we asked, "Are the proportions of afflicted individuals the same in all three groups?" This research question would involve comparison of [4] ______________. If the research question were, "Is gender associated with the disease condition?" we would examine the relationship between [5] ______________. If the research question were, "Are the weights of the subjects associated with their disease status?" we would examine the relationship between [6] ______________. If the research question were, "Are the weights and the concentrations of some hormone related in the afflicted people?" we would examine [7] ______________. If the question were, "Can we predict whether a person will develop the illness by looking at selected laboratory tests?" we would be predicting [8] ______________. Finally, the question, "Can we induce the illness in animals by administering some blood components from the afflicted patients?" would be answered by [9] ______________.

STUDENT QUESTIONNAIRE

Please complete the following questionnaire by filling in the blank or circling the most appropriate answer.

1. Sex: (1) Female (2) Male
2. Age: _______
3. Class status:
   (1) Freshman (2) Sophomore (3) Junior (4) Senior (5) Graduate
4. Height in inches: _______
5. Weight in pounds: _______
6. Pulse (take your pulse for 15 seconds, multiply by 4, and record the number): _______
7. Run in place for one minute and then take your pulse again.

   Pulse after running in place for one minute: _______
8. Smoking cigarettes:
   (1) I never smoked. (2) I quit smoking. (3) I smoke. (4) Other
9. Daily physical activity level:
   (1) Minimal (2) Moderate (3) Vigorous
10. Daily stress level:
    (1) Minimal (2) Moderate (3) Heavy
11. I have at least three friends or relatives with whom I can discuss the positive and negative events of my day.
    (1) True (2) False
12. I take at least one five-minute "time out" per day to be quiet.
    (1) True (2) False
13. I would rate my overall health as:
    (1) Excellent (2) Good (3) Average (4) Below average
14. I generally keep the thermostat at my residence set at _______ °F.
15. What is your opinion about national health insurance?

    ____

    ____

Research Questions in the Context of Observational Research

Consider the student questionnaire we first presented in Chapter 2. (We reproduce it on page 392 so you can readily refer to it. We looked at the questionnaire data cards and wrote research questions that correspond to each of the categories given in Figure 12.1. In Activity 12.1, use our questions as examples and write research questions that correspond to other items in the questionnaire. It is helpful to be conscious of the type of variable (categorical or continuous ratio) you are using in the question and to visualize the table or graph your question may potentially generate.

ACTIVITY 12.1

Formulating research questions: observational research

Research Tools	Our Question	Your Question
Simple descriptions (Note that AGE is a quantitative variable. You may depict this in a histogram or a frequency polygon.)	What is the age composition of the statistics class?	____________________
Comparison with values from the literature (Select a ratio variable and a parameter such as mean for your hypothesis.)	The mean age of the students at the university is known to be twenty-three years. Does the sample from these statistics classes have the same mean?	____________________
Comparison of two groups	Is there a difference between the mean age of males and females in the statistics class?	____________________
Comparison of three or more groups	Are the ages of people in the different activity groups different?	____________________
Determining association between two categorized variables	Are activity level and perceived health level associated, or are activity and health level independent of each other?	____________________

Determining association between one categorized and one continuous variable	Are activity level and weight related?	______________
Determining association between two continuous variables	Is there a relationship between height and weight?	______________
Predicting one variable from another	Can we predict a person's weight if we know his or her height?	______________

Research Questions in the Context of Experimental Research

You will learn about diverse research questions and how researchers formulate them by reading journal articles. Here we present one question that refers to the faculty exercise data (FED).

Activity 12.2

Formulating research questions: experimental research

Refer to FED data and write a research question (as you did in Activity 12.1).

Research Tool	Our Question	Your Question
Determining the effect of an intervention on a response variable	What happens to the body fat content of the subjects when they exercise with high intensity for four months?	______________

Hypotheses

The word hypothesis in layperson's language means "guess." Researchers, however, use it in a special sense: A **hypothesis** is a testable proposition about a population characteristic, the relationship between variables, or differences between characteristics of two or more populations. The proposition is a hunch or an educated guess, and it usually arises from a study of the existing literature or the researcher's experience.

Steps in Formulating and Testing Hypotheses

A research question arises in the mind of a researcher when she or he observes a phenomenon that is puzzling. The question may arise when at work, when reading a journal, or when just relaxing. We will illustrate the steps in formulating a hypothesis by working through a hypothetical example.

Formulating a Question or Problem Statement

You notice that some people claim that they gain weight even when they eat moderately, whereas others claim that they can eat as much as they want without gaining weight. This leads you to formulate the question, "Is there a connection between the amount of food that is eaten and weight gain?"

One way to answer this question would be to seek the opinion of an authority on the subject; reading literature is another way; conducting an experiment is a third way. Whichever approach you choose, you must begin with an educated guess, which is your hypothesis: When people eat more, they gain weight.

Conducting an Experiment

Subjects You decide to conduct a preliminary experiment using rats as subjects. You choose to divide the rats into two groups, the experimental group and the control group. There will be ten rats in each group.

Stimulus Variable The stimulus variable is the amount of food. You decide to make no changes in the diet of the control group, which has been shown over time to keep the rats at a stable weight. You then must decide on an amount of food to feed to the experimental group.

Response Variable The response variable will be the measured weight. You must decide how weight will be measured: what units will be used (grams or ounces), what instruments will be used to weigh the subjects, and the specific procedure to be followed to measure weight. This procedure will, you hope, ensure standardization of all measurements.

Other Factors Other factors that must be considered are such things as the amount of daily exercise the rats get, hereditary tendency to gain weight, type of food eaten, and basal metabolic rate. As much as a researcher may try, however, it is impossible to consider everything at once. So, we control for these factors by randomly assigning subjects from a homogeneous population to control and experimental groups. We may choose a large group of rats whose weights are approximately the same (within one standard deviation of the mean weight, for instance) and then choose our experimental and control groups randomly from these rats. The researcher's interest is in diet and weight only. Nothing else will be measured.

Formulating Statistical Hypotheses

A statistical hypothesis explicitly states which parameter will be included in the hypothesis. Many hypotheses are possible, depending on the original purpose of the experiment.

Null Hypothesis Let us stipulate that our interest is in testing whether there is a change in the mean weight of the rats in the experimental group. We assume that if the increased amount of food caused no significant weight gain in the experimental group, these rats would have the same mean weight as the control group at the end of the experimental period. One way of phrasing this hypothesis is to state that there will be no difference in the weight gained between the two groups. A hypothesis stated this way (i.e., that some variable is zero) is called a **null hypothesis**. The null hypothesis may be stated in the following equivalent ways: "The mean weight gain of the two groups of rats, fed different amounts of food, will be *the same*," or, "There will be *no difference* between the mean weight gain of the two groups of rats when fed different amounts of food." Note that the crucial elements of the original educated guess are still present in this hypothesis. Defining the two groups, the variable to be measured (weight gain), and the statistic to be computed (mean gain in weight), clarifies the question and makes the statistical analysis of the experimental results feasible.

In symbols the null hypothesis is stated as follows:

$$H_0\colon \mu_1 = \mu_2 \quad \text{or} \quad \mu_1 - \mu_2 = 0$$

where H_0 is an abbreviation for null hypothesis and μ_1 and μ_2 are the means of weight gained by the rats in the experimental and control groups, respectively. Note that we used the Greek letter μ to indicate that the hypothesis is about the *population* means. The null hypothesis is an **explicit statement** involving the **parameter** (population attribute) of interest, which declares that there is **no difference** between two or more parameters, or no difference between a parameter and a published value, or no difference between distributions.

Alternative Hypothesis The **alternative hypothesis** is the opposite of the null hypothesis. In other words, the alternative hypothesis says that there *will* be a difference in mean weight gain. The alternative hypothesis could be stated as "The mean weight gain of the two groups of rats, fed different amounts of food, will not be the same," or, "There will be a difference between the mean weight gain of the two groups of rats when fed different amounts of food."

In symbols, the alternative hypothesis is expressed as

$$H_a\colon \mu_1 \neq \mu_2 \quad \text{or} \quad \mu_1 - \mu_2 \neq 0$$

where H_a is an abbreviation for "alternative hypothesis" and μ_1 and μ_2 are the means of the weight gained by the rats in the experimental and control groups, respectively. ($\neq$ means "not equal to.")

Actually, alternative hypotheses come in different kinds. The one stated above is an example of a two-sided hypothesis. A one-sided hypothesis would be in one of the following forms:

$$H_a\colon \mu_1 < \mu_2 \quad \text{or} \quad H_a\colon \mu_1 > \mu_2$$

(The symbol $>$ means "greater than" and the symbol $<$ means "less than.") A one-sided hypothesis is used when the other alternative is assumed to be impossible or of no interest. For example, a one-sided hypothesis might be appropriate for this experiment, since we expect that, if anything, the experimental group will either gain the same amount of weight as the control group (in which case the null hypothesis is satisfied) or gain more weight than the control group (in which case the alternative hypothesis is satisfied). We do not consider it possible that the experimental group will gain less weight, and if at the end of the experiment they were found to have gained (somewhat) less we would take that to be evidence of the null hypothesis. The difference in weight gain in the two groups would, in this instance, be attributed to factors other than the experimental variable, the difference in diet.

The process of transforming a research question into a hypothesis helps clarify analysis of the data. Table 12.1 summarizes some of the important factors in the experiment we have been considering. The stimulus or treatment variable is the amount of food. Rats in the experimental group will receive more food than rats in the control group. The response variable must be a measure of weight gain or loss. In this case, the logical choice for this variable will be the mean weight of the rats at the end of the experiment minus the mean weight of the same rats at the beginning. These means will be computed for both groups of rats and statistical methods will be used to decide whether the mean weight gained by the experimental group is sufficient to reject the null hypothesis that the mean weight gained in the two groups was the same.

TABLE 12.1 Factors in the Experiment as They Relate to the Research Question

Element of the Question	Element of the Hypothesis or Experiment	Comment
Interest in people	Experiment on *rats*	Will we be able to generalize the results of our experiment to humans? We will probably have to do experiments on humans, also.
Amount of food	*Different quantities* for the two groups	When we start the actual experiment, we have to specify the exact quantities and the composition of the food, feeding schedules, etc.
Weight gain	*Mean difference* in weight for each group	Notice that we chose the mean as the appropriate measure of change.
Is there a difference?	Weight gain is the *same* for both groups or there is *no difference*. (You may reject this hypothesis based on the experimental results and conclude that there is a difference.)	We formulate a null hypothesis that there will be no difference between the means of the weight gained by the rats in the two groups.

ACTIVITY 12.3

Steps in formulating and testing hypotheses

Literature Example 12.1 is from an article entitled "Effects of Mental Practice on Rate of Skill Acquisition." Refer to the abstract and answer the questions below as a way of reviewing the steps in formulating hypotheses.

Step	Example and Questions	
A research question arises in the mind of a researcher when she or he observes a phenomenon that is puzzling.	A researcher reads in the literature that mental practice of a novel skill helps the acquisition of that skill. Write the question or problem statement this researcher wished to investigate.	1 ________
One way to answer this question would be to seek the opinion of an authority on the subject; reading the literature is another way; conducting an experiment is a third way.	The researcher found five references in the literature that substantiated her hunch. In the article she states, "No research was found to document whether mental practice can accelerate EMG (electromyography) changes found following physical practice." Therefore she decided to conduct an experiment. What was the contribution of the literature review to this research effort?	2 ________
Initially a researcher formulates an educated guess.	Write a layperson's version of the guess the researcher developed.	3 ________
Conducting an experiment: choosing subjects	In the case of effect of diet on weight we used rats. We cannot get rats to do mental activity, however. What kind of subjects were chosen?	4 ________

Choosing the stimulus variable	The humans the researcher selected needed to have a new task.	What task was given the subjects? 5 ____________________
Choosing the research design	What is the research design of this study?	6 ____________________
	What are the two treatments?	____________________ ____________________
Response variable	What is the response variable?	7 ____________________
Other factors	Some of the subjects may be in exceptionally good physical condition or have a high degree of manual dexterity. How does the experimental design control for such factors?	8 ____________________
Statistical hypotheses: the null hypothesis	We need to translate the "guess" into a statistical hypothesis. Rate of skill acquisition is of interest. What null hypothesis might this researcher have developed? Express it as an equation. Regression analysis is usually done to determine b, the slope of the regression equation, which represents the rate of acquisition of the skill; a t test is then used to test whether the rates in the two groups are the same. The symbol for population slope is β.	9 ____________________
Alternative hypothesis	Express the researcher's alternative hypothesis.	10 ____________________

LITERATURE EXAMPLE 12.1

Effects of Mental Practice on Rate of Skill Acquisition

Joyce R Maring

The purpose of this study was to investigate the effectiveness of mental practice in increasing the rate of skill acquisition during a novel motor task. Twenty-six subjects were randomly assigned to two groups. The Control Group (n = 13) performed only physical practice; the Experimental Group (n = 13) performed both mental and physical practice. The task was to toss, by flexing the elbow, a Ping-Pong ball held in a cup on a forearm splint to a target. The biceps brachii muscle and the long and lateral heads of the triceps brachii muscle were monitored electromyographically to determine any changes occurring during skill acquisition. The Experimental Group's accuracy improved at a significantly greater rate than that of the Control Group. In addition, the Experimental Group demonstrated changes in timing variables that led to a more efficient movement. These changes included a decrease in time from the onset of muscle activity to peak activity and an increase in the time elapsed from the onset of agonist contraction to the onset of antagonist contraction. These results suggest that mental practice may be an important tool in facilitating the acquisition of a new motor skill. [Maring JR: Effects of mental practice on rate of skill acquisition. Phys Ther 70:165–172, 1990]

Hypotheses in the Context of Observational Research

ACTIVITY 12.4

Formulating hypotheses: observational research

In this activity we have translated each research question we developed in Activity 12.1 into a hypothesis. Do the same with each research question you developed in that activity (page 393).

Research Question	Our Hypothesis	Your Hypothesis (corresponding to the research question you wrote in Activity 12.1)
What is the age composition of the statistics class?	No hypothesis is needed here; this question calls for a description.	______
The mean age of the students at the university is known to be twenty-three years. Does the sample from these statistics classes have the same mean?	The mean age of the students in the statistics classes is twenty-three years. (The null hypothesis here is actually asserting that the population from which the sample was drawn, statistics students, is identical age-wise to the population of all students at this university. The parameter of interest is the mean age. Compare the content of this hypothesis with the three components of a null hypothesis discussed earlier.)	______
Is there a difference between the mean age of males and females in the statistics class?	There is no difference in the mean age of males and females in the statistics class. Or, the mean age of males and females in the statistics class is the same. (Compare the	______

	content of this hypothesis with the three components of the null hypothesis.	
Are the ages of people in the different activity groups different?	The mean age of subjects in all activity groups is the same.	
Are activity level and perceived health level associated, or are activity and health level independent of each other?	Activity level and perceived health level are independent of each other. Or, there is no association between activity level and perceived health level. (Variables are ACTIVITY and HEALTH; χ^2 measures independence or association; the hypothesis asserts that the variables are independent.)	
Are activity level and weight related?	Activity level and weight are not related.	
Is there a relationship between height and weight?	There is no relationship between height and weight. Or, height and weight are not associated. (Variables are HEIGHT and WEIGHT; the correlation coefficient measures independence or association; the hypothesis asserts that the variables are independent.)	
Can we predict a person's weight if we know his or her height?	Weight cannot be predicted from height. (Variables are HEIGHT and WEIGHT; the regression coefficient measures prediction capability; the hypothesis asserts that the weight cannot be predicted from height.)	

Examples from the Literature

Some authors include the research questions and hypotheses explicitly when they publish articles about their research. Literature Example 12.2 is from one such paper, "Breastfeeding and Employment." Read the abstract and the research questions and complete Activity 12.5.

ACTIVITY 12.5

Research questions and related hypotheses in an observational study

Refer to the first question the researchers addressed.

What are the two comparison groups? 1 ______________________

____

____

What is the explanatory variable? 2 ______________________

What are the response variables? 3 ______________________

____

State the hypothesis. 4 ______________________

____

____

____

Refer to the second question the researchers addressed.

What are the two comparison groups? 5 ______________________

____

____

What is the explanatory variable? 6 ______________________

What is the response variable? 7 ______________________

State the hypothesis. 8 ______________________

____

Refer to the third question the researchers addressed.

What is the explanatory variable? 9 ______________________

What are the four response variables? 10 ______________________

____

State the hypothesis. 11

LITERATURE EXAMPLE 12.2

MARGARET H. KEARNEY, RNC, MS

LINDA CRONENWETT, RN, PhD, FAAN

Breastfeeding and Employment

Breastfeeding problems, outcomes, and satisfaction of married, well-educated first-time mothers who returned to work within six months postpartum were compared to those of mothers with the same characteristics who stayed at home. Mothers who planned to work after giving birth anticipated and experienced shorter durations of breastfeeding than did those who planned to remain at home. Breastfeeding experiences and satisfaction among working mothers differed little from the experiences and satisfaction of their nonworking counterparts; however, employment prior to two months postpartum exerted some negative effects on breastfeeding outcomes.

Accepted: March 1991

JOGGN
Volume 20 Number 6
November/December 1991

In the United States, more than half of the married mothers of infants younger than one year are employed outside the home.[1] In response to a surge in the number of new mothers who return to work, guidelines have proliferated regarding breastfeeding while employed.[2–4] Perhaps because new mothers returning to employment is a recent demographic change,[2] recommendations for the management of breastfeeding and employment have been derived primarily from clinical experience, retrospective surveys, and anecdotal reports. No studies to date have examined, within a homogeneous sample, the effects on breastfeeding of returning to work versus staying at home.

Data collected as part of a larger study were analyzed to address the following questions:

1. Do pregnant women who plan to return to work after delivery and those who plan not to work differ in their prenatal attitudes toward breastfeeding or planned duration of breastfeeding?
2. Do mothers who return to employment between two weeks and six months postpartum or who put in more hours per week after returning to work experience more breastfeeding problems during that time than do mothers who remain at home or work fewer hours?
3. Is there a relationship between employment status at six months postpartum and duration of breastfeeding, support for breastfeeding, satisfaction with breastfeeding, or perception of infant temperament?

Hypotheses in the Context of Experimental Research

ACTIVITY 12.6

Formulating hypotheses: experimental research

In this activity we have translated the research question we developed in Activity 12.2 into a hypothesis. Do the same with the research question you developed in that activity (page 394).

Our question	Our hypothesis	Your hypothesis
What happens to the body fat content of the faculty when they exercise with high intensity for four months?	There is no difference in the body fat of the faculty before the exercise program and after the exercise program. H_0: $\mu_1 = \mu_2$ or $\mu_1 - \mu_2 = 0$ where μ_1 and μ_2 are the means of body fat of the population before the experimental exercise and after the experimental exercise.	____________ ____________ ____________ ____________

Examples from the Literature

Literature Example 12.3, "Infant Responses to Saline Instillations and Endotracheal Suctioning," lists three hypotheses. Read the abstract and answer the questions in Activity 12.7.

ACTIVITY 12.7

Hypotheses in experimental research

What is the overall objective of the study?

1 ____________________
____
____

Refer to the first hypothesis formulated by the researchers.

What are the two treatments? 2 ______________________

____

____

What are the response variables? 3 ______________________

State the research question. 4 ______________________

____

____

Refer to the second hypothesis formulated by the researchers.

What is the independent variable? 5 ______________________

What are the response variables? 6 ______________________

State the research question. 7 ______________________

____

Refer to the third hypothesis formulated by the researchers.

What is the independent variable? 8 ______________________

What are the response variables? 9 ______________________

State the research question. 10 ______________________

____

____

LITERATURE EXAMPLE 12.3

DAVID R. SHORTEN, RN, BScN, MN

PAUL J. BYRNE, MB, CHA

ROBERT L. JONES, MD, PhD

Infant Responses to Saline Instillations and Endotracheal Suctioning

The study examined the effects of endotracheal suction with and without saline instillation on neonates with respiratory distress. In a completely counterbalanced factorial-within-subjects design, 27 intubated neonates were randomly assigned to two orders of presentation of treatment conditions. Heart rate and blood pressure were continuously recorded throughout both treatment conditions. The ratio of arterial oxygen tension to alveolar oxygen tension was used to assess oxygenation. Results indicated that clinically stable newborns tolerated instillations of 0.25–0.5 ml. The suctioning protocol used in this study minimized changes in infants' heart rates and blood pressures.

Accepted: April 1991

JOGNN
Volume 20 Number 6
November/December 1991

Hypotheses

Three hypotheses were formulated:

1. Receiving a saline instillation with tracheal suctioning will cause significant changes in heart rate (HR), blood pressure (BP), and the ratio of arterial oxygen tension to alveolar oxygen tension (a/APO_2), compared to receiving tracheal suctioning alone (main effect of condition).
2. There will be significant changes in HR, BP, and a/APO_2 over the 13 time periods (main effect of time).
3. There will be significant changes in HR and BP during or after the treatment period (interaction effect of condition by time).

Activity 12.8

Writing hypotheses implied by tables

Literature Example 12.4 is from an article entitled "Self-Reported Eating Disorders of Black, Low-Income Adolescents: Behavior, Body Weight Perceptions, and Methods of Dieting." Read the abstract and the following quotes from the body of the article. Answer the questions about each quote.

Quote	Question	Hypothesis
"Demographics for students who thought they had an eating disorder are similar to the overall population."	What are the two groups?	1 ______
	What are the groups being compared on?	2 ______
	What null hypothesis is suggested by the quote?	3 ______
"Students were asked to indicate attitudes, feelings related to dieting, food, and feelings about self, which are characteristics closely related to anorexia nervosa and bulimia nervosa. . . . Food or eating-related characteristics not related to existence of an eating disorder were few, as evidenced by results of the chi square tests of significance."	What are the response variables?	4 ______
	What is the explanatory variable?	5 ______
	What null hypothesis is suggested by the quote?	6 ______
"In every case where a dependency relative to eating disorder was evident, those reporting a disorder were more likely to possess an eating-related behavior than those who reported no eating disorder."	What is the response variable?	7 ______
	What is the explanatory variable?	8 ______
	What null hypothesis is suggested by the quote?	9 ______

"In comparing the same nonfood-related behavioral characteristics with the presence of an eating disorder for males and females, 'feel ineffective as a person' was significantly related for both genders. . . ."	What is the nonfood-related variable?	10 ______
	What are the explanatory variables?	11 ______
	Write two null hypotheses, one for males and the other for females.	12 ______
"'Hate being less than best,' 'parents have high goals for me,' and 'desire to do everything perfectly' were only significantly related to females."	What are the nonfood-related variables?	13 ______
	What are the explanatory variables?	14 ______
	Write two null hypotheses, one for males and the other for females.	15 ______
"No significant difference existed between these attitudes and behaviors between age groups or between males and females."	What are the nonfood-related variables?	16 ______
	What are the explanatory variables?	17 ______
	What null hypotheses are suggested by the quote?	18 ______

LITERATURE EXAMPLE 12.4

Self-Reported Eating Disorders of Black, Low-Income Adolescents: Behavior, Body Weight Perceptions, and Methods of Dieting

Margaret Balentine, Kathleen Stitt, Judith Bonner, Louise Clark

ABSTRACT: *This study identified black, low-income adolescents who thought they had anorexia nervosa or bulimia nervosa, identified behaviors common to those who thought they had the disorders, and compared their actual and perceived body weight as well as methods of dieting. Of the 1,930 students in grades 7-12 who participated, about 12% thought they might have an eating disorder. These students also reported having food-related behaviors similar to individuals with eating disorders and which differed significantly from those reported by their peers. Gender was more likely to affect food behavior than age. Though most students were within their expected weight range, those with self-reported eating disorders perceived themselves to be heavier more often than their peers and when their actual weights were compared, they were more likely to weigh more. Of those who had dieted to lose weight, fasting was the most frequently reported restrictive method used.* (J Sch Health. 1991;61(9):392-396)

Table 1
Demographic Description of Low-Income, Black Adolescents

	n = 1,930 Total Population		n = 226 Self-Reported Eating Disorders	
Item	**n**	**%**	**n**	**%**
School setting				
Middle	1,046	54.2	127	56.2
Senior	884	45.8	99	43.8
Age				
11	17	0.9	3	1.3
12	242	12.5	32	14.2
13	443	23.0	47	20.8
14	412	21.3	47	20.8
15	324	16.8	35	15.5
16	219	11.3	22	9.7
17	172	8.9	25	11.1
18	81	4.2	10	4.4
19	20	1.0	5	2.2
Gender				
Male	940	48.7	97	42.9
Female	990	51.3	129	57.1
Child nutrition program participation				
Lunch	1,656	86.2	188	83.6
Breakfast	581	30.2	65	29.0

Table 2
Food-Related Attitudes and Behaviors of Low-Income, Black Adolescents Who Reported Having Eating Disorders

Attitudes/Behaviors	Percentage "Always or Sometimes"					
	Age Group			Gender		
	(n=125) ≤14 Yr	(n=95) >14 Yr	P	(n=96) Male	(n=124) Female	P
Eat sweets without feeling nervous	72.8[1]	76.8[1]	.500	71.9[1]	76.6[1]	.428
Eat when upset	41.6[1]	45.9	.523	38.6	47.2[1]	.204
Stuff with food	54.5	66.3	.078	61.3	58.4	.669
Think about dieting	48.8[1]	52.6[1]	.576	30.5[1]	65.6[1]	.000[2]
Feeling guilty after overeating	55.2[1]	49.0[1]	.360	47.3[1]	56.3[1]	.187
Terrified of gaining	64.3[1]	55.9[1]	.211	51.7[1]	67.2[1]	.021[2]
Exaggerate importance of weight	60.5[1]	63.5[1]	.653	56.5[1]	65.6[1]	.171
Gone on eating binges	54.8[1]	58.9[1]	.543	66.7[1]	48.8[1]	.008[2]
Desire to be thinner	65.6[1]	75.7[1]	.105	62.5[1]	75.8[1]	.047[2]
Think about bingeing	46.8[1]	53.2[1]	.349	56.8[1]	43.9	.059
Eat moderately with others, but stuff when alone	56.8	53.9	.670	58.0	53.7	.527
If gain, worry that will continue	62.0[1]	60.7[1]	.845	46.3[1]	72.8[1]	.000[2]
Think about vomiting to lose weight	20.6[1]	38.3[1]	.004[2]	23.9[1]	31.4[1]	.223
Eat or drink away from friends	33.6[1]	29.8[1]	.551	37.9[1]	27.5[1]	.103

[1]*Identifies significantly different attitudes and behaviors ($p \leq .05$) between those reporting either one or both eating disorders and those reporting neither.*
[2]*Identifies significant differences ($p \leq .05$) between age groups or gender.*

Table 3
Nonfood-Related Attitudes and Behaviors of Low-Income, Black Adolescents Who Reported Having Eating Disorders

Attitudes/Behaviors	Percentage "Always or Sometimes"					
	Age Group			Gender		
	(n=125) ≤14 Yr	(n=95) > 14 Yr	P	(n=96) Male	(n=124) Female	P
Feel ineffective as a person	53.2	54.9[1]	.803	54.3[1]	53.6[1]	.919
Family expects outstanding performance	85.5	79.8	.267	85.0	81.6	.510
Try not to disappoint parents	85.6	85.4	.963	82.3	88.0	.226
Hate being less than best	88.1	80.0	.095	83.3	85.6[1]	.645
Parents have high goals for me	90.5	84.6	.187	90.4	86.2[1]	.341
Desire to do everything perfectly	80.0[1]	78.2	.762	73.7	83.6[1]	.074
Have extremely high goals	92.0	91.6	.915	87.5	95.1	.051

[1]*Identifies significantly different attitudes and behaviors ($p \leq .05$) between those reporting either one or both eating disorders and those reporting neither.*

Self-Assessment

Tasks:	How well can you do it? *Poorly*				*Very well*	Page
1. Given the research context and a description of variables, formulate research hypotheses.	1	2	3	4	5	401
2. Formulate hypotheses regarding means.	1	2	3	4	5	401
3. Formulate hypotheses regarding correlation.	1	2	3	4	5	402
4. Formulate hypotheses regarding prediction.	1	2	3	4	5	402
5. Generate an example illustrating all the steps that culminate in a statistical hypothesis.	1	2	3	4	5	395

Answers for Chapter 12

Answers for Figure 12.1: Question numbers correspond exactly to the category numbers in Figure 12.1.

Answers for Activity 12.3: 1) Does mental practice help in the acquisition of a skill? 2) Some work on this question has already been done. But measuring EMG to demonstrate the changes in muscle activity secondary to the acquisition of the skill is a different way of documenting the skill acquisition. 3) Mental practice helps a person acquire a skill. 4) Human subjects, perhaps volunteers 5) The actual task was throwing a ping pong ball.

6) O_1 X O_2
O_3 O_4

The control group performed only physical practice; the experimental group performed both mental and physical practice. 7) Rate of improvement in accuracy; changes in timing variables 8) Random assignment of subjects to the groups controls for other factors. 9) $\beta_1 = \beta_2$, where β_1 is the rate (slope of a line) for the experimental group and β_2 is the rate for the control group. 10) $\beta_1 \neq \beta_2$. $\beta_1 < \beta_2$ is also a possibility.

Answers for Activity 12.5: Group I: mothers who plan to return to work after delivery; Group II: mothers who do not plan to work 2) Work after delivery 3) Prenatal attitudes toward breastfeeding and planned duration of breastfeeding 4) Pregnant women who plan to return to work after delivery and those who plan not to work do not differ in their prenatal attitudes toward breastfeeding or planned duration of breastfeeding. 5) Group I: mothers who return to work; Group II: mothers who remain at home or work fewer hours 6) Work status 7) Number of breastfeeding problems 8) Mothers in Group I and Group II have similar numbers of breastfeeding problems. 9) Employment status 10) Duration of breastfeeding; satisfaction with breastfeeding; support for breastfeeding; perception of infant temperament 11) There is no relationship between the employment status and duration of breastfeeding, satisfaction with breastfeeding, support for breastfeeding, or perception of infant temperament.

Answers for Activity 12.7: 1) To examine the effects of endotracheal suctioning with and without saline instillation on neonates with respiratory distress 2) Treatment I: saline instillation with tracheal suctioning; Treatment II: tracheal suctioning alone 3) HR, BP, a/APO_2 4) Do treatments one and two differ in their effect on HR, BP, and a/APO_2? 5) Time 6) HR, BP, a/APO_2 7) Will there be significant changes in HR, BP, a/APO_2 over the time periods? 8) Time (during and after treatment period) 9) HR and BP 10) Will there be significant changes in HR and BP during or after the treatment period?

Answers for Activity 12.8: 1) Students who thought they had an eating disorder and the overall population 2) Demographics such as gender, age, etc. 3) Students who thought they had an eating disorder and students in the overall population are similar on demographics. 4) Attitudes, feelings related to dieting, food, and feelings about self 5) Perceived existence of the eating disorder 6) Those reporting eating disorders and those reporting no disorder have the same characteristics. 7) Eating-related behavior 8) Reporting a disorder, reporting no eating disorder 9) Those reporting a disorder and those who reported no eating disorder manifest the same eating-related behavior. 10) Feeling ineffective as a person 11) Presence and absence of a perceived eating disorder; males and females 12) The presence or absence of a perceived eating disorder among males is not related to feeling ineffective as a person. The presence or absence of a perceived eating disorder among females is not related to feeling ineffective as a person. 13) Hating being less than best; parents having high goals; desiring to do everything perfectly 14) Presence or absence of a perceived eating disorder; males and females 15) The presence or absence of a perceived eating disorder among males is not related to hating being less than best, parents having high goals, and desiring to do everything perfectly. The presence or absence of a perceived eating disorder among females is not related to hating being less than best, parents having high goals, and desiring to do everything perfectly. 16) Attitudes and behaviors 17) Age groups; males and females 18) Age groups have similar non-food related attitudes and behaviors; males and females have similar non-food related attitudes and behaviors.

CHAPTER

13

Distribution of Sample Means

Prelude

"I never knew words could be so confusing," Milo said to Tock as he bent down to scratch the dog's ear.

"Only when you use a lot to say a little," answered Tock.

— Norton Juster
The Phantom Toll Booth, p. 44

Chapter Outline

Self-Assessment

Tasks:	**How well can you do it?** *Poorly*				*Very well*	**Page**
1. State the components of the Central Limit Theorem.	1	2	3	4	5	416
2. Classify events as likely or unlikely, given cutoff points.	1	2	3	4	5	425
3. Distinguish the two types of error in testing a hypothesis.	1	2	3	4	5	427
4. State and explain the four essential assumptions made in statistical analyses of data, relating to:						
(a) The composition of any given measurement	1	2	3	4	5	436
(b) The distribution of population measurements	1	2	3	4	5	437
(c) The equivalence of measures of variation	1	2	3	4	5	438
(d) The shapes of fitted lines	1	2	3	4	5	439
5. Comment on transformations to induce linearity and normality.	1	2	3	4	5	441

Introduction

In Chapter 4 we saw how to sample from a population, but what can we do with a sample once we've taken it? One thing we can do is infer from a sample certain characteristics of the population from which it was drawn. In this chapter we will discuss how we use sample statistics to estimate population parameters. We will also discuss the use of samples in testing hypotheses. Each research question we considered in Chapter 12 and its corresponding hypotheses may be tested with the help of samples. Regardless of the experimental design, we are certain to use samples when we test a hypothesis. The basis for making statistical inferences will be the theoretical distributions discussed in Chapter 11.

The Central Limit Theorem

Many populations have normal or approximately normal distributions. Even when a population itself is not normally distributed though, the means of samples from that population are approximately normally distributed. The mathematical theorem that asserts this fact is called the **Central Limit Theorem.** This theorem is extremely useful, because it tells us how to estimate population parameters from sample statistics.

Components of the Central Limit Theorem

The Central Limit Theorem states the following:

1. The distribution of means, $\bar{x}$, of all possible samples of size n from a given population approximates a normal distribution. The approximation becomes more precise as n increases.
2. The mean of the means of all possible samples of size n from a population, denoted $\mu_{\bar{x}}$, is equal to the population mean, μ.
3. The mean of the variances of all possible samples equals the variance of the population, σ^2.
4. The variance of the means of all possible samples of size n is equal to

$$\sigma_{\bar{x}}^2 = \frac{\sigma^2}{n}.$$

5. The standard deviation of the means of all possible samples of size n is equal to

$$\sigma_{\bar{x}} = \frac{\sigma}{\sqrt{n}}.$$

An Artificial Example to Demonstrate the Central Limit Theorem

We will work with a very small population from which we can take all possible samples. Assume that the population consists of only the four numbers 10, 8, 6, and 4. We will take samples of size two ($n = 2$). To do this, we will draw one number, copy it, replace the number, and draw another one. We will do this in all possible ways, so that we will have some samples of two with the same number repeated (e.g., 8, 8). The order of the numbers drawn is important, too, so, for example, we count 8, 6 and 6, 8 as two different samples. Activity 13.1 includes all possible samples.

ACTIVITY 13.1

All possible samples from a known population

Compute $\bar{x}$, s^2, and s for each of the samples shown using the following formulas:

$$\bar{x} = \frac{\Sigma x_i}{n} \qquad s^2 = \frac{\Sigma (x_i - \bar{x})^2}{n - 1} \qquad s = \sqrt{s^2}$$

Then compute

$$\mu = \frac{\Sigma x_i}{N}, \qquad \sigma^2 = \frac{\Sigma (x_i - \mu)^2}{N}, \qquad \text{and } \sigma = \sqrt{\sigma^2}$$

for the population. (See Chapter 7 if you need help using the formulas.)

↓ **Population→**	**10**	**8**	**6**	**4**
10	10, 10 $\bar{x}$ = ______ s^2 = ______ s = ______	10, 8 $\bar{x}$ = ______ s^2 = ______ s = ______	10, 6 $\bar{x}$ = ______ s^2 = ______ s = ______	10, 4 $\bar{x}$ = ______ s^2 = ______ s = ______
8	8, 10 $\bar{x}$ = ______ s^2 = ______ s = ______	8, 8 $\bar{x}$ = ______ s^2 = ______ s = ______	8, 6 $\bar{x}$ = ______ s^2 = ______ s = ______	8, 4 $\bar{x}$ = ______ s^2 = ______ s = ______
6	6, 10 $\bar{x}$ = ______ s^2 = ______ s = ______	6, 8 $\bar{x}$ = ______ s^2 = ______ s = ______	6, 6 $\bar{x}$ = ______ s^2 = ______ s = ______	6, 4 $\bar{x}$ = ______ s^2 = ______ s = ______
4	4, 10 $\bar{x}$ = ______ s^2 = ______ s = ______	4, 8 $\bar{x}$ = ______ s^2 = ______ s = ______	4, 6 $\bar{x}$ = ______ s^2 = ______ s = ______	4, 4 $\bar{x}$ = ______ s^2 = ______ s = ______

Population: 10, 8, 6, 4. μ = ______, σ^2 = ______, σ = ______
Sample size: $n = 2$. Number of samples is 16.

ACTIVITY 13.2

Verification of the components of the Central Limit Theorem (using the artificial population from Activity 13.1)

Components of the Central Limit Theorem		Questions
The distribution of the means of all possible samples from a given population approximates a normal distribution.	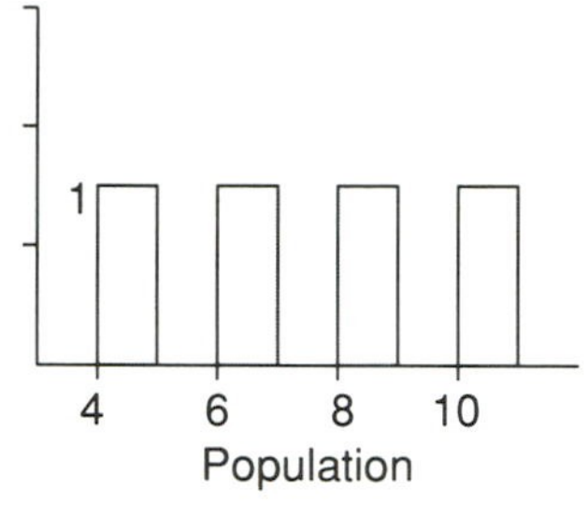	The population shown in the graph is [1]normal/not normal.
	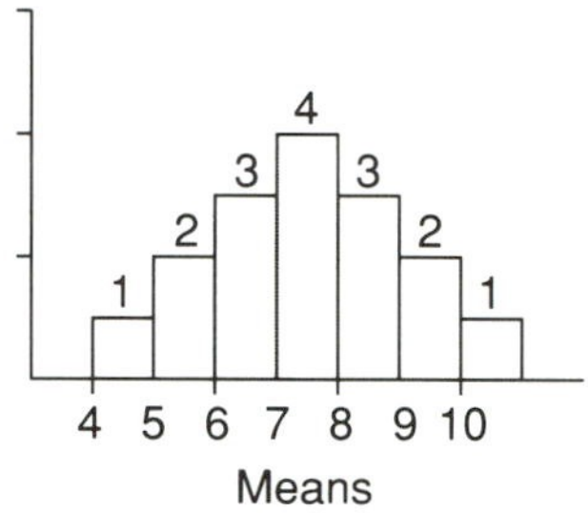	The means of all possible samples [2]approximate/do not approximate a normal distribution.
The mean of the means of all possible samples from a population is equal to the population mean (μ).	$\mu = 7$ $\sigma^2 = 5$ $\sigma = 2.24$ Mean of all possible $\bar{x} = 7$	The mean of all possible $\bar{x}$ [3]does/does not equal μ.
The mean of the variances of all possible samples equals the variance of the population, σ^2.	Variances of all sixteen possible samples: 0, 0, 0, 0, 2, 2, 2, 2, 2, 2, 8, 8, 8, 8, 18, 18; mean $s^2 =$ [4]_______	The mean of all possible s^2 [5]does/does not equal σ^2.
The variance of the means of all possible samples is equal to ($\sigma_{\bar{x}}^2 = \sigma^2/n$), where n is the sample size.	Means of all sixteen possible samples: 4, 5, 5, 6, 6, 6, 7, 7, 7, 7, 8, 8, 8, 9, 9, 10	The variance of the means of all possible samples $\sigma_{\bar{x}}^2$ [7]is/is not equal to σ^2/n.

This is a population, not a sample, so you use $N = 16$ in the denominator:

$$\sigma^2_{\bar{x}} = \frac{\Sigma (x_i - \mu)^2}{N} = 2.5$$

$\sigma^2 = 5$
$n = 2$

$\frac{\sigma^2}{n} =$ [6] ________

The standard deviation of the means of all possible samples is equal to $\sigma_{\bar{x}} = \sigma/\sqrt{n}$, where n is the sample size.

$\sigma^2_{\bar{x}} = 2.5$

$\sigma_{\bar{x}} =$ [8] ________

$\sigma^2 = 5$

$n = 2$

$\sigma = 2.24$

$\frac{\sigma}{\sqrt{n}} =$ [9] ________

$\sigma_{\bar{x}}$ [10]is/is not equal to $\sigma/\sqrt{n}$.

We rarely know σ. s is usually the best estimate of σ. So we usually use $s_{\bar{x}} = s/\sqrt{n}$ as the standard deviation of the means of all possible samples. It is called the **standard error** and is often denoted $s_{\bar{x}}$ or SE, to distinguish it from standard deviation, denoted s or SD.

The Central Limit Theorem Applied to Real Data

ACTIVITY 13.3

Twenty random samples from PREVO2 *observations from FED*

Following are the PREVO2 observations from the faculty exercise data (FED):

29.5	28.7	29.6	34.2	40.7	27.6	24.6	23.3	36.5	23.3	25.7
28.6	36.1	28.7	23.0	30.3	27.0	26.0	32.8	33.3	26.9	29.4
37.6	39.1	30.5	40.8	21.4	40.8	48.0	30.0			

The number in the population = 30; the mean = 31.13; the st. dev = 6.312; the variance = 39.84. Twenty random samples of size $n = 10$ were taken from the PREVO2 data (with replacement*). Complete the activity to see how the Central Limit Theorem applies to real data.

---

* When we take relatively small samples from very large populations, replacement is not necessary. When the population is small, we need to sample with replacement for the Central Limit Theorem to work.

Components of the Central Limit Theorem

The distribution of the means of all possible samples from a given population approximates a normal distribution

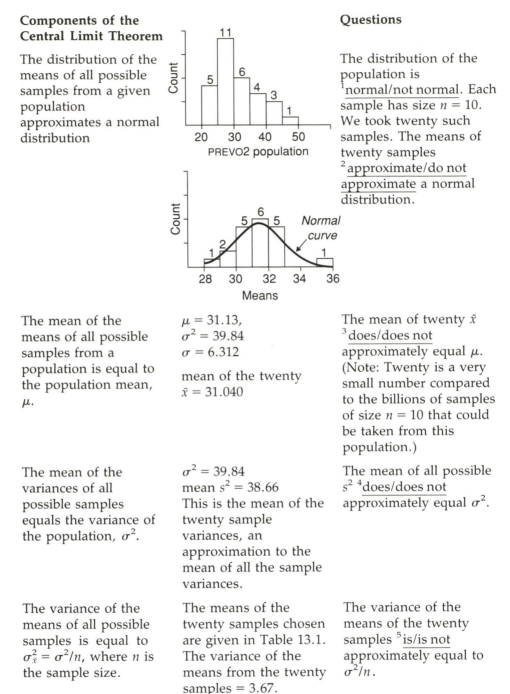

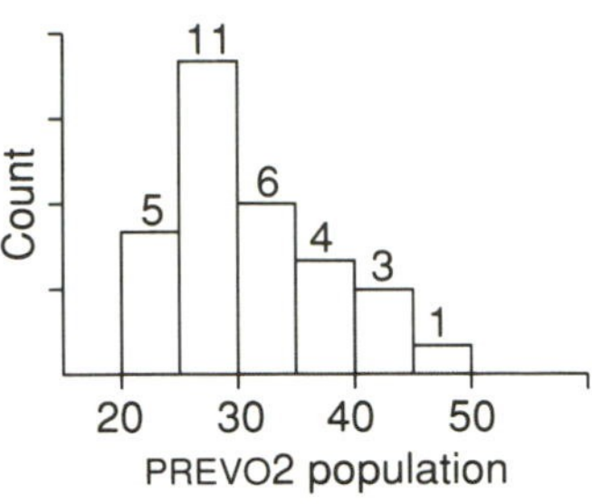

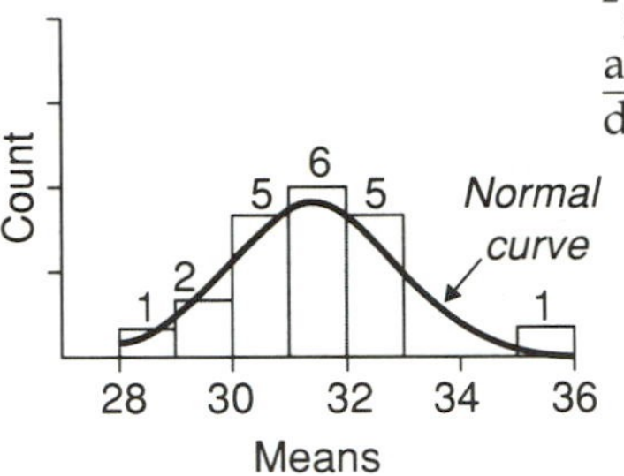

Questions

The distribution of the population is [1]normal/not normal. Each sample has size $n = 10$. We took twenty such samples. The means of twenty samples [2]approximate/do not approximate a normal distribution.

The mean of the means of all possible samples from a population is equal to the population mean, μ.

$\mu = 31.13$,
$\sigma^2 = 39.84$
$\sigma = 6.312$

mean of the twenty $\bar{x} = 31.040$

The mean of twenty $\bar{x}$ [3]does/does not approximately equal μ. (Note: Twenty is a very small number compared to the billions of samples of size $n = 10$ that could be taken from this population.)

The mean of the variances of all possible samples equals the variance of the population, σ^2.

$\sigma^2 = 39.84$
mean $s^2 = 38.66$
This is the mean of the twenty sample variances, an approximation to the mean of all the sample variances.

The mean of all possible s^2 [4]does/does not approximately equal σ^2.

The variance of the means of all possible samples is equal to $\sigma_{\bar{x}}^2 = \sigma^2/n$, where n is the sample size.

The means of the twenty samples chosen are given in Table 13.1. The variance of the means from the twenty samples = 3.67.

$\sigma^2 = 39.84$

$n = 10$

$$\frac{\sigma^2}{n} = 3.984$$

The variance of the means of the twenty samples [5]is/is not approximately equal to σ^2/n.

The standard deviation of the means of all possible samples is $\sigma_{\bar{x}} = \sigma/\sqrt{n}$, where n is the sample size

The standard deviation of the means computed from the twenty samples = 1.968.

$\sigma = 6.312$

$n = 10$

$\sigma/\sqrt{n} = 1.996$

The standard deviation of the means computed from twenty samples [6]is/is not equal to $\sigma/\sqrt{n}$ (to one-decimal-place accuracy).

TABLE 13.1 Random Samples from PREVO2 Observations

	Sample Number									
	1	2	3	4	5	6	7	8	9	10
1	26.9	30.5	27.0	27.6	27.0	23.0	29.4	40.8	30.3	21.4
2	23.3	24.6	24.6	23.3	23.3	25.7	23.3	37.6	29.6	32.8
3	29.4	27.6	40.8	32.8	39.1	29.6	28.7	24.6	26.0	21.4
4	26.0	40.8	29.6	23.3	40.7	34.2	29.4	28.7	29.5	23.0
5	28.7	39.1	28.6	28.7	21.4	40.7	34.2	36.5	36.1	24.6
6	29.5	30.5	37.6	27.6	30.0	26.0	36.1	30.3	30.0	30.0
7	34.2	34.2	24.6	40.8	40.7	32.8	37.6	23.0	27.0	40.8
8	30.3	40.8	29.4	25.7	40.8	30.0	28.7	29.6	39.1	34.2
9	27.6	30.0	28.7	29.5	23.3	28.7	28.7	34.2	40.8	21.4
10	40.7	48.0	30.5	21.4	39.1	40.8	36.5	36.5	40.8	32.8
Mean	29.7	34.6	30.1	28.1	32.5	31.2	31.3	32.2	32.9	28.2
Var.	23	53	27	31	69	36	21	34	33	47
S. Dev	4.83	7.31	5.23	5.61	8.30	6.04	4.59	5.86	5.71	6.82

	Sample Number										
	11	12	13	14	15	16	17	18	19	20	21
1	40.8	33.3	40.8	36.1	28.6	36.5	27.0	23.3	26.9	29.6	
2	29.6	28.6	28.7	23.0	36.1	29.5	30.3	37.6	26.0	24.6	
3	32.8	32.8	23.3	32.8	25.7	37.6	48.0	30.5	27.6	25.7	
4	29.5	40.7	29.4	30.5	29.5	39.1	21.4	29.6	36.5	30.3	
5	29.4	40.8	30.5	32.8	21.4	40.8	21.4	29.4	26.0	28.7	
6	36.5	23.3	48.0	28.7	29.6	48.0	29.4	34.2	23.3	39.1	
7	29.4	33.3	23.0	40.8	37.6	29.6	33.3	21.4	28.6	27.0	
8	33.4	23.0	30.0	33.3	23.0	28.6	27.6	39.1	32.8	28.7	
9	24	23.3	28.7	39.1	36.5	21.4	29.5	24.6	28.7	23.3	
10	4.85	29.5	23.3	24.6	29.6	39.1	36.5	23.3	37.6	34.2	
Mean	33.4	30.9	30.6	32.2	29.8	35.0	30.4	29.3	29.4	29.1	
Var.	24	44	64	33	31	59	60	39	22	22	
S. Dev	4.85	6.62	8.03	5.74	5.58	7.68	7.74	6.21	4.72	4.69	

Twenty random samples of size ten were taken with replacement from the PREVO2 observations on thirty faculty. Each column in Table 13.1 gives the data for a different sample. The twenty samples are all [1] the same as/different from each other. Means, variances, and standard deviations are [2] the same/different. In practice, we would take only one sample and base our conclusions on that sample. Here we are taking twenty samples for pedagogical purposes to illustrate the Central Limit Theorem. Take your own sample from the FED and record the values of PREVO2 as sample 21 in Table 13.1. Compute mean, variance, and standard deviation, and record them in the appropriate blanks.

Table 13.2 gives the mean, standard deviation, and variance of the mean for each of the twenty samples taken from the population of thirty PREVO2 observations on thirty faculty.

TABLE 13.2 Mean, Standard Deviation, and Standard Error of the Mean for Each of the Twenty Random Samples from PREVO2 Observations

	n	Mean	St. Dev.	SE Mean	Variance
1	10	29.66	4.83	1.53	23.3289
2	10	34.61	7.31	2.31	53.4361
3	10	30.14	5.23	1.65	27.3529
4	10	28.07	5.61	1.77	31.4721
5	10	32.54	8.30	2.62	68.8900
6	10	31.15	6.04	1.91	36.4816
7	10	31.26	4.59	1.45	21.0681
8	10	32.18	5.86	1.85	34.3396
9	10	32.92	5.71	1.81	32.6041
10	10	28.24	6.82	2.16	46.5124
11	10	33.39	4.85	1.53	23.5225
12	10	30.86	6.62	2.09	43.8244
13	10	30.57	8.03	2.54	64.4809
14	10	32.17	5.74	1.81	32.9476
15	10	29.76	5.58	1.77	31.1364
16	10	35.02	7.68	2.43	58.9824
17	10	30.44	7.74	2.45	59.9076
18	10	29.30	6.21	1.96	38.5641
19	10	29.40	4.72	1.49	22.2784
20	10	29.12	4.69	1.48	21.9961

No. of Samples	*n*	Mean	St. Dev.	Variance
20	10	31.040	1.968	3.67

Number in Population	Mean	St. Dev.	Variance
30	31.13	6.312	39.84

ACTIVITY 13.4

Implications of the Central Limit Theorem

Components of the Central Limit Theorem

The means of all possible samples from a given population have an approximately normal distribution.

Purpose of this component

The means of all possible samples approximately fit the normal distribution. So we can find the relative position of a computed sample mean, $\bar{x}$, among all the means.

Even when the population distribution is known to be asymmetric, we can safely use the normal tables for the distribution of $\bar{x}$.

Questions

Mark the approximate location of each of the sample means, $\bar{x}$, from Table 13.2 on the graph below.

Means closer to the mean (31) of the population occur [1]more/less frequently than those farther away from 31. Mark the mean from your sample on the figure. How many standard errors away is your sample mean from 31?

$$\frac{\bar{x} - 31}{2} = \_\_\_\_\_\_\_\_.$$

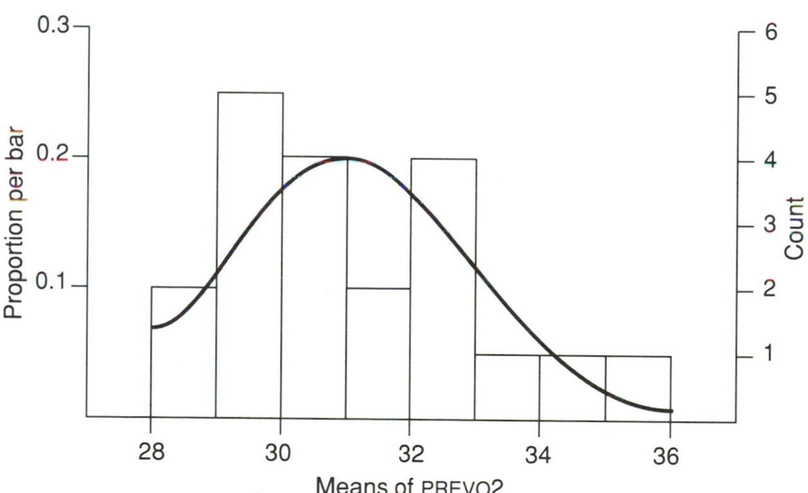

The mean of the means of all possible samples from a population is equal to the population mean, μ.

This component of the theorem indicates that the sample mean is an unbiased statistic. An **unbiased statistic** is defined as one for which the mean of its

Suppose we develop a hypothesis that states that $\mu_{PREVO2} = 40$. We take one sample and obtain $\bar{x} = 31$. Because $\bar{x}$ is an unbiased estimate of μ, we can assume that sampling distribution is equal to the parameter (in this case μ). We can use the sample mean to estimate μ and to test hypotheses about μ.

this sample mean will provide evidence whether the sample could have come from a population with $\mu = 40$.

Mark the hypothesized $\mu = 40$ in the graph you used earlier. 40 is [2]likely/unlikely to be the true population mean. Notice where you indicated the mean from your sample in the figure. How many standard errors away is your sample mean from the actual mean, 31? If all you had to go on was this one sample mean, you would conclude that your sample is [3]likely/unlikely to have come from a population with mean = 31.

The mean of the variances of all possible samples equals the variance of the population, σ^2.

This component of the theorem indicates that s^2 is an unbiased statistic. We can use it to estimate σ^2.

The variance of the means of all possible samples is equal to $(\sigma_{\bar{x}}^2 = \sigma^2/n)$, where n is the sample size.

This component of the theorem provides a way of computing variance of the distribution of means of all possible samples if we know σ^2.

The standard deviation of the means of all possible samples, $\sigma_{\bar{x}}$, is equal to $\sigma/\sqrt{n}$, where n is the sample size.

This component of the theorem would provide us with a way of computing the standard deviation of the means of all possible samples (the sampling distribution) if we knew σ. But because we almost never know σ, we

The distributions shown in the graph below have [4]the same/a different mean and [5]the same/different standard deviations. Here $n = 10$. In the distribution of means, standard error is [6]reduced/increased by a factor of approximately 1/3 ($\approx 1/\sqrt{n}$).

must use s as the best estimate of σ and compute the standard error, or

$$SE_{\bar{x}} = s_{\bar{x}} = \frac{s}{\sqrt{n}}$$

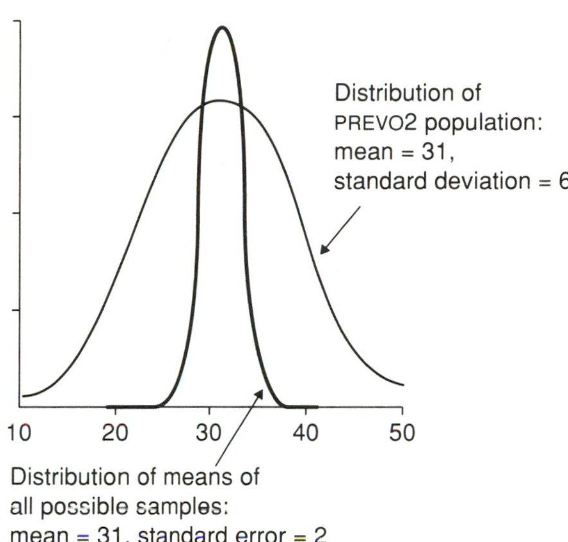

Classifying Events as Likely and Unlikely

Figure 13.1 shows plots of the probability distributions of heart rates of faculty members before and after an exercise program. Below are data on the heart rates of the faculty participants that correspond to the graphs in Figure 13.1.

	Heart Rate (HR)	
	PRE	POST
n of Cases	17	17
Minimum	64.0	59.0
Maximum	107.0	87.0
Mean	84.0	73.5
Standard Dev.	10.7	8.6

The mean heart rates and standard deviations seem to differ. This is reflected in the probability distributions in Figure 13.1, which gives both histograms and normal curves. The normal curve has 95 percent of the area within two standard deviations of the mean. (That is, 95 percent of the observations are between mean − 2 SD and mean + 2 SD.) The PRE heart rate distribution shows a [1] greater/lesser spread than the POST heart rate distribution. For the PREHR data,

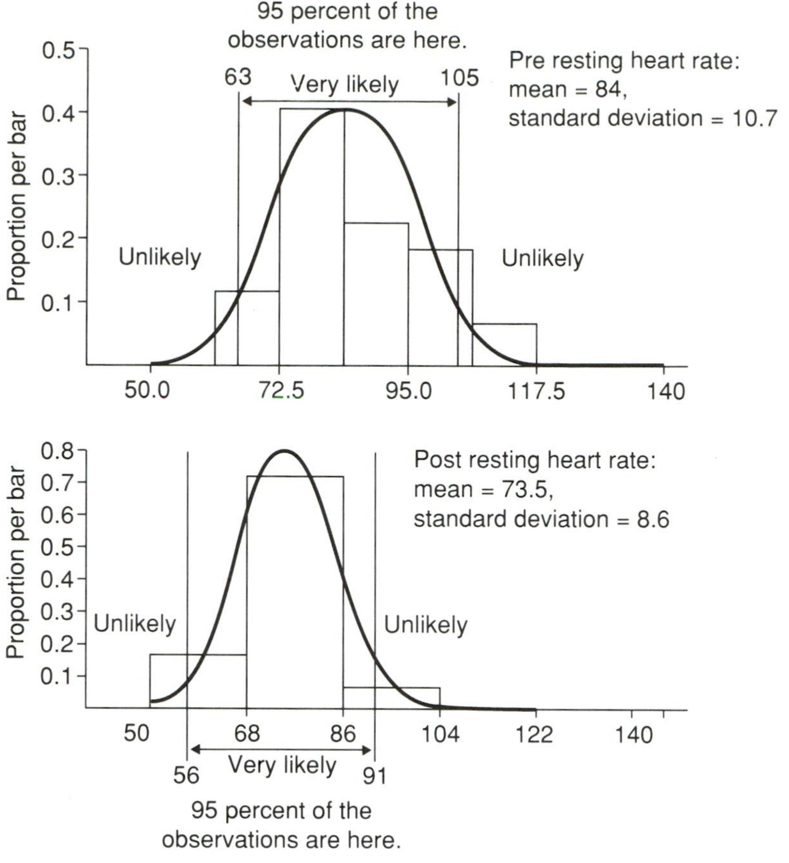

FIGURE 13.1 Pre and Post Resting Heart Rates (PREHR and POSTHR) of Seventeen Faculty Members

95 percent of the observations are between 63 and 105, approximately; for the POSTHR data, 95 percent are between 56 and 91, approximately.

ACTIVITY 13.5

Classifying events as likely and unlikely

Suppose we separate out the cards for the seventeen faculty members who exercised from the faculty exercise data (FED) cards and draw one card. Which interval is the pre heart rate very likely to be in? [1]Below 63/63–105 Which interval is the pre heart rate unlikely to be in? [2]Below 63/63–105 Previously we decided that the cutoff points for determining whether a value is likely or unlikely are the mean ±2 SD (approximately). This is because we want to have about 95 percent of the observations included in the "very likely" category. That is, we are labeling an event that happens 95 percent of the time as very likely and one that happens 5 percent of the time as unlikely. (Someone else might be inclined to choose a different percent and get different cutoffs.)

Consider the post heart rate graph in Figure 13.1. Which post heart rates are we very likely to draw from the deck? [3]Below 56/56–91 Which post heart rates are we unlikely to get [4]Below 56/57–91 Again, the cutoff points for determining whether a value is likely or unlikely are the mean ±2 SD. This is because, as with the pre observations, we want to have 95 percent of the post observations included in the "very likely" category.

Probability of Being Correct or Incorrect When Rejecting a Hypothesis

We have been treating the FED cases as if they constituted the entire population, and we have been drawing samples trying to estimate means and standard deviations. For example, we took a sample from FED and computed its mean. Then we asked whether this sample could have come from the FED population. Of course we already knew that it *did* come from this population, because we chose it from the FED. But suppose we did not know where it came from. Suppose that we knew that the mean of the FED PREVO2 data was 31. We then tested ten individuals and wanted to know if they were a sample from that population. We might ask this, for example, if our original FED population came from a western university, our sample of ten came from an eastern university, and we were trying to determine if this geographical factor made a difference. Suppose that the sample chosen had a mean of 35.02. (We know that one of the samples from the FED actually has this mean.) This sample mean is slightly more than two standard deviations away from 31, the mean of the population. We would conclude that it is unlikely that the sample is typical of the western faculty. (Yet we have seen that it's possible to get that high a mean even from the western data. It's possible, but unlikely.) Thus if the mean from our eastern sample was 35.02 we would conclude that it's unlikely these data came from a population with a mean of 31 and SE of 2, and as we've seen, this conclusion could be wrong.

When we move from this contrived world into the real world where we don't know the parameters, such as the mean of the population, we use similar reasoning. We compare our sample statistics with the theoretical distribution and decide whether such a sample is likely to have come from a population with specified μ and SE. Our task in statistical inference is either to estimate a parameter such as μ or to test a value that we guess to be the population mean (the hypothesized mean).

A hypothesis is a testable proposition, for example, about a population characteristic, or a relationship of variables, or the differences between characteristics of two or more populations. The proposition is a hunch or an educated guess and arises from the literature or the researcher's total experience. On the basis of a sample we may reject or accept the hypothesis we sought to test. Whether we accept or reject the hypothesis, there is always the possibility of being wrong. We would like to have a known low probability of being wrong (to control for error). When we make a decision about the hypothesis and we know the probability of being wrong, we are said to have controlled for error.

We will never know the real population characteristics with certainty. What we test is a null hypothesis, and we either reject or accept it. When we test, we may commit one of two types of error: (1) α error, or Type I error, the probability of rejecting a true hypothesis, or (2) β error, or Type II error, the probability of accepting a false hypothesis. Figure 13.2 illustrates the two types of error.

Hypotheses are tested by comparing the sample statistic to the hypothesized value of the parameter using the distribution (t, F, or χ^2) that is appropriate for that parameter. Assuming the null hypothesis is true, we use the appropriate distribution to estimate the probability, α, of obtaining a value "as bad or worse" than the statistic obtained from our sample. This probability, α, can be thought of as the measure of how likely it would be to obtain a sample with the value obtained given that the null hypothesis is true.

Researchers often report the value of α and leave rejection or acceptance of the hypothesis to the reader. Other researchers choose a critical value of α (usually .05 or smaller) before sampling and then reject the hypothesis if the probability turns out to be greater than the chosen critical value.

The Central Limit Theorem tells us that the sampling distribution of $\bar{x}$ (sample mean) has an approximately normal distribution. Other statistics such as proportions, correlation coefficients, variances, etc., also have sampling distributions. The technique for using these sampling distributions is basically the same as what we have illustrated for testing means using the normal distribution.

The curve in Figure 13.3 is a normal curve with mean 31.2 and standard deviation 1.7. (Since the variable is a mean, the appropriate value to be used for the standard deviation is the standard error of the mean.) It portrays forty samples of PREVO2 values taken from the FED.

The cutoff points we set for the likelihood of an event's occurrence depend upon what chance we are willing to take of being wrong. In Activity 13.5 when we decided on an interval containing 95 percent of the means, we excluded 2.5 percent in each tail of the curve. If we were to decide on a similar interval and select a sample at random from the FED PREVO2 data, it is entirely possible that the sample mean would be below 27.8 or above 34.6. In that case an event we

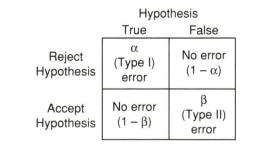

	Hypothesis: True	Hypothesis: False
Reject Hypothesis	α (Type I) error	No error $(1 - \alpha)$
Accept Hypothesis	No error $(1 - \beta)$	β (Type II) error

FIGURE 13.2 Alpha (α, or Type I) and Beta (β, or Type II) Errors

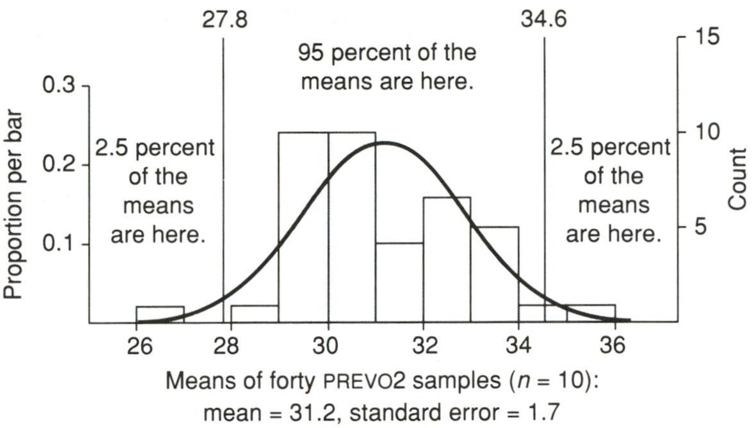

Figure 13.3 Means of Forty Samples from PREVO2 Values

called unlikely would have occurred. (Only two out of the forty sample means actually fall outside the 95 percent interval.) Unlikely events can and do happen. Thanks to the Central Limit Theorem, we can say that this unlikely event occurs 5 percent of the time, in contrast to 95 percent of the time when it does not occur.

Suppose we are testing the hypothesis that the population mean is actually 31. We draw a sample and happen to get one of the unlikely samples with a mean larger than 34.6 or smaller than 27.8. Based on this sample mean, we would reject the hypothesis that the population mean is 31—and we would be wrong. This kind of error is called an alpha (α) error. (Of course, we would probably never know that we had made this error. We would have every reason to believe that our sample was typical.) Using this protocol for testing, we can expect to be wrong in this way 5 percent of the time. We say that we have tested at the 5 percent level of significance, or we have set $\alpha = .05$. This is the meaning of the α in the phrase "alpha error."

Suppose that instead of 95 percent, we use a 90 percent interval (28.3–34.1). In this case we will have 10 percent of the total area in the tails; 10 percent is the [1] _______ error. Commonly chosen alphas are 10 percent, 5 percent, and 1 percent. These are chosen by the researcher in advance of collecting data, based on previous knowledge of the research context and convention. The most commonly chosen α is 5 percent, probably because this number corresponds to an interval of approximately two (actually 1.96) standard deviations around the mean.

Take one sample of $n = 10$ PREVO2 values. Compute the mean and see where this falls in Figure 13.3. Did you get an unlikely or a likely sample? If the sample was an unlikely sample, what kind of error occurred? [2] _______

Examples from the Literature

ACTIVITY 13.6

Rejection or acceptance of a hypothesis based on the probability values

Rejection or acceptance of a hypothesis depends on the α level (chance of being wrong when we reject a null hypothesis) set by the researcher. Let us assume $\alpha = 0.05$. Literature Example 13.1 is from the article "Effects of an Exercise Program on Sick Leave Due to Back Pain." (Note: In Table 2, the p value is roughly the same thing as α.) Answer the questions below.

What are the two response variables? 1 __________

Diagram the research design. 2 __________

What is the explanatory variable? 3 __________

Write a hypothesis regarding the difference between control and exercise groups in sick days attributable to back pain (difference between period 2 and period 1). 4 __________

Assume α is set at 0.05. Do you reject or accept the hypothesis? 5 Reject/Accept

Write a sentence describing the result of this portion of the study. 6 __________

Write a hypothesis regarding the difference between control and exercise groups in episodes of back pain (difference between period 2 and period 1). 7 __________

Assume α is set at 0.05. Do you reject or accept the hypothesis? 8 Reject/Accept

Write a sentence describing the result of this portion of the study. 9 __________

LITERATURE EXAMPLE 13.1

Effect of an exercise program on sick leave due to back pain

Karin M. Kellett, David A. Kellett, Lena A. Nordholm

Physical Therapy, 1991, vol. 71, pp. 283–293

The purposes of this study were to evaluate the effect of a weekly exercise program on short-term sick leave (<50 days) attributable to back pain and to determine whether changes in absenteeism were related to changes in cardiovascular fitness. Subjects were randomly assigned to an exercise group ($n = 58$) and a control group ($n = 53$). Sick leave attributable to back pain was determined in the intervention period of $1\frac{1}{2}$ years and a comparable $1\frac{1}{2}$-year period prior to the study. In the exercise group, the number of episodes of back pain and the number of sick-leave days attributable to back pain in the intervention period decreased by over 50%. Absenteeism attributable to back pain increased in the control group. The decrease in sick leave in the exercise group was not accompanied by any change in cardiovascular fitness. Suggestions for establishing exercise programs are given. [Kellett KM, Kellett DA, Nordholm LA. Effects of an exercise program on sick leave due to back pain. *Phys Ther.* *1991;71:283–293.*]

Results

The exercise group and the control group were first compared to ascertain the equivalence of the two groups prior to intervention. There were no significant differences between the two groups for the variables age, BMI, or cardiovascular fitness. Table 2 shows that the groups did not differ significantly for either sick days attributable to back pain or episodes of back pain prior to intervention (period 1). The groups were equivalent with respect to sex distribution (i.e., women composed 30% of each group) and the ratio between managerial and shop floor workers (e.g., 1:2).

Sick Leave

Thirty-two percent of the exercise group took sick leave because of back pain during period 1, and 21.6% took sick leave during period 2. Twenty-seven percent of the control group took sick leave because of back pain in period 1, and 29% took sick leave during period 2. Sick-leave data for the two groups and the two data-collection periods are presented in Table 2.

Table 2 shows a significant difference between the 51.2% decrease in sick days attributable to back pain in the exercise group and the 65% increase in the control group. A similar significant difference in change scores is noted for the other sick-leave variable, episodes of back pain (Tab. 2). From the data presented in Table 2, the changes in each group were examined separately using paired t tests. In the exercise group, the number of sick days attributable to back pain decreased ($t = 2.29$, df = 36, $p < .05$), but there was no significant change in the number of episodes of back pain ($t = 1.79$, df = 36, $p <$ NS). The control group did not change significantly in either sick days attributable to back pain ($t = -1.34$, df = 47, $p =$ NS) or episodes of back pain ($t = -1.18$, df = 47, $p =$ NS). Thus, it appears that the significant difference in change scores on the sick-days variable can mainly be attributed to changes in the exercise group. With respect to the sickness episode variable, the difference in change scores was produced because the groups changed in different directions, even though each such change was not significant.

TABLE 2 Comparison of Means, Standard Deviations, and Results of *t* Tests of Sick-Leave Variables for Exercise and Control Groups

Variable	Exercise Group (n = 37) $\bar{x}$	SD	Range	Control Group (n = 48) $\bar{x}$	SD	Range	t	df	p
Sick days attributable to back pain									
Period 1	5.59	12.54	0.0–65.00	2.50	5.98	0.0–32.00	1.50	83	NS
Period 2	2.73	7.63	0.0–41.00	4.13	9.32	0.0–40.00	−.73	83	NS
Difference	2.86	7.62		−1.63	8.40		2.55	83	<.02
Episodes of back pain									
Period 1	0.54	0.93	0.0–3.00	0.33	0.60	0.0–2.00	1.27	83	NS
Period 2	0.27	0.61	0.0–3.00	0.52	1.07	0.0–4.00	−1.27	83	NS
Difference	0.27	0.96		−0.19	1.10		2.01	83	<.05

Literature Example 13.2 is from the article "Employment Status and Heart Disease Risk Factors in Middle-Aged Women: The Rancho Bernardo Study." Answer the following questions about the article.

F statistics are given in the last column of Table 4. The authors used asterisks (*) to indicate the significance levels.

Risk factors significant at <.10 were

1 _______, _______, and _______.

Risk factors significant at <.05 were

2 _______ and _______.

What α levels did the authors choose in Table 4?

3 _______, _______

χ^2 is used in the context of dichotomous populations (yes or no events). In Table 3, which variable is tested using χ^2?

4 _______

What α levels did the authors report in Table 3?

5 _______

Suppose we use α = .05. Which are the significant risk factors in Table 3?

6 _______, _______,

_______, _______,

and _______

LITERATURE EXAMPLE 13.2

Employment Status and Heart Disease Risk Factors in Middle-Aged Women: The Rancho Bernardo Study

Donna Kritz-Silverstein, PhD, Deborah L. Wingard, PhD, and Elizabeth Barrett-Connor, MD

Background. In recent years, an increasing number of women have been entering the labor force. It is known that in men, employment is related to heart disease risk, but there are few studies examining this association among women.

Methods. The relation between employment status and heart disease risk factors including lipid and lipoprotein levels, systolic and diastolic blood pressure, fasting and postchallenge plasma glucose and insulin levels, was examined in 242 women aged 40 to 59 years, who were participants in the Rancho Bernardo Heart and Chronic Disease Survey. At the time of a follow-up clinic visit between 1984 and 1987, 46.7% were employed, primarily in managerial positions.

Results. Employed women smoked fewer cigarettes, drank less alcohol, and exercised more than unemployed women, but these differences were not statistically significant. After adjustment for covariates, employed women had significantly lower total cholesterol and fasting plasma glucose levels than unemployed women. Differences on other biological variables, although not statistically significant, also favored the employed women.

Conclusions. Results of this study suggest that middle-aged women employed in managerial positions are healthier than unemployed women. (*Am J Public Health.* 1992;82:215–219)

TABLE 2—Age-Adjusted Comparisons[a] of Employed and Unemployed Women Aged 40–59 on Physical and Life-style Variables: Rancho Bernardo, Calif, 1984–1987 (n = 242)

Variable	Unemployed (n = 113)	Employed (n = 129)	*F* or χ^2
	Mean	Mean	
Body mass index (kg/m^2)	24.9	24.4	0.72
Waist-hip ratio	77.7	76.9	0.84
Alcohol (mL/wk)	90.9	75.6	1.02
	Percent	Percent	
Using estrogen (% yes)	35.1	35.3	0.01
Regular exercise (% yes)	22.2	29.8	1.48
Smoking (% yes)	39.3	33.5	0.67

[a]Comparisons were adjusted for age by analysis of covariance for the continuous variables of body mass index, waist-hip ratio, alcohol consumption, and by the Mantel-Haenzel Extension Test for the categorical variables of estrogen use, exercise and cigarette smoking.

TABLE 3—Age-Adjusted Comparisons[b] of Employed and Unemployed Women Aged 40–59 on Heart Disease Risk Factors: Rancho Bernardo, Calif, 1984–1987 (n = 242)

Risk Factor	Unemployed (n = 113)	Employed (n = 129)	*F* or χ^2
Cholesterol (mg/dL)	226.8	216.6	4.63**
High-density lipoprotein (mg/dL)	68.4	68.0	0.03
Low-density lipoprotein (mg/dL)	135.5	128.5	2.24
Triglycerides (mg/dL)[a]	97.7	85.1	3.78**
Fasting plasma glucose (mg/dL)	99.1	95.4	3.46*
Postchallenge plasma glucose (mg/dL)	120.6	114.3	1.43
Fasting insulin (μ/dL) (n = 216)[a]	12.3	11.2	1.61
Postchallenge insulin (μ/dL) (n = 205)[a]	66.1	53.7	3.82**
Systolic blood pressure (mm Hg)	123.9	118.8	5.24**
Diastolic blood pressure (mm Hg)	76.2	74.9	1.16
Hypertension (% yes)	27.6	12.8	7.02***

[a]Statistics computed on $\log_{10}$ triglycerides, $\log_{10}$ fasting insulin, and $\log_{10}$ nonfasting insulin, respectively.
[b]Age-adjusted comparisons of hypertension rates were computed with the Mantel-Haenzel Extension Test. Age-adjusted comparisons on all other risk factors were computed with analysis of covariance.
$*P < .07$, $**P < .05$, $***P < .01$.

TABLE 4—Mean Differences between Employed and Unemployed Women in Heart Disease Risk Factors after Adjustment for Possible Confounders:[a] Rancho Bernardo, Calif, 1984–1987

Risk Factor	Difference[b]	F
Cholesterol (mg/dL)	−11.8	4.57**
High-density lipoprotein (mg/dL)	0.9	0.10
Low-density lipoprotein (mg/dL)	−7.9	2.19
Triglycerides (mg/dL)	−1.1	1.88
Fasting plasma glucose (mg/dL)	−4.5	5.78**
Postchallenge plasma glucose (mg/dL)	−9.5	3.05*
Fasting plasma insulin (μ/dL)	−1.0	0.22
Postchallenge plasma insulin (μ/dL)	−1.2	2.03
Systolic blood pressure (mm Hg)	−3.8	2.80*
Diastolic blood pressure (mm Hg)	−0.7	0.28
Hypertension (% yes)	−10.1	3.49*

[a]Possible confounders are age, body mass index, estrogen use, alcohol consumption, cigarette smoking, regular exercise, marital status, education, and social class.
[b]Difference is the beta weight obtained from the multiple regression analysis in which unemployed = 1 and employed = 2.
*$P < .10$, **$P < .05$.

Literature Example 13.3 is from the article "Relationship between Body Mass Indices and Measures of Body Adiposity." Answer the questions below.

Consider two variables: W/H and skinfold fat %. The authors give correlation coefficients (r). Any correlation coefficient above .17 may be considered to be significant at the level $\alpha = .05$.

Are all the reported correlation coefficients significant at the level $\alpha = .05$? [1]Yes/No

Hypothesis: There is no relationship between the pairs of variables under consideration (e.g., Hydrostatic Fat % versus W/H^2).
Check to see if any of the confidence intervals (given in parentheses) contain 0, the hypothesized correlation coefficient. If the hypothesized correlation coefficient (0) is not contained in the reported confidence interval, we may reject the hypothesis.

For which pairs of variables would you **accept** the hypothesis of no relationship? [2]All/None

LITERATURE EXAMPLE 13.3

Relationship between Body Mass Indices and Measures of Body Adiposity

DENNIS A. REVICKI, PHD, AND RICHARD G. ISRAEL, EDD

Abstract: We examined the relationship between various body mass indices (BMIs), skinfold measures, and laboratory measures of body fat in 474 males aged 20–70 years. Evaluations included height, weight, skinfold thickness, and hydrostatic measurements of adiposity. The weight-height ratio (W/H), Quetelet index (W/H^2), Khosla-Lowe index (W/H^3), and Benn index (W/H^P) were calculated. The correlations among the various BMIs were high, ranging from 0.91 to 0.99, and all were strongly correlated with weight (rs = 0.81 – 0.98), while only W/H^2 (r = –.03) and W/H^P (r = –.01) were not correlated with height. The W/H^2 and W/H^P had the strongest correlation with hydrostatic and skinfold measurements, although all the BMIs were significantly correlated with these measurements. Results suggest that the Benn index and the Quetelet index are equally valid estimates of body fat in respect to their relationship with hydrostatic measures. (*Am J Public Health* 1986; 76:992–994.)

TABLE 4—Correlations, Partial Correlations Adjusting for Age, and 95% Confidence Intervals between Four Body Mass Indices, Skinfold Fat Per Cent and Hydrostatic Fat Per Cent

	Body Mass Indices			
Body Fat Measures	W/H	W/H^2	W/H^3	W/H^P
Correlations (Confidence Intervals)				
Skinfold Fat %	.75 (.71, .79)	.76 (.72, .80)	.73 (.68, .77)	.76 (.72, .80)
Hydrostatic Fat %	.70 (.65, .75)	.71 (.66, .75)	.69 (.64, .74)	.71 (.66, .75)
Partial Correlations (Confidence Intervals)				
Skinfold Fat %	.70 (.65, .74)	.74 (.70, .78)	.72 (.67, .76)	.72 (.67, .76)
Hydrostatic Fat %	.52 (.45, .59)	.58 (.51, .64)	.58 (.51, .64)	.54 (.47, .60)

Assumptions, Models, and Reality

Scientists construct **models** to describe and explain the phenomena that they are investigating. These models are based on **assumptions**, which are reasonable beliefs that many people hold. It is important to identify these assumptions explicitly. It is also important to remember that a model is not a complete and accurate description of reality. It is a simplified explanation of some chosen aspects of the world, whose purpose is to help us understand those aspects and make predictions.

There are a few essential assumptions in statistics. They relate to (1) the composition of any given measurement (x_i or y_i); (2) the distribution of the population of measurements (x_i or y_i); (3) equivalence of measures of variation ($s_1^2 = s_2^2$); and (4) the shapes of fitted curves (linear or curvilinear). These assumptions are melded together in different ways to form models. In this section we will discuss some of these assumptions and provide illustrations of how they are used. In the following discussion we are concerned with a collection of values of a particular variable, and we call each specific individual value of this variable a measurement.

Assumptions about the Composition of Any Given Measurement

Any given measurement may be thought of as being composed of the population mean (μ) and a random element (ϵ). The relationship is expressed succinctly in the following formula:

$$y = \mu + \epsilon$$

μ is the mean of the population of y and ϵ is the error term.* μ is a constant (for any given population) and ϵ varies with each member of the population. For example, the measurement $y_i = 28.7$ from the PREVO2 data may be seen as being composed of $\mu = 31.13$ and $\epsilon = -2.96$. ϵ varies from individual to individual for many reasons such as diet, heredity, exercise, etc. Many of the factors are not known. ϵ is called a random variable. The model we are using here is expressed by the equation above. Because ϵ varies randomly, y will also vary randomly.

The above model makes two assumptions:

1. The equation applies to all members of the population.
2. The values of ϵ are independent, i.e., the error term of one member of the population cannot be predicted from the error term of another member. The variation from member to member is truly random.

Note: ϵ has mean = 0 and standard deviation σ, where σ is the standard deviation of y. This can be demonstrated by some fairly easy algebraic computations.

---

* Don't think of error as meaning "mistake" in this context. The error term simply gives the deviation from the mean. There's nothing wrong with deviating from the mean—it's not a mistake to deviate from the mean.

Activity 13.7

Composition of PREVO2 *observations as determined using assumption #1*

$y = \mu + \epsilon$

$y_i = 31.13 - 5.13$

Compute y_i when $\mu = 31.13$ and $\epsilon = -5.13$

1 ________

$28.7 = 31.13 + \epsilon$

Compute ϵ

$\epsilon = y_i - \mu$

2 ________

We computed all the values of ϵ for the PREVO2 data (listed below) and computed the mean, μ, and the standard deviation, σ, for y_i. We also computed the mean and standard deviations of ϵ.

ϵ [3]does/does not have mean = 0 and standard deviation = 6.42.

−1.63	−2.43	−1.53	3.07
9.57	−3.53	−6.53	−7.83
5.37	−7.83	−5.43	−2.53
4.97	−2.43	−8.13	−0.83
−4.13	−5.13	1.67	2.17
−4.23	−1.73	6.47	7.97
−0.63	9.67	−9.73	9.67
16.87	−1.13		

	n	Mean	Standard Dev.
PREVO2 (y_i)	30	31.13	6.42
(PREVO2 − μ) = ϵ	30	0.00	6.42

An Assumption About the Distribution of the Population of Measurements

We often assume that a population is normally distributed. Activity 13.2 pointed out that even when this assumption is flawed the means of all possible samples will approximate a normal distribution. If we are testing means and the sample size is reasonably large, we are safe to assume the sample means are distributed normally, regardless of how the original data were distributed.

In cases in which we wish to determine whether a population is normally distributed but have no theoretical basis for this determination, it is possible to test the assumption. Most statistical software provides a way to do this. A

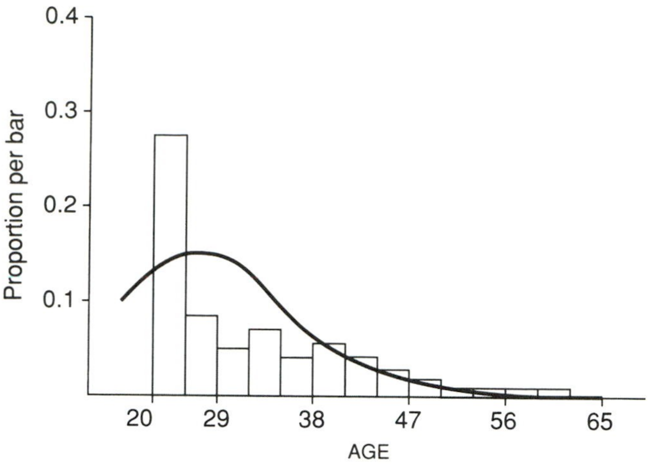

FIGURE 13.4 Histogram of Student Ages

normal quantile plot is obtained by forming a sorted list of the y_i (the data values being tested for normality). Each observation is regarded as a quantile (percentile) of the distribution, and these are plotted against the corresponding quantiles of the standard normal distribution. That is, each observed value is plotted against the z-score that has the same fraction of the normal distribution below it. A straight line indicates that the observations are normally distributed.

Figure 13.4 shows the distribution of student ages. The population size was $n = 153$. We chose this example because the data are highly skewed. Figure 13.5 shows the normal quantile plot for these data. As you can see, the points

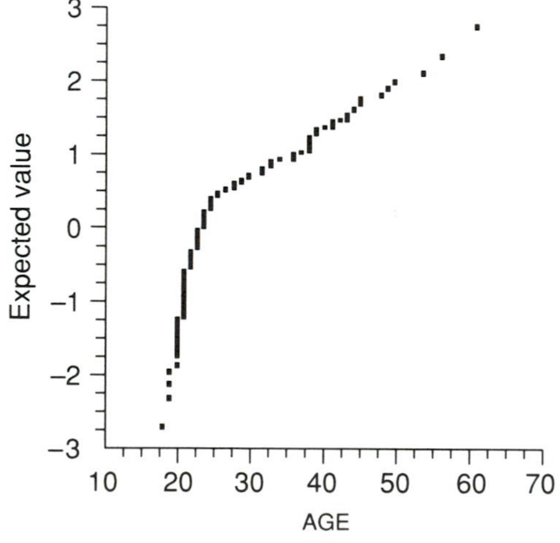

FIGURE 13.5 Normal Quantile Plot of Student Ages

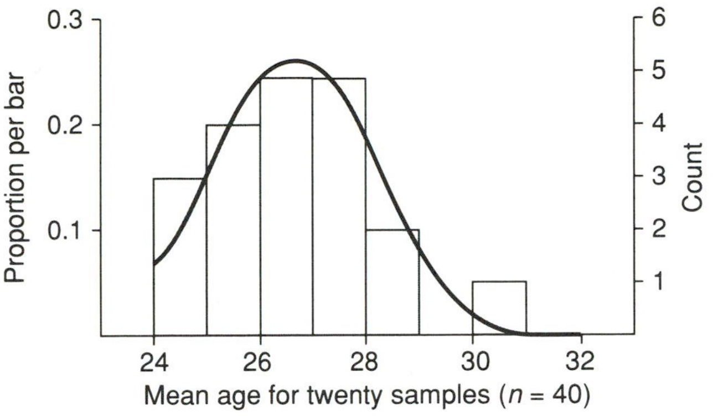

FIGURE 13.6 Histogram of Mean Age for twenty Samples of Size 40

do not fall on a straight line. The next two graphs, Figures 13.6 and 13.7, are similar graphs for twenty sample means taken from this population of ages. The sample size, $n = 40$, is reasonably large, and as you can see, these data fit the normal distribution fairly well as shown by the fact that the points in Figure 13.7 do fall fairly close to a straight line.

An Assumption About Equivalence of Measures of Variation ($s_1^2 = s_2^2$)

Normal populations can differ from each other in their means and/or standard deviations. When we test to see if the means are equal, we frequently assume that the standard deviations are equal. But are they? We have seen that when the units of measurement of two distributions differ, standard deviations are not directly comparable. So we need to compute the coefficient of variation (see

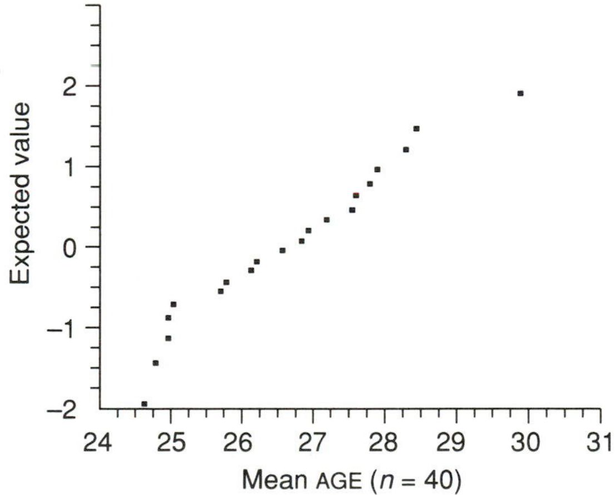

FIGURE 13.7 Normal Quantile Plot for Mean Age, Samples of Size 40

Chapter 8). Even when both distributions have the same units, the standard deviations may still look different because of variation in the samples. (Table 13.2, page 422) illustrates this point. Note that the standard deviations of twenty samples from the same population are [1] the same/different. For these twenty samples, $s_1^2/s_2^2 * \text{df}$ has a χ^2 distribution. That distribution can be used to test whether the variation in the standard deviations we see is likely to be caused by random fluctuation or whether it is more likely that the standard deviations of the samples are actually different.

An Assumption About the Shapes of Fitted Curves (Linear versus Curvilinear)

When we attempt to predict y from known x, we perform regression analysis, which is a method for fitting a straight line to the data in question. For example, suppose we are interested in looking at the relationship betweeen pre-weight (PREWT) of the faculty and the percent body fat content (PREFAT%). These data

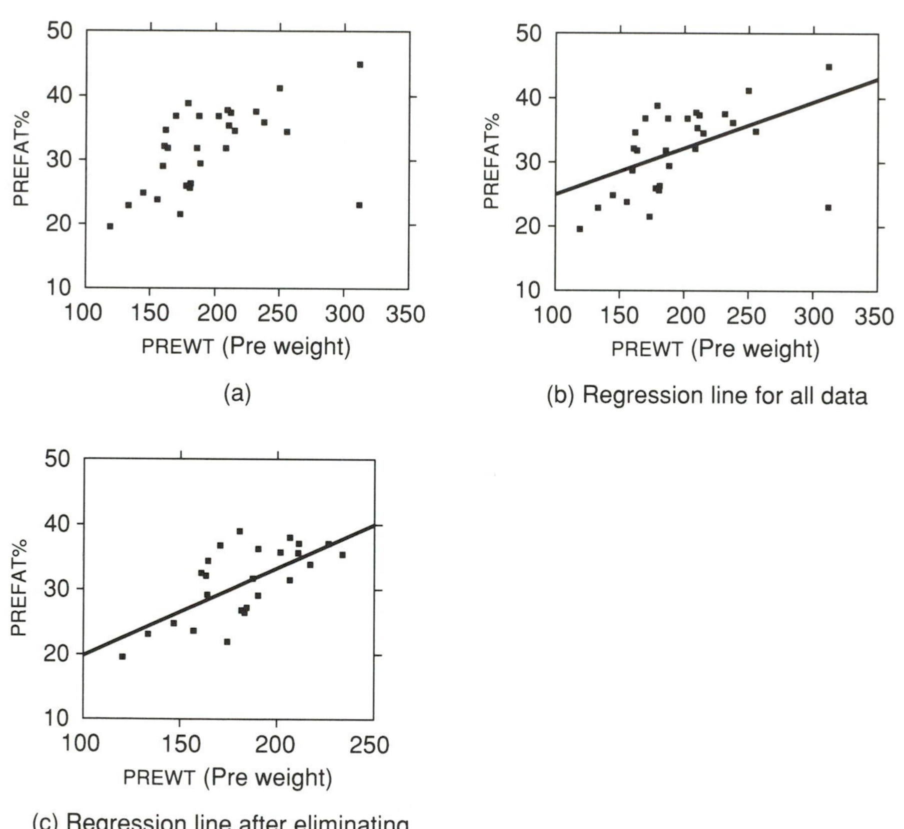

FIGURE 13.8

are plotted in part (a) of Figure 13.8. Are we justified in assuming a straight-line relationship between them? Part (b) of the figure shows the line that best fits the data points using the method of regression analysis. Is it a reasonable fit? To us it does not seem reasonable. Up to the weight of 250 pounds, a case can perhaps be made for a linear relationship. The assumption of linearity can be checked by eliminating the three data points above 250 and plotting the remaining points, as shown in part (c). Still there are many data points between 175 and 200 that are "too far" from the line. We need to check that linearity is actually present before we analyze data using methods that assume a linear relationship.

Transformations to Induce Linearity and Normality

Sometimes when two variables are not linearly related, the logarithm of one of the variables will be linearly related to the other variable. Or, it may be that the square root of the values of the one will be linearly related to the other. These kinds of relationships can be discovered by applying a mathematical transformation to the observations. Mathematical transformations change the values of a variable by applying a rule to all values of that variable. Taking the logarithm of each observation is known as applying a **log transformation**. This transformation is commonly used in biological work. Taking the square root of every observation is known as applying a **square root transformation**. Taking the z-score of every observation is known as applying a **transformation into z-scores**. One feature of many transformations is that one can recover the original observations by using the inverse of the mathematical operation. The choice of transformation depends upon the kind of data, and one is guided in this choice by the literature or by experience.

Self-Assessment

Tasks:	**How well can you do it?**					**Page**
	Poorly				*Very well*	
1. State the components of the Central Limit Theorem.	1	2	3	4	5	416
2. Classify events as likely or unlikely, given cutoff points.	1	2	3	4	5	425
3. Distinguish the two types of error in testing a hypothesis.	1	2	3	4	5	427
4. State and explain the four essential assumptions made in statistical analyses of data, relating to:						
(a) The composition of any given measurement	1	2	3	4	5	436
(b) The distribution of population measurements	1	2	3	4	5	437

(c) The equivalence of measures of variation	1	2	3	4	5	438
(d) The shapes of fitted lines	1	2	3	4	5	439
5. Comment on transformations to induce linearity and normality.	1	2	3	4	5	441

Answers for Chapter 13

Answers for Activity 13.1:

↓ **Population**→	**10**	**8**	**6**	**4**
10	10, 10 $\bar{x}$ = **10** s^2 = **0** s = **0**	10, 8 $\bar{x}$ = **9** s^2 = **2** s = **1.4**	10, 6 $\bar{x}$ = **8** s^2 = **8** s = **2.8**	10, 4 $\bar{x}$ = **7** s^2 = **18** s = **4.2**
8	8, 10 $\bar{x}$ = **9** s^2 = **2** **s = 1.4**	8, 8 $\bar{x}$ = **8** s^2 = **0** s = **0**	8, 6 $\bar{x}$ = **7** s^2 = **2** s = **1.4**	8, 4 $\bar{x}$ = **6** s^2 = **8** s = **2.8**
6	6, 10 $\bar{x}$ = **8** s^2 = **8** s = **2.8**	6, 8 $\bar{x}$ = **7** s^2 = **2** s = **1.4**	6, 6 $\bar{x}$ = **6** s^2 = **0** s = **0**	6, 4 $\bar{x}$ = **5** s^2 = **2** s = **1.4**
4	4, 10 $\bar{x}$ = **7** s^2 = **18** s = **4.2**	4, 8 $\bar{x}$ = **6** s^2 = **8** s = **2.8**	4, 6 $\bar{x}$ = **5** s^2 = **2** s = **1.4**	4, 4 $\bar{x}$ = **4** s^2 = **0** **s = 0**

Population: $\mu = \mathbf{7}$, $\sigma^2 = \mathbf{5}$, $\sigma = \mathbf{2.24}$

Answers for Activity 13.2: 1) Not normal 2) Approximate; but in this case not a very good approximation of a normal distribution because the sample size, 2 is so small. 3) Does 4) 5 5) Does 6) 2.5 7) Is 8) 1.58 9) 1.58 10) Is

Answers for Activity 13.3: 1) Not Normal 2) Approximate 3) Does 4) Does 5) Is 6) Is

Answers for p. 422: 1) Different from 2) Different

Answers for Activity 13.4: 1) More 2) Unlikely 3) Likely, but you never can tell, as unlikely events do sometimes happen by chance. You may have gotten a sample with an unusually large or unusually small sample mean. 4) The same 5) Different 6) Reduced

Answer for p. 425: 1) Greater

Answers for Activity 13.5: 1) 63–105 2) Below 63 3) 56–91 4) Below 56

Answer for p. 429: 1) Alpha (α) 2) Alpha (α) error

Answers for Activity 13.6: 1) Sick days attributable to back pain; episodes of back pain 2) $\begin{matrix} O_1 & X & O_2 \\ O_3 & & O_4 \end{matrix}$ 3) Exercise 4) The mean number of sick days attributable to back pain is the same for exercise group and control group: $(O_2 - O_1) - (O_4 - O_3) = 0$. 5) Reject 6) Table 2 shows a significant difference between the . . . decrease in sick days attributable to back pain in the exercise group and the . . . increase in the control group. 7) The mean difference in episodes of back pain are the same for exercise group and control group: $(O_2 - O_1) - (O_4 - O_3) = 0$. 8) Reject 9) A similar significant difference in change scores is noted for the other sick-leave variable, episodes of back pain. (See the results section of article.)

Answers for Literature Example 13.2: 1) Postchallenge plasma glucose, systolic blood pressure, hypertension 2) Cholesterol, fasting plasma glucose 3) $<.10$ and $<.05$ 4) Hypertension 5) $<.07$, $<.05$, $<.01$ 6) Cholesterol, triglycerides, postchallenge insulin, systolic blood pressure, hypertension

Answers for Literature Example 13.3: 1) Yes 2) None

Answers for Activity 13.7: 1) 26 2) −2.43 3) Does

Answer for p. 440: 1) Different

CHAPTER

14

Venturing into the Unknown: From Sample Statistics to Population Parameters

Prelude

"Now will you tell me where we are?" asked Tock as he looked around the desolate island.

"To be sure," said Canby; "you're on the Island of Conclusions. Make yourself at home. You're apt to be here for some time."

"But how did we get here?" asked Milo, who was still a bit puzzled by being there at all.

"You jumped, of course," explained Canby. "That's the way most everyone gets here. It's really quite simple: every time you decide something without having a good reason, you jump to Conclusions whether you like it or not. It's such an easy trip to make that I've been here hundreds of times." . . .

As he spoke, at least eight or nine more people sailed onto the island from every direction possible.

"Well, I'm going to jump right back," announced the Humbug, who took two or three practice bends, leaped as far as he could, and landed in a heap two feet away.

"That won't do at all," scolded Canby, helping him to his feet. "You can never jump away from Conclusions." . . .

. . . "But from now on I'm going to have a very good reason before I make up my mind about anything. You can lose too much time jumping to Conclusions."

— Norton Juster
The Phantom Toll Booth, pp. 168, 170

Chapter Outline

Self-Assessment

Tasks:	**How well can you do it?**					**Page**
	Poorly				*Very well*	
1. Transform raw data into *z*-scores and *t*-scores.	1	2	3	4	5	447
2. Estimate and interpret a confidence interval for the mean when all the essential information is given.	1	2	3	4	5	450
3. Estimate and interpret a confidence interval for a proportion when all the essential information is given.	1	2	3	4	5	454
4. Use a confidence interval for making decisions about μ.	1	2	3	4	5	464
5. Test a hypothesis about the mean of a single population when all the essential data are given, using three different methods.	1	2	3	4	5	464
6. Test hypotheses about means of two dependent populations when all the essential data are given, using three different methods.	1	2	3	4	5	478
7. Test hypotheses about means of two independent populations when all the essential data are given, using three different methods.	1	2	3	4	5	494

8. Test hypotheses about means of three or more groups using analysis of variance techniques.	1	2	3	4	5	501
9. Estimate the sample size needed to obtain a sample proportion within ±.05 of the population proportion with a 95 percent confidence level when the population size is known.	1	2	3	4	5	520
10. Estimate the sample size needed to obtain a sample mean within a given number of standard deviations of the population mean for a given confidence level.	1	2	3	4	5	522

Introduction

Venturing into the statistical unknown is neither dangerous nor benign—it just is. As we have seen, statistics rarely deals with certainties, only probable or improbable occurrences.

We have learned to describe observed data with numbers, charts, and tables. We even guessed at the relationships between variables by plotting graphs or constructing two-way and three-way tables. In this chapter we will learn to use known sample statistics to estimate population parameters.

Histograms of population distributions take a multitude of shapes, as Figure 14.1 suggests. Such diversity hampers any attempts at prediction or estimation of population characteristics from samples. So we look for mathematical models that adequately describe populations for given purposes. Some populations cannot be described with simple mathematical models. However, many popula-

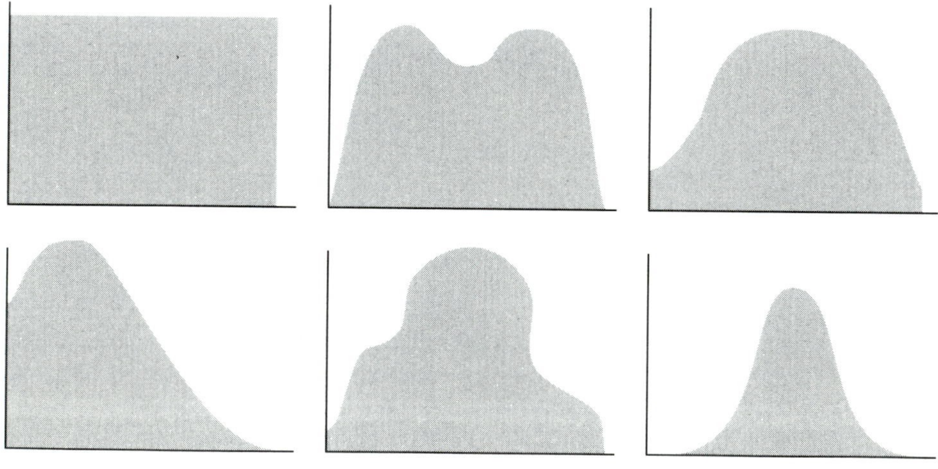

Figure 14.1 Some Possible Distributions of Populations

tions are adequately described by known mathematical models. The normal distribution (also known as the Gaussian distribution, or z-distribution), the Poisson distribution, and the binomial distribution are some examples. You already learned to find values from tables of distributions of statistics such as z, t, χ^2, and F in Chapter 11.

If each population has its own distribution it would seem to be a hopeless task to generalize and describe unknown populations by observing or gathering data from relatively small samples. Through a fortunate coincidence, the normal distribution—described by Abraham DeMoivre in 1733, later by Pierre Simon de Laplace and Karl Friedrich Gauss—approximates many population distributions. Sir Francis Galton was the first to use the normal distribution in the health context.

In Chapter 11 we showed how the z-distribution (normal distribution) table allows us to calculate area under the curve (probability or relative frequency) if z is known. What is z? z is a contrived measure that takes into account an observation and the standard deviation and mean of a population. Converting measurements into z-scores is called **standardization**. You practiced calculating z in Chapter 7. Any measurement can be converted to its corresponding z-score, or **standardized score**, which measures the distance between the observation and the population mean in units of population standard deviation.

ACTIVITY 14.1

Standardization of measurements to z-scores

Convert the measurements below into z-scores using the formula

$$z = \frac{x_i - \mu}{\sigma}$$

Name of variable	μ	σ	Measurements of Interest	z-scores
Heart rate	87	13	74, 100	(74 − 87)/13 = [1] ______ (100 − 87)/13 = ______
Respiratory quotient	1.2	0.1	1.0, 1.4	[2] ______
Resting diastolic blood pressure	80	8	72, 88	[3] ______
% body fat obtained by Lomen formula	32	6	26, 32	[4] ______
Minute ventilation	132	32	100, 148	[5] ______
Age	50	7.5	42.5, 57.5	[6] ______
Weight in pounds	190	45	100, 235	[7] ______
Height in inches	70	3	68.5, 71.5	[8] ______

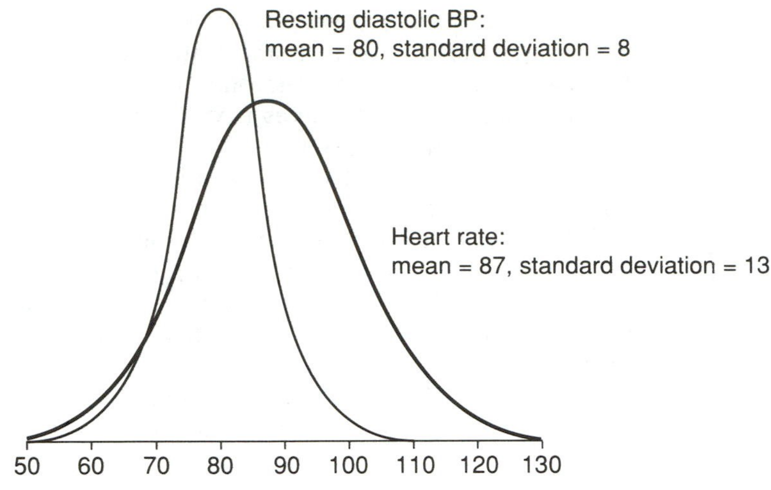

FIGURE 14.2

The variables in Activity 14.1 are all measured in different units and have different means and standard deviations. *z*-scores have no units. When we convert *z*-scores back, the measurements acquire units.

Figure 14.2 shows the normal distributions of resting diastolic blood pressure and heart rate. Draw lines indicating the means for these two populations, and also shade the area between the *z*-scores you calculated in Activity 14.1.

The populations shown in Figure 14.2 differ from each other in mean as well as standard deviation. We see a [1] greater/lesser spread in heart rate than in resting diastolic blood pressure because the standard deviation of heart rate is higher. The mean of heart rate is larger than the mean of blood pressure, so we see a [2] _______ toward the right in the normal distribution for heart rate relative to the distribution for blood pressure.

The seven graphs in Figure 14.3 depict normal distributions for seven of the variables mentioned in Activity 14.1. From the *z*-distribution table (Appendix B, page 690), we found that *z* for the middle 95 percent of the cases is 1.96. We rounded it to 2. Mark the 95 percent area in each of the distributions. The first one is done for you as an example.

Now extend the shading to 99 percent of the area. The heart rate distribution is shown in Figure 14.4 as an example.

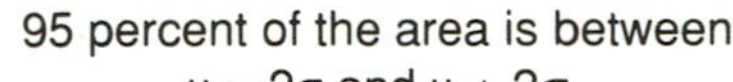
95 percent of the area is between
μ − 2σ and μ + 2σ

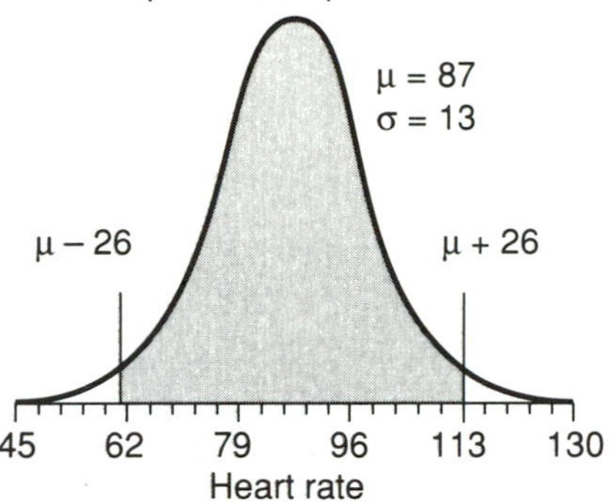
μ = 87
σ = 13
μ − 26
μ + 26
45 62 79 96 113 130
Heart rate

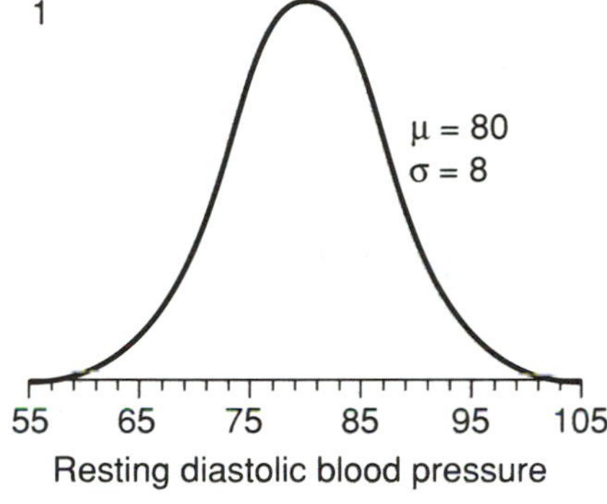
1
μ = 80
σ = 8
55 65 75 85 95 105
Resting diastolic blood pressure

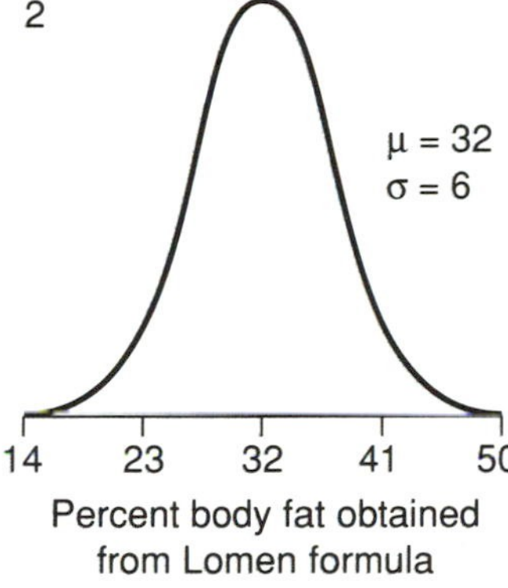
2
μ = 32
σ = 6
14 23 32 41 50
Percent body fat obtained
from Lomen formula

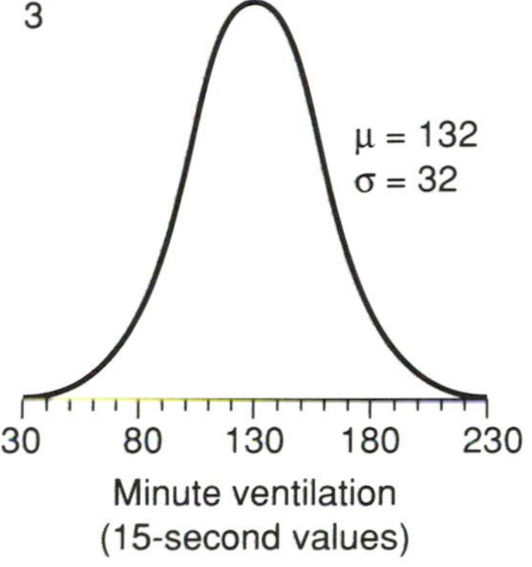
3
μ = 132
σ = 32
30 80 130 180 230
Minute ventilation
(15-second values)

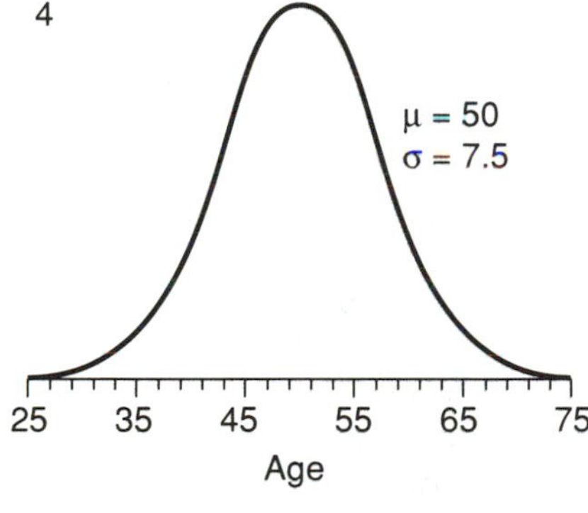
4
μ = 50
σ = 7.5
25 35 45 55 65 75
Age

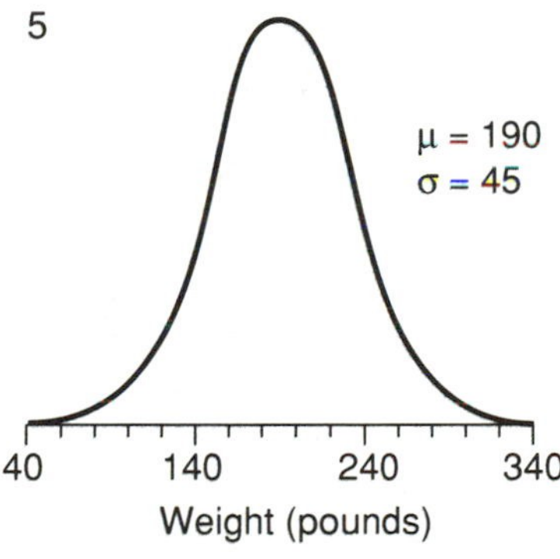
5
μ = 190
σ = 45
40 140 240 340
Weight (pounds)

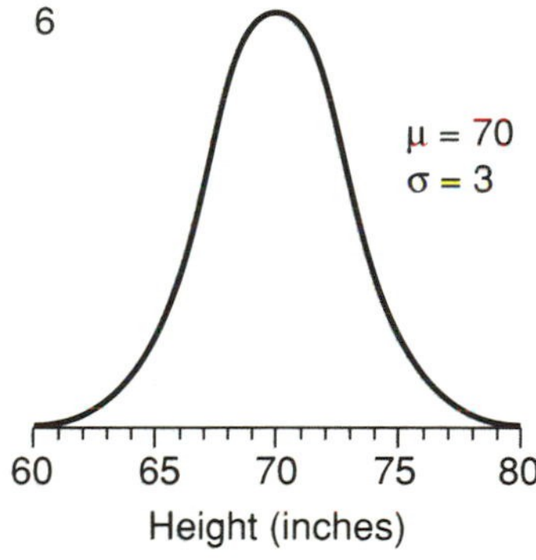
6
μ = 70
σ = 3
60 65 70 75 80
Height (inches)

FIGURE 14.3

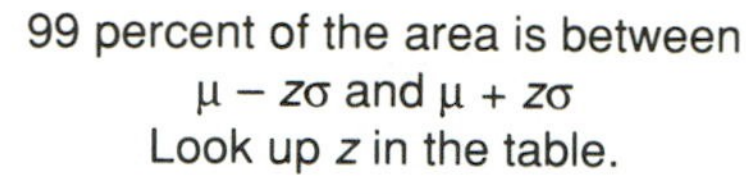

99 percent of the area is between
μ − zσ and μ + zσ
Look up z in the table.

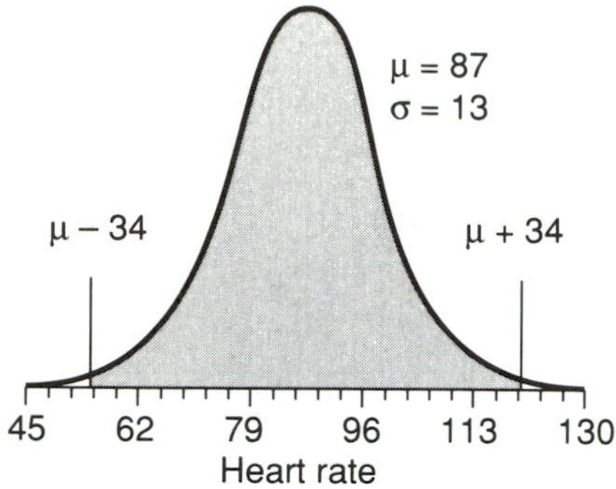

FIGURE 14.4

Estimating Parameters: Confidence Intervals

Confidence Interval for Ratio Data

We are interested in learning about the population by taking a sample. How do we do this? We know that sample means are distributed with mean μ and standard deviation $\sigma/\sqrt{n}$. We know neither μ nor σ, so we use the best estimates from the sample, $\bar{x}$ and s. We also know that when we take one sample the probability of obtaining a sample with $\bar{x}$ close to μ is [1] higher/lower than the probability of getting one far away from μ. So we assume that the mean of our single sample, $\bar{x}$, is close to μ. $\bar{x}$ is the best single-number estimate for μ. But we can do better than this.

As we saw in Chapter 13, the Central Limit Theorem tells us that the mean of the sample means of all possible samples of size n is the population mean. Furthermore, the deviation of the sample means from the population mean as measured by the standard deviation can be made very small by making the sample size large enough. This is so because of $\sqrt{n}$ in the denominator. In addition, sample means are distributed approximately normally, and the approximation gets better as sample size increases. These facts not only allow us to use the sample mean as an estimate of the population mean but also allow us to put error bounds on the estimate. We are able to give the estimate as an interval, a range of values, instead of just a single number. Furthermore, the Central Limit Theorem provides a means to quantify how likely it is that the true mean will actually be in the interval.

A **confidence interval (C.I.)** is a range of values that includes the parameter with known probability, called the confidence level. The confidence level represents the probability (or approximate probability) that a sample will actually have the value of the parameter in the confidence interval. For μ, the population mean, confidence intervals are calculated using the sample means and their standard error,

$$\text{C.I.} = \left(\bar{x} - t\frac{s}{\sqrt{n}}\right) \text{ to } \left(\bar{x} + t\frac{s}{\sqrt{n}}\right)$$

where t represents the statistic obtained from the t-distribution with $(n-1)$ degrees of freedom for the chosen level of confidence, $(1-\alpha)$. Confidence intervals are usually expressed in mathematical interval notation. For example, the interval that begins at 4.5 and ends at 5.3 would be denoted (4.5, 5.3). This confidence interval can also be given in the form 4.9 ± 0.4, where 4.9 is the midpoint of the interval (the mean) and 0.4 is half the width of the interval.

In practice we obtain exactly one sample and calculate the confidence interval from it. To help you understand the concept of confidence intervals, we will calculate confidence intervals for forty samples of size $n = 10$ obtained from the PREVO2 data and see how many actually cover μ. Note that for $n = 10$, df = 9 and $t_{95} = 2.262$. For the first sample (numbered 0), C.I. $= 32.11 \pm 2.262 * 1.74 =$ (28.18, 36.04). The data from all of the calculations appear in Table 14.1. In Figure 14.5, each horizontal line represents the 95 percent confidence interval for a different sample. The sample number appears to the left of every fifth line. Which of the C.I.s does not include μ, i.e., 31? [1] ________. What percentage of the C.I.s does not include μ? [2] ________. When

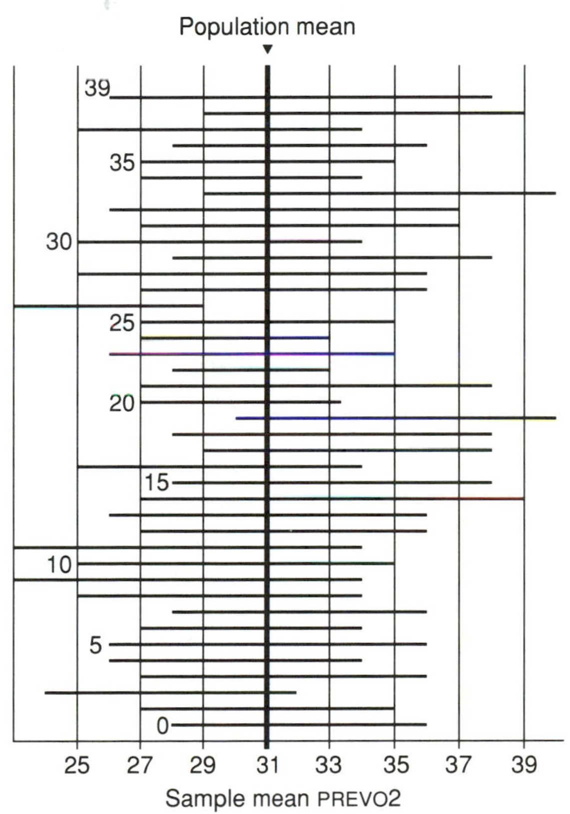

FIGURE 14.5 95 Percent Confidence Intervals for PREVO2 Samples

TABLE 14.1 Confidence Intervals for the Means of Forty Samples of Size Ten

	n	Mean	St Dev	SE Mean	95.0 Percent C.I.
0	10	32.11	5.49	1.74	(28.18, 36.04)
1	10	30.99	5.75	1.82	(26.87, 35.11)
2	10	28.57	5.46	1.73	(24.67, 32.47)
3	10	31.59	6.67	2.11	(26.82, 36.36)
4	10	29.67	5.60	1.77	(25.66, 33.68)
5	10	30.66	6.93	2.19	(25.70, 35.62)
6	10	30.62	5.10	1.61	(26.97, 34.27)
7	10	32.53	5.11	1.62	(28.87, 36.19)
8	10	29.41	5.78	1.83	(25.27, 33.55)
9	10	29.06	6.70	2.12	(24.27, 33.85)
10	10	29.96	6.58	2.08	(25.25, 34.67)
11	10	29.24	7.20	2.28	(24.09, 34.39)
12	10	31.72	6.46	2.04	(27.10, 36.34)
13	10	30.84	7.15	2.26	(25.73, 35.95)
14	10	33.11	8.00	2.53	(27.38, 38.84)
15	10	33.10	7.58	2.40	(27.67, 38.53)
16	10	29.74	6.02	1.90	(25.43, 34.05)
17	10	33.51	6.27	1.98	(29.02, 38.00)
18	10	32.93	7.23	2.29	(27.76, 38.10)
19	10	35.11	6.94	2.19	(30.14, 40.08)
20	10	29.59	4.31	1.36	(26.51, 32.67)
21	10	32.08	7.72	2.44	(26.55, 37.61)
22	10	30.55	4.09	1.29	(27.63, 33.47)
23	10	30.33	5.85	1.85	(26.15, 34.51)
24	10	29.59	4.15	1.31	(26.62, 32.56)
25	10	30.55	5.52	1.75	(26.60, 34.50)
26	10	26.61	3.27	1.03	(24.27, 28.95)
27	10	31.74	6.07	1.92	(27.40, 36.08)
28	10	30.33	8.03	2.54	(24.58, 36.08)
29	10	33.14	7.08	2.24	(28.07, 38.21)
30	10	29.76	6.18	1.95	(25.34, 34.18)
31	10	32.21	7.19	2.28	(27.06, 37.36)
32	10	32.08	8.06	2.55	(26.31, 37.85)
33	10	34.47	7.57	2.39	(29.05, 39.89)
34	10	30.60	5.22	1.65	(26.87, 34.33)
35	10	30.93	5.59	1.77	(26.93, 34.93)
36	10	32.10	6.15	1.94	(27.70, 36.50)
37	10	29.71	6.36	2.01	(25.16, 34.26)
38	10	33.80	6.70	2.12	(29.01, 38.59)
39	10	31.77	8.16	2.58	(25.93, 37.61)

we compute a confidence interval, it is likely to include the population parameter with probability $1 - \alpha$, the prescribed confidence level. In this case we set the level of confidence at 95 percent, so we would expect 95 percent of the confidence intervals to contain $\mu = 31$. To compute the confidence intervals, we used df = 9 because the sample size is 10. Consequently, we found the t-value in the table for $1 - \alpha = .95$ with df = 9 to be $t =$ [1] _______.

Since we computed 95 percent confidence intervals, we would have expected 95 percent of the forty samples to have included $\mu = 31$ and 5 percent not to have included this value. However, 97.5 percent included 31 and 2.5 percent did not include 31. What are the reasons for the discrepancy? One is that use of the t-distribution in this context is only accurate when the samples are drawn from a population that has a normal distribution. Another reason is that forty is a rather small fraction of the samples of size $n = 10$ that can be chosen from this population. Perhaps, had we taken several hundred samples, we would have captured the population mean in a percentage much closer to 95 percent. Notice also that 5 percent of 40 is 2. Instead of one confidence interval out of forty not containing the mean, we would have expected two. We were only off by one sample!

ACTIVITY 14.2

Computation of a confidence interval

Take a random sample of ten heart rates from the FED and compute the confidence interval for the pre heart rate (PREHR). Choose your own α and complete the table below. We are assuming for purposes of this exercise that the thirty cases represent the population.

Variable	$1 - \alpha$	n	df	t	Sample mean	Standard deviation	SE	C.I.
PREHR								

Does the C.I. you computed for the mean heart rate include the FED mean heart rate ($\mu_{HR} = 87$)? If you are one of the few people whose C.I. did not contain 87, the actual mean heart rate, you got one of the possible atypical samples. If one hundred people carried out this activity, each choosing 95 percent for α, we would expect about five of them to get confidence intervals that didn't include 87.

Estimate confidence intervals for the following variables from the questionnaire data.

Variable	$1 - \alpha$	n	t	Sample Mean	Sample Standard Deviation	$SE_{\bar{x}}$	C.I ($\bar{x} - t * SE_{\bar{x}}$, $\bar{x} + t * SE_{\bar{x}}$)
Age	99%	51	2.7*	25.2	6.9	.96	[1] (______, ______)
Weight	95%	51	2	66.4	4.3	.60	[2] (______, ______)
Height	99%	51	2.7	140.5	33.7	4.7	[3] (______, ______)
Pulse I	95%	51	2	75.9	11.3	1.6	[4] (______, ______)
Pulse II	95%	51	2	98	17.1	2.4	[5] (______, ______)

* The required df, $n - 1 = 50$, is not given in the table, so you read t for the next smaller df that is given, in this instance df = 40.

Confidence Intervals for Proportions

Our interest is in learning about a population by taking a sample. Suppose we know that about 65 percent of our statistics class is female ($n = 51$). In future statistics classes what percentage is likely to be women? Notice that the population we are now discussing does not yet exist, because we are talking about the future. We cannot draw a random sample from future populations. An appropriate procedure would be to make sure that this statistics class was chosen at random from all the statistics classes offered during the past two or three semesters. Another possibility would be to check that students are assigned to statistics classes more or less at random, and that no gender bias is operating in the choosing of statistics classes. For this discussion we will assume we have done these things.

We want to estimate the percentage of females in future statistics classes, assuming that the future will be pretty much like the past. Consider the following facts:

1. Each time we take a sample from a population we get [6] a different/the same sample.
2. We can calculate the standard deviation of a proportion, and from it the standard error (SE) of the proportion.
3. The sample proportion (p) is the best estimate of the population proportion (P). In the case of proportions we may use the normal distribution for estimating confidence intervals, provided both np and $n * (1 - p)$ are larger than 5.

The confidence interval (C.I.) for a proportion is the range of values C.I. = $(p - z * \text{SE}, p + z * \text{SE})$, where z is obtained from the normal table for the chosen level of confidence $(1 - \alpha)$.

What is the dispersion for proportions? For populations with twenty or more cases the following formula is reasonably accurate:

$$\text{SE} = \sqrt{\frac{p(1-p)}{n}}$$

So, for our example,

$$\text{SE} = \sqrt{\frac{.65(1-.65)}{51}} \approx .07$$

The 95 percent confidence interval for the proportion of females is calculated as follows (we rounded the z-value to 2):

$$(p = 0.65 \qquad \text{SE} = 0.07 \qquad z = 2 \qquad n = 51)$$

$$\begin{aligned} \text{C.I.} &= (p - z * \text{SE}, \quad p + z * \text{SE}) \\ &= [0.65 - (2 * 0.07), \; 0.65 + (2 * 0.07)] \\ &= (0.65 - 0.14, \quad 0.65 + 0.14) \\ &= (0.51, \quad 0.79) \end{aligned}$$

This means that we expect the percent of females in statistics classes to be between 51 percent and 79 percent. Our estimate may be wrong approximately 5 percent of the time.

We found that the 95 percent confidence interval for the percentage of females runs from 51 percent to 79 percent. If this interval is too big for our purposes, i.e., if we would like a closer estimate, there are two ways that we can decrease the size of the confidence interval: we can increase the sample size or we can decrease the level of confidence. For example, the 80 percent confidence interval goes from [1] _______ to [2]_______. When we say that the 95 percent confidence interval runs from 51 percent to 79 percent, just what is it that we are 95 percent confident of? Basically we are saying the procedure will produce a confidence interval that actually contains the parameter (in this case the true proportion of female students in statistics classes) 95 percent of the time. Since some approximations have been used in arriving at this result, we can, in fact, only be sure of capturing the true parameter *approximately* 95 percent of the times that we use these methods.

ACTIVITY 14.3

Computation of confidence intervals for proportions

Estimate the confidence intervals for the following variables from the questionnaire data ($n = 51$).

Variable	Sample p	$(1-\alpha)$ α is specified by the researcher	z	Variance $p(1-p)$	Standard error $SE = \sqrt{\frac{p(1-p)}{n}}$	C.I. $(p - z * SE,$ $p + z * SE)$
SEX	Fem 0.65	95%	2	0.23	0.07	(.51, .79)
TIMEOUT	Yes 0.67	90%	1 ______	2 ______	3 ______	4 ______
FRIENDS	Yes 0.9	99%	5 ______	6 ______	7 ______	8 ______

The main subject of this chapter is estimation and comparison of means. Figure 14.6 shows that when we compare means of two groups we use a

Description	Comparison of Group Means	Determinimg Relationships Among Variables: Association	Determinimg Relationships Among Variables: Prediction	Determining the Effect of Intervention
Simple descriptions (1)	Comparison with values given in the literature (2) *t-test* Comparison of two groups (3) *t-test* Comparison of three or more groups (4) *Analysis of variance, or ANOVA*	Two categorized variables (5) *Chi-square* One categorized variable and one continuous variable (6) *Correlation coefficient* Continuous data (7) *Correlation coefficient*	The value of one variable from the value of another (8) *Regression analysis*	Doing experiments (9) *t-test* *Analysis of variance, or ANOVA* *Chi-square* *Correlation coefficient* *Regression analysis*

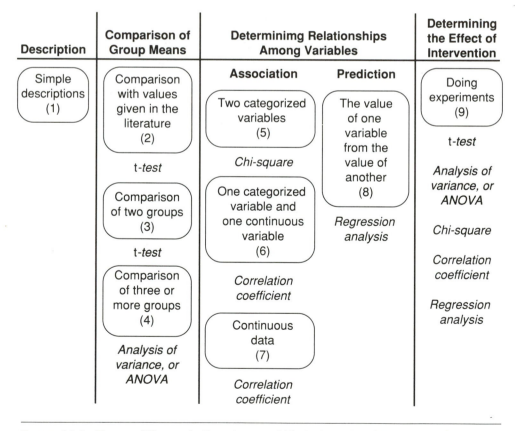

FIGURE 14.6 Types of Research Questions and Their Associated Tests

[1] ______________ and when we compare means of three or more groups we use [2] ______________ and the *F*-distribution.

Examples from the Literature

Literature Example 14.1 is from the article entitled "Safety Belt and Helmet Use Among High School Students—United States, 1990." Read the article and answer the questions below:

Use of a safety belt while riding in cars and use of a helmet while riding motorcycles and bicycles are important for safety. What are the year 2000 objectives for helmet use for motorcycles and bicycles?

Motorcycle [1] ______________________

Bicycle ______________________

What percent of students used safety belts in 1990?

[2] ______________________

In the space to the right, plot 95 percent confidence intervals for seat belt use by males, females, and total, and comment on how they overlap.

[3]

According to the authors, "Use of safety belts did not vary significantly by sex, race/ethnicity, or grade." Examine the confidence intervals and confirm these conclusions.

[4] ______________________

"White students (59.8%) were significantly more likely than Hispanic students (39.3%) to wear motorcycle helmets. Helmet use by black students (55.9%) was not significantly different from helmet use by either white or Hispanic Students."

Could you have come to these conclusions by reading numbers from the figure alone?

[5] Yes/No

What statistical tool clarifies this statement?

____

____

What in your opinion will contribute to achievement of the year 2000 objectives?

LITERATURE EXAMPLE 14.1

Safety-Belt and Helmet Use Among High School Students—United States, 1990

MMWR. *1992;41:111-114*

DURING 1988, injuries were the leading cause of death among persons aged 15-19 years in the United States. More than half (53%) of these deaths were motor-vehicle related, including crashes involving bicycles and motorcycles with motor vehicles (CDC, unpublished data, 1988). Among persons aged 15-19 years, motor-vehicle-related injuries are the leading contributor to hospital and emergency department medical costs associated with injuries.[1] This article presents 1990 self-reported data from U.S. students in grades 9-12 regarding the prevalence of three behaviors that reduce the risk for injuries from motor-vehicle crashes—safety-belt use, motorcycle-helmet use, and bicycle-helmet use.

The national school-based Youth Risk Behavior Survey (YRBS) is a component of CDC's Youth Risk Behavior Surveillance System, which periodically measures the prevalence of priority health-risk behaviors among youth through representative national, state, and local surveys.[2] The 1990 YRBS used a three-stage sample design to obtain a representative sample of 11 631 students in grades 9-12 in the 50 states, the District of Columbia, Puerto Rico, and the Virgin Islands. Students were asked how often they wore safety belts when riding in a car or truck driven by someone else and how often they wore a helmet when riding a motorcycle or a bicycle.

Less than one fourth (24.3%) of all students in grades 9-12 "always" used safety belts when riding in a car or truck driven by someone else. An additional 23.0% of students used safety belts "most of the time," and 13.4% reported "never" using safety belts. Use of safety belts did not vary significantly by sex, race/ethnicity, or grade.

Male students (44.8%) were significantly more likely than female students (23.6%) to ride motorcycles, and white students (37.3%) were significantly more likely than Hispanic (24.8%) or black students (18.9%) to ride motorcycles. Among students who rode motorcycles, 57.9% wore motorcycle helmets "always" or "most of the time." White students (59.8%) were significantly more likely than Hispanic students (39.3%) to wear motorcycle helmets. Helmet use by black students (55.9%) was not significantly different from helmet use by either white or Hispanic students. Use of motorcycle helmets did not vary significantly by sex or grade.

Male students (67.3%) were significantly more likely than female students (53.4%) to ride bicycles; 2.3% of students wore bicycle helmets "always" or "most of the time." Use of bicycle helmets did not vary significantly by sex, race/ethnicity, or grade.

TABLE. Percentage of high school students who reported "always" using safety belts when riding in a car or truck driven by someone else, by sex, race/ethnicity, and grade—United States, Youth Risk Behavior Survey, 1990*

	Female		Male		Total	
Category	**%**	**(95% CI†)**	**%**	**(95% CI)**	**%**	**(95% CI)**
Race/Ethnicity						
White	27.1	(±5.5)	22.5	(±5.1)	24.8	(±4.9)
Black	23.3	(±5.1)	20.7	(±9.4)	22.4	(±6.5)
Hispanic	25.4	(±5.5)	17.8	(±4.9)	21.8	(±4.1)
Grade						
9th	21.1	(±3.9)	19.7	(±5.1)	20.7	(±3.9)
10th	25.7	(±6.7)	22.5	(±4.9)	24.1	(±5.1)
11th	27.5	(±4.9)	22.5	(±5.3)	25.1	(±4.3)
12th	32.3	(±5.3)	23.1	(±6.3)	27.3	(±5.1)
Total	**26.4**	**(±4.3)**	**22.1**	**(±4.3)**	**24.3**	**(±3.9)**

*Unweighted sample size=11,631 students.
†Confidence interval.

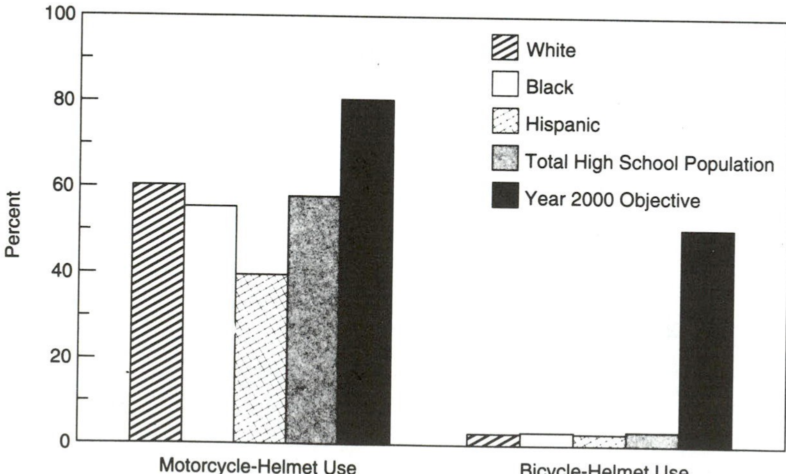

*Helmet use "most of the time" or "always" among students who rode motorcycles or bicycles.
†Unweighted sample size=11,631 students.

FIGURE.—Percentage of high school students reporting use of motorcycle or biclycle helmets,* by race/ethnicity—United States, Youth Risk Behavior Survey, 1990†

Reported by: Div of Injury Control, National Center for Environmental Health and Injury Control; Div of Adolescent and School Health, National Center for Chronic Disease Prevention and Health Promotion, CDC.

It used to be common practice in journal articles to report data in the form $\bar{x} \pm s$. More recently, articles began giving the somewhat more appropriate $\bar{x} \pm SE_{\bar{x}}$. (This form is more appropriate because the standard error is the correct measure of dispersion of sample means.) However, this formula tends to imply that the deviation is precisely one standard error. Actually, $\bar{x}$ can deviate by any number of standard errors from μ. Therefore, the current trend is to report a mean as a specific confidence interval. Literature Example 14.2 gives the data in something like the second form by giving the mean of the differences followed by the standard error in parentheses. (Note that the variable of interest, in this case, is the difference (Diff) of the two variables, BRFS and FCPS.)

Literature Example 14.2 is from the article entitled "The Behavioral Risk Factor Survey and the Stanford Five-City Project Survey: A Comparison of Cardiovascular Risk Behavior Estimates." Read the abstract to learn about the study.

Standard error of the mean,

$$SE = \frac{s}{\sqrt{n}}$$

where n = sample size, is reported for each mean difference in this article. When we compute confidence intervals around means or test hypotheses about means, the standard error of the mean is the appropriate standard deviation to use.

In this context SE is [1]appropriate/inappropriate.

A confidence interval is useful because it gives probable bounds on the population mean. Furthermore, using confidence intervals, we can test hypotheses directly without any further computations.

Give 95 percent confidence intervals for:
% obese BMI for all cities
$-11.2 \pm 1.96 * SE =$

([2] _______, _______)

% cholesterol for all cities
$.11 \pm 1.96 * SE =$

([3] _______, _______)

Write null hypotheses about BMI and cholesterol (in symbols).

[4] ____________________________

____

Hypotheses can be tested using confidence intervals. If the 95 percent confidence interval contains the hypothesized mean, for instance, then you can be 95 percent sure that the study or experiment has given no statistically significant evidence to reject the null hypothesis. Do you reject or accept the hypotheses?

BMI: $\mu_1 - \mu_2 = 0$

[5] Reject/Accept

Cholesterol: $\mu_1 - \mu_2 = 0$

[6] Reject/Accept

LITERATURE EXAMPLE 14.2

The Behavioral Risk Factor Survey and the Stanford Five-City Project Survey: A Comparison of Cardiovascular Risk Behavior Estimates

Christine Jackson, PhD, Darius E. Jatulis, MS, and Stephen P. Fortmann, MD

Background. Nearly all state health departments collect Behavioral Risk Factor Survey (BRFS) data, and many report using these data in public health planning. Although the BRFS is widely used, little is known about its measurement properties. This study compares the cardiovascular risk behavior estimates of the BRFS with estimates derived from the physiological and interview data of the Stanford Five-City Project Survey (FCPS).

Method. The BRFS is a random telephone sample of 1588 adults aged 25 to 64; the FCPS is a random household sample of 1512 adults aged 25 to 64. Both samples were drawn from the same four California communities.

Results. The surveys produced comparable estimates for measures of current smoking, number of cigarettes smoked per day, rate of ever being told one has high blood pressure, rate of prescription of blood pressure medications, compliance in taking medications, and mean total cholesterol. Significant differences were found for mean body mass index, rates of obesity, and, in particular, rate of controlled hypertension.

Conclusions. These differences indicate that, for some risk variables, the BRFS has limited utility in assessing public health needs and setting public health objectives. A formal validation study is needed to test all the risk behavior estimates measured by this widely used instrument. (*Am J Public Health.* 1992;82:412–416)

As shown in Table 5, the surveys produced comparable estimates of mean total cholesterol.

TABLE 5—Behavioral Risk Factor Survey (BRFS) and Five-City Project Survey (FCPS): Estimates for Body Mass Index (BMI) and Cholesterol

	BMI			% Obese BMI			Cholesterol mmol/L		
	BRFS (1559)	FCPS (1501)	Diff (SE)	BRFS (1559)	FCPS (1501)	Diff (SE)	BRFS (149)	FCPS (1470)	Diff (SE)
All	24.2	25.7***	−1.5 (.16)	14.1	25.3***	−11.2 (1.4)	5.18	5.07	.11 (.12)
City 1	23.8	25.4***	−1.6 (.31)	11.6	21.8***	−10.2 (2.6)	4.97	5.08	−.11 (.16)
City 2	24.8	25.8**	−1.0 (.32)	17.8	25.2*	−7.4 (3.0)	4.92	5.15	−.23 (.19)
City 3	24.6	26.4***	−1.8 (.33)	15.5	31.6***	−16.0 (3.0)	5.53	5.11	.42 (.32)
City 4	23.7	25.1***	−1.4 (.34)	11.2	22.6***	−11.4 (2.9)	5.28	4.91	.37 (.26)
All adjusted	24.3[a]	25.8***		13.7[b]	24.7***		—	—	

Note. SE = standard error.
[a]adjusted for sex.
[b]adjusted for sex and education.
*$P \le .05$, **$P \le .01$, ***$P \le .001$.

Testing Hypotheses

When we draw a sample to calculate sample statistics (such as $\bar{x}$), our sample may turn out to be typical or atypical. If $\bar{x}$ is from a typical sample, then it is close to the population parameter and hence is a good approximation of μ, whereas if $\bar{x}$ is from an atypical sample, $\bar{x}$ is far away from μ and hence is a poor approximation of μ. Although we never know whether a sample is typical or atypical, we can, at least, quantify the probability that we will err because the sample was atypical. When we estimate the mean using a 95 percent confidence interval, for instance, we are acknowledging that we expect to be wrong 5 percent of the time because of obtaining an atypical sample. In testing a scientific hypothesis using statistical testing procedures, we use a similar method to quantify the chance of sampling error when we choose to reject, for example, the null hypothesis and accept the alternative hypothesis. The best we can do is quantify the probability of error, because as we saw in Chapter 13, whether we accept or reject the hypothesis we may be wrong. The kinds of errors that we are talking about are not mistakes; they are unavoidable sampling errors that result from the possibility of choosing an atypical sample.

When we test a hypothesis, we may commit one of two types of errors shown in Figure 14.7: α error, or Type I error, the probability of rejecting a true hypothesis; or β error, or Type II error, the probability of accepting a false hypothesis. As the probability of Type I error decreases, the probability of Type II error increases, and vice versa. In most cases, a researcher is trying to show that there is a significant difference (between two means, for instance); that is, the researcher hopes to obtain statistically significant evidence that a null hypothesis is false.

A researcher can increase the probability of rejecting a null hypothesis by testing at a lower level (increasing α), but the cost of this is to make more likely the rejection of a true null hypothesis.

α is often set at .05, but there is nothing magic about this number. It has become traditional in some disciplines to use it, but one should consider the consequences of committing α or β error, decide which is more damaging, and set the level of α accordingly rather than sticking blindly with $\alpha = .05$.

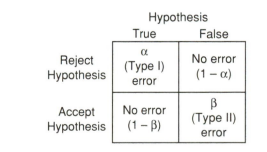

	Hypothesis	
	True	False
Reject Hypothesis	α (Type I) error	No error ($1 - \alpha$)
Accept Hypothesis	No error ($1 - \beta$)	β (Type II) error

FIGURE 14.7 Alpha (Type I) and Beta (Type II) Errors

Hypotheses are tested by comparing a sample statistic to the hypothesized value of the parameter using the distribution (t, F, or χ^2) that is appropriate for that parameter. Assuming the hypothesis is true, we use the appropriate distribution to estimate the probability of obtaining a value as "bad" as or "worse" than the statistic obtained from our sample. This probability, α, is the measure of how likely it would be to obtain a sample with the sample value given that the null hypothesis is true.

Suppose we are testing a hypothesis that some parameter (for instance, a mean) is zero. Would we reject the hypothesis if the sample statistic turned out to be non-zero but very small (0.0000001, for instance)? Probably not. In fact, it would be very rare for the sample statistic to be exactly the hypothesized value of the population parameter, even if the hypothesis were true. Hypothesis testing gives us a method for deciding rationally how much leeway to allow in the statistic's variation from the hypothesized parameter. We use a cut-off value such that if the statistic is enough more than that value it's very unlikely that the parameter is equal to zero.

One other consideration about hypothesis testing should be noted: It is possible to use a hypothesis test that is too powerful, i.e., one that detects a difference that is statistically significant but not of practical significance. Too much accuracy can actually be harmful! For example, a researcher might prove that the average score achieved by men on a certain test is different from those achieved by women, and that this difference is statistically significant. In fact, however, the difference might only amount to a few points (say one or two points on a 300-point test) and therefore be of no consequence whatsoever.

Two-Tailed and One-Tailed Tests

Consider Figure 14.8. When the rejection region includes both tails of the distribution, as shown in part (a), the test is called a two-tailed test. When the

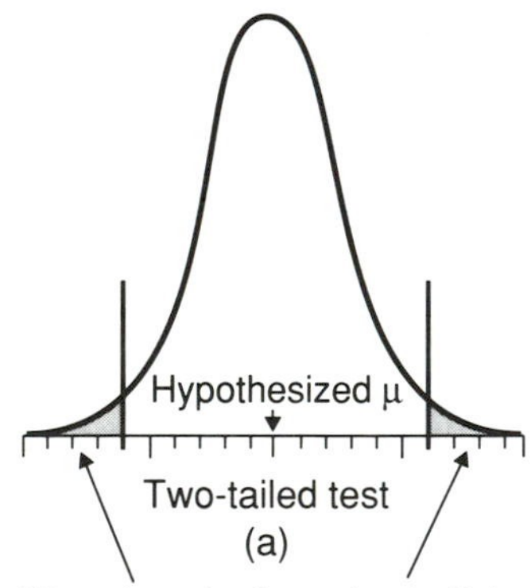

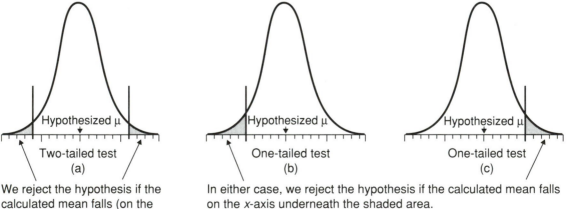

We reject the hypothesis if the calculated mean falls (on the x-axis) underneath either of the shaded areas.

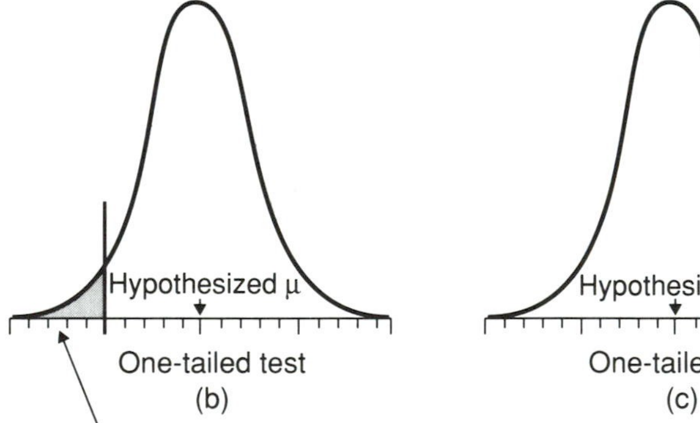

In either case, we reject the hypothesis if the calculated mean falls on the x-axis underneath the shaded area.

FIGURE 14.8

rejection region includes one tail only, as in parts (b) and (c), the test is called a one-tailed test.

Suppose you suspect that doing exercise will increase body fat. Your alternative hypothesis will then be $\mu_2 > \mu_1$, i.e., mean body fat after exercise will be greater than mean body fat before. What if, after exercise, the exercise group has *lost* a significant amount of body fat? In this case you certainly can't count this as evidence of the alternative hypothesis, so you must accept the null hypothesis. When you decided to use a one-sided test, you were effectively saying that the chance of seeing a decrease in body fat was small or of no interest to you. Some people express one-sided tests like those in Figure 14.8 in the form

$$H_0: \mu_2 \leq \mu_1$$
$$H_a: \mu_2 > \mu_1$$

including the unlikely or uninteresting case in the null hypothesis.

ACTIVITY 14.4

One- or two-tailed tests of hypotheses

Depending upon the interest of the researcher, she or he may write three types of alternative hypotheses. Suppose a researcher considering the faculty exercise data wants to develop a hypothesis about body fat. The null hypothesis will be that there is no difference in body fat before and after exercising. The researcher hopes to reject the null hypothesis in favor of an alternative hypothesis. The three forms for alternative hypotheses and the corresponding tests are listed below.

Alternative hypothesis	Type of test	Comment
There is a difference in the body fat (of the faculty who exercised) before and after the exercise.	Two-tailed test: You split α into two halves. If α is 5 percent, the shaded areas will have 2.5 percent each. See part (a) of Figure 14.8.	The alternative hypothesis states that there is a difference in body fat before and after exercising (i.e., the difference is not zero), so we reject the null hypothesis in favor of the alternative hypothesis if the difference is enough less than zero or enough more than zero. Therefore we set up the rejection region on [1] one/both side(s). Hence it is a [2] one/two tailed test.

After doing exercise, body fat will decrease.	One-tailed test: α will be located on the lower, or left, end of the normal curve; the shaded area is equal to α ($\alpha = 5$ percent). See part (b) of Figure 14.8.	The alternative hypothesis states that body fat will decrease after exercising, so we accept the null hypothesis if body fat increases or stays the same. We set the rejection region on the 3 lower/upper side. Therefore it is a 4 one/two tailed test.
After doing exercise, body fat will increase.	One-tailed test: α will be located on the upper, or right, end of the normal curve; the shaded area is equal to α ($\alpha = 5$ percent). See part (c) of Figure 14.8.	The alternative hypothesis states that body fat will increase, so we accept the null hypothesis if body fat decreases or stays the same. We set the rejection region on the 5 lower/upper side. Therefore it is a 6 one/two tailed test.

Three Methods of Testing Hypotheses

Method I: Confidence Interval Method

Use of Confidence Intervals for Testing Hypotheses about Means When we have a C.I., we know that it is highly likely that interval will contain μ, the population mean. Another way of viewing it is to say that any number outside the C.I. is not likely to be the value of μ. Figure 14.9 illustrates this idea. Note that according to this method we would have rejected the hypothesis that $\mu =$

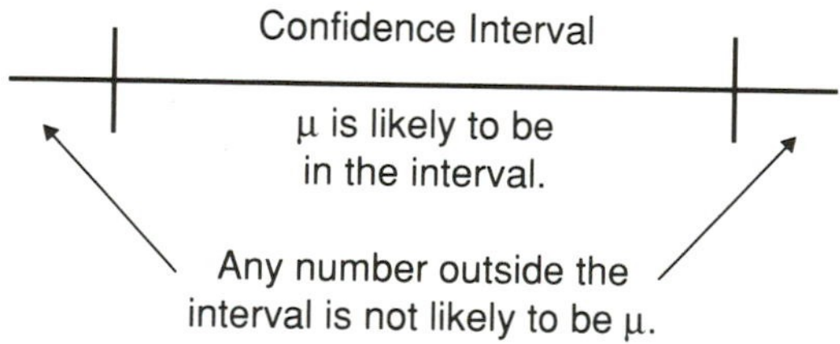

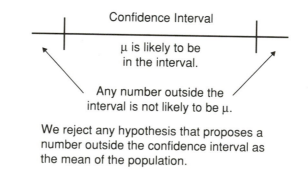

FIGURE 14.9

31, the true mean for PREVO2, for one of the forty samples listed in Table 14.1, for example, and hence would have been wrong. This method allows us to reject many possible values for μ instead of one hypothesized μ. Therefore some statisticians believe that this is the most practical way to test population values.

Use of Confidence Intervals for Testing Hypotheses about Proportions Hypotheses about the proportion (percentage) of a population having a certain property (for example, the proportion of "yes" responses to a question in a questionnaire) are tested in much the same way that hypotheses about means are tested, provided the sample size is large enough. (For most practical purposes, the sample size may be assumed to be large enough, provided that np and $n(1-p)$ are both 5 or larger. Methods for testing hypotheses when the sample size is not this large involve use of the Binomial Distribution; such methods are beyond the scope of this book.)

The confidence interval for a proportion is computed much like the confidence interval for the mean; however, the formula for the standard error of the proportion,

$$\mathrm{SE} = \sqrt{\frac{p(1-p)}{n}}$$

is used. Also, the normal distribution is used instead of the t-distribution for determining the factor that multiplies the standard error. The C.I. is the interval $(p - z * \mathrm{SE},\ p + z * \mathrm{SE})$, where p is the proportion of the sample having the property in question. Any hypothesis that postulates a proportion outside the confidence interval is rejected, and any hypothesis postulating a proportion inside the confidence interval is accepted.

Method II: Critical Means and Proportions Method

The set of all numbers that fall underneath the shaded regions in Figure 14.8 is called the **critical region**. To use this method, you compute the boundary points of the critical region, whether two-sided or one-sided (depending on the hypothesis) and reject the null hypothesis if the sample mean ($\bar{x}$) or the sample proportion falls in the critical region. The critical region is the set of numbers for which the null hypothesis is unlikely to be true.

Use of Critical Means and Proportions with a Two-Tailed Test The boundaries of the critical region for a two-tailed test are obtained by calculating the critical means:

$$\text{Critical means} = \mu_{\text{hyp}} \pm t * \mathrm{SE},$$

where μ_{hyp} is the hypothesized mean. The t-value is obtained from the t-table (Table 11.7) for the appropriate degrees of freedom. You reject the null hypothesis, $\mu = \mu_{\text{hyp}}$, if the calculated mean falls underneath either of the shaded areas shown in part (a) of Figure 14.8 (page 462). Note: The dark vertical bars are at $\mu_{\text{hyp}} \pm t * \mathrm{SE}$, the critical means.

Critical proportions are computed in much the same way, making the modifications described in the previous section. The boundaries of the critical region for a two-tailed test are obtained by calculating the critical proportions:

$$\text{Critical proportions} = p_{\text{hyp}} \pm z * \text{SE}$$

where p_{hyp} is the hypothesized proportion and

$$\text{SE} = \sqrt{\frac{p_{\text{hyp}}(1 - p_{\text{hyp}})}{n}}.$$

Here, the standard error is computed using the hypothesized value of p rather than the sample value as in the computation of a confidence interval, because we are assuming it is the true proportion. The z-value is obtained from the normal table (Table 11.4) for the appropriate value of $\alpha/2$. For $\alpha = .05$, you look up z in the normal table for left area = .025. The closest area in the table is .02275, which corresponds to $z = 2.0$.

In using the critical value methods, the critical values are computed *in advance* of taking any data. Tampering with the critical values based on the data, i.e., setting these values so as to obtain a desired result, is called "data snooping" and is dishonest, because it defeats the purpose of hypothesis testing.

Critical Means with a One-Tailed Test on the Lower End

$$\begin{aligned} H_0&: \mu = \mu_{\text{hyp}} \text{ or } \mu \geqslant \mu_{\text{hyp}} \\ H_a&: \mu < \mu_{\text{hyp}} \\ \text{Critical mean} &= \mu_{\text{hyp}} - t * \text{SE} \end{aligned}$$

where μ_{hyp} is the hypothesized mean. For a one-tailed test the entire critical region is on one side. (Since α is the probability that the sample statistic, in this case $\bar{x}$, will be in the critical region given that the null hypothesis is true, you do not divide α by two in a one-tailed test.) You look up t for α in the right-hand part of Table 11.7 if the critical region is to be on the lower end.* You reject the null hypothesis, $\mu = \mu_{\text{hyp}}$, if the calculated mean falls underneath the shaded area. Proportions are handled similarly, by finding z for α in the normal table.

Critical Means with a One-Tailed Test on the Upper End

$$\begin{aligned} H_0&: \mu = \mu_{\text{hyp}} \text{ or } \mu \leqslant \mu_{\text{hyp}} \\ H_a&: \mu > \mu_{\text{hyp}} \\ \text{Critical mean} &= \mu_{\text{hyp}} + t * \text{SE} \end{aligned}$$

where μ_{hyp} is the hypothesized mean. For a one-tailed test the entire critical region is on one side. You look up t for α in the left-hand part of Table 11.7 if the critical region is to be on the upper end. You reject the null hypothesis if the calculated mean falls underneath the shaded area. Proportions are handled similarly by looking up z for left area = α in the normal table.

---

* Note that Tables 11.7 and 11.8 only give t-values for $\alpha = .05$ and .01. For other significance levels you will need to find more complete tables or use a computer statistical package.

Method III: Critical *t*- and *z*-Values Method

To test a mean by this method, you find the critical t, whether two-sided or one-sided (depending on the hypothesis), and reject the hypothesis if the sample t falls in the critical region. For a proportion, you find the critical z and reject the hypothesis if the sample value falls in the critical region.

Critical *t*-Values Method with a Two-Tailed Test For a two-tailed test, if $\alpha = .05$ the critical t-values are obtained from Table 11.8. Next, the data are collected, and then t is calculated using the following formula:

$$t = \frac{\bar{x} - \mu_{\text{hyp}}}{\dfrac{s}{\sqrt{n}}}$$

Part (a) of Figure 14.10 illustrates this method.

For a proportion the critical z-values are found in the normal table (Table 11.4, page 371, or Table B-2, page 690) and the sample z is calculated using the formula

$$z = \frac{p - p_{\text{hyp}}}{\sqrt{\dfrac{p_{\text{hyp}}(1 - p_{\text{hyp}})}{n}}}$$

where p is the proportion in the sample having the property and p_{hyp} is the hypothesized proportion.

Critical *t*-Values Method with a One-Tailed Test on the Lower Tail For a one-tailed test the entire critical region is on one side. So we look up the critical t for α in the right-hand part of Table 11.8 if the critical region is to be on the

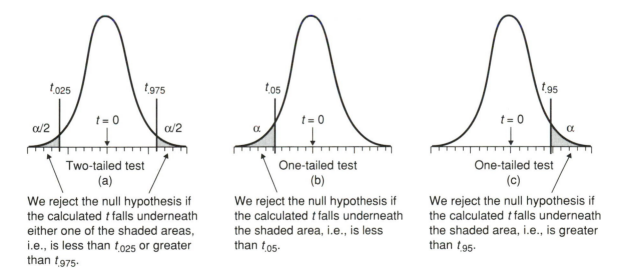

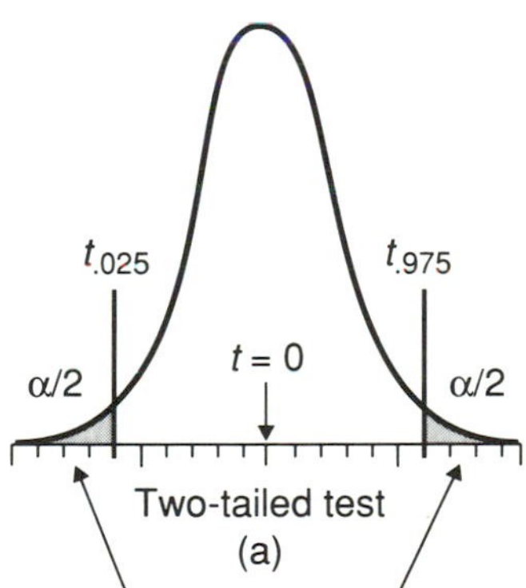

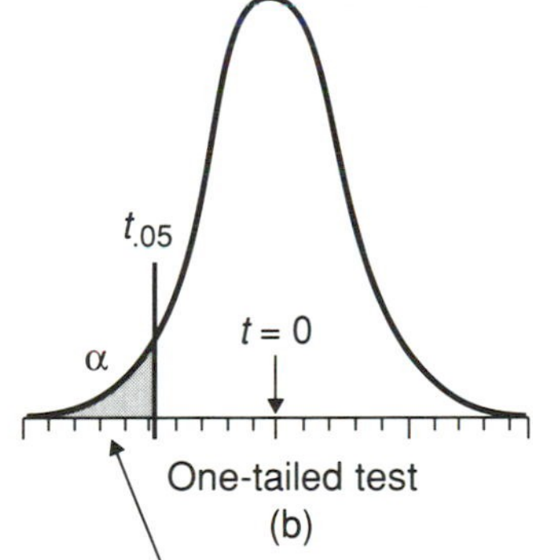

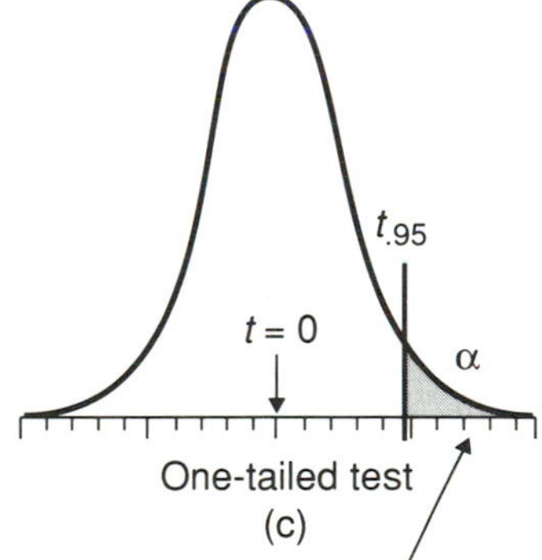

FIGURE 14.10

lower end. As before, after the data have been collected, t is calculated using the formula

$$t = \frac{\bar{x} - \mu_{\text{hyp}}}{\frac{s}{\sqrt{n}}}$$

Part (b) of Figure 14.10 illustrates this method. Proportions are tested in the same way as for two-tailed tests except that the critical z is the z-value from the normal table corresponding to α.

Critical t-Values Method with a One-Tailed Test on the Upper Tail For a one-tailed test with a critical region on the upper end, you look up the critical t for α in the right-hand part of Table 11.7. As before, t is calculated using the formula

$$t = \frac{\bar{x} - \mu_{\text{hyp}}}{\frac{s}{\sqrt{n}}}$$

See part (c) of Figure 14.10. Proportions are tested in the same way as for two-tailed tests, except that the critical z is the z-value from the normal table corresponding to right area $= \alpha$.

Designs and Analyses of Data

ACTIVITY 14.5

Testing a hypothesis about a hypothesized parameter (mean) using one sample

This activity will illustrate the use of each of the three testing methods discussed in the previous section by applying them to one of the research designs discussed in Chapter 9.

Recall the research design symbolized O_1 (see Chapter 9, page 326). This design is usually analyzed by giving summary statistics (mean, median, mode, standard deviation, range, quartiles, etc.) and by displaying the distribution in a table or graph. Results from such a study can also be compared to results from similar studies as reported in the literature. For example, you might compare results from your sample to a published study done with an extremely large sample to determine whether your sample is from a population that is significantly different from the large group with respect to some important studied characteristic.

For example, suppose that a medical study done on a random sample of several thousand middle-aged men found that their mean systolic blood pressure was 130. You wish to compare male faculty to see if their blood pressure is the same. Develop a hypothesis about systolic blood pressure about male faculty members. You will use the FED as your sample to test the hypothesis.

Write the null hypothesis in words and in symbols:

1 ____

____

Write the alternative hypothesis in words and in symbols:

____

____

Testing the Hypothesis: Method I, Confidence Interval Method

Sample size: $n = 26$
Sample mean: $\bar{x} = 124.7$
Sample st. dev.: $s = 14.6$
Std. error of mean: $SE_{\bar{x}} = 2.9$

Use a confidence level $(1 - \alpha)$ of 95 percent and calculate C.I. $= x \pm t * SE =$ [2](_______). Does the confidence interval contain the hypothesized mean, $\mu =$ 130, for systolic blood pressure? [3] Yes/No Do you accept or reject the null hypothesis? [4] Accept/Reject

Testing the Hypothesis: Method II, Critical Means Method (Two-Tailed Test) Find the critical means by calculating $\mu_{hyp} \pm t * SE$, where μ_{hyp} is the hypothesized mean. Critical means = [5] _______ and _______. Is the sample mean between the critical means or does it fall in the critical region? [6] Between means/In critical region Do you accept or reject the null hypothesis? [7] Accept/Reject

Testing the Hypothesis: Method III, Critical *t* Method (Two-Tailed Test) Find the critical t-values for $\alpha = .05$ and df = 25 in the right-hand part of Table 11.8 (page 377) or Table B-4 (page 692). The critical values are [8] _______ and _______. Now calculate the t-value for the sample mean using the formula

$$t = \frac{\bar{x} - \mu_{hyp}}{\frac{s}{\sqrt{n}}}$$

The calculated $t =$ [9] _______. Is the calculated t between the critical t-values or does it fall in the critical region? [10] Between critical t-values/In critical region Do you accept or reject the null hypothesis? [11] Accept/Reject

You will often see statistical significance reported in terms of a P-value, a concept similar to significance level, α. The P-value is the probability of getting a value of the statistic (in this case, t) as extreme or more extreme than the value obtained from the sample, assuming that the null hypothesis is true. You will see P-values reported in some of the Literature Examples given in this book, although sometimes they are labeled with a lowercase p. In this chapter we use uppercase P for P-values to distinguish this concept from proportions, for which we use lowercase p.

Tables such as Table 11.8 are not detailed enough to enable you to find P-values; however, hypotheses can be tested using computer programs, which do give P-values.

Both the *P*-value and α are graphically represented by areas in the tail of the distribution. The *P*-value gives more information than α, however, because the *P*-value is the smallest significance level at which the data are significant. For example, if α was set at .05 but the *P*-value turned out to be .0004, we would reject the hypothesis. With this *P*-value we would also reject the hypothesis at the .005, the .001, or even the .0005 levels—in fact, at any level down to .0004. Thus, a *P*-value gives more information than a statement such as, "The hypothesis was rejected at the .05 level of significance."

For our sample of systolic blood pressures of FED participants, we found $t = 1.82$. Our computer program tells us that the *P*-value for this t is .076. This is the probability of getting a sample with $t < -1.82$ or $t > +1.82$. We chose α to be .05. Since .076 is greater than .05, we [12] accept/reject the null hypothesis.

Examples from the Literature

The articles that follow all tested hypotheses about proportions (percentages of a population that have a certain property) rather than hypotheses about means. Methods of testing such hypotheses were discussed earlier. Literature Example 14.3 is from the article entitled "Patient, Provider and Hospital Characteristics Associated with Inappropriate Hospitalization." Read the abstract to learn about the study.

Suppose we formulate the following general hypothesis: There is no difference among the standardized percentages in each category.

Write a specific hypothesis for the variable Years from Licensing.

[1] ________________

____

____

If the confidence interval covers 0, we accept the hypothesis.

We [2] reject/accept the hypothesis about the Years from Licensing variable.

Check all the confidence intervals in the third column of Table 2 to see which C.I. do not cover 0.

The confidence interval for these variables do not cover 0:

[3] ________________

Does the C.I. related to Board Certified cover 0?

[4] Yes/No

Read the authors' conclusion. Is your conclusion the same?

[5] Yes/No

LITERATURE EXAMPLE 14.3

Patient, Provider and Hospital Characteristics Associated with Inappropriate Hospitalization

ALBERT L. SIU, MD, MSPH, WILLARD G. MANNING, PHD, AND BERNADETTE BENJAMIN, BS

To determine the relation between patient and provider characteristics and inappropriate hospital use, we examined adult nonpregnancy hospitalizations from a randomized trial of health insurance conducted in six sites in the United States. Appropriateness of inpatient treatment was based on medical record review; patient characteristics on sociodemographic, economic, and health status; and provider characteristics on descriptors of physician practice and hospital facilities.

Twenty-seven percent of admissions attended by physicians licensed for more than 15 years were judged inappropriate, compared to 20 percent for younger physicians. Admissions were more likely to be inappropriate if the patient was female (27 percent compared with 18 percent). Controlling for patient and provider characteristics reduces but does not eliminate the differences in the appropriateness of inpatient care across the study's six sites. Differences in available provider and patient characteristics do not account for geographic differences in inappropriate hospitalization in this study. (*Am J Public Health* 1990; 80:1253–1256.)

TABLE 2—Inappropriate Days by Provider and Patient Characteristics

Characteristics	Standardized Percentage	Difference (95% CI) versus Initial Category
Specialty		
General/family	39.7	—
Internal Medicine	36.6	−3.1 (−14.9, 8.7)
Surgery	30.0	−9.7 (−20.3, 1.0)
Obstetrics/gynecology	33.3	−6.4 (−19.8, 7.1)
Solo Practice		
Yes	34.4	—
No	32.4	−2.0 (−10.6, 6.5)
Board-Certified		
Yes	31.6	—
No	39.2	7.6 (−1.0, 16.3)
Years from Licensing		
≥15	39.0	—
<15	29.9	−9.1 (−17.0, −1.2)
JCAH Accredited		
Yes	33.7	—
No	41.2	7.5 (−6.6, 21.6)
Teaching Status		
Yes	36.5	—
No	33.9	−2.6 (−16.3, 11.0)
Public Hospital		
Yes	30.3	—
No	34.9	4.6 (−10.9, 20.0)
Sex		
Male	23.5	—
Female	42.0	18.5 (10.9, 26.1)

For unstandardized comparisons, the likelihood of a hospital day being inappropriate is increased if the physician was not board certified or was licensed more than 15 years ago, if the hospital was nonpublic, and if the patient was female. The standardized estimates (Table 2) are similar to those for inappropriate admissions although years from licensing and patient gender are more strongly associated and physician specialty less strongly associated.

Literature Example 14.4 is from the article entitled "Crop Duster Aviation Mechanics: High Risk for Pesticide Poisoning." Read the abstract to learn about the study.

Write a null hypothesis regarding the variable Poisoned.	1 ______________________________
Test the hypothesis using the given confidence interval.	Do you 2 reject/accept the null hypothesis?
Which variables have confidence intervals containing zero?	3 ______________________________
Read the authors' conclusions. In light of the confidence intervals they furnished, should you agree with their interpretation, in general?	4 Yes/No

Consider the last two sentences in the abstract. Research can be a powerful tool to influence governmental policies. Can you think of another instance of such action by a governmental agency to protect workers?

LITERATURE EXAMPLE 14.4

Crop Duster Aviation Mechanics: High Risk for Pesticide Poisoning

Rob McConnell, MD, A. Feliciano Pacheco Antón, MD, and Ralph Magnotti, PhD

A cross-sectional medical survey was conducted among 63 Nicaraguan aviation mechanics exposed to organophosphate and other toxic pesticides. Thirty-one (49 percent) reported having been acutely poisoned on the job. Also, seven of 14 novice mechanics, with less than one year on the job, reported that they had been poisoned. Thirty-eight (61 percent) had cholinesterase levels below the lower limit of normal, including three workers with levels less than 20 percent of the lower limit of normal. Risk factors for low cholinesterase included recent hire and recent poisoning. Workers did not use protective equipment, nor were there facilities for bathing on site. As a result of this survey, the government has prohibited the mixing and loading of pesticides at this airport and requires the washing of planes prior to maintenance work; coveralls and thin, pesticide impermeable gloves are to be issued to mechanics handling pesticide-contaminated parts. Closed system mixing and loading systems have been installed at satellite airstrips. (*Am J Public Health* 1990; 80:1236–1239.)

TABLE 2—Mean Cholinesterase and Frequency of Low Cholinesterase among Aviation Mechanics by Selected Risk Factors, Chinandega, Nicaragua, December 1987

Risk Factor	Number with Low Cholinesterase (%)	Relative Risk (95% CI)	Mean Cholinesterase	Difference of Means (95% CI)
Poisoned in the previous year (N – 12)	11 (92)	1.7 (1.1, 2.6)	1.5 IU	−2.0 (−1.3, −2.7)
Not poisoned in the previous year (N = 50)	27 (54)		3.5 IU	
Less than 1 year in job (N = 14)	13 (93)	1.8 (1.2, 2.7)	2.0 IU	−1.4 (−0.7, −2.1)
One or more year in job (N = 48)	25 (52)		3.4 IU	
Has not received pesticide safety training at airport (N = 45)	30 (67)	1.5 (0.91, 2.5)	2.9 IU	−0.7 (−0.07, −1.3)
Has received training (N = 16)	7 (44)		3.6 IU	
Changes work clothes less often than daily (N = 47)	25 (53)	0.61 (0.40, 0.93)	3.2 IU	0.6 (1.3, −0.1)
Changes work clothes daily (N = 15)	13 (87)		2.6 IU	
More than 2 symptoms compatible with cholinesterase inhibitors (N = 23)	12 (52)	0.78 (0.50, 1.2)	3.0 IU	−0.1 (0.6, −0.8)
Two or fewer symptoms compatible with cholinesterase inhibitors (N = 38)	26 (67)		3.1 IU	

Cholinesterase Measurements

We evaluated the following risk factors for depressed cholinesterase levels among these workers: recent poisoning, recent hire, no training about safe use of pesticides, less than daily change of clean work clothes, and symptoms compatible with organophosphate insecticide poisoning. In addition, the mean cholinesterase for workers with each risk factor was compared with the mean for all workers without the risk. Almost all workers poisoned in the previous year and almost all mechanics hired within the previous year had low cholinesterase values at the time of the screening (see Table 2). The presence or absence of more than two symptoms compatible with organophosphate insecticide poisoning was not reflected either in higher prevalence of low cholinesterase values (52 percent compared with 67 percent among those with two symptoms or less) or in lower mean cholinesterase levels.

Paradoxically, there was a higher prevalence of low cholinesterase among mechanics who reported a daily change of clean work clothes (87 percent) compared with mechanics who changed work clothes less often (53 percent).

There was a slightly lower prevalence of low cholinesterase levels among workers who had received some previous training about the safe handling of pesticides, and the mean cholinesterase level among workers who had received training (3.6 IU) was higher than among workers without any training (2.9 IU). However, this beneficial effect of training was confounded by the high exposure tasks of recently hired workers, who were also less likely to have received training. After controlling for a dummy variable representing recent hire (within the last year) in a multiple regression procedure, there was still a slight beneficial average effect on cholinesterase of having received training, but the effect was reduced (regression coefficient = 0.41 I.U.; 95% CI = −0.28, 1.13).

Literature Example 14.5 is from the article entitled "Gender Differences in Cigarette Smoking and Quitting in a Cohort of Young Adults." Read the abstract to learn about the study.

Are there gender differences in the barriers and pressures? To find out, check Table 4, which gives 95 percent confidence intervals. For each statement check to see if the confidence intervals for men and women overlap. Checking to see whether confidence intervals overlap is not an accurate method for testing the hypothesis that two parameters are equal. It should only be used as a rough indication. Usually, however, when the confidence intervals do not overlap, there will be statistically significant evidence that the two groups are different, i.e., evidence to reject the null hypothesis, which says they are the same. Even when the intervals overlap, it is often possible to reject the null hypothesis. In this exercise, however, we will accept the null hypothesis when the intervals overlap and reject it when they don't. For example, for the statement "If I quit smoking, I would probably gain a lot of weight," the C.I.s do not overlap. Therefore, we reject the null hypothesis that there is no difference between the percentages of men and women who agree with the statement.

"If I quit smoking, I would feel tense and irritable."	The C.I.s [1] overlap/do not overlap; therefore we [2] reject/accept the null hypothesis.
"I enjoy smoking too much to quit."	The C.I.s [3] overlap/do not overlap; therefore we [4] reject/accept the null hypothesis.
"If I quit smoking, it would be hard to go out with friends who smoke."	The C.I.s [5] overlap/do not overlap; therefore we [6] reject/accept the null hypothesis.
"If I got tense and irritable from not smoking, it would make the people I'm close to unhappy."	The C.I.s [7] overlap/do not overlap; therefore we [8] reject/accept the null hypothesis.
"A doctor has strongly urged me to quit smoking."	The C.I.s [9] overlap/do not overlap; therefore we [10] reject/accept the null hypothesis.
"The people I am closest to would like me to quit smoking."	The C.I.s [11] overlap/do not overlap; therefore we [12] reject/accept the null hypothesis.

LITERATURE EXAMPLE 14.5

Gender Differences in Cigarette Smoking and Quitting in a Cohort of Young Adults

Phyllis L. Pirie, PhD, David M. Murray, PhD, and Russell V. Luepker, MD

Background. Smoking among young women is associated with a variety of negative health outcomes. Gender specific influences on smoking, quitting and attempting to quit are hypothesized to occur and may have implications for cessation programs.

Methods. Telephone surveys were conducted in a large (n = 6,711) cohort of young men and women (average age 19.2 years) which was first established in 1979 and has been resurveyed several times since then. Questions concerned smoking, successful and unsuccessful attempts to quit, withdrawal symptoms during quit attempts, and concerns about quitting.

Results. More women than men reported current smoking (26.5 vs 22.6 percent), but quitting attempts, successful and unsuccessful, were equally common. Withdrawal symptoms were reported equally, except for wanting to eat more than usual and weight gain, both of which were reported more often by women than men. Women smokers reported substantially more concern about weight gain if they quit smoking (57.9 vs 26.3 percent expressing concern).

Conclusions. Targeted programs are needed to address issues of concern to young women smokers, particularly fear of gaining weight. (*Am J Public Health* 1991;81:324–327)

TABLE 4—Percent Agreement with Statements Reflecting Barriers and Pressures to Quit Smoking Reported by Young Adult Current Smokers, by Gender

Barriers/Pressures	Men (95% CI) n = 779	Women (95% CI) n = 863
Barriers to Quitting—%		
If I quit smoking, I would probably gain a lot of weight	26.3 (23.2, 29.4)*	57.9 (54.6, 61.2)
If I quit smoking, I would feel tense and irritable	62.1 (58.7, 65.5)	67.7 (64.6, 70.8)
I enjoy smoking too much to quit	50.4 (46.9, 53.9)	50.9 (45.6, 54.2)
If I quit smoking, it would be hard to go out with friends who smoke	32.7 (29.4, 36.0)	46.5 (43.2, 49.8)
If I got tense and irritable from not smoking, it would make the people I'm closest to unhappy	54.7 (51.2, 58.2)	59.2 (55.9, 62.5)
Pressures to Quit—%		
A doctor has strongly urged me to quit smoking	20.2 (17.4, 23.0)	31.5 (28.4, 34.6)
The people I'm closest to would like me to quit smoking	63.5 (60.1, 66.9)	70.6 (67.6, 73.6)

*95% confidence interval.

Simple proportions and their 95 percent confidence intervals are reported for categorical variables, and means and their confidence intervals for continuous variables. All analyses were conducted using SAS.[19]

Results

Finally, we examined various beliefs about the social and physical aspects of quitting smoking which may act as barriers to quitting, as well as potential social pressures to quit (Table 4). Women were more likely to agree with two of the "barriers" items: "If I quit smoking, I would probably gain a lot of weight" and, "If I quit smoking, it would be hard to go out with friends who smoke." Women were also more likely to agree that they felt social pressure to quit, reporting that, "A doctor has strongly urged me to quit smoking" and, "The people I'm closest to would like me to quit smoking."

Literature Example 14.6 is from the article entitled "Brand Logo Recognition by Children Aged 3 to 6 Years: Mickey Mouse and Old Joe the Camel." Read the abstract to learn the purpose of the study.

Notice the black bars that correspond to recognition of Old Joe the Camel in the figure.

Percentage of recognition of Old Joe the Camel [1] does/does not increase with the child's age.

Suppose we formulate the following hypothesis: The recognition rate for the Disney Channel logo and Old Joe logo is the same for 3-, 4-, and 5-year-olds. We stipulate $\alpha = .05$. Notice the P-values given on the top of the bars in the figure. On the basis of these probabilities, do you reject or accept the hypothesis?

[2] Reject/Accept

What type of sampling procedure did the investigators use?

3 ______________________________

How would you take your sample if you were to repeat this study?

4 ______________________________

____

____

____

LITERATURE EXAMPLE 14.6

Brand Logo Recognition by Children Aged 3 to 6 Years: Mickey Mouse and Old Joe the Camel

Paul M. Fischer, MD; Meyer P. Schwartz, MD; John W. Richards, Jr, MD; Adam O. Goldstein, MD; Tina H. Rojas

Objective

Little is known about the influence of advertising on very young children. We, therefore, measured product logo recognition by subjects aged 3 to 6 years.

Design

Children were instructed to match logos with one of 12 products pictured on a game board. Twenty-two logos were tested, including those representing children's products, adult products, and those for two popular cigarette brands (Camel and Marlboro).

Setting

Preschools in Augusta and Atlanta, Ga.

Participants

A convenience sample of 229 children attending preschool.

Results

The children demonstrated high rates of logo recognition. When analyzed by product category, the level of recognition of cigarette logos was intermediate between children's and adult products. The recognition rates of The Disney Channel logo and Old Joe (the cartoon character promoting Camel cigarettes) were highest in their respective product categories. Recognition rates increased with age. Appproximately 30% of 3-year-old children correctly matched Old Joe with a picture of a cigarette compared with 91.3% of 6-year-old children.

Conclusion

Very young children see, understand and remember advertising. Given the serious health consequences of smoking, the exposure of children to environmental tobacco advertising may represent an important health risk and should be studied further. (*J.A.M.A.* 1991;266:3145–3148)

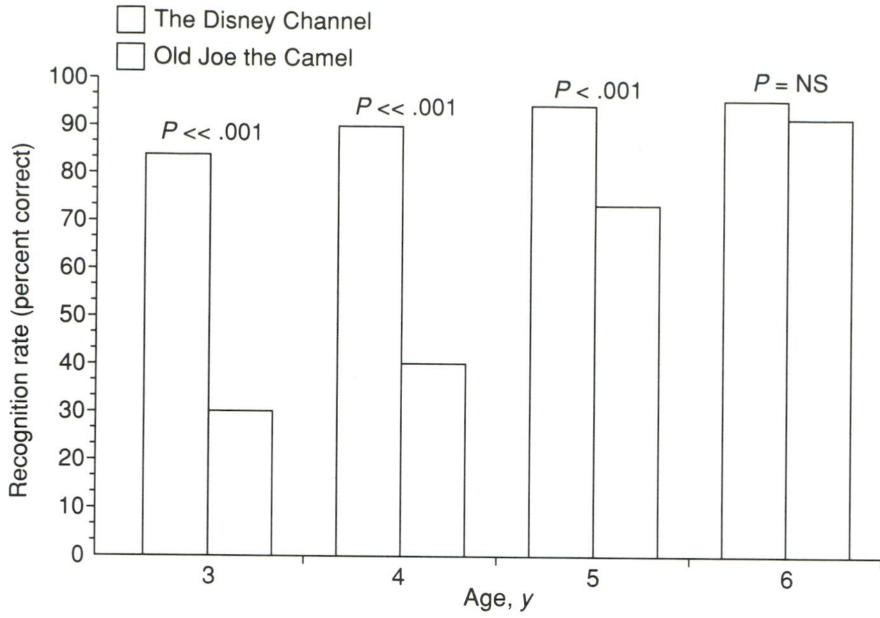

Logo Recognition Rates for the Disney Channel and Old Joe the Camel by Subject Age

ACTIVITY 14.6

Testing a hypothesis about pre-test and post-test (dependent) data

In Chapter 9 (page 327) we discussed the design $O_1 \quad X \quad O_2$. Recall that because there is no control group this design leaves some doubt as to whether an observed change in the response variable was caused by the intervention. Nevertheless, this design may be used for testing hypotheses about a change in the response variable. (Sometimes it's just not possible to have a control group.)

As an example, we will consider the hypothesis that low-intensity exercise affects heart rate, and we will use the low-intensity group (PARTICIP = 3) from FED as the sample. We will reconsider this example later in this chapter when we study comparing data from an experimental group to a control group. For now, we will create an artificial example by stipulating that there was no control group. (In fact, there was one.)

The hypothesis might be $\mu_1 = \mu_2$, or $\mu_1 - \mu_2 = 0$. Another way of expressing this hypothesis would be to say that the mean of the variable $x_1 - x_2$, denoted $\mu_{x_1 - x_2}$, equals zero. These three statements are algebraically equivalent, but the first, $\mu_1 = \mu_2$, is generally used to suggest that μ_1 and μ_2 are independent of each other. It is necessary to make this distinction because standard error is computed differently when two samples are drawn independently. In this study, pre- and post-test data are taken from the same group of people, so the pre and post observations are dependent observations. Post-test heart rate depends not only on the effect of exercise but also on the pre-test heart rate. (The two observations are called paired data.) Therefore, in this study it is appropriate to subtract the pre observation from the post observation for each subject and compute the mean of the differences. We are testing the hypothesis that the mean of the differences is zero. Below you will do so using each of the three testing methods discussed earlier in the chapter.

Testing the Hypothesis: Method I, Confidence Interval Method We denote the low-intensity exercise pre- and post-test differences by $d = x_1 - x_2$.

Sample size: $n = 8$
Sample mean difference: $\bar{d} = -8.75$
St. dev. differences: $s = 10.61$
St. error mean difference: $SE_{\bar{d}} = 3.75$

Use a confidence interval $(1 - \alpha)$ of 95 percent and calculate C.I. $= \bar{d} \pm t * SE_{\bar{d}} =$ [1] (_______). Does the interval contain 0, the hypothesized value of $\mu_{x_1 - x_2}$? [2] Yes/No Do you reject or accept the null hypothesis? [3] Accept/Reject

Testing the Hypothesis: Method II, Critical Means Method (Two-Tailed Test) Find the critical means by calculating the hypothesized mean μ_{hyp}, using the formula $\mu_{\text{hyp}} \pm t * SE_{\bar{d}}$. For $\alpha = .05$ and df = 7, $t =$ [4] _______. The critical means are [5] _______ and _______. Is the calculated mean between the critical means or does it fall in the critical region? [6] Between critical means/In critical region Do you accept or reject the null hypothesis? [7] Accept/Reject

Testing the Hypothesis: Method III, Critical *t* Method (Two-Tailed Test) Find the critical *t*-values for $\alpha = .05$ and df = 7 in the right-hand part of Table 11.8. The critical *t*-values are [8] _______ and _______. The calculated *t* comes from the formula

$$t = \frac{\bar{x} - \mu_{\text{hyp}}}{\frac{s}{\sqrt{n}}}$$

with $\mu_{\text{hyp}} = 0$. The calculated $t =$ [9] _______. Is the calculated *t* between the critical *t*-values or does it fall in the critical region? [10] Between *t*-values/In critical region The *P*-value for $t = -2.33$ is .052. Given that we chose $\alpha = .05$, do you accept or reject the null hypothesis? [11] Accept/Reject

Examples from the Literature

Literature Example 14.7 is from the article entitled "Influence of High Voltage Pulsed Direct Current on Edema Formation Following Impact Injury." Read the abstract to learn about the study.

Write null hypotheses regarding the effect of treatment.

During treatments:

1 ______________________________

____

____

Between treatments:

2 ______________________________

____

____

Look up critical t for df = 19 and $\alpha = .05$.

3 ______________________________

Based on the t-values given in the table for treated limbs, do you reject or accept the null hypothesis for each item?

During treatments: 4 Reject/Accept the hypothesis
Between treatments: 5 Reject/Accept the hypothesis

Based on the p-values given in the table do you reject or accept each null hypothesis for untreated limbs?

During treatment intervals:
6 Reject/Accept
Between treatment intervals:
7 Reject/Accept

Read the last two sentences of the results. Can we accept the authors' conclusions?

8 Yes/No

Look at the figure. Vertical lines depict standard errors (SE).

Treated limbs have consistently 9 lower/higher volume than untreated limbs over a period of 17 hours.

On the basis of this experiment would you recommend HVPC to a friend, should the need for treatment of edema arise? What are your reasons?

10 Yes/No

LITERATURE EXAMPLE 14.7

Influence of High Voltage Pulsed Direct Current on Edema Formation Following Impact Injury

Josette A Bettany
Dale R Fish
Frank C Mendel

Edema results in pain and may lead to reduced functional mobility. High voltage pulsed direct current (HVPC) has recently been advocated for edema control. The purpose of our study was to determine the effect of HVPC on edema formation in frogs. Hind limbs of 20 anesthetized frogs were injured by dropping a 450-g weight onto the plantar aspects of the feet. One hind limb of each frog was randomly selected to receive continuous 120-Hz HVPC at voltages 10% lower than those needed to evoke muscle contraction. Four 30-minute treatments were administered at 1.5-hour intervals beginning 10 minutes after trauma. Limb volumes were measured by water displacement. An analysis of variance for repeated measures and a Newman-Keuls post hoc *test were used to determine the significance of treatment effects. The HVPC significantly* ($p < .01$) *reduced edema formation. We hypothesize that HVPC may also be effective in controlling edema formation after impact injuries in humans. [Bettany JA, Fish DR, Mendel FC: Influence of high voltage pulsed direct current on edema formation following impact injury. Phys Ther 70:219–224, 1990]*

To test the hypothesis that treatment effects would be achieved only during the administration of HVPC, we used Student's paired t tests to assess differences ($p < .05$) between treated and untreated limb volume changes during and between treatment periods.

Table 2. *Results of Student's Paired* t *tests for Changes in Limb Volumes During and Between Treatment Intervals*

	df	x-y[a]	*t*	*p*[b]
Treated limbs				
During treatments	79	1.04	1.45	.153
Between treatments	59	1.67	2.24	.029
Untreated limbs				
During treatment intervals	79	3.10	6.36	.001
Between treatment intervals	59	0.85	1.76	.042

[a] x-y represents mean change in limb volume (in milliliters per kilogram) from beginning to end of associated time interval.

[b] $p < .05$.

Results

High voltage pulsed direct current significantly reduced edema formation in treated limbs as compared with untreated limbs (Tab. 1, Fig. 2). The Scheffé *post hoc* analysis revealed that volumes of treated limbs were significantly less ($p < .01$) than volumes of untreated limbs from the end of the first treatment to the last volume measurement at 17 hours posttrauma. Table 2 shows that mean volumes of treated limbs did not increase significantly during HVPC administration, but did increase significantly between treatments. Mean volumes of untreated limbs increased significantly during time intervals corresponding to treatment and rest periods (Tab. 2).

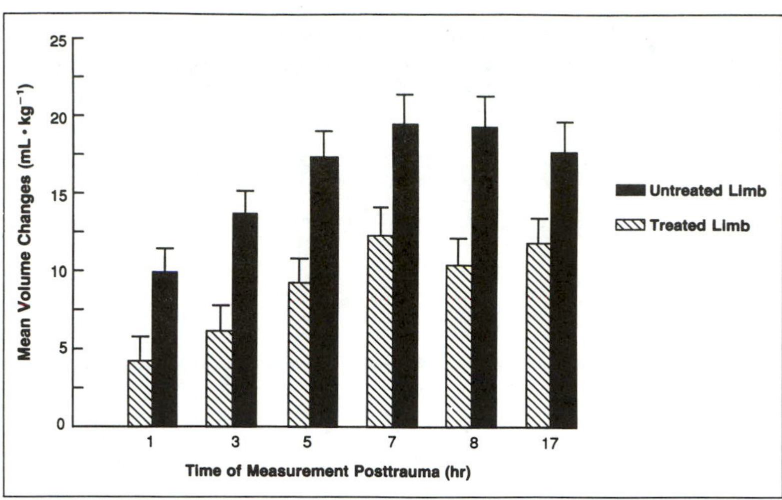

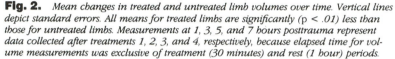

Fig. 2. *Mean changes in treated and untreated limb volumes over time. Vertical lines depict standard errors. All means for treated limbs are significantly* ($p < .01$) *less than those for untreated limbs. Measurements at 1, 3, 5, and 7 hours posttrauma represent data collected after treatments 1, 2, 3, and 4, respectively, because elapsed time for volume measurements was exclusive of treatment (30 minutes) and rest (1 hour) periods.*

Literature Example 14.8 is from the article entitled "Calcium supplementation on normotensive and hypertensive pregnant women." Read the abstract to learn about the study.

The article notes that "Paired difference t tests were used to determine whether changes between beginning and ending blood pressure measurements and calcium values were significant." This means that the authors subtracted the beginning reading from the ending reading for each person and used the differences as the variable. They calculated the t-value for this variable. This is the test we have been considering.

The paired difference t-test is used when the observations are [1] dependent/independent.

A difference of −0.02 was found to be statistically significant, i.e., it's unlikely to have gotten so small a difference if the hypothesis that the difference is zero is true. Does this mean that this difference is necessarily significant from the medical point of view?

[2] Yes/No

The table gives beginning and ending means of each variable measured and the standard deviations of the observations. Is this enough information to compute the t-values and hence determine the actual p-values?

[3] Yes/No

LITERATURE EXAMPLE 14.8

Calcium supplementation on normotensive and hypertensive pregnant women

Kathy B. Knight and Robert E. Keith

Abstract

Normotensive and hypertensive pregnant women participated in a study to determine the effects of calcium supplementation on blood pressure. Subjects were randomly assigned to control or supplemented groups (1000 mg Ca/d). Blood pressure and serum total and ionic calcium were measured during the 20-wk supplementation period. Calcium supplementation had a significant lowering effect on diastolic blood pressure over the course of the study in the hypertensive group only. The hypertensive control subjects' mean serum ionic calcium value decreased significantly ($P < 0.05$) over the course of the experiment. A significant ($P < 0.05$) inverse relationship was observed between dietary calcium intake and blood pressure ($r = -0.386$ for systolic pressure and -0.359 for diastolic pressure). *Am J Clin Nutr* 1992;55:891–5.

Statistical analysis

Means and standard deviations of initial and final total serum calcium and ionic calcium concentrations, systolic and diastolic blood pressure measurements, and dietary calcium were calculated. The change in these indices over time was also calculated. *T* tests were performed to determine significant differences between hypertensive control subjects and supplemented subjects and between normotensive control subjects and supplemented subjects. Paired difference *t* tests were used to determine whether changes between beginning and ending blood pressure measurement and calcium values were significant. Correlation coefficients were used to examine possible relationships between total serum and ionic calcium and blood pressure as well as calcium intake vs initial blood pressure. All values were considered significant at $P < 0.05$ (20).

Serum calcium

As seen in Table 4, subjects randomly assigned to the normotensive supplemented group had a slightly higher mean initial total serum calcium concentration than did the normotensive control group. No other significant differences in mean total serum calcium and serum ionic calcium concentrations were found.

However, when differences between beginning and ending ionic and total serum calcium were calculated, hypertensive control subjects had a significant ($P < 0.05$) decrease in ionic calcium. No significant changes in serum or ionic calcium were seen in the normotensive or hypertensive supplemented groups. No significant correlations between total and ionic serum calcium and blood pressure were found.

TABLE 4 Ionic and Total Serum Calcium for Normotensive and Hypertensive Control and Calcium-Supplemented Pregnant Women*

Group	Beginning (12 wk)	Ending (32 wk)	Difference†
		mmol/L	
Normotensive control (n = 15)			
Ionic calcium	1.02 ± 0.13	1.01 ± 0.10	−0.01
Total serum calcium	2.40 ± 0.02	2.40 ± 0.02	0.00
Normotensive supplemented (n = 15)			
Ionic calcium	1.01 ± 0.10	1.03 ± 0.08	0.02
Total serum calcium	2.42 ± 0.03‡	2.41 ± 0.03	−0.01
Hypertensive control (n = 10)			
Ionic calcium	0.99 ± 0.11	0.97 ± 0.12	−0.02§
Total serum calcium	2.43 ± 0.02	2.43 ± 0.02	0.00
Hypertensive supplemented (n = 10)			
Ionic calcium	1.01 ± 0.05	1.00 ± 0.05	−0.01
Total serum calcium	2.42 ± 0.03	2.42 ± 0.02	0.00

* $\bar{x}$ ± SD.
† Ending value minus beginning value.
‡ Significantly different from normotensive control serum calcium, $P < 0.05$.
§ Significant difference between beginning and ending values, $P < 0.05$.

Literature Example 14.9 is from the article entitled "Concordance for Dyslipidemic Hypertension in Male Twins." Read the abstract to learn about the study.

Look at the bars in the figure. Which pairs look different?

[1] First Examination/Military Induction/ Adult Gain in Body-Mass Index

The authors indicate $p = .01$ for two of the pairs. Assume that the hypothesis is that twins with and without dyslipidemic hypertension have the same mean BMI. Also assume $\alpha = .05$. For each variable, indicate whether you would accept or reject this hypothesis.

First Examination: [2] Reject/Accept the hypothesis

Military Induction: [3] Reject/Accept the hypothesis

Adult Gain in Body-Mass Index: [4] Reject/Accept the hypothesis

State the results of the research in your own words.

[5] ________________________________

____

____

LITERATURE EXAMPLE 14.9

Concordance for Dyslipidemic Hypertension in Male Twins

Joseph V. Selby, MD; Beth Newman, PhD; Jose Quiroga, MD; Joe C. Christian, MD, PhD; Melissa A. Austin, PhD; Richard R. Fabsitz, MA

Sixty cases of dyslipidemic hypertension were identified in the 1028 middle-aged, white, male twin participants in the first examination of the National Heart, Lung, and Blood Institute Twin Study (1969 to 1973). The prevalence of dyslipidemic hypertension was similar by zygosity but proband concordance was three times greater in monozygotic than dizygotic twins (0.44 [seven concordant and 18 discordant pairs] vs 0.14 [two concordant and 24 discordant pairs]), suggesting a genetic effect on the condition. Low high-density lipoprotein cholesterol level was the most common lipid abnormality in concordant pairs. Mortality from ischemic heart disease was significantly higher in individuals with dyslipidemic hypertension. Obesity and glucose intolerance were closely associated with the syndrome. Moreover, within the 18 discordant monozygotic twin pairs, the twins with dyslipidemic hypertension had gained significantly more weight as adults and were significantly heavier than their unaffected cotwins. Thus, although genetic factors may influence development of dyslipidemic hypertension, nongenetic, potentially modifiable aspects of obesity are also closely related to expression of this clinically important syndrome.

(*JAMA*. 1991;265:2079-2084)

A matched-pair t test was used in the 18 discordant monozygotic twin pairs to test for a possible association of nongenetic (ie, behavioral or environmental) aspects of obesity with expression of the syndrome.

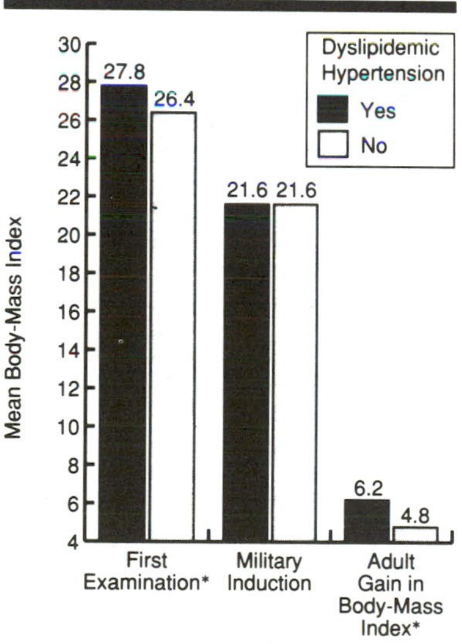

Fig 1.—Association of obesity and adult weight gain with dyslipidemic hypertension within 18 discordant monozygotic twin pairs. Asterisk indicates $P = .01$. P values are for matched intrapair t tests.

RESULTS

Findings at the First Examination

There were 18 monozygotic twin pairs discordant for dyslipidemic hypertension. In these pairs, the twins with dyslipidemic hypertension had a significantly higher mean BMI than their nonaffected cotwins (Fig 1). Mean BMI at military induction did not differ at all within these 18 pairs. The entire difference in BMI at the first examination was due to a greater gain in BMI during adult life in the twin who developed dyslipidemic hypertension.

Literature Example 14.10 is from the article entitled "Aspirin Effects on Mortality and Morbidity in Patients With Diabetes Mellitus." Read the abstract to learn about the study.

Read the section on statistical methods. We use t for testing the differences in samples. Differences in proportions are tested using the binomial distribution. When the sample sizes are large, as they were in this study, the normal distribution provides a good approximation to the binomial distribution. This is why the normal table is used for testing proportions. Thanks to computers, the practice of approximating binomials with normals is becoming less common. We expect that it will eventually disappear completely.

Look up critical values of z for $\alpha = .05$.

1 _______ to _______

Which z-value in the last column of the table is larger in size than the critical z-value?

2 _______

Which variable does this z-value correspond to?

____

Read the section on Total Cause-Specific Mortality. The authors comment about all cardiovascular causes. Can we accept the authors' conclusions, based on the evidence presented?

3 Yes/No

LITERATURE EXAMPLE 14.10

Aspirin Effects on Mortality and Morbidity in Patients With Diabetes Mellitus

Early Treatment Diabetic Retinopathy Study Report 14

ETDRS Investigators

Objectives.—This report presents information on the effects of aspirin on mortality, the occurrence of cardiovascular events, and the incidence of kidney disease in the patients enrolled in the Early Treatment Diabetic Retinopathy Study (ETDRS).

Study Design.—This multicenter, randomized clinical trial of aspirin vs placebo was sponsored by the National Eye Institute.

Patients.—Patients (N=3711) were enrolled in 22 clinical centers between April 1980 and July 1985. Men and women between the ages of 18 and 70 years with a clinical diagnosis of diabetes mellitus were eligible. Approximately 30% of all patients were considered to have type I diabetes mellitus, 31% type II, and in 39% type I or II could not be determined definitely.

Intervention.—Patients were randomly assigned to aspirin or placebo (two 325-mg tablets once per day).

Main Outcome Measures.—Mortality from all causes was specified as the primary outcome measure for assessing the systemic effects of aspirin. Other outcome variables included cause-specific mortality and cardiovascular events.

Results.—The estimate of relative risk for total mortality for aspirin-treated patients compared with placebo-treated patients for the entire study period was 0.91 (99% confidence interval, 0.75 to 1.11). Larger differences were noted for the occurrence of fatal and nonfatal myocardial infarction; the estimate of relative risk was 0.83 for the entire follow-up period (99% confidence interval, 0.66 to 1.04).

Conclusions.—The effects of aspirin on any of the cardiovascular events considered in the ETDRS were not substantially different from the effects observed in other studies that included mainly nondiabetic persons. Furthermore, there was no evidence of harmful effects of aspirin. Aspirin has been recommended previously for persons at risk for cardiovascular disease. The ETDRS results support application of this recommendation to those persons with diabetes at increased risk of cardiovascular disease.

(*JAMA*. 1992;268:1292-1300)

Statistical Methods.—Comparisons of outcome measures expressed as proportions of events were made with the two-sample test of equality of proportions. Comparisons of continuous variables were based on the two-sample z test of equality of means.[26]

Total and Cause-Specific Mortality

The number and percentage of deaths among patients in each group, shown in Table 2, are based on total follow-up. The percentages of patient deaths in the aspirin and placebo groups were nearly equal (18.3% and 19.7%, respectively). The majority of deaths in both groups were classified as cardiovascular by the Mortality and Morbidity Classification Committee: 244 (71.8%) of the 340 deaths in the aspirin group and 275 (75.1%) of the 366 deaths in the placebo group. The patients who died of cardiovascular causes represented 13.1% and 14.8% of the aspirin and placebo groups, respectively. Only a few patients died of cerebrovascular disease, and there was no difference between the treatment groups. About 2% of the deaths could not be classified or were attributed to unknown causes.

Table 2.—Deaths by Cause and Nonfatal Events

	Aspirin, No. (%) (N=1856)	Placebo, No. (%) (N=1855)	z
Death—all causes	340 (18.3)	366 (19.7)	−1.10
All cardiovascular	244 (13.1)	275 (14.8)	−1.47
Sudden coronary death	47 (2.5)	67 (3.6)	−1.91
Cerebrovascular	25 (1.3)	25 (1.3)	0.00
All noncardiovascular	87 (4.7)	85 (4.6)	0.15
Cancer	16 (0.9)	14 (0.8)	0.37
Other noncardiovascular	71 (3.8)	71 (3.8)	0.00
Cause of death unknown	9 (0.5)	6 (0.3)	0.78
Fatal or nonfatal myocardial infarction	241 (13.0)	283 (15.3)	−1.99
Fatal or nonfatal stroke	92 (5.0)	78 (4.2)	1.10
Amputation	88 (4.7)	93 (5.2)	−0.61
Hypertension	1393 (75.1)	1356 (73.1)	1.36
Any of the following	239 (12.9)	227 (12.2)	0.59
Renal transplantation	144 (7.8)	146 (7.9)	−0.13
Renal dialysis	36 (1.9)	33 (1.8)	0.36
Candidate for dialysis	101 (5.4)	100 (5.4)	0.07
Death from kidney failure	21 (1.1)	18 (1.0)	0.48
Cardiovascular death, nonfatal myocardial infarction, or stroke	350 (18.9)	379 (20.4)	−1.21
All deaths, nonfatal myocardial infarction, or stroke	439 (23.7)	463 (25.0)	−0.93

Literature Example 14.11 is from the article entitled "Ventricular Arrhythmias in Patients Undergoing Noncardiac Surgery." Read the abstract to learn about the study.

Look at the p-values in the figure and name the groups for which the difference is not significant at $\alpha = .05$.

1 ______________________________

____

Which items are significant at $\alpha = .05$?

2 ______________________________

____

____

____

LITERATURE EXAMPLE 14.11

Ventricular Arrhythmias in Patients Undergoing Noncardiac Surgery

Brian O'Kelly, MB, MRCPI, FRCPC; Warren S. Browner, MD, MPH; Barry Massie, MD; Julio Tubau, MD; Long Ngo, MS; Dennis T. Mangano, PhD, MD; for the Study of Perioperative Ischemia Research Group

Objective.—To determine the incidence, clinical predictors and prognostic importance of perioperative ventricular arrhythmias.

Design.—Prospective cohort study (Study of Perioperative Ischemia).

Setting.—University-affiliated Department of Veterans Affairs Medical Center, San Francisco, Calif.

Subjects.—A consecutive sample of 230 male patients, with known coronary artery disease (46%) or at high risk of coronary artery disease (54%), undergoing major noncardiac surgical procedures.

Measurements.—We recorded cardiac rhythm throughout the preoperative (mean=21 hours), intraoperative (mean=6 hours), and postoperative (mean=38 hours) periods using continuous ambulatory electrocardiographic monitoring. Adverse cardiac outcomes were noted by physicians blinded to information about arrhythmias.

Main Results.—Frequent or major ventricular arrhythmias (>30 ventricular ectopic beats per hour, ventricular tachycardia) occurred in 44% of our patients: 21% preoperatively, 16% intraoperatively, and 36% postoperatively. Compared with the preoperative baseline, the severity of arrhythmia increased in only 2% of patients intraoperatively but in 10% postoperatively. Preoperative ventricular arrhythmias were more common in smokers (odds ratio [OR], 4.1; 95% confidence interval [CI], 1.2 to 15.0), those with a history of congestive heart failure (OR, 4.1; 95% CI, 1.9 to 9.0), and those with electrocardiographic evidence of myocardial ischemia (OR, 2.2; 95% CI, 1.1 to 4.7). Preoperative arrhythmias were associated with the occurrence of intraoperative and postoperative arrhythmias (OR, 7.3; 95% CI, 3.3 to 16.0, and OR, 6.4; 95% CI, 2.7 to 15.0, respectively). Nonfatal myocardial infarction or cardiac death occurred in nine men; these outcomes were not significantly more frequent in those with prior perioperative arrhythmias, albeit with wide CIs (OR, 1.6; 95% CI, 0.4 to 6.2).

Conclusion.—Almost half of all high-risk patients undergoing noncardiac surgery have frequent ventricular ectopic beats or nonsustained ventricular tachycardia. Our results suggest that these arrhythmias, when they occur without other signs or symptoms of myocardial infarction, may not require aggressive monitoring or treatment during the perioperative period.

(*JAMA*. 1992;268:217-221)

Data Analysis

For each patient, rates of ventricular ectopic beats and ventricular tachycardia were calculated by dividing the number of ventricular ectopic beats and the number of episodes of ventricular tachycardia by the number of hours analyzed in that period. Differences between the periods were assessed by analysis of variance (Bonferroni multiple comparison method) over all periods, then by paired *t* tests. Differences in the proportion of

RESULTS

Incidence of Ventricular Arrhythmias

Ventricular arrhythmias were common, occurring in 101 patients (44%) during the perioperative monitoring period (Fig 1). None was accompanied by symptoms of palpitations, syncope, angina, or pulmonary congestion. Few arrhythmias were noted on routine ECG monitoring. For example, although 36 patients had intraoperative ventricular arrhythmias, only four (11%) were clinically detected.

Preoperatively, 39 patients (17.0%) had greater than 30 ventricular ectopic beats per hour, 21 (9.1%) had ventricular tachycardia, and 49 (21.3%) had either arrhythmia. Unadjusted for the length of monitoring, the incidence of ventricular arrhythmias declined to 15.7% during surgery (P=.07). During the intraoperative period, a total of 30 episodes of ventricular tachycardia occurred in 17 patients, of which seven (23%) occurred within 30 minutes of tracheal intubation. A greater proportion of patients (36.1%) had ventricular arrhythmias postoperatively (Fig 2), at least in part because of a longer mean monitoring period.

Testing the Hypothesis: Method I, Confidence Interval Method

	Experimental group (low-intensity exercise) Post − Pre	Control group (no exercise) Post − Pre
Sample size	$n_1 = 8$	$n_2 = 7$
Mean difference	$\bar{x}_1 = -8.7$	$\bar{x}_2 = -4.1$
St. dev. of difference	$s_1 = 10.6$	$s_2 = 14.3$
Variance of difference	$s_1^2 = 112.36$	$s_2^2 = 204.5$
Confidence level $(1 - \alpha) = 95\%$		

$$SE_{\bar{x}_1 - \bar{x}_2} = \sqrt{\frac{s_1^2}{n_1} + \frac{s_2^2}{n_2}}$$

$SE_{\bar{x}_1 - \bar{x}_2} =$ [1] ____________

Calculate the 95 percent confidence interval for $\mu_1 - \mu_2$ using the formula C.I. $= (\bar{x}_1 - \bar{x}_2) \pm t * SE_{\bar{x}_1 - \bar{x}_2}$. Note: For rough calculations, the degree of freedom for t can be taken to be the smaller of $n_1 - 1$ and $n_2 - 1$. For this computation use this value for df. This choice of df estimates the t-statistic very conservatively (that is, it makes the confidence interval larger than it really is).

Actually, the distribution of the difference of two means is only approximately a t-distribution, and the value of df that provides the best approximation is usually not an integer. Most statistical computer software computes the t-statistic using the non-integer value of df for this test. It is also worth noting that this test assumes that the two populations are normally distributed, though this assumption is rarely checked in practice.

Confidence interval for $\mu_1 - \mu_2 =$ [2] ____________. Does the interval contain 0, the hypothesized value of $\mu_1 - \mu_2$? [3] Yes/No Do you accept or reject the null hypothesis? [4] Accept/Reject

Testing the Hypothesis: Method II, Critical Means Method (Two-Tailed Test) Find the critical means by calculating $(\mu_1 - \mu_2) \pm t * SE_{\bar{x}_1 - \bar{x}_2}$, where the hypothesized difference $\mu_1 - \mu_2 = 0$. For $\alpha = .05$ and df $= 6$, $t =$ [5] ______. The critical differences of means are [6] ______. $\bar{x}_1 - \bar{x}_2 =$ [7] ______. Does the calculated difference in experimental and control means fall between the critical

means or does it fall in the critical region? [8] Between means/In critical region
Do you reject or accept the null hypothesis? [9] Accept/Reject

Testing the Hypothesis: Method III, Critical t Method (Two-Tailed Test) Calculate critical t-values using the formula

$$t = \frac{(\bar{x}_1 - \bar{x}_2) - (\mu_1 - \mu_2)}{\mathrm{SE}_{\bar{x}_1 - \bar{x}_2}}$$

The critical t-values are [10] _______. The calculated $t =$ [11] _______. Is the calculated t between the critical t-values or does it fall in the critical region? [12] Between t-values/In critical region The probability of such a t-value occurring if the hypothesis is true is .49. Given the α value we chose, do you accept or reject the hypothesis? [13] Accept/Reject

Even using the less conservative (non-integer) value for df, we find that we still cannot reject the null hypothesis. Notice, however, the small size of the samples. A test with such small samples is rarely powerful enough to reject a null hypothesis unless it is grossly false, i.e., unless the difference of the means is quite large.

Examples from the Literature

Literature Example 14.12 is from the article entitled "Home Apnea Monitoring and Disruptions in Family Life: A Multidimensional Controlled Study." Read the abstract to learn about the study.

Having a high-risk infant in the family can be detrimental to the mother's health. For this study the sample consists of ninety-three mothers. Let us stipulate α to be .05. Look up the critical t-values at which you reject the hypotheses.

Critical t given in the table for df = 92 and $\alpha = .05$ (two-tailed test) [1] _______

Notice t-values are given in the footnote to the table: −3.8 for infant gestation age and −2.23 for satisfaction with social support.

[2] Do you reject/accept the null hypothesis for each of these variables?

Notice the P-values given for each of the above variables.

On the basis of P-values do you [3] reject/accept the null hypothesis for each variable?

These two groups are different from each other.

This is [4] a dependent/an independent t-test.

LITERATURE EXAMPLE 14.12

Home Apnea Monitoring and Disruptions in Family Life: A Multidimensional Controlled Study

Elizabeth Ahmann, ScD, RN, Louise Wulff, RN, ScD, and Robert G. Meny, MD

TABLE 2—Means for Case and Comparison Mothers on Potentially Confounding Variables

	Case Mothers		Comparison Mothers	
	Mean	SD	Mean	SD
Infant gestational age at birth, mo	35.5	4.6	37.6*	3.0
Difficult temperament, mean score[a]	11.2	4.0	10.2	3.7
Maternal age, y	26.7	5.5	27.4	6.5
No. of children at home	1.8	0.8	1.6	0.8
Years married or living together	5.1	3.8	5.2	4.5
Satisfaction with social support, mean score[b]	54.1	7.4	56.3**	5.5

[a]Scale developed for this study; conceptually based on fussy-difficult scale of Bates' Infant Characteristics Questionnaire (1979), but modified (with permission) for use in telephone interviewing.
[b]Crnic's Social Support measure, satisfaction scale: a higher score implies greater satisfaction.
*$t = -3.6$, $P = .0004$.
**$t = -2.23$, $P = .0268$.

We used data from telephone interviews and mailed questionnaires to examine 12 aspects of family life among 93 families with infants considered at high risk for sudden infant death syndrome and on home apnea monitors and a matched comparison group with infants not requiring monitoring. Using logistic regression to control confounding variables, we found that case mothers were at an increased risk of poor health, but we found no other significant differences in family life between the two groups. (*Am J Public Health.* 1992;82:719–722)

Results

Case and comparison groups differed on only three of many potentially confounding variables examined: gestational age, days hospitalized at birth, and satisfaction with social support (Tables 1 and 2).

Among mothers, gestational age confounded the relationship between case-comparison status and one dependent variable, relationship with baby. Satisfaction with social support was a confounder in relation to several of the dependent variables: parental health, parental sense of competence, parental role restriction, relationship with spouse, marital satisfaction, and social isolation. No other study variables were confounders among mothers, and none were confounders among fathers.

Literature Example 14.13 is from the article entitled "Comparisons of Hospital Care for Patients With AIDS and Other HIV-Related Conditions." Read the abstract to learn the purpose of the study.

Look at the row in the table that shows "No. of patients per hospital."

Calculate CV (see Chapter 8, page 267):

AIDS [1] _______

Other HIV illness [2] _______

Total [3] _______

Look at the ranges for this variable.

When we consider the ranges and CVs we conclude that there is quite a bit of variation in the values. We do not have any information about possible outliers. In such circumstances, [4] mean/median may be a desirable choice for a measure of center.

Paired t-test comparisons give $p < .001$.

We [5] reject/accept all the hypotheses that state equality of mean measurements for all the stated variables in the table.

LITERATURE EXAMPLE 14.13

Comparisons of Hospital Care for Patients With AIDS and Other HIV-Related Conditions

Dennis P. Andrulis, PhD, MPH; Virginia Beers Weslowski, MPA; Elizabeth Hintz, MHSA; Audrey Wright Spolarich, MPA

Objective.—To compare utilization and financing of inpatient care for persons with the acquired immunodeficiency syndrome (AIDS) (as defined by the Centers for Disease Control) and those with "other HIV [human immunodeficiency virus]-related illness."

Design.—A mailed survey of the members of five national organizations representing public, teaching, children's, community, and Catholic hospitals. The survey requested information on demographics, service utilization, costs, and financing of care for AIDS and other HIV patients. Statistical analysis using paired *t* tests was conducted to evaluate differences between group means for AIDS and other HIV patients. Differences among categorical variables were evaluated by calculation of proportions and compared using χ^2 tests.

Participants.—Five hundred eighteen of 1158 hospitals surveyed responded to the AIDS and other HIV portions of the survey.

Results.—Three hundred twenty-five hospitals reported treating at least one other HIV patient. These 325 hospitals treated 30% of all AIDS patients (16 213) estimated to have been alive during 1988, and provided care to over 11 000 other HIV patients. Service utilization by other HIV patients was found to comprise a substantial portion of the total HIV burden and related costs, representing 35% of all HIV-related admissions, 29% of all inpatient costs, and 35% of all inpatient losses. Demographic and mode of exposure analysis indicated that other HIV patients were more likely than AIDS patients to be intravenous drug users, female, nonwhite, and to have no source of public or private coverage for their health care.

Conclusions.—Our results demonstrate that accounting for only the utilization of services by persons with AIDS as defined by the Centers for Disease Control will understate significantly the total burden of the HIV epidemic on hospitals. The results suggest that the expanded definition proposed by the Centers for Disease Control would incorporate a large hospitalized HIV population.

(*JAMA*. 1992;267:2482-2486)

To examine differences in the care of persons with AIDS and OHIV, we conducted statistical analyses using SPSS-PC+ software (SPSS Inc, Chicago, Ill). Differences between AIDS and OHIV were evaluated by calculations of means, standard deviations, and statistically compared by paired *t* tests. Differences among categorical variables were evaluated by calculation of proportions and statistically compared by contingency table analyses using χ^2 tests.

RESULTS

Utilization

Responding hospitals (n=325) reported 28 000 admissions for persons with CDC-defined AIDS and nearly 15 000 admissions for persons with other HIV-related conditions in 1988. These admissions accounted for nearly 700 000 days of inpatient care. Utilizing these services were 16 213 distinct AIDS inpatients and 11 077 distinct inpatients with OHIV representing a total burden of approximately 27 290 distinct HIV-related patients, an average of 84 individuals per hospital (Table 1). (This number overcounts the actual number of distinct cases. Patients who were admitted at least once during the year as an OHIV patient and were also admitted at least once as an AIDS patient would be counted twice. This will result in an understatement of per patient utilization statistics.) People with OHIV accounted for about one third of all HIV-related admissions and slightly fewer than one third of the hospital days. The differences in the mean number of admissions per facility (86.2 for AIDS and 45.7 for OHIV) were statistically significant as were the differences in the mean number of patient and inpatient days.

Table 1.—Hospital Utilization: Acquired Immunodeficiency Syndrome (AIDS) and Other Human Immunodeficiency Virus (HIV) Illness*

Variable†	AIDS	Other HIV Illness	Total
No. of patients	16 213	11 077	**27 290**
Per hospital (mean±SD)	49.9±100	34.1±76	84.0±136
Range	0-750	1-605	1-1033
No. of admissions	28 024	14 866	**42 890**
Per hospital (mean±SD)	86.2±172	45.7±97	132.0±211
Range	0-1280	1-685	2-1426
Inpatient care, d	460 379	214 422	**674 801**
Per hospital (mean±SD)	1416.6±3415	659.8±1848	2076.3±4196
Range	0-37 483	1-18 972	7-39 460
Length of stay, d (mean±SD)	16.4±8	14.4±12	15.7±7
Range	1-68	1-181	3-60
Days per patient per year, mean	28.4	19.4	24.7
Admissions per patient per year, mean	1.7	1.3	1.6

*Includes 325 hospitals, mean number of beds per hospital, 491; includes 17 hospitals that treated HIV patients, but no patients with AIDS, as defined by the Centers for Disease Control, Atlanta, Ga.

†Paired *t*-test comparisons of AIDS and other HIV illness resulted in significant differences at $P<.001$ for all variables.

Testing Hypotheses About Three or More Groups Using Analysis of Variance (ANOVA)

We present a brief introduction to analysis of variance (ANOVA), a subject we discuss in greater detail in Chapter 17. We used a *t*-test when we compared means of two groups. Now we wish to test means of three or more groups using analysis of variance. Suppose the hypothesis states that there is no difference between the means of the groups. Assuming it is true we may ignore the fact that we have three groups and compute the sum of squares, which is the numerator of the variance formula

$$\text{Mean sum of squares} = s^2 = \frac{\Sigma\,(x_i - \bar{\bar{x}})^2}{n - 1}$$

In this case x_i represents a single observation and $\bar{\bar{x}}$ represents the "grand mean," or the mean of all the observations. The numerator in the above formula gives us the total sum of squares (total SSq). Observations in each group will vary. So we can compute the sum of squares for each group and add them up. This sum is called within-groups sum of squares (within SSq).

$\Sigma\,(x_i - \bar{x}_1)^2$ Group I sum of squares

$\Sigma\,(x_i - \bar{x}_2)^2$ Group II sum of squares

$\Sigma\,(x_i - \bar{x}_3)^2$ Group III sum of squares

Sum of the above = within sum of squares

The sum of squares between the means (between SSq) can be obtained by subtracting within SSq from total SSq. When the group means are all pretty much the same size, between SSq will be small, whereas if the group means vary a great deal (i.e., the differences between them are large), between SSq will be large. The total sum of squares is separated into within SSq and between SSq, and these components are used for testing the hypothesis of equal means for three or more groups.

Total sum of squares = Between SSq + Within SSq

Activity 14.8

Calculation of total, within, and between sums of squares

Suppose a study involves three groups, and the hypothesis is that there is no difference in the means of the three groups. We present a small set of fictitious data to illustrate the calculations.

Data Set:

Group 1: 3, 3, 4, 2, 3, 3

Group 2: 5, 4, 5, 6, 7, 3

Group 3: 8, 8, 7, 8, 9, 2

The numbers of observations in each group are represented by n_i (that is, n_1, n_2, n_3).

To begin the calculations we compute the numerator of the variance formula, the sum of squares (SSq):

$$\text{Variance} = s^2 = \frac{\Sigma (x_i - \bar{\bar{x}})^2}{n - 1}$$

$$\text{SSq} = \sum (x_i - \bar{\bar{x}})^2$$

There are [1] _______ observations in each group and [2] _______ observations in the entire data set.

Entire data set:

3, 3, 4, 2, 3, 3   5, 4, 5, 6, 7, 3

8, 8, 7, 8, 9, 2

The total number of observations $n = n_1 + n_2 + n_3$.

We analyze the total SSq as follows: Total SSq = Between SSq + [3] _______ SSq

The mean of all observations, the grand mean, or GM, is

$$\bar{\bar{x}} = \frac{\Sigma x_i}{n}$$

x_i = each observation in the data set
n = total number of observations in all three data sets

$\bar{\bar{x}}$ = [4] _______

Mean for Group 1 = $\bar{x}_1$ [5] _______

Mean for Group 2 = $\bar{x}_2$ [6] _______

Mean for Group 3 = $\bar{x}_3$ [7] _______

Calculate total SSq.

Total SSq = $\sum (x_i - \bar{\bar{x}})^2$ = [8] _______

Now we consider each group separately and compute SSq by taking each observation in the group and subtracting the mean for that group, squaring, and summing. Within SSq = Group 1 SSq + Group 2 SSq + Group 3 SSq

Group 1 SSq [9] _______

Group 2 SSq [10] _______

Group 3 SSq [11] _______

Within SSq = [12] _______

Total sum of squares = Between SSq + Within SSq
Therefore, Between SSq = Total Sum of Squares − Within SSq

Between SSq = [13] _______

Another way to calculate between SSq:
Between SSq $= \sum n_i(\bar{x}_i - \bar{\bar{x}})^2$

Group 1 $= n_1(\bar{x}_1 - \bar{\bar{x}})^2 =$ [14] ________

Group 2 $= n_2(\bar{x}_2 - \bar{\bar{x}})^2 =$ [15] ________

Group 3 $= n_3(\bar{x}_3 - \bar{\bar{x}})^2 =$ [16] ________

Between SSq = [17] ________

The number of degrees of freedom for between SSq = between df = number of groups − 1.
The number of degrees of freedom for within SSq $= \Sigma\,(n_i - 1) = (\Sigma\, n_i) -$ number of groups

$$\text{Between mean SSq} = \frac{\text{Between SSq}}{\text{Between df}}$$

$$\text{Within mean SSq} = \frac{\text{Within SSq}}{\text{Within df}}$$

Between df = [18] ________

Within df = [19] ________

Between mean SSq = [20] ________

Within mean SSq = [21] ________

If there is no difference in the means, mean SSq of both between and within estimate the variance in the observations. If there is a difference, mean SSq of between will be greater. The ratio of these two mean SSq has an F-distribution.

$$F = \frac{\text{Mean SSq between groups}}{\text{Mean SSq within groups}}$$

$F =$ [22] ________

Look up $F_{\text{between df, within df}}$ in Table 11.10 for $\alpha = .05$. If the computed value exceeds the value from the table, we reject the hypothesis that the means are equal; otherwise we accept it.

$F_{\text{between df, within df}} =$ [23] ________

The calculated F is [24] smaller/larger than the F-value obtained from the table. Therefore we [25] accept/reject the hypothesis that the means of the three groups are equal.

Table 14.2 shows one of the formats in which analysis of variance data are presented.

TABLE 14.2 Analysis of Variance Data as Displayed by Minitab Computer Program

SOURCE	DF	SS	MS	F	p
FACTOR	2	48	24		
ERROR	15	44	2.93	8.18	.004
TOTAL	17	92			

Source = source of variation; Factor = between groups variation; Error = within groups variation; DF = degrees of freedom; SS = sum of squares; MS = mean sum of squares; p is the p-value for such F. Note that you cannot obtain this p-value from the tables given in this book; the computer calculated it.

It is instructive to draw confidence intervals. We use the pooled standard deviation for calculating the C.I.s.

$$\text{Pooled standard deviation} = S_p = \sqrt{\text{Error MS}}$$

$$\text{POOLED STDEV} = 1.713$$

Choose $1 - \alpha = 95\%$ and read t from the table for df = 15. We get $t = 2.131$. Use the confidence interval formula to compute the confidence intervals and write them in the table below:

$$\text{Confidence interval: } (\bar{x} - t * \text{SE}, \bar{x} + t * \text{SE})$$

	n	**Mean**	**St Dev**	$\text{SE} = \frac{S_p}{\sqrt{n}}$	Confidence Interval
Group 1	6	3.000	0.632	.7	1 ______________
Group 2	6	5.000	1.414	.7	2 ______________
Group 3	6	7.000	2.530	.7	3 ______________

Plot the confidence intervals on the graph below:

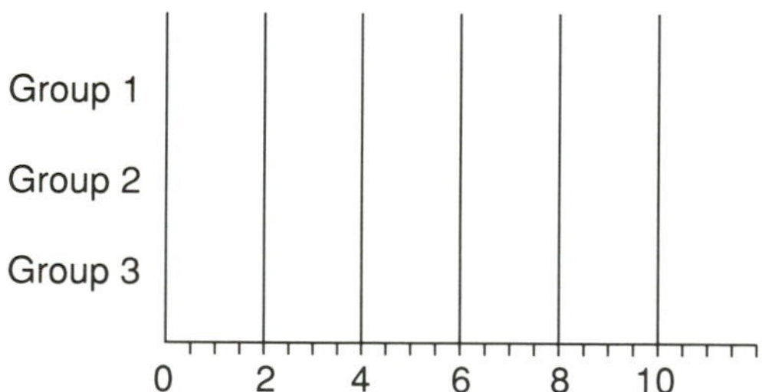

How much overlap is there in the confidence intervals? [4]Considerable/Some/None

We first illustrated analysis of variance using a fictitious data set. Now we will give examples using data from the faculty exercise data. The study used the following design:

Randomly assigned Group I:	High intensity (HI)	O_1	X_{HI}	O_2
Randomly assigned Group II:	Low intensity (LI)	O_3	X_{LI}	O_4
Randomly assigned Group III:	Control	O_5		O_6

We have three groups and six sets of observations, two sets on each group. Our interest is in the difference in the subjects' weights. When we obtain the three differences we will have three sets of data (Table 14.3). A *t*-test is useful for comparing two group means, but not three—which is why we will use analysis of variance.

First we may check by plotting the confidence intervals, using the information in Table 14.3. The intervals are listed below and graphed in Figure 14.11.

	n	Mean	St Dev	SE Mean	95.0% C.I.
Control	7	2.57	4.61	1.74	(−1.70, 6.84)
Low Intensity	8	−2.12	6.27	2.22	(−7.37, 3.12)
High Intensity	9	1.00	4.85	1.62	(−2.73, 4.70)

The confidence intervals of the weights overlap quite a bit, but the means look different. According to the means, those in the control and high-intensity groups gained a little weight and those in the low-intensity group lost a little. But are these differences more than one would expect to see had there been no exercise intervention? In other words, is the variation between the means

TABLE 14.3 Pre and Post Weight Differences (POSTWT − PREWT) for Three Exercise Groups

	Control	Low Intensity (LI)	High Intensity (HI)
1	1	−7	1
2	7	3	0
3	2	6	7
4	0	2	8
5	−4	−14	−7
6	10	−3	−4
7	2	−2	4
8		−2	−1
9			1

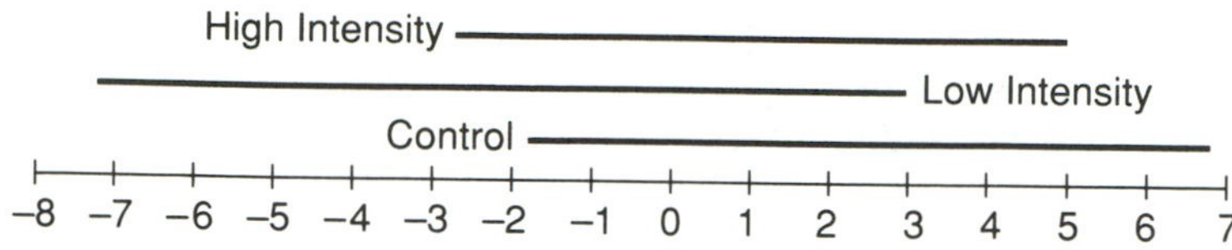

FIGURE 14.11 Individual 95 Percent Confidence Intervals for POSTWT − PREWT

more than the variation within the samples? Visual inspection of the confidence intervals suggests that the between means variation may not be more than the within sample variation. In performing analysis of variance we separate the variation according to its source. This separation of the variation into different categories, e.g., within each sample and between means of each sample is at the heart of the analysis of variance method.

Note: The confidence intervals in the example just above were not computed in quite the same way as the confidence intervals in Activity 14.8. In that activity, the pooled standard deviation was used, but just above we used the individual standard deviations. Also, df in Activity 14.8 was the number of degrees of freedom for the error, or within groups variation. Here we used the sample size minus 1 as df for each t. What allowed us to use the pooled standard deviation in Activity 14.8 was the assumption that the standard deviations of the three groups were equal, an assumption that is necessary to perform analysis of variance. If the assumption is that each group has the same standard deviation, it makes sense to pool the samples so that we estimate this common standard deviation with the biggest possible set of data.

In the example above, we hadn't done the analysis of variance computations, so we didn't know the pooled standard deviation, and it wasn't worth computing for this very rough analysis. Instead, we chose to compare the confidence intervals as we would if we only had two groups and we had no reason to believe the standard deviations were the same.

Since t gets smaller as df increases, using the smaller df will, if anything, overestimate the size of the confidence interval. Since we were only doing a rough analysis in the example above, it seemed appropriate to overestimate in this way. If this procedure had resulted in confidence intervals that did not overlap or overlapped only slightly, we would then have proceeded to perform the analysis of variance computations, and there would have been a good chance that we'd have detected a significant difference. This method of comparing was appropriate in this case because we were fishing to see if it was worth pursuing further analysis. In the following example (Activity 14.9), we will already have done the analysis of variance, having assumed the standard deviations of the three groups to be equal. We, therefore, will use the pooled standard deviation and the df for the within (error) SSq.

Although graphs like the one in Figure 14.11 are found in the literature, we want to make it clear that the methods just described here are useful only for rough comparisons. In many cases a graph will show overlap between the bars whereas an analysis of variance test will give significant evidence against equality of the means. ANOVA is a much more sensitive test.

ACTIVITY 14.9

Calculation of total, within, and between sums of squares for the faculty weight data

The hypothesis is that there is no difference in the means of the three groups: $\mu_1 = \mu_2 = \mu_3$. The variables being studied are the differences between PREWT and POSTWT of each group. The means of each difference are:

Control	μ_1
Low-Intensity	μ_2
High-Intensity	μ_3

The data were given in Table 14.3 and are repeated below. As before, the numbers of observations in each group are represented by n_i (that is, n_1, n_2, n_3).

Entire data set:

1, 7, 2, 0, −4, 10, 2;
−7, 3, 6, 2, −14, −3, −2, −2;
1, 0, 7, 8, −7, −4, 4, −1, 1

The total number of observations is $n = n_1 + n_2 + n_3$.

To begin the calculations we compute the numerator of the variance formula, the sum of squares (SSq):

$$\text{Variance} = s^2 = \frac{\Sigma (x_i - \bar{\bar{x}})^2}{n-1}$$

$$\text{SSq} = \sum (x_i - \bar{\bar{x}})^2$$

There are [1] _______, _______, and _______ observations in each group and [2] _______ observations in the entire data set.

We analyze the total SSq and separate it into Total SSq = Between SSq + [3]_______ SSq.

The mean of all observations, the grand mean, or GM, is

$$\bar{\bar{x}} = \frac{\Sigma x_i}{n}$$

x_i = each observation in the data set
n = total number of observations in all three data sets

$\bar{\bar{x}}$ = [4]_______
Mean for Group 1 = $\bar{x}_1$ [5] _______
Mean for Group 2 = $\bar{x}_2$ [6] _______
Mean for Group 3 = $\bar{x}_3$ [7] _______

Calculate total SSq.

Total SSq = $\sum (x_i - \bar{\bar{x}})^2$ = [8] _______

Now we consider each group separately and compute SSq by taking each observation in the

Group 1 SSq [9]_______
Group 2 SSq [10] _______
Group 3 SSq [11] _______

group and subtracting the mean for that group, squaring, and summing. Within SSq = Group 1 SSq + Group 2 SSq + Group 3 SSq	Within SSq = [12] _______
Total sum of squares = Between SSq + Within SSq Therefore, Between SSq = Total sum of squares − Within SSq	Between SSq = [13] _______
Another way to calculate between SSq: Between SSq = $\Sigma\, n_i(\bar{x}_i - \bar{\bar{x}})^2$	Group 1 = $n_1(\bar{x}_1 - \bar{\bar{x}})^2$ = [14] _______ Group 2 = $n_2(\bar{x}_2 - \bar{\bar{x}})^2$ = [15] _______ Group 3 = $n_3(\bar{x}_3 - \bar{\bar{x}})^2$ = [16] _______ Between SSq = [17] _______
The number of degrees of freedom for between SSq = between df = number of groups − 1. The number of degrees of freedom for within SSq = $\Sigma\,(n_i - 1) = (\Sigma\, n_i)$ − number of groups.	Between df = [18] _______ Within df = [19] _______
$$\text{Between mean SSq} = \frac{\text{Between SSq}}{\text{Between df}}$$	Between mean SSq = [20] _______
$$\text{Within mean SSq} = \frac{\text{Within SSq}}{\text{Within df}}$$	Within mean SSq = [21] _______
If there is no difference in the means, mean SSq of both between and within estimate the variance in the observations. If there is a difference, mean SSq of between will be greater. The ratio of these two mean SSq has an *F*-distribution.	
$$F = \frac{\text{Mean SSq between groups}}{\text{Mean SSq within groups}}$$	F = [22] _______
Look up $F_{\text{between df, within df}}$ in Table 11.10 for $\alpha = .05$. If the computed value exceeds the value from the table, we reject the hypothesis that the means are equal; otherwise we accept it.	$F_{\text{between df, within df}}$ = [23] _______ The calculated *F* is [24] smaller/larger than the *F*-value obtained from the table. Therefore we [25] accept/reject the hypothesis that the means of the three groups are equal.

TABLE 14.4 Differences between Post- and Pre-Tests (POSTWAT − PREWAT) for Three Groups (control, low-intensity exercise, and high-intensity exercise)

	Control	Low Intensity (LI)	High Intensity (HI)
1	0	40	41
2	19	41	62
3	−18	42	43
4	0	20	83
5	21	41	83
6	19	62	64
7	12	41	42
8		39	53
9			64

Enter the results from Activity 14.9 in the analysis of variance table below.

Source of Variation	df	SSq	Mean SSq	F
Between	2	1 ______	2 ______	3 ______
Within	21	4 ______	5 ______	
Total	23	6 ______		

The calculated value of F was 1.55, which is less than the value from the table, 3.47. So we have no reason to reject the hypothesis that $\mu_1 = \mu_2 = \mu_3$. As we pointed out earlier, the confidence intervals overlap each other. F is likely to be small under such conditions.

Next we will consider a situation in which only two of three confidence intervals overlap. It will be instructive to compare the confidence intervals and the analysis of variance results to the results we arrived at using weight data. Table 14.4 lists pre and post energy expenditure values for subjects in the three groups. Again, the hypothesis we want to test is that there is no difference in the means of the three groups ($\mu_1 = \mu_2 = \mu_3$). Calculate the confidence intervals for each group and plot the C.I.s:

	n	Mean	St Dev	$SE = \frac{S_p}{\sqrt{n}}$	t	95% C.I.
Control	7	7.57	14.32	7 ______	2.080	(8 ______, ______)
Low Intensity	8	40.75	11.26	9 ______	2.080	(10 ______, ______)
High Intensity	9	59.44	16.26	11 ______	2.080	(12 ______, ______)

S_p = Pooled standard deviation = 14.19

Note: Since we assume that all three groups have the same standard deviation, it makes sense to estimate it with s_p. The pooled standard deviation is the best estimate of the standard deviation of the population because it makes use of the biggest available set of data. Plot the confidence intervals in the figure below.

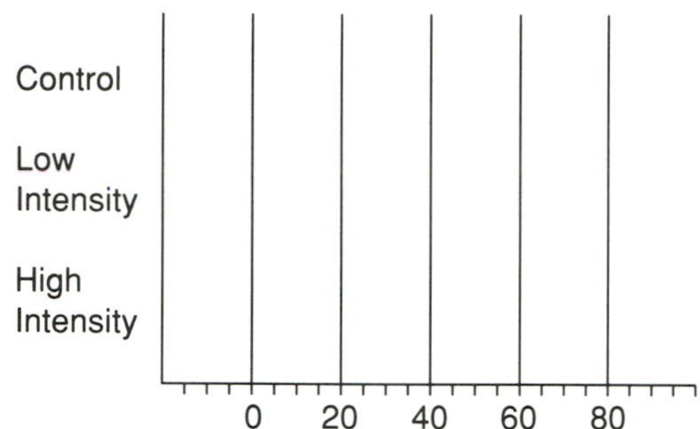

Which confidence intervals overlap? [1] Control and Low Int./Control and High Int./Low Int. and High Int.

As Figure 14.12 shows, the control group confidence interval does not overlap the experimental group confidence interval at all. So the control group is very likely actually different from the intervention groups. There is a little overlap between low intensity and high intensity exercise groups. This difference could be statistically significant. The confidence interval indicates the variation within the sample. The differences between the sample means are much larger than the variation found within the observations in each sample. We can make this statement because we have separated the analysis of variance into two categories: (1) variation within each sample and (2) variation between the means of the samples.

Let us look at the analysis of variance for differences between the three groups. We looked at the confidence intervals and means and decided that there is no overlap between the control group confidence interval and those of the other groups. Apparently the exercise had an effect on the WAT capacity. Consider the analysis of variance table.

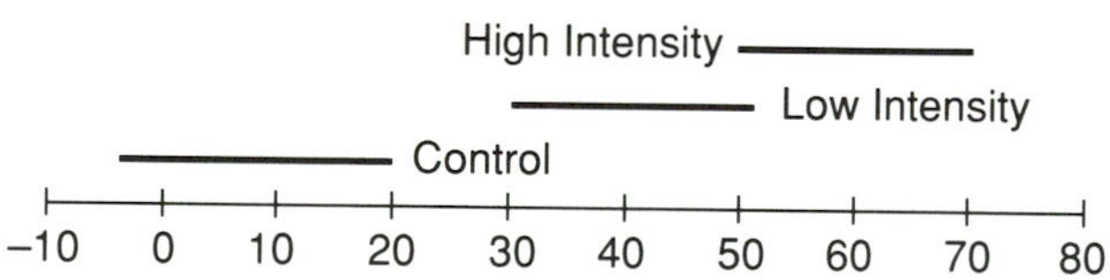

FIGURE 14.12 Individual 95 Percent Confidence Intervals for POSTWAT − PREWAT

Source	df	SSq	Mean SSq	F	p
Between	2	10,680	5,340	26.50	0.000
Within	21	4,231	201		
TOTAL	23	14,912			

F is 26.5, and it would be quite unlikely to get such a large F if the hypothesis of equality were true. So we reject the hypothesis $\mu_1 = \mu_2 = \mu_3$. F for the weight differences was 1.5. F for the watt differences is 26.5. Relate the confidence interval plots to the F-values of the watt differences and the weight differences.

____

____

____

Examples from the Literature

Literature Example 14.14 is from the article entitled "Isolated Systolic Hypertension and Subclinical Cardiovascular Disease in the Elderly." Read the abstract to learn about the study.

Read the paragraph preceding the results section at the bottom of the page. The authors transformed the data for the distributions that were skewed. The tests, in general, assumed a normal distribution. What transformation did the authors use?

1 ____________________

What probability did the authors list that relates to the following statement: "The prevalence of hypertension increased with age in both men and women."

2 ________

This number is statistically significant because it is less than the usually used $\alpha =$ 3 ________.

Name the three groups being compared.

4 ____________________

____

List the significant variables (risk factors) shown in Table 1.

5 ____________________

LITERATURE EXAMPLE 14.14

Isolated Systolic Hypertension and Subclinical Cardiovascular Disease in the Elderly

Initial Findings From the Cardiovascular Health Study

Bruce M. Psaty, MD, PhD; Curt D. Furberg, MD, PhD; Lewis H. Kuller, MD, DrPH; Nemat O. Borhani, MD; Pentti M. Rautaharju, MD, PhD; Daniel H. O'Leary, MD; Diane E. Bild, MD, MPH; John Robbins, MD, MHS; Linda P. Fried, MD, MPH; Cheryl Reid, MD

Objective.—To assess the association between isolated systolic hypertension (ISH) and subclinical disease in adults aged 65 years and above.

Design.—Medicare eligibility lists were used to obtain a representative sample of 5201 community-dwelling elderly persons for the Cardiovascular Health Study, a National Heart, Lung, and Blood Institute–sponsored cohort study of risk factors for coronary heart disease and stroke. In this cross-sectional analysis of baseline data, we excluded 3012 participants who were receiving antihypertensive medications, had clinical cardiovascular disease, or had a diastolic blood pressure of at least 90 mm Hg.

Main Outcome Measures.—For electrocardiogram: myocardial infarction, left ventricular hypertrophy, and left ventricular mass as measures of myocardial damage and strain; for echocardiography: left ventricular mass, fractional shortening, and Doppler flow velocities as measures of cardiac systolic and diastolic function; and for carotid sonography: carotid arterial intima-media thickness as a measure of atherosclerosis.

Results.—Among the 2189 men and women in this analysis, 195 (9%) had ISH (systolic blood pressure, ≥160 mm Hg) and 596 (23%) had borderline ISH (systolic blood pressure, 140 to 159 mm Hg). Systolic blood pressure was associated with myocardial infarction by electrocardiogram (P=.02). Borderline and definite ISH were strongly associated with left ventricular mass (P<.001). While there was little association with cardiac systolic function, borderline and definite ISH were associated with cardiac diastolic function (P<.001). Isolated systolic hypertension was also strongly associated with increased intima-media thickness of the carotid artery (P<.001).

Conclusions.—While cohort analyses of future repeated measures will provide a better assessment of risk, both borderline and definite ISH were strongly related to a variety of measures of subclinical disease in elderly men and women.

(*JAMA.* 1992;268:1287-1291)

We used SPSS-PC for data analysis.[20] Techniques included analysis of variance and logistic regression.[21,22] When distributions were skewed, we used the log function to normalize them. Data in the tables are untransformed; but for skewed distributions, statistical tests were performed on the log-transformed data. All P values represent two-sided tests.

RESULTS

Of the 5201 participants, we excluded 16 who did not have both a DBP and an SBP and 2923 who had clinical cardiovascular disease or who were taking antihypertensive medications. Figure 1 shows the prevalence of hypertension among the remaining 2262 participants according to 5-year age groups. The prevalence increased with age in both men and women (P<.0001). The proportion with borderline ISH was slightly higher in women (22.7%) than in men (20.9%), as was the prevalence of ISH (8.7% for women, and 8.5% for men). The distributions of hypertension type differed significantly between the gen-

Table 1.—Mean Levels of Selected Covariates According to Type of Systolic Hypertension

Covariate	Type of Hypertension: None (n=1489)	Borderline Isolated Systolic (n=496)	Isolated Systolic (n=195)	*P*
Age, y	71.4	73.0	74.9	<.001
Male, %	39.8	38.9	40.0	.936
Systolic blood pressure, mm Hg	121.1	147.9	171.6	<.001
Diastolic blood pressure, mm Hg	66.7	73.5	75.7	<.001
Height, cm	165.2	163.4	162.8	<.001
Weight, kg	70.1	71.7	68.6	.015
Waist circumference, cm	91.5	94.4	91.1	<.001
Cholesterol, mmol/L	5.58	5.70	5.75	.007
High-density lipoprotein cholesterol, mmol/L	1.46	1.45	1.51	.297
Fasting glucose, mmol/L	5.70	5.92	5.94	.002
History of hypertension, %	10.3	21.6	35.4	<.001
History of diabetes, %	5.7	7.1	5.1	.477
Current smoker, %	14.5	11.5	11.3	.144
Ever-smoker, %	54.1	47.6	51.8	.040

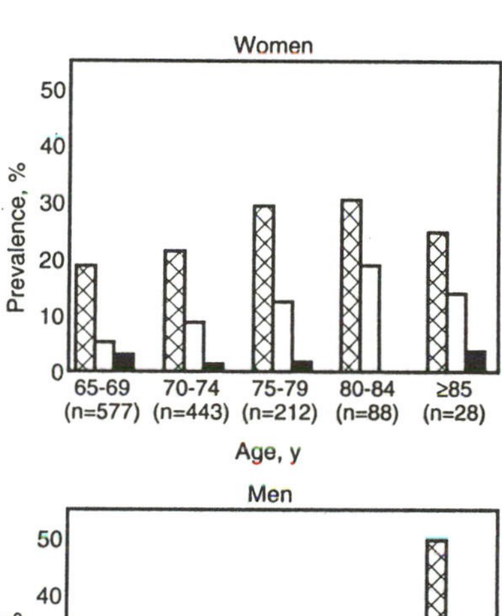

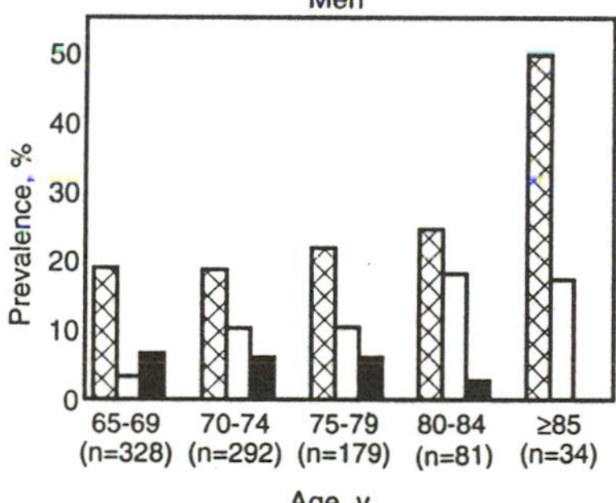

Fig 1.—Prevalence of borderline isolated systolic hypertension, isolated systolic hypertension, and diastolic hypertension according to age and gender among subjects who were free of clinical cardiovascular disease and not taking antihypertensive medications.

ders ($P<.0001$). Diastolic hypertension was uncommon; so we also excluded the 47 men and the 26 women who had diastolic hypertension.

The subsequent analyses focused on the 1322 women and the 867 men who were not receiving antihypertensive medications, were free of clinical cardiovascular disease, and had a DBP of less than 90 mm Hg. Table 1 lists the major risk factors for cardiovascular disease according to type of systolic hypertension. Most of the differences among groups were significant. Both fasting blood glucose and cholesterol levels, for instance, were higher among those with ISH.

Literature Example 14.15 is from the article entitled "Work-Site Nutrition Intervention and Employees' Dietary Habits: The Treatwell Program." Read the abstract to learn about the study.

How were the samples for the intervention and control groups selected? 1 ______

Consider mean total dietary fat. Note the age groups and corresponding *F*- and *p*-values. Write the hypothesis in symbols. 2 ______

If $\alpha = .05$, do you accept or reject the hypothesis? 3 Reject/Accept

What other variables are significant? 4 ______

______

Read the results section. Check the *F*-values and *p*-values in the table. Are the values in the table in consonance with the authors' conclusions? 5 Yes/No

What are the implications of this study? 6 ______

______

______

LITERATURE EXAMPLE 14.15

Work-Site Nutrition Intervention and Employees' Dietary Habits: The Treatwell Program

Glorian Sorensen, PhD, MPH, Diane M. Morris, PhD, RD, Mary K. Hunt, MPH, RD, James R. Hebert, ScD Donald R. Harris, PhD, Anne Stoddard, ScD, and Judith K. Ockene, PhD

In a randomized, controlled study of the Treatwell work-site nutrition intervention program, which focused on promoting eating patterns low in fat and high in fiber, 16 work sites from Massachusetts and Rhode Island were recruited to participate and randomly assigned to either an intervention or a control condition. The intervention included direct education and environmental programming tailored to each work site; control work sites received no intervention. A cohort of workers randomly sampled from each site was surveyed both prior to and following the intervention. Dietary patterns were assessed using a semiquantitative food frequency questionnaire. Adjusting for work site, the decrease in mean dietary fat intake was 1.1% of total calories more in intervention sites than in control sites ($P < .005$). Mean changes in dietary fiber intake between intervention and control sites did not differ. This study provides evidence that a work-site nutrition intervention program can effectively influence the dietary habits of workers. (*Am J Public Health.* 1992;82:877–880)

Results

Table 4 presents the relationship of these employee characteristics to total intake of dietary fat and dietary fiber (geometric mean), measured at baseline. As shown in the table, total dietary fat varied significantly by age, sex, and BMI; total dietary fiber varied significantly by age, sex, and education.

TABLE 4—Mean Baseline Level of Dietary Fat and Dietary Fiber by Employee Characteristics and Results of ANOVA Testing of Differences in Means

Employee Characteristic	Mean Total Dietary Fat (% cal.)	F Test (*df*)	*P* Value	Geometric Mean Total Dietary Fiber[a]	F Test (*df*)	*P* Value
Age						
17–35	34.1			12.56		
36–50	35.0			12.41		
51–75	33.3	7.86 (2, 1827)	<.01	14.19	9.96 (2, 1827)	<.01
Education						
High school	34.1			12.16		
Some college	33.9			13.32		
College graduate	34.2	0.37 (2, 1887)	.69	13.46	6.99 (2, 1887)	<.01
Ethnicity						
White	34.2			12.89		
Nonwhite	30.4	0.32 (1, 1908)	.57	16.14	0.01 (1, 1908)	.94
Sex						
Male	33.7			13.47		
Female	34.4	4.28 (1, 1901)	.04	12.37	12.15 (1, 1901)	<.01
Body Mass Index						
Underweight	34.2			11.95		
Acceptable weight	33.8			13.07		
Moderately overweight	33.9			13.13		
Obese	35.5			13.35		
Morbidly obese	37.6	2.50 (4, 1829)	.04	12.09	1.32 (4, 1829)	.26

Note. ANOVA = analysis of variance; *df* = degrees of freedom.
[a]Geometric mean for total dietary fiber intake derived by exponentiating the log value.

Literature Example 14.16 is from the article entitled "Influence of High Voltage Pulsed Direct Current on Edema Formation Following Impact Injury." Read the abstract to learn about the study.

Look at Table 1 and pay particular attention to the *F*-statistics and the corresponding *p*. The values of *p* for both group and time are less than .05. Do we accept or reject the null hypothesis for the two variables, group and time?

[1] Accept/Reject

SS indicates sum of squares, which corresponds to the numerator in the variance formula.

What does mean sum of squares correspond to?

[2] variance/standard deviation

Mean sum of squares of error is an estimate of the variance. If there is no group effect, MS for group and error would be similar. If there is an effect, the group MS will be considerably larger. The proportion of group MS and error MS is *F*.

Compute the proportion of time MS to error MS: [3] ____________________

____

Group X Time indicates the interaction between time and treatment.

Is *F* corresponding to Group X Time significant?

[4] Yes/No

Read the results paragraph. Does the treatment seem effective for the frog limbs?

[5] ________

LITERATURE EXAMPLE 14.16

Influence of High Voltage Pulsed Direct Current on Edema Formation Following Impact Injury

Josette A Bettany
Dale R Fish
Frank C Mendel

Edema results in pain and may lead to reduced functional mobility. High voltage pulsed direct current (HVPC) has recently been advocated for edema control. The purpose of our study was to determine the effect of HVPC on edema formation in frogs. Hind limbs of 20 anesthetized frogs were injured by dropping a 450-g weight onto the plantar aspects of the feet. One hind limb of each frog was randomly selected to receive continuous 120-Hz HVPC at voltages 10% lower than those needed to evoke muscle contraction. Four 30-minute treatments were administered at 1.5-hour intervals beginning 10 minutes after trauma. Limb volumes were measured by water displacement. An analysis of variance for repeated measures and a Newman-Keuls post hoc *test were used to determine the significance of treatment effects. The HVPC significantly* ($p < .01$) *reduced edema formation. We hypothesize that HVPC may also be effective in controlling edema formation after impact injuries in humans. [Bettany JA, Fish DR, Mendel FC: Influence of high voltage pulsed direct current on edema formation following impact injury. Phys Ther 70:219–224, 1990]*

Data Analysis

The hypothesis that limb volumes would increase less in treated limbs than in untreated limbs was tested by a one-way analysis of variance (ANOVA) for repeated measures. Differences were accepted as significant at the .05 level.

Results

High voltage pulsed direct current significantly reduced edema formation in treated limbs as compared with untreated limbs (Tab. 1, Fig. 2). The Scheffé *post hoc* analysis revealed that volumes of treated limbs were significantly less ($p < .01$) than volumes of untreated limbs from the end of the first treatment to the last volume measurement at 17 hours posttrauma.

Table 1. *One-Way Analysis of Variance for Repeated Measures—Effects of Treatment and Time on Changes in Limb Volumes*

Source	*df*	SS	MS	*F*	p^a
Group	1	2540.67	2540.67	12.53	.001
Error	34	6893.60	202.75		
Time	5	2166.14	433.23	25.60	.001
Group × time	5	44.29	8.86	0.52	.758
Error	170	2876.57	16.92		

$^a p < .05$.

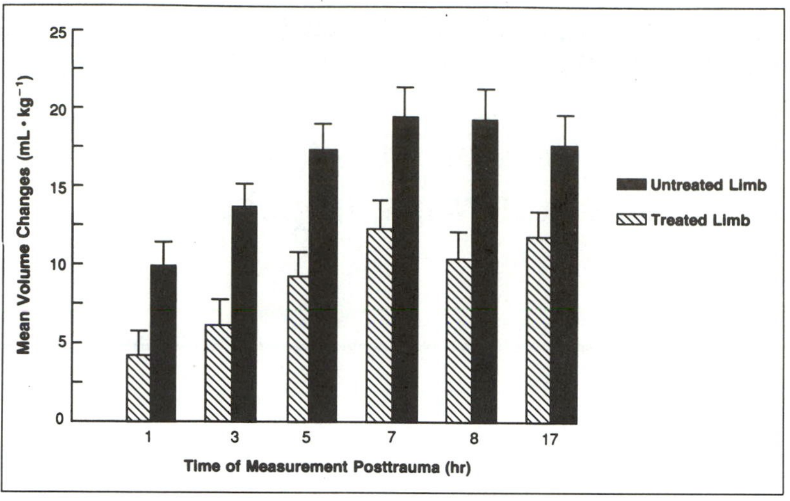
Mean Volume Changes (mL • kg⁻¹)
25
20
15
10
5
0
1
3
5
7
8
17
Time of Measurement Posttrauma (hr)
Untreated Limb
Treated Limb

Some Sample Size Considerations

When planning an experiment it is extremely desirable to consider in detail the statistical analyses that will be used before any data are collected. One question that must be addressed is what size samples to take. Sampling is frequently the most expensive aspect of an experiment or a study, especially when human subjects are involved, and sample size is often limited by the funds available. When animal subjects are used and the intervention is painful or damaging to the animal, you also want to use as few subjects as possible for ethical reasons. On the other hand, you want to use enough subjects that the results of the experiment or study can be significant. This means that a small confidence interval is desirable. As you have seen, the size of the confidence interval is dependent on standard deviation, sample size, and confidence level.

A preliminary computation of sample size based on the considerations in Table 14.5 is often helpful. Occasionally, a preliminary experiment is necessary to obtain a rough estimate of standard deviation. Use information from Table 14.5 and the top part of the figure on the following page and add appropriate arrows for the case in which sample size decreases.

Table 14.5

Factor	Effect on Sample Size	Comment
Variance (σ^2)	The larger the variance, the larger the sample size needed to estimate the variance to a given degree of accuracy.	If all subjects were of exactly the same height ($\sigma^2 = 0$), a sample of $n = 1$ would be sufficient to estimate the population height. The more variation in the sample, the larger the variance and hence the larger the confidence interval.
Significance level, or α error	The smaller the α level, the bigger the sample size needed to achieve that level.	α error quantifies our willingness to be wrong when we reject a hypothesis. As α error gets smaller, the sample size required to obtain significant results becomes larger.
Power of the test $(1 - \beta)$	The larger the required power, the bigger the sample size.	Power is the probability of being correct when the hypothesis is true.
The difference between the sample statistic we may obtain and the hypothesized parameter	As the difference we are willing to accept increases, the sample size decreases.	The difference indicates how close our estimates need to be. The greater the difference we are willing to accept, the smaller the sample size needs to be.

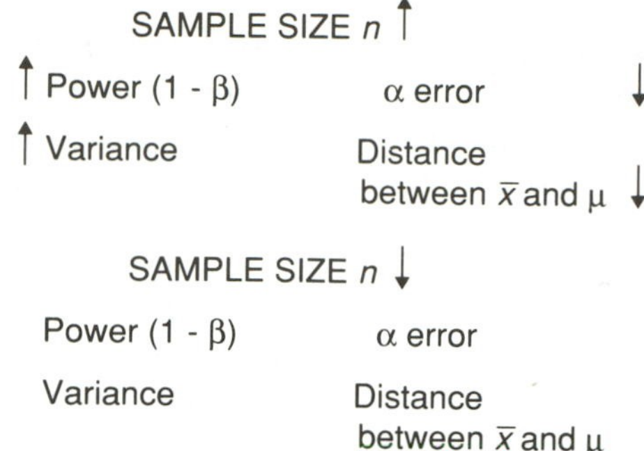

Sample size calculations can be complicated, and researchers generally consult with a specialist when figuring appropriate sample sizes. However, there are tables that are very helpful for estimating sample size. Table 14.6 is for estimating sample size to obtain a 95 percent confidence level that a sample proportion will be within ±.05 of the population proportion. Table 14.7 can be used to estimate sample size to achieve a given probability $1 - \alpha$ that a sample mean falls within $\pm k\sigma$ of the population mean for various values of k. The activities that follow illustrate the use of these tables.

ACTIVITY 14.10

Estimating sample size for a proportion

Use Table 14.6 to determine the sample size needed to obtain a proportion, p, within ±.05 of the population proportion with a 95 percent level of confidence. The first two are done for you as examples:

Population Size	Sample Size
10	10
30	28
45	1 ________
60	2 ________
100	3 ________
150	4 ________
300	5 ________
500	6 ________
1,000	7 ________
3,000	8 ________
20,000	9 ________
100,000	10 ________

TABLE 14.6 Sample Size Needed to Estimate Sample Proportion to Within ±.05 of Population Proportion at 95 Percent Confidence Level

N	*S*	*N*	*S*	*N*	*S*
10	10	220	140	1200	291
15	14	230	144	1300	297
20	19	240	148	1400	302
25	24	250	152	1500	306
30	28	260	155	1600	310
35	32	270	159	1700	313
40	36	280	162	1800	317
45	40	290	165	1900	320
50	44	300	169	2000	322
55	48	320	175	2200	327
60	52	340	181	2400	331
65	56	360	186	2600	335
70	59	380	191	2800	338
75	63	400	196	3000	341
80	66	420	201	3500	346
85	70	440	205	4000	351
90	73	460	210	4500	354
95	76	480	214	5000	357
100	80	500	217	6000	361
110	86	550	226	7000	364
120	92	600	234	8000	367
130	97	650	242	9000	368
140	103	700	248	10000	370
150	108	750	254	15000	375
160	113	800	260	20000	377
170	118	850	265	30000	379
180	123	900	269	40000	380
190	127	950	274	50000	381
200	132	1000	278	75000	382
210	136	1100	285	100000	384

Note: *N* is population size; *S* is sample size.
Source: Reprinted with permission from Krejcie, R.V., and Morgan, D.W., 1970. "Determining Sample Size for Research Activities." *Educational and Psychological Measurement* 30, pp. 607–610.

ACTIVITY 14.11

Estimating sample size for a mean

Use Table 14.7 to answer the following questions. For a 99 percent confidence level that $\bar{x}$, the sample mean, falls between $\mu - .5\sigma$ and $\mu + .5\sigma$, we would need a sample size of $n = 27$. This number is found at the intersection of the

TABLE 14.7 Sample Size Needed to Estimate Population Mean

Values are for the necessary sample size for the probability $1 - \alpha$ that the sample mean $\bar{x}$ of a random sample from a normally distributed population will fall between $\mu - k\sigma$ and $\mu + k\sigma$. (For most naturally occurring populations the restriction to normality is not required if the sample size is larger than 5.) For example, for a probability $1 - \alpha = .99$ that $\bar{x}$ will fall between $\mu - .5\sigma$ and $\mu + .5\sigma$, a sample of size $n = 27$ would be required.

k	.50	.60	.70	.80	.90	.95	.99	.999
.01	4,543	7,090	10,733	16,436	27,061	38,416	66,357	108,306
.05	182	284	430	658	1,083	1,537	2,655	4,333
.10	46	71	108	165	271	385	664	1,084
.15	21	32	48	74	121	171	295	482
.20	12	18	27	42	68	97	166	271
.25	8	12	18	27	44	62	107	174
.30	6	8	12	19	31	43	74	121
.40	3	5	7	11	17	25	42	68
.50	2	3	5	7	11	16	27	44
.60	2	2	3	5	8	11	19	31
.70	1	2	3	4	6	8	14	23
.80		2	2	3	5	7	11	17
.90		1	2	3	4	5	9	14
1.00			2	2	3	4	7	11
1.25			1	2	3	2	5	7
1.50				1	2	2	3	5
1.75					1	2	3	4
2.00						1	2	3
2.50							2	2
3.00							1	2
3.25								2
3.50								1

Source: Reprinted by permission of the publisher from Wilfrid J. Dixon and Frank J. Massey, Jr., *Introduction to Statistical Analysis*, 4th edition. (New York: McGraw-Hill, Inc., 1983).

.5 row and the .99 column. Suppose $\sigma = 2$ and we want $\bar{x}$ to be between $\mu - .5$ and $\mu + .5$ with a confidence level of 95 percent. We would need a sample size of 62. To see this, we notice that $k\sigma = .5$, i.e., $2k = .5$, so $k = .5/2 = .25$. We find the answer at the intersection of the .25 row and the .95 column. For a 90 percent confidence level that $\bar{x}$ is between $\mu - .2\sigma$ and $\mu + .2\sigma$, we need a sample size of [1] _______. For a 99 percent confidence level, with $\sigma = .5$, that $\bar{x}$ is between $\mu - .1$ and $\mu + .1$, we need a sample size of [2] _______. Suppose the sample size was 25 and the sample mean, $\bar{x}$, turned out to be 10.5. The population standard deviation is known to be .5. Our hypothesis says that the population mean, μ, is 10.0. Can we accept this hypothesis at the $\alpha = .05$ level, using a two-sided test? [3] _______

Self-Assessment

Tasks:	How well can you do it?					Page
	Poorly				*Very well*	
1. Transform raw data into z-scores and t-scores.	1	2	3	4	5	447
2. Estimate and interpret a confidence interval for the mean when all the essential information is given.	1	2	3	4	5	450
3. Estimate and interpret a confidence interval for a proportion when all the essential information is given.	1	2	3	4	5	454
4. Use a confidence interval for making decisions about μ.	1	2	3	4	5	464
5. Test a hypothesis about the mean of a single population when all the essential data are given, using three different methods.	1	2	3	4	5	464
6. Test hypotheses about means of two dependent populations when all the essential data are given, using three different methods.	1	2	3	4	5	478
7. Test hypotheses about means of two independent populations when all the essential data are given, using three different methods.	1	2	3	4	5	494
8. Test hypotheses about means of three or more groups, using analysis of variance techniques.	1	2	3	4	5	501

9. Estimate the sample size needed to obtain a sample proportion within ±.05 of the population proportion with a 95 percent confidence level when the population size is known.	1	2	3	4	5	520
10. Estimate the sample size needed to obtain a sample mean within a given number of standard deviations of the population mean for a given confidence level.	1	2	3	4	5	522

Answers for Chapter 14

Answers for Activity 14.1: 1) −1, +1 2) −2, +2 3) −1, +1 4) −1, 0 5) −1, +0.5 6) −1, +1 7) −2, +1 8) −0.5, +0.5

Answers for p. 448: 1) Greater 2) Shift

Answers for Figure 14.3: The shading on the first graph is between 61 and 113. 1) 64–96 2) 20–44 3) 68–196 4) 35–65 5) 100–280 6) 64–76

Answers for Figure 14.4: The shading in Figure 14.4 is between 53 and 121. (We rounded z to 2.6). Your shading in Figure 14.3 should now be between the following values: 1) 59.2–100.8 2) 16.4–47.6 3) 48.8–215.2 4) 30.5–69.5 5) 73–307 6) 62.2–77.8

Answer for p. 450: 1) Higher

Answers for p. 451: 1) Confidence interval for sample number 26 2) 2.5%

Answer for p. 453: 1) 2.262

Answers for p. 454: 1) (22.6, 27.8) 2) (65.2, 67.6) 3) (127.8, 153.2) 4) (72.7, 79.1) 5) (93.2, 102.8) 6) Different

Answers for p. 455: These can be computed exactly the same way the C.I. above was computed, except that $z = 1.28$. You then get 1) 56% 2) 74%

Answers for Activity 14.3: 1) 1.6 2) .221 3) .066 4) (.56, .78) 5) 2.6 6) .09 7) .04 8) (.79, 1.01)

Answers for p. 457: 1) t-distribution or t-test 2) Analysis of variance

Answers for Literature Example 14.1: 1) 80%; 55% 2) 24.3 3) See figure below. They all overlap each other. 4) The overlapping confidence intervals confirm this statement. 5) No; confidence intervals

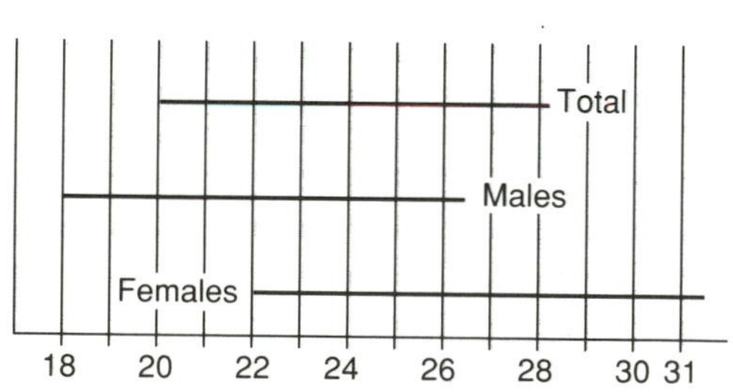

Answers for Literature Example 14.2: 1) Appropriate 2) (−13.94, −8.45) 3) (−.13, .35) 4) $\mu_1 - \mu_2 = 0$; $\mu_1 - \mu_2 = 0$ 5) Reject the BMI null hypothesis, as the C.I. does not include 0. 6) Accept the cholesterol null hypothesis, as the C.I. includes 0.

Answers for Activity 14.4: 1) Both 2) Two 3) Lower 4) One 5) Upper 6) One

Answers for Activity 14.5: 1) The mean systolic blood pressure from the sample is the same as that reported in the literature for the general population, i.e., 130; H_0; $\mu = 130$. Alternative hypothesis: The mean systolic blood pressure from the sample is not the same as that reported in the literature for the general population; $H_a \neq 130$. 2) $124.7 \pm 2.06 * 2.9 = (118.7, 130.7)$ 3) Yes 4) Accept 5) $130 \pm 2.06 * 2.9 = (124, 136)$ 6) Between means 7) Accept 8) −2.06 and +2.06 9) −1.82 10) Between critical *t*-values 11) Accept 12) Accept

Answers for Literature Example 14.3: 1) There is no difference in percentages of inappropriate hospitalization between physicians licensed more than fifteen years ago and those licensed less than fifteen years ago. 2) Reject 3) Years from Licensing, Sex 4) Yes 5) No; −1 to 16.3 covers 0. This means that the difference is *not* statistically significant.

Answers for Literature Example 14.4: 1) The difference in the mean cholinesterase between those who were poisoned and not poisoned is 0; $\mu_1 - \mu_2 = 0$. 2) Reject 3) Changes work clothes less than daily; number of symptoms 4) Yes, you should agree.

Answers for Literature Example 14.5: 1) Overlap 2) Accept 3) Overlap 4) Accept 5) Do not overlap 6) Reject 7) Overlap 8) Accept 9) Do not overlap 10) Reject 11) Do not overlap 12) Reject

Answers for Literature Example 14.6: 1) Does 2) Reject 3) Convenience sample 4) The results of the convenience sample can only be trusted to be accurate for the sample itself. A random sample would provide statistically more reliable results because those results could be generalized to the population from which the sample was drawn.

Answers for Activity 14.6: 1) (−17.62, 0.12) 2) Yes 3) Accept; this value is very close to the rejection region, however. Further research with a larger sample would be warranted. 4) 2.365 5) −8.87 and +8.87 6) Between critical means 7) Accept 8) ±2.365 9) $(-8.75 - 0)/3.75 = -2.33$ 10) Between *t*-values 11) Accept

Answers for Literature Example 14.7: For each limb, limb volumes are the same at the beginning and at the end of the treatment interval. 2) For each limb, limb volumes are the same at the end of one treatment interval and the beginning of the next. 3) 2.093 4) Accept 5) Reject 6) Reject 7) Reject 8) Yes 9) Lower 10) It might be advisable to look for studies that document use of HVPC on humans. Animal experiments do not directly lead to application in humans.

Answers for Literature Example 14.8: 1) Dependent 2) No, it depends on the judgment of clinicians. 3) No, the standard deviation of the differences is not given and cannot be computed from the given information. As a matter of fact, the standard deviations given are irrelevant for these purposes.

Answers for Literature Example 14.9: 1) First Examination and Adult Gain in Body-Mass Index 2) Reject 3) Accept 4) Reject 5) There is a possible relationship between nongenetic aspects of obesity and the syndrome of dyslipidemic hypertension.

Answers for Literature Example 14.10: 1) Our tables give ± 2 as the closest approximation. More accurate tables give ± 1.96. 2) −1.99, corresponding to fatal or nonfatal myocardial infarction 3) Yes

Answers for Literature Example 14.11: 1) Preoperative and intraoperative ventricular arrhythmias 2) Preoperative and postoperative ventricular arrhythmias; intraoperative and postoperative ventricular arrhythmias

Answers for Activity 14.7: 1) 6.6 2) (−20.7, 11.5) 3) Yes 4) Accept 5) 2.45 6) ±16.1 7) −4.6 8) Between means 9) Accept 10) ±2.45 11) (−4.6 − 0)/6.6 = .7 12) Between *t*-values 13) Accept

Answers for Literature Example 14.12: 1) −2, +2 2) Reject 3) Reject 4) Independent

Answers for Literature Example 14.13: 1) 200% 2) 223% 3) 162% 4) Median 5) Reject

Answers for Activity 14.8: 1) 6 2) 18 3) Within 4) 5 5) 3 6) 5 7) 7 8) 92 9) 2 10) 10 11) 32 12) 44 13) 48 14) 24 15) 0 16) 24 17) 48 18) 2 19) 15 20) 48/2 = 24 21) 44/15 = 2.93 22) 8.18 23) 3.68 24) Larger 25) Reject

Answers for p. 504: 1) (1.5, 4.5) 2) (3.5, 6.5) 3) (5.5, 8.5) 4) Some

Answers for Activity 14.9: Your answers may differ slightly because of roundoff error. 1) 7, 8, 9 2) 24 3) Within 4) .42 5) 2.6 6) −2.13 7) 1 8) 677.8 9) 127.71 10) 274.88 11) 188 12) 590.6 13) 87.2 14) 32.45 15) 51.69 16) 3.06 17) 87.2 18) 2 19) 21 20) 43.6 21) 28.1 22) 1.55 23) 3.47 24) Smaller 25) Accept

Answers for p. 509: 1) 87.2 2) 43.6 3) 1.55 4) 590.6 5) 28.1 6) 677.8 7) 5.36 8) Approximately (−4, 19) 9) 5.02 10) Approximately (30, 51) 11) 4.73 12) Approximately (50, 69)

Answers for p. 510: 1) Low int. overlaps high int. slightly

Answers for p. 511: If the means are close, between SSq will be relatively smaller than within SSq, so *F* will be smaller. If the means are farther apart between SSq will be relatively larger than within SSq, so *F* will be larger.

Answers for Literature Example 14.14: 1) Log transformation 2) $P < .0001$ 3) .05 4) Type of hypertension: None, Borderline isolated systolic, and isolated systolic 5) All the variables are significant at the .05 level except Male %, High-density lipoprotein, History of diabetes, and Current Smoker.

Answers for Literature Example 14.15: 1) Randomly 2) $\mu_1 = \mu_2 = \mu_3$ 3) Reject 4) Sex, Body Mass Index 5) Yes 6) A work-site nutrition intervention program can influence workers' dietary habits.

Answers for Literature Example 14.16: 1) Reject 2) Variance 3) Time MS/Error MS = $F = 25.6$ 4) No; $p > .05$ 5) Yes

Answers for figure on p. 520: Power and variance arrows should point down; α error and distance arrows should point up.

Answers for Activity 14.10: 1) 40 2) 52 3) 80 4) 108 5) 169 6) 217 7) 278 8) 341 9) 377 10) 384

Answers for Activity 14.11: 1) 68 2) 166 3) No. For $n = 25$ and $1 - \alpha = .95$, we see from the table that $k = .40$. So $\mu + k\sigma = 10 + (.40 * .5) = 10.2$. This is the upper bound for the 95 percent confidence interval, so 10.5 is outside that interval. The probability is less than .05/2 = .025 that we would get a sample mean larger than 10.2, so the probability that we would get a sample mean as large as 10.5 is less than .025.

CHAPTER

15

Venturing into the Unknown: Association and Prediction

Prelude

As the cheering continued, Rhyme leaned forward and touched Milo gently on the arm.

"They're shouting for you," she said with a smile.

"But I could never have done it," he objected, "without everyone else's help."

"That may be true," said Reason gravely, "but you had the courage to try; and what you can do is often simply a matter of what you *will* do."

"That's why," said Azaz, "there was one very important thing about your quest that we couldn't discuss until you returned."

"I remember," said Milo eagerly. "Tell me now."

"It was impossible," said the king, looking at the Mathemagician.

"Completely impossible," said the Mathemagician, looking at the king.

"Do you mean—" stammered the bug, who suddenly felt a bit faint.

"Yes, indeed," they repeated together; "but if we'd told you then, you might not have gone—and, as you've discovered, so many things are possible just as long as you don't know they're impossible."

— Norton Juster
The Phantom Toll Booth, p. 247

Chapter Outline

Self-Assessment

Tasks:	How well can you do it?					Page
	Poorly				*Very well*	
1. Interpret the correlation coefficient, r.	1	2	3	4	5	530
2. Interpret r^2, $(1 - r^2)$.	1	2	3	4	5	558
3. Formulate and test hypotheses about r using critical values of r and confidence intervals.	1	2	3	4	5	541
4. Estimate a confidence interval for a population correlation coefficient.	1	2	3	4	5	541
5. Predict y when x is known, on the basis of a line drawn using the least squares method.	1	2	3	4	5	550
6. Interpret an analysis of variance table for regression.	1	2	3	4	5	556
7. Interpret a regression equation.	1	2	3	4	5	551

Introduction

We have analyzed the SQD and FED data from different perspectives. We described the variables with single-number summaries, displayed them in tables and graphs, tested hypotheses about the effectiveness of interventions, and estimated population parameters. In this chapter we will explore the relationship between variables and learn how to predict the values of one variable from those of another. Such relationships are not readily discernible from the raw data, and we need to use a variety of statistical tools to understand different aspects of the data.

How do we know which relationships to explore? Researchers generally have some idea what they want to study from their past experience and from reading the literature. Sometimes a research direction is a shot in the dark, especially if one is working in an uncharted field. Often a choice of hypotheses or variables is dictated by special circumstance. Our motivation in choosing variables and relationships to use for examples in this book has been based on our knowledge and experience in teaching statistics to college students. We have been more concerned with finding examples that are pedagogically revealing than with finding examples that constitute actual cutting-edge research challenges. The examples from the literature that we have presented, however, give a true picture of the sort of questions that researchers actually ask.

Figure 15.1 outlines the various tests associated with different types of research questions. Consider the many variables in the student questionnaire data. We may wish to explore the association between WEIGHT and HEIGHT, SMOKE and HEALTH, SEX and SMOKE, AGE and WEIGHT, or STRESS and TIMEOUT. Notice that some of these variables are quantitative (continuous) and others are qualitative (categorical). Different kinds of relationship measures have been devised to deal with variables measured in these different ways. If both variables are categorized (or categorical) variables, we use [1]______________. If both variables are continuous, we use [2]______________. For predicting the value of one variable based on the value of another, we use [3]______________ analysis. Some statisticians contend that these different methods for measuring relationships are a result of an effort to ease computation and most are descendants of the correlation coefficient. So we will use the correlation coefficient in all cases as a preliminary screening device for determining relationships between variables. We will discuss in detail the Pearson correlation coefficient, named after Karl Pearson (1852–1936), the English statistician who also developed the χ^2 statistic.

Association of Continuous Variables: Correlation

The central question that we wish to consider when dealing with continuous variables is, "How does the variation in one variable, x_i, affect the variation in another, y_i?" It is assumed that the values of the two variables are linked in some logical way, for instance, temperature and pressure both taken at the same time, or height and weight of the same person.

One difficulty that surfaces immediately is that of units. Two continuous

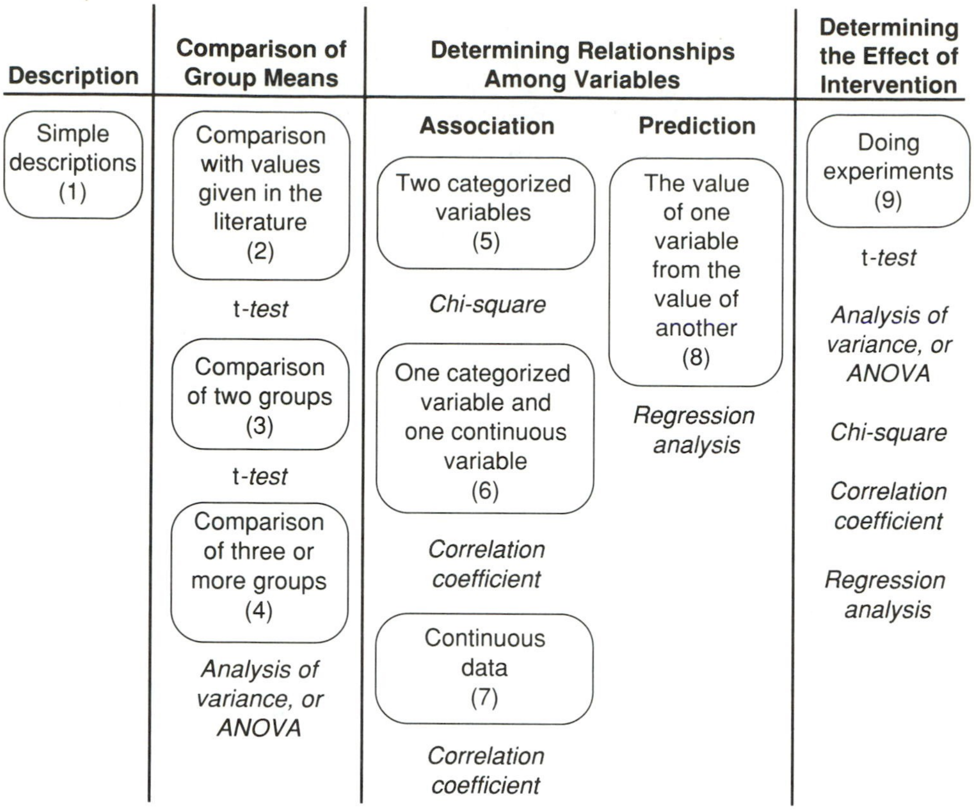

Description

Simple descriptions (1)

Comparison of Group Means

Comparison with values given in the literature (2)

t-*test*

Comparison of two groups (3)

t-*test*

Comparison of three or more groups (4)

Analysis of variance, or ANOVA

Determining Relationships Among Variables

Association

Two categorized variables (5)

Chi-square

One categorized variable and one continuous variable (6)

Correlation coefficient

Continuous data (7)

Correlation coefficient

Prediction

The value of one variable from the value of another (8)

Regression analysis

Determining the Effect of Intervention

Doing experiments (9)

t-*test*

Analysis of variance, or ANOVA

Chi-square

Correlation coefficient

Regression analysis

FIGURE 15.1 Types of Research Questions and Their Associated Tests

variables, such as PREWAT (y_i) and PREVO2 (x_i) in the FED, are measured in different units. In order to compare their co-variation we must first transform them into dimensionless quantities, taking the mean and standard deviations of both x and y into account. The z-score provides such a dimensionless quantity. Multiplying the respective z-scores of two variables by one another and averaging the products give a relatively simple measure of association called the **correlation coefficient**. The correlation coefficient is usually denoted r and is defined as the mean of the products of the z-scores of two variables,

$$r = \frac{\sum z_x * z_y}{n}$$

where x and y are subscripts referring to the x and y variables and n signifies the number of paired observations in the sample.

Interpretation of the Correlation Coefficient

The correlation coefficient (r) actually measures the strength of the *linear* relationship between two variables. When both z_x and z_y are large, their product will be large, increasing r; when both are small, their product will be small, decreasing r. If one variable consistently gets smaller as the other gets larger, the z-score of the one will be negative when the z-score of the other is positive, and therefore the product will be negative. In this case, the correlation coefficient will become negative. r varies between -1 and $+1$. A correlation coefficient of 0 indicates no linear relationship; -1 and $+1$ indicate a perfect linear relationship. With a perfect linear relationship, the points on a graph of the two variables fall on a straight line. If the value of y increases as the value of x increases, as shown in part (c) of Figure 15.2, the coefficient will be $+1$; if the value of y decreases as the value of x increases, as in part (a) of Figure 15.2, the coefficient will be -1.

There are several different conditions that can give a correlation coefficient of 0. Even when most of the values approximate a linear relationship, one outlier can render the correlation coefficient very small or even 0, as in part (d) of Figure 15.2. If we find that an outlier is heavily influencing the correlation coefficient, it is a good idea to double-check that value and perhaps eliminate it from the analysis, provided that further analysis gives strong reason to believe that it is an erroneous value.

The correlation coefficient only measures the strength of linear (straight-line) relationships. It is possible for x and y to be strongly related and yet have a correlation coefficient of 0, as shown in part (e) of Figure 15.2. The correlation

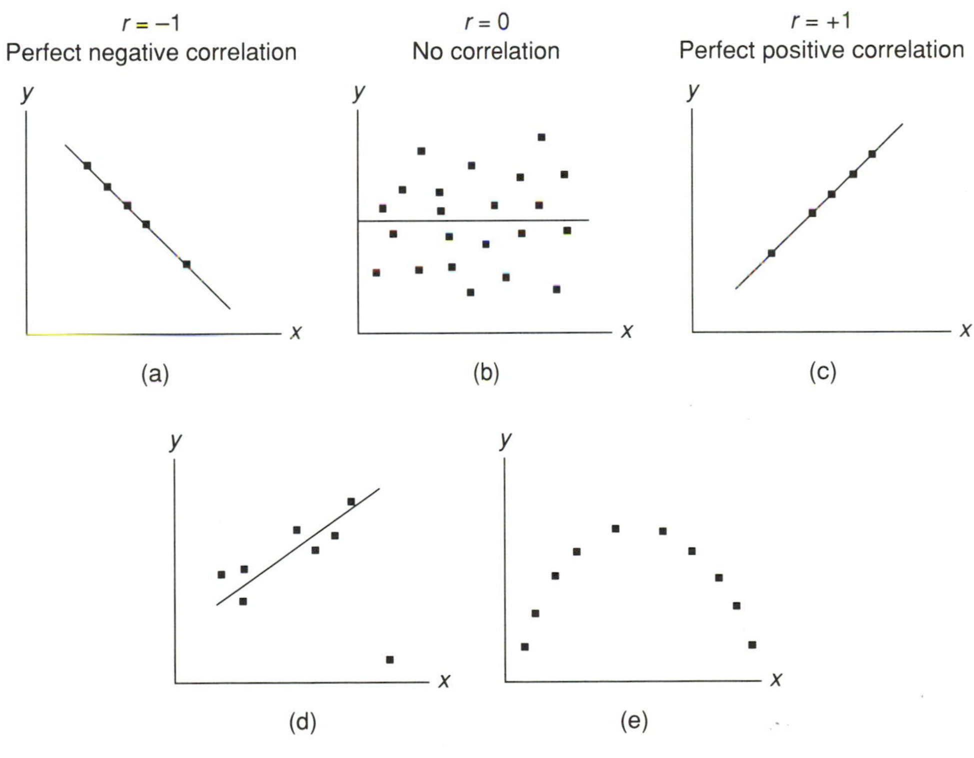

FIGURE 15.2 Scatter Plots Depicting Different Extreme Correlation Coefficients

coefficient may underestimate the strength of a curvilinear relationship. The correlation coefficient measures only the [1]linear/curvilinear relationship. So it is always a good idea to plot the variables to check for a non-linear relationship. If the data points are distributed in other than a linear fashion, the curvilinear association between them is not measured by the correlation coefficient, so it is inappropriate to use the correlation coefficient as a measure of association in that case.

When r has a positive value, the variables are said to have **positive** correlation. In such cases, when one variable increases, the paired variable [2] increases/decreases. When r has a negative value, the variables are said to have a **negative correlation**; in that case, when one variable increases the paired variable will [3] increase/decrease. When $r = 0$ the two variables do not vary with each other. Extreme values also may make the correlation coefficient equal to [4] _______. If we compute a correlation coefficient for data that have a curvilinear distribution we may obtain a correlation coefficient of [5] _______ because the correlation coefficient is designed to measure [6] linear/curvilinear relationships only.

Examples from the Faculty Exercise Data

Figure 15.3 shows a plot of pre maximum heart rate and pre heart rate percent values taken from the faculty exercise data. When pre maximum heart rate increases, the paired variable pre heart rate % [7]increases/decreases. The correlation coefficient is [8]positive/negative. Although these points do not all fall on the same straight line, they seem to cluster around a line. Since the correlation

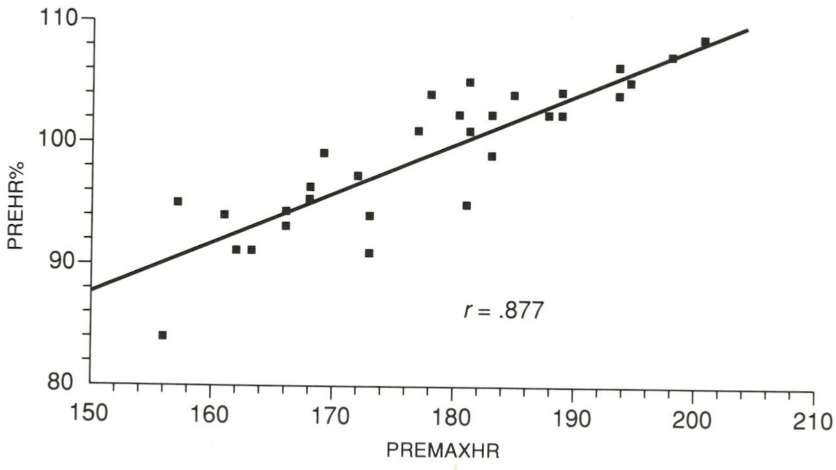

FIGURE 15.3 Scatter Plot and Correlation of Pre Maximum Heart Rate (PREMAXHR) and Pre Heart Rate % (PREHR%)

coefficient is .877, a number fairly close to 1.00, these variables are said to be highly correlated. The variations from linearity result from unknown causes, but the relationship between the variables is approximately a [1]linear/curvilinear relationship.

For the two variables graphed in Figure 15.4, r has a negative value, so the variables are said to have negative correlation. When one increases, the paired variable decreases. The correlation coefficient for the data in the figure is [2]________. Here we see that the points are more dispersed than the points in Figure 15.3. The variables are negatively correlated and are not as strongly correlated as in the previous example (0.664 is smaller than 0.887). It is difficult to judge the amount of dispersion from this kind of graph, because changes in scale used on the axes can disperse the points visually by different amounts.

Figure 15.5 graphs the same variables shown in the previous two figures, but with the variable values transformed into their z-scores.

Consider the relationship between blood pressure and heart rate graphed in Figure 15.6. Does pre resting systolic blood pressure (PRERESSYS) increase as the pre heart rate (PREHR) increases? [3]Yes/No For any given systolic blood pressure the heart rate takes on a wide range of values. Heart rate and systolic blood pressure do not seem to have any relationship to each other.

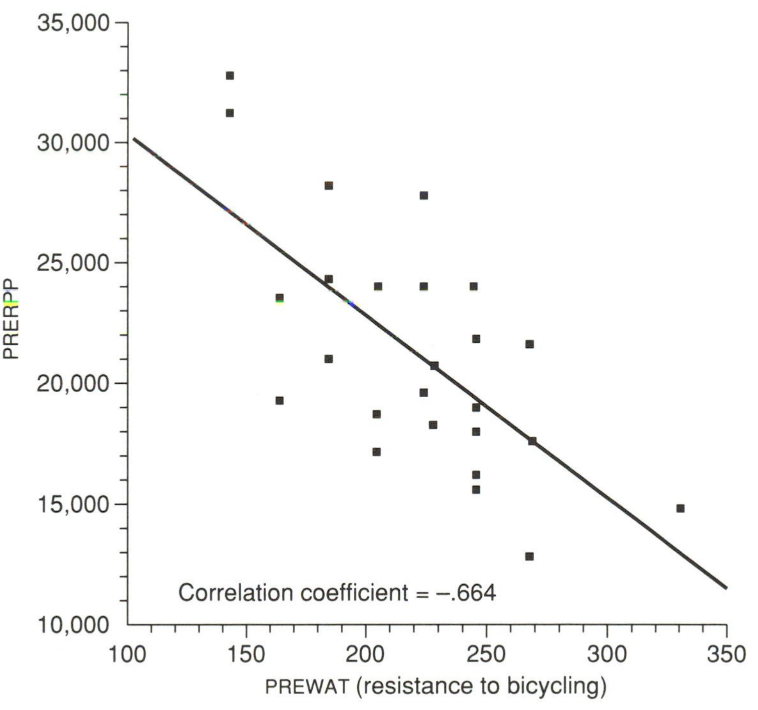

FIGURE 15.4 Scatter Plot and Correlation Coefficient of PREWAT and PRERPP

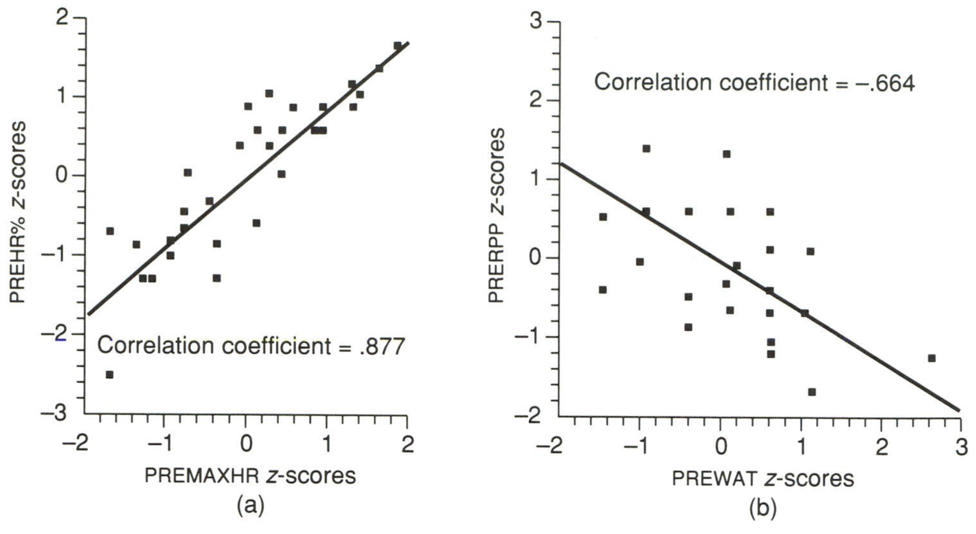

FIGURE 15.5 Scatter Plots of Variables with Standardized Values

Covariance of Two Variables Versus Variance of One Variable

You are familiar with variance and standard deviation. Both x and y have standard deviations that may be computed by using the formulas

$$s_x = \sqrt{\frac{\Sigma\,(x - \bar{x})^2}{n - 1}} \qquad s_y = \sqrt{\frac{\Sigma\,(y - \bar{y})^2}{n - 1}}$$

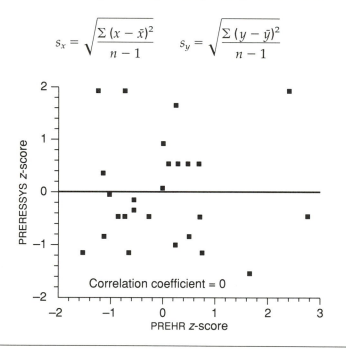

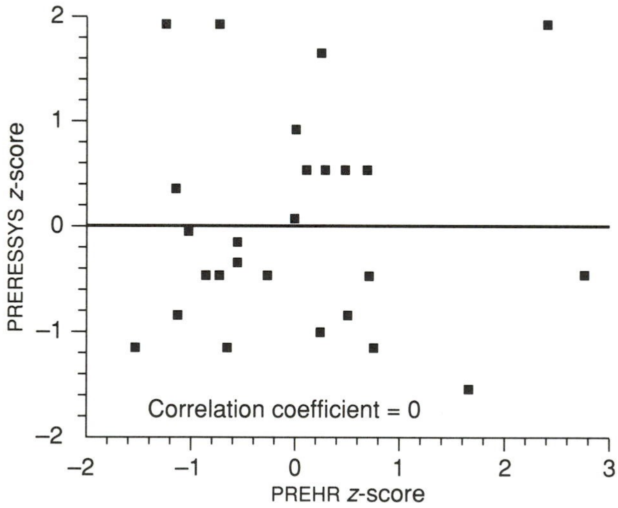

FIGURE 15.6 Scatter Plot of PREHR and PRERESSYS Using Standardized Values (Correlation Coefficient = 0)

where s_x is the standard deviation of x and s_y is the standard deviation of y. We want to know how x and y vary together. This is measured by the sample covariance, defined by the following formula

$$\text{Sample covariance } s_{xy} = \sum \frac{(x - \bar{x})(y - \bar{y})}{n - 1}$$

Compare the expression in the numerator of the sample covariance with the expressions in the numerators of the formulas for variance of x and y.

$$s_x^2 = \frac{\Sigma\,(x - \bar{x})(x - \bar{x})}{n - 1} \qquad s_y^2 = \frac{\Sigma\,(y - \bar{y})(y - \bar{y})}{n - 1}$$

Instead of squaring the deviation, as in the case of one variable (x or y), we [1]____________ the deviations of x and y together and sum the products.

To illustrate how the sample covariance behaves, we have plotted PREMAXHR against PREHR% in Figure 15.7. Note that the graph is divided into four quadrants separated by lines representing $\bar{x}$ and $\bar{y}$. Quadrant 1 contains all the points representing x and y values [2]smaller/greater than the means of x and y. Quadrant 3 contains all the points representing x and y values [3]smaller/greater than the means of x and y. Quadrant 2 contains all the points representing x values [4]smaller/greater than the mean of x and y values [5]smaller/greater than the mean of y. Quadrant 4 contains all the points representing x values [6]smaller/greater than the mean of x and y values [7]smaller/greater than the mean of y.

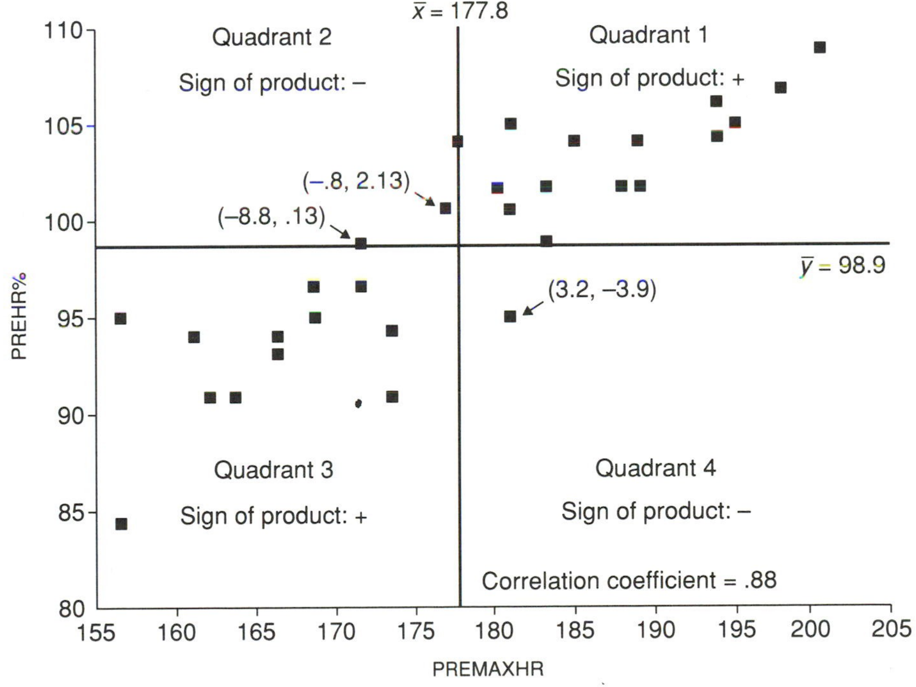

FIGURE 15.7 Scatter Plot of PREMAXHR and PREHR%

Activity 15.1

Determining the quadrant in which points fall

In the rightmost column of the table below, circle the number of the quadrant in which each point falls. Refer to Figure 15.7.

Row	PREMAXHR (x)	PREHR% (y)	$(x - \bar{x})$	$(y - \bar{y})$	$(x - \bar{x})(y - \bar{y})$	Quadrant
1	169	99	−8.7800	0.1300	−1.141	1 2 3 4
2	180	102	2.2200	3.1300	6.949	1 2 3 4
3	163	91	−14.7800	−7.8700	116.319	1 2 3 4
4	195	105	17.2200	6.1300	105.559	1 2 3 4
5	183	99	5.2200	0.1300	0.679	1 2 3 4
6	189	104	11.2200	5.1300	57.559	1 2 3 4
7	162	91	−15.7800	−7.8700	124.189	1 2 3 4
8	181	105	3.2200	6.1300	19.739	1 2 3 4
9	173	91	−4.7800	−7.8700	37.619	1 2 3 4
10	168	95	−9.7800	−3.8700	37.849	1 2 3 4
11	177	101	−0.7800	2.1300	−1.661	1 2 3 4
12	189	102	11.2200	3.1300	35.119	1 2 3 4
13	194	106	16.2200	7.1300	115.649	1 2 3 4
14	178	104	0.2200	5.1300	1.129	1 2 3 4
15	166	93	−11.7800	−5.8700	69.149	1 2 3 4
16	168	96	−9.7800	−2.8700	28.069	1 2 3 4
17	173	94	−4.7800	−4.8700	23.279	1 2 3 4
18	157	84	−20.7800	14.8700	308.999	1 2 3 4
19	161	94	−16.7800	−4.8700	81.719	1 2 3 4
20	198	107	20.2200	8.1300	164.389	1 2 3 4
21	157	95	−20.7800	−3.8700	80.419	1 2 3 4
22	201	109	23.2200	10.1300	235.219	1 2 3 4
23	185	104	7.2200	5.1300	37.039	1 2 3 4
24	181	95	3.2200	−3.8700	−12.461	1 2 3 4
25	172	97	−5.7800	−1.8700	10.809	1 2 3 4
26	188	102	10.2200	3.1300	31.989	1 2 3 4
27	183	102	5.2200	3.1300	16.339	1 2 3 4
28	166	94	−11.7800	−4.8700	57.369	1 2 3 4
29	194	104	16.2200	5.1300	83.209	1 2 3 4
30	181	101	3.2200	2.1300	6.859	1 2 3 4

Calculation of Sample Covariance and Correlation Coefficient

Sample covariance is dependent upon the units in which x and y are measured. Covariance is standardized by dividing it by the product of the standard deviations of x and y, which gives the correlation coefficient. The correlation coefficient r_{xy} is a standardized measure of association: $r_{xy} = s_{xy}/s_x * s_y$. For the data in Activity 15.1,

$$\text{Sum of } (x - \bar{x})(y - \bar{y}) = 1{,}878$$

$$\text{Degrees of freedom } (n - 1) = 29$$

$$s_x = 12.47$$

$$s_y = 5.92$$

$$s_{xy} = 64.7$$

The correlation coefficient for the data is therefore

$$r_{xy} = \frac{s_{xy}}{s_x s_y} = {}^{1}\_\_\_\_\_\_\_\_.$$

This is the same correlation coefficient that we introduced earlier (denoted r). The two formulas can be shown to be mathematically equivalent.

Consider Figure 15.8. The graph in part (a) has an outlier (an extreme value) in quadrant 4. Without it the values of x and y have a strong correlation. When the outlier is included, the correlation coefficient is [2]________. For the

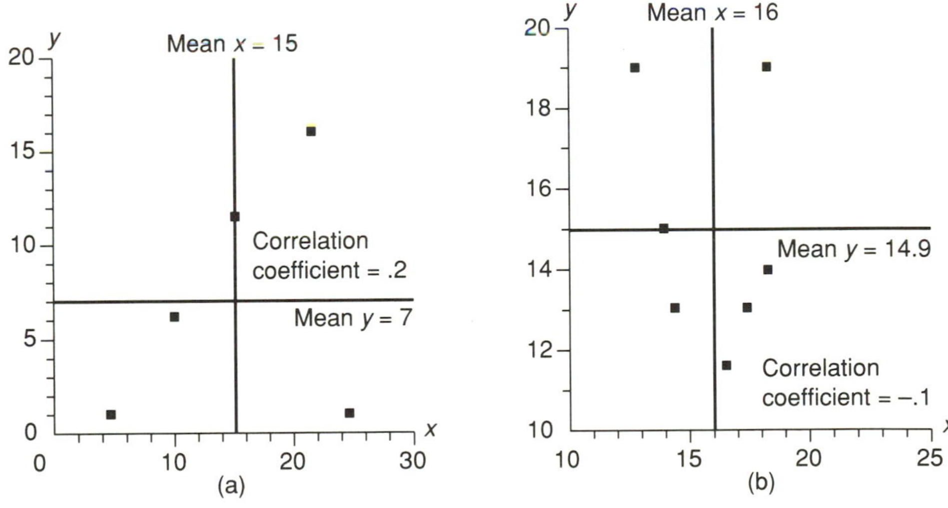

FIGURE 15.8 The Influence of Outliers and Curvilinear Distribution on Correlation Coefficients

outlier, the value of $(x - \bar{x})(y - \bar{y})$ is negative, and hence (since all the other terms are positive) the outlier causes the correlation coefficient to be underestimated. We see that outliers in quadrants 2 and 4 can cancel out the products with positive values in quadrants 1 and 3 and result in underestimation of the correlation coefficient.

The data in part (b) have a correlation coefficient of [1]_______ because x and y have a curvilinear relationship. You can see how the plus and minus values of the products in the covariance cancel each other out. Even when there is a strong relationship between the variables we can obtain $r = 0$, because the relationship is non-linear. We must keep in mind that the correlation coefficient measures the strength of the linear relationship only; if the relationship is non-linear, the correlation coefficient is not meaningful.

Equivalent Versions of the Correlation Coefficient Formula

As noted, the correlation coefficient is a measure of the strength of a linear association. Plotting tells us whether its use is appropriate. The graph will tell us if the data points are distributed in a curvilinear fashion.

$$\text{Correlation coefficient, } r_{xy} = s_{xy}/s_x * s_y$$

For the data in Activity 15.1, we calculated

$$r_{xy} = \frac{s_{xy}}{s_x s_y} = \frac{64.7}{12.47 * 5.92} = .877$$

An equivalent formula for the correlation coefficient is

$$r = \frac{\Sigma\,\{(x_i - \bar{x})(y_i - \bar{y})\}}{\sqrt{\Sigma\,(x_i - \bar{x})^2(y_i - \bar{y})^2}}$$

In this formula you should readily recognize covariance in the numerator and standard deviations of x and y in the denominator. (The $n - 1$ factors cancel out.)

We also defined r using the formula

$$r = \frac{\Sigma\, z_x * z_y}{n}$$

where x and y are subscripts referring to the x and y variables and n signifies the number of paired observations in the sample.* This formula is mathematically equivalent to the other correlation coefficient formulas. You may use whichever formula makes the meaning of r intuitively accessible to you. Please review the calculation of r given in Chapter 7.

Estimation of the Population Correlation Coefficient, ρ

The population correlation coefficient ρ (rho) is estimated by the confidence interval. Unlike C.I. for means, the computation of C.I. for the population correlation coefficient is cumbersome. So we satisfy ourselves by reading the C.I. of ρ from the diagram of confidence belts for ρ in Figure 15.9.

1. Locate the value of r that you calculated on the horizontal axis at the bottom or top of the diagram.
2. Follow the vertical line that is closest to the calculated r to the two symmetrical curved lines (above and below) that correspond to your sample size.
3. At the points where the vertical r line intersects each of the sample size curves, find the closest horizontal lines and follow them to the vertical ρ axis. Read the upper and lower C.I. estimates on the vertical axis.

For example, if $r = .6$ and $n = 50$, the C.I. will extend from .4 to .75. You determine this by following the steps listed above.

1. Find 0.6 (sample correlation coefficient) on the horizontal axis.
2. Follow the vertical line either down or up. Find the intersections of this vertical line with the sample size curves for $n = 50$ (two different places). Follow the nearest horizontal lines from the intersection points over to the vertical scale, which gives the endpoints of the 95 percent confidence interval for the population correlation coefficients.
3. Read the endpoints of the confidence interval as 0.4 and 0.75.

If the size of your sample is not given on the curves, choose the sample size that is smaller than and closest to your sample size. The smaller sample size gives a larger confidence interval, and it is preferable to overestimate the size of a confidence interval than to underestimate it.

---

* Note that this formula assumes that the z-scores were computed using the parameters σ and μ rather than sample statistics. The difference is that we use N in the denominator when calculating σ instead of $n - 1$ as we do when we calculate s. The larger n is, of course, the less difference it makes whether you divide by the population size or by one less than the population size.

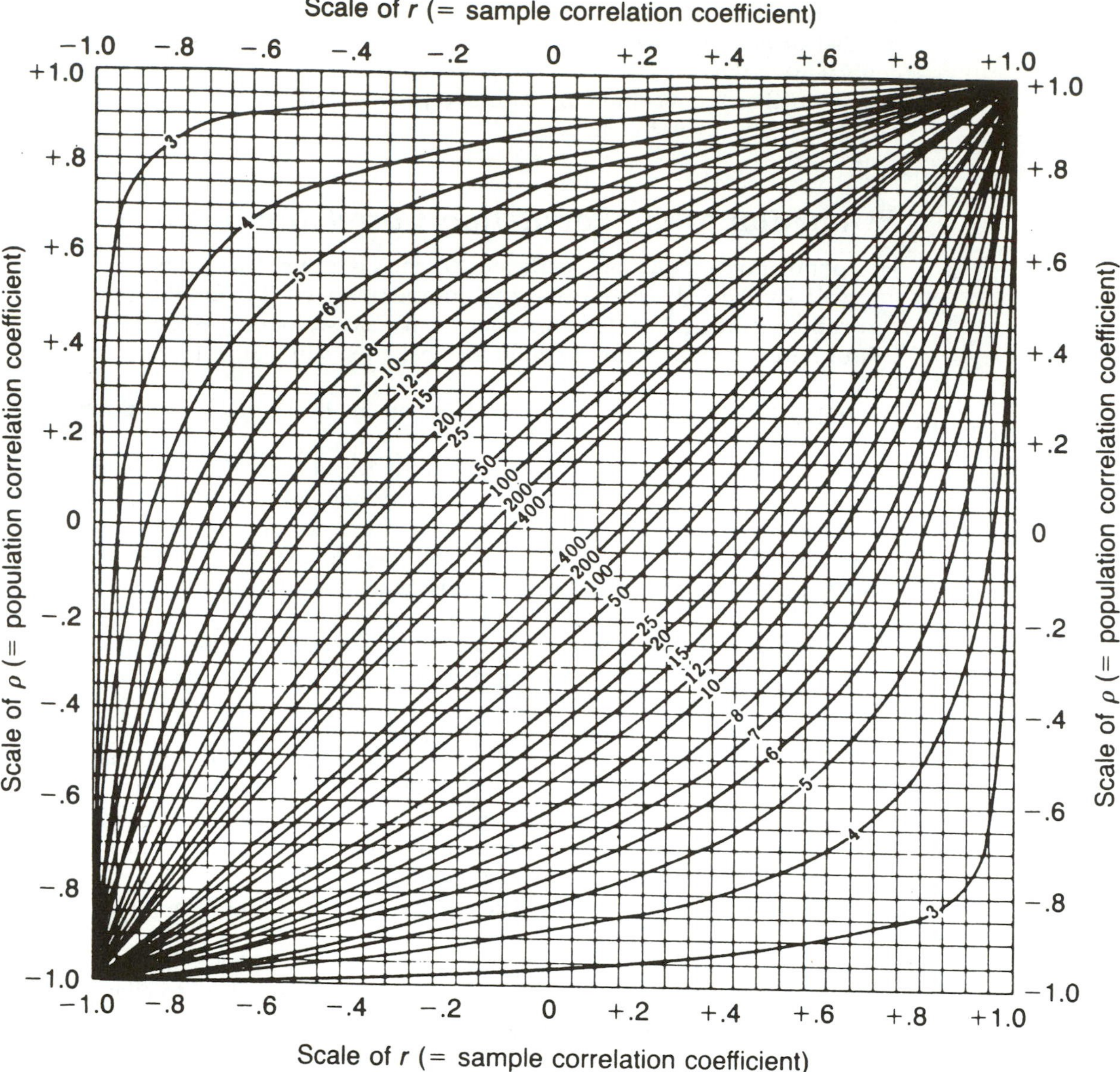

FIGURE 15.9 Confidence Belts for the Correlation Coefficient ($1 - \alpha = .95$)

ACTIVITY 15.2

Calculation of confidence intervals for correlation coefficients

Use Figure 15.9 to find the following confidence intervals. The variable names are given in the first column and the sample correlation coefficient is given in the second column. The sample size was 30. Since no confidence belts for $n =$ 30 are shown in the figure, you should use the curves for $n = 25$.

Variables $n = 30$	**Correlation coefficient,** r	**Approximate confidence interval**
PREMAXHR PREHR%	0.88	[1](_______, _______)
PREHR PRERESSYS	0.00	[2](_______, _______)
PREMAXHR PREVO2	0.37	[3](_______, _______)
PREWT PREFAT%	0.52	[4](_______, _______)

Tests for the Correlation Coefficient

We will discuss three ways of testing hypotheses about r:

1. Read the confidence interval from the confidence belts diagram and see whether the hypothesized ρ is covered.
2. Look up the critical value of r in a table and compare the calculated r values with the critical r.
3. Compute a t-value using the calculated r and compare it with the critical t-value from the t-table.

We have seen that when testing hypotheses about means, if the C.I. does not cover the hypothesized μ, we reject the null hypothesis. Similarly, if the confidence belt does not cover the hypothesized ρ, we will reject the null hypothesis and conclude that the computed r is significant. Conversely, any null hypothesis about a value of ρ between the confidence limits will be accepted. For the following activity, read the confidence interval from the confidence belts diagram in Figure 15.9 and indicate whether the hypothesized $\rho = 0$ is covered.

Activity 15.3

Testing hypotheses using confidence intervals

Variables $n = 30$	Correlation coefficient, r	Approximate confidence interval	Does the confidence interval cover hypothesized ρ? ($\rho = 0$)	Reject or accept the hypothesis?
PREMAXHR PREHR%	0.88	(.8, 1)	[1]Yes/No	[2]R/A
PREHR PRERESSYS	0.00	(−.4, +.4)	[3]Yes/No	[4]R/A
PREMAXHR PREVO2	0.37	(−.1, .65)	[5]Yes/No	[6]R/A
PREWT PREFAT%	0.52	(.15, .75)	[7]Yes/No	[8]R/A

A second way to test hypotheses about r is to look up the critical value of r in a table and compare the calculated r-values with the critical value. Suppose we are testing the hypothesis that there is no linear association between two variables, i.e., H_0: $\rho = 0$. Even when there is no linear correlation at all, it is possible, in fact likely, that we will compute a correlation coefficient that is not zero, though it will be close to zero. As we saw with means, there is some variation introduced because of sampling error. Different samples would have different values for r, most of them quite small, even if the population, in fact, had $\rho = 0$. What we need is an indication of which values of r are least likely to occur if the null hypothesis is true. Fortunately it is easy to find a table of critical values of r. Table 15.1 is an example of such a table. The critical values vary with the desired α and the sample size.

TABLE 15.1 Percentage Points, Distribution of the Correlation Coefficient, When $\rho = 0$

$P[r \leq \text{tabular value, assuming } \rho = 0] = 1 - \alpha$

ν	$\alpha = 0.05$ $2\alpha = 0.1$	0.025 0.05	0.01 0.02	0.005 0.01	0.0025 0.005	0.0005 0.001	ν	$\alpha = 0.05$ $2\alpha = 0.1$	0.025 0.05	0.01 0.02	0.005 0.01	0.0025 0.005	0.0005 0.001
1	0.9877	$0.9^{2}692$	$0.9^{3}507$	$0.9^{3}877$	$0.9^{4}692$	$0.9^{6}877$	16	0.400	0.468	0.543	0.590	0.631	0.708
2	0.9000	0.9500	0.9800	$0.9^{2}000$	$0.9^{2}500$	$0.9^{2}000$	17	0.389	0.456	0.529	0.575	0.616	0.693
3	0.805	0.878	0.9343	0.9587	0.9740	$0.9^{2}114$	18	0.378	0.444	0.516	0.561	0.602	0.679
4	0.729	0.811	0.882	0.9172	0.9417	0.9741	19	0.369	0.433	0.503	0.549	0.589	0.665
5	0.669	0.754	0.833	0.875	0.9056	0.9509	20	0.360	0.423	0.492	0.537	0.576	0.652
6	0.621	0.707	0.789	0.834	0.870	0.9249	25	0.323	0.381	0.445	0.487	0.524	0.597
7	0.582	0.666	0.750	0.798	0.836	0.898	30	0.296	0.349	0.409	0.449	0.434	0.554
8	0.549	0.632	0.715	0.765	0.805	0.872	35	0.275	0.325	0.381	0.418	0.452	0.519
9	0.521	0.602	0.685	0.735	0.776	0.847	40	0.257	0.304	0.358	0.393	0.425	0.490
10	0.497	0.576	0.658	0.708	0.750	0.823	45	0.243	0.288	0.338	0.372	0.403	0.465
11	0.476	0.553	0.634	0.684	0.726	0.801	50	0.231	0.273	0.322	0.354	0.384	0.443
12	0.457	0.532	0.612	0.661	0.703	0.780	60	0.211	0.250	0.295	0.325	0.352	0.408
13	0.441	0.514	0.592	0.641	0.683	0.760	70	0.195	0.232	0.274	0.302	0.327	0.380
14	0.426	0.497	0.574	0.623	0.664	0.742	80	0.183	0.217	0.257	0.283	0.307	0.357
15	0.412	0.482	0.558	0.606	0.647	0.725	90	0.173	0.205	0.242	0.267	0.290	0.338
							100	0.164	0.195	0.230	0.254	0.276	0.321

$\alpha = 1 - F(r$ given ν and assuming $\rho = 0)$ is the upper-tail area of the distribution of r appropriate for use in a single-tail test. For a two-tail test, 2α must be used. If r is calculated from n paired observations, enter the table with $\nu = n - 2$. For partial correlations enter with $\nu = n - k - 2$, where k is the number of variables held constant.

Activity 15.4

Testing hypotheses using the critical r method

For a two-tailed test with $\alpha = .05$ and $n = 32$, $\nu = n - 2 = 30$ the critical value of r is 0.349. If the calculated value is greater than this tabulated r, we conclude that it is unlikely to have gotten this large a value of r if $\rho = 0$. Therefore, we reject the hypothesis $\rho = 0$. Indicate whether you would accept or reject the hypothesis $\rho = 0$ for each pair of variables below.

Variables $n = 30$	Correlation coefficient, r	Critical r	Reject or accept the hypothesis?
PREMAXHR PREHR%	0.88	±0.36	[1]R/A
PREHR PRERESSYS	0.00	±0.36	[2]R/A
PREMAXHR PREVO2	0.37	±0.36	[3]R/A
PREWT PREFAT%	0.52	±0.36	[4]R/A

The third way to test hypotheses about r is to compute a t-value using the calculated r and compare it with the critical t-value from the t-table. We use the t-statistic with $n - 2$ degrees of freedom to decide about the significance of the calculated statistic, r. The formula for computing t in this context is

$$t = \frac{r - \rho_{\text{hyp}}}{\sqrt{\dfrac{1 - r^2}{n - 2}}}$$

where r = correlation coefficient, hypothesized $\rho = 0$, and standard deviation of r =

$$\sqrt{\frac{1 - r^2}{n - 2}} \quad \text{or} \quad t = r * \sqrt{\frac{n - 2}{1 - r^2}}$$

Figure 15.10 shows that for a two-tailed test when $\alpha = .05$ and df = 28, t = [1]_______. (Note: For calculating the correlation coefficient, df = $n - 2$.) The t-value for a one-tailed test with alternate hypothesis $\rho < 0$ when $\alpha = .05$ with df = 28 is [2]_______. The t-value for a one-tailed test with alternative hypothesis $\rho > 0$ when $\alpha = .05$ with df = 28 is [3]_______. (See Table 11.7, page 376, or Table B-3, page 691.)

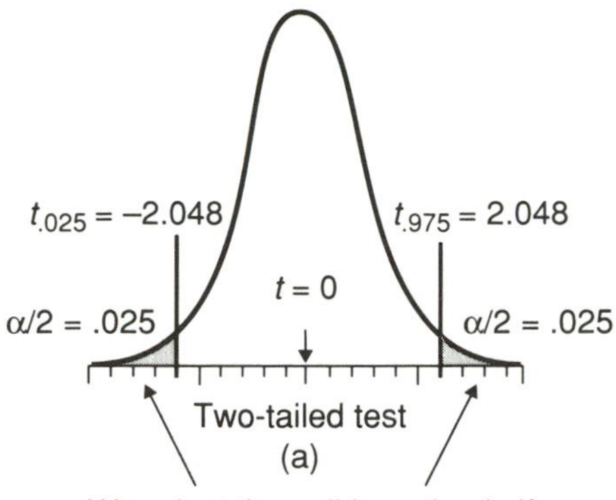

We reject the null hypothesis if the calculated t falls underneath either of the shaded areas.

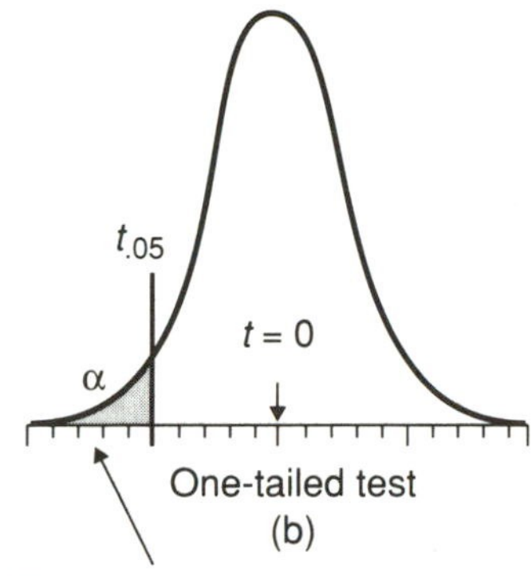

We reject the null hypothesis if the calculated t falls underneath the shaded area.

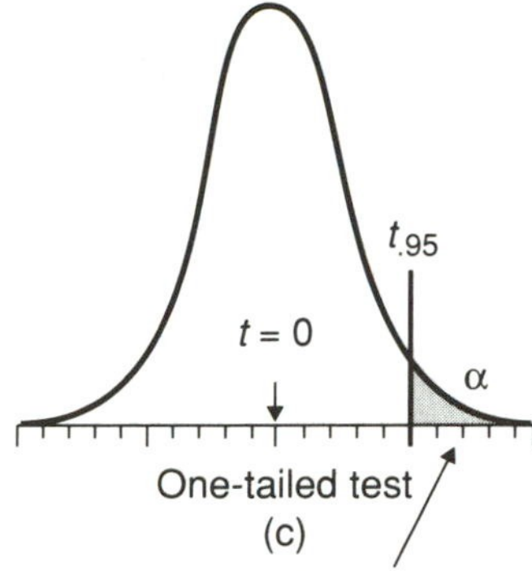

We reject the null hypothesis if the calculated t falls underneath the shaded area.

FIGURE 15.10

ACTIVITY 15.5

Testing hypotheses using the critical t method

Variables $n = 30$	Correlation coefficient, r	Calculated t	Use a two-tailed test. Do you reject or accept the hypothesis?
PREMAXHR PREHR%	0.88	1 _______	2 R/A
PREHR PRERESSYS	0.00	3 _______	4 R/A
PREMAXHR PREVO2	0.37	5 _______	6 R/A
PREWT PREFAT%	0.52	7 _______	8 R/A

Suppose we are considering more than two variables, and we are interested in seeing if any of the variables are related to each other. As a preliminary screening device we can compute all the correlation coefficients, see which ones are statistically significant, and choose those for further investigation. Some statistical software allows the user to plot many variables at a time. Activity 15.6 and Figure 15.11 present the output of one software package. Based on the preliminary results we will select interesting variables to compare and produce larger plots to investigate the relationships further.

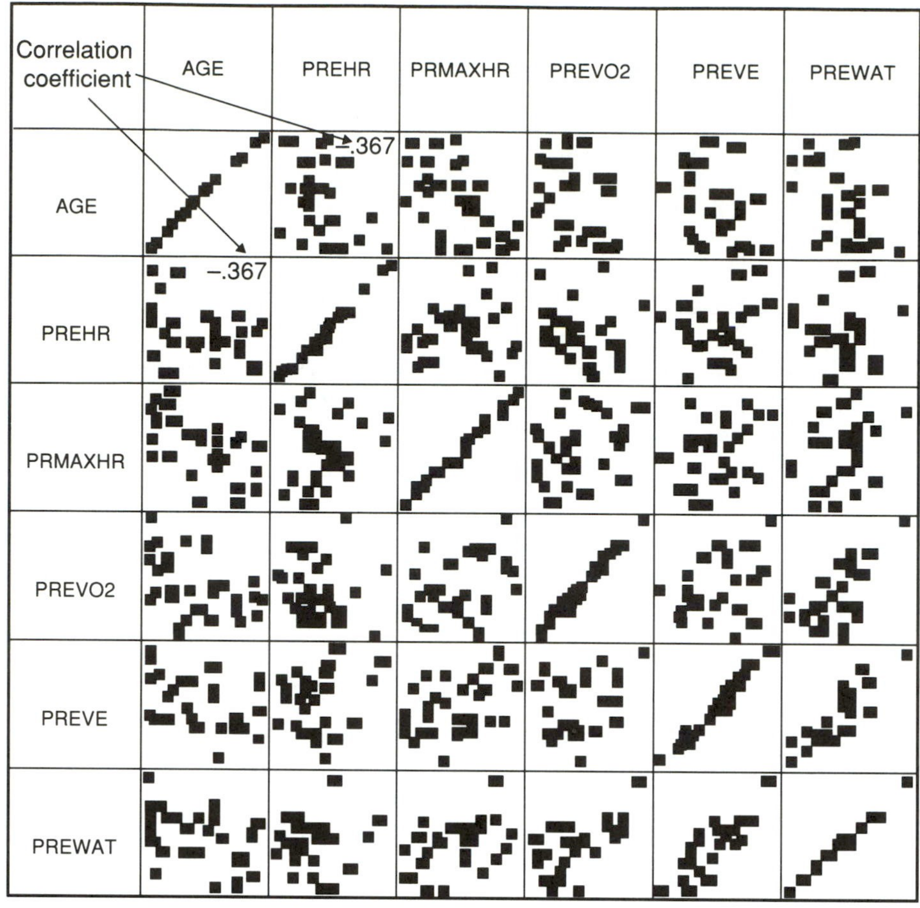

FIGURE 15.11 Scatter Plots of Selected Variables as Aids to Selecting Variables with a Linear Relationship

ACTIVITY 15.6

Testing several correlation coefficients for significance using the critical r method

In the table on the facing page, circle all the correlation coefficients that are greater than +0.36 or less than −0.36.

Copy the correlation coefficients from the table onto the corresponding scatter plots in Figure 15.11. We gave AGE versus PREHR as an example. The correlation coefficient for these two variables is statistically significant. Does the scatter plot show a linear relationship? These small graphs serve well as preliminary screening devices, as does the table of correlation coefficients. In practice, first we compute correlation coefficients for all the variables of interest.

Correlation Coefficients for All Pretest Variables in FED Correlated with Each Other

	AGE	PREHR	PREMAXHR	PREVO2	PREVE	PREWAT	PREHR%	PRERQ	PREFAT%	PREWT	PREHT	PRERESHR	PRRESSYS	PRRESDIA
PREHR	−0.367													
PREMAXHR	−0.638	0.334												
PREVO2	−0.394	−0.191	0.367											
PREVE	−0.463	0.219	0.396	0.421										
PREWAT	−0.454	0.081	0.419	0.636	0.812									
PREHR%	−0.265	0.256	0.877	0.178	0.307	0.283								
PRERQ	−0.216	0.066	0.277	0.087	0.128	−0.198	0.282							
PREFAT%	−0.077	−0.083	−0.035	−0.340	0.221	0.101	0.012	−0.069						
PREWT	−0.077	0.331	−0.037	−0.460	0.460	0.316	0.015	−0.311	0.520					
PREHT	0.025	0.059	0.058	−0.169	0.439	0.499	0.125	−0.437	0.379	0.730				
PRERESHR	−0.019	0.439	0.252	−0.290	0.018	0.047	0.255	−0.252	−0.017	0.408	0.485			
PRRESSYS	0.372	0.000	−0.086	−0.387	0.045	−0.090	0.117	−0.084	0.195	0.399	0.366	0.322		
PRRESDIA	0.120	0.310	0.185	−0.315	0.352	0.152	0.369	−0.048	0.219	0.599	0.581	0.398	0.583	
PREREPP	0.305	0.063	−0.006	−0.423	−0.519	−0.664	0.102	0.210	−0.116	−0.232	−0.207	0.332	0.399	0.229

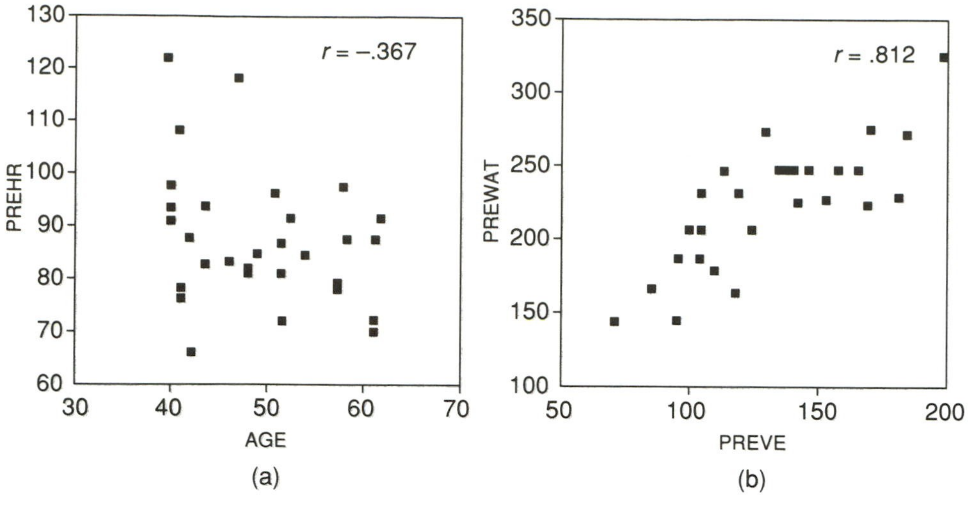

FIGURE 15.12 Scatter Plots with Significant Correlation Coefficients

Then we look at all the scatter plots to see which ones appear linear. Next, we test to see if the correlation coefficients are statistically significant.

Two of the scatter plots from Figure 15.11 are reprinted enlarged in Figure 15.12, along with the correlation coefficients. Graph (a) tends to [1]refute/substantiate the moderate correlation suggested by $r = -.367$. Note, for instance, the almost vertical line of points for age 40. Even though this r is statistically significant, we are strongly suspicious of a linear relationship between AGE and PREHR. The correlation coefficient of PREVE and PREWAT shows a strong correlation, and nothing in the graph leads us to [2]refute/substantiate this relationship. The relationship may be curvilinear rather than linear, but the size of the correlation coefficient (.812) suggests that the amount of curvature is not very great. Use the scatter plots in Figure 15.11 to determine which of the other variables seem to have a linear relationship with PREWAT. [3] ________

Association and Causal Relationships

Does significant correlation between two variables x and y imply a causal relationship between the variables? No. Other conditions need to be fulfilled before we can conclude that x causes y.

1. The correlation coefficient of x and y needs to be statistically significant. For example, there is a strong correlation between average number of cigarettes smoked and incidence of lung cancer across many diverse population groups.
2. x must precede y. Smoking precedes lung cancer. You couldn't say that smoking caused a person's lung cancer if she didn't start smoking until after she contracted lung cancer.

3. There should not be another variable z that causes both x and y. Some scientists contend that life-style and genetics may be other factors that influence both lung cancer and smoking. Other related studies refute this claim.
4. There should be experimental evidence that conclusively links x to y. Since we cannot perform these experiments on human subjects, scientists attached animals to smoking machines and a significant number of animals developed lung cancer.
5. The finding that x causes y needs to be consistent with all other related results. Toxic substances have been shown to cause tumors. The hypothesis that some active toxic ingredients of smoke cause uncontrolled proliferation of cells is consistent with the literature. Prospective studies that followed smokers and non-smokers also demonstrated much higher incidence of lung cancer in the smokers. Life-long non-smokers living with smokers are also known to have an increased incidence of lung cancer.

The criteria listed above are illustrated in Figure 15.13.

We would expect a high correlation between the number of sales of fans and air-conditioners and the amount of ice cream sold, but does that mean that fan sales cause ice cream sales or ice cream sales cause fan sales? Obviously not. In this case there is a third factor (rising temperature) that causes both ice cream sales and fan sales to increase.

Prediction: Linear Regression

Once we find that two variables have a statistically significant linear relationship (significant r), we wonder if we can predict y by knowing x. We introduce the subject by plotting height in inches vs. the same height in centimeters in Figure 15.14. These two variables will have a perfect correlation, $r = +1$. All the points will fall on the same straight line, and that line can be described by an equation of the form $y = a + bx$ where a represents the intercept of the line with the

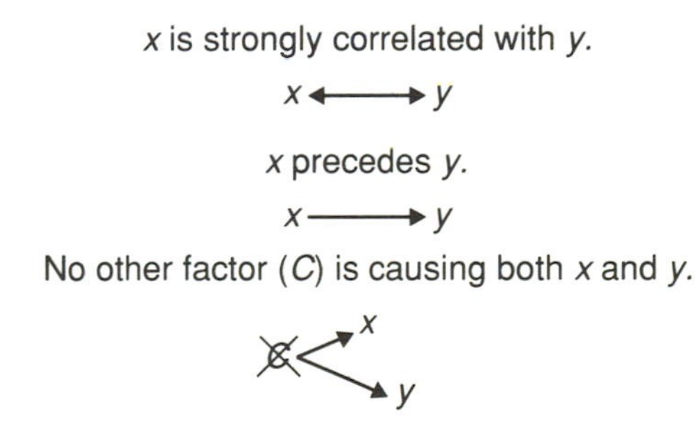

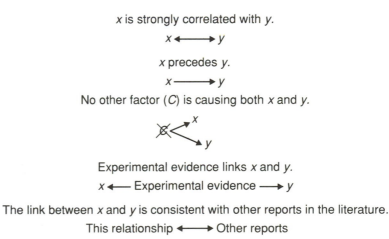

Figure 15.13 Criteria for Asserting a Causal Relationship

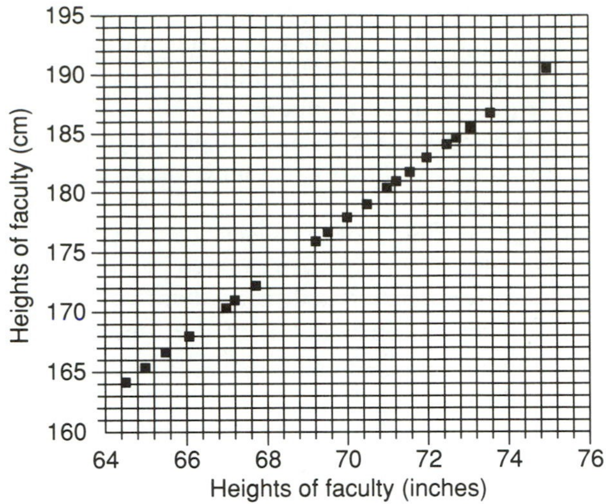

FIGURE 15.14 Heights of Faculty Measured in Inches and Centimeters

vertical axis ($x = 0$) and b is the slope of the line. x is the height in inches and y is the height in centimeters. Any equation of the form $y = a + bx$ graphs as a straight line.

The particular example in Figure 15.14 is highly artificial; it is extremely rare to encounter variables that have a perfect linear correlation ($r = +1$ or $r = -1$). Usually, even when there is a linear relationship between the variables, there will be some random variation from perfect linearity. With such data, the problem is to choose a straight line that best fits the data points, i.e., to find the straight line that is "closest" to most of the points.

Consider the relationship between PREHR% and PREMAXHR graphed in Figure 15.15. Note that the correlation coefficient is .88, which is statistically significant. The data points seem to be linearly associated. Draw a straight line that seems to fit the points approximately. What informal criterion did you use to draw the line?

Regression Analysis for Prediction

Clearly we can draw many straight lines to approximate the line "connecting" the points in Figure 15.15. How do we draw the best line? The concept of "best" straight line can be defined in many ways, but there is one definition that is almost universally accepted by statisticians. This particular concept of "best" straight line uses a method called **least squares approximation.** When we use this method, the sum of squared vertical distances between each data point and the point on the line corresponding to that particular x will be minimum when compared to any other line. The line arrived at by this method is called the **regression line** and the method of arriving at it is called **linear**

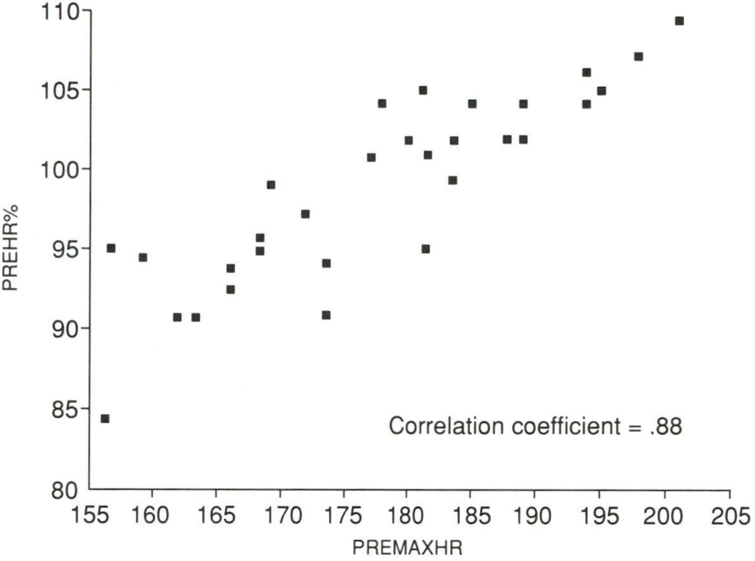

FIGURE 15.15 PREMAXHR VS. PREHR%

regression. Figure 15.16 illustrates a regression line. (The terms residual component, total SSq component, and regression component given in the figure will be explained later.)

The equation of the regression line is of the form

$$y' = a + bx$$

where y' is the value of the response variable, y, predicted by the regression

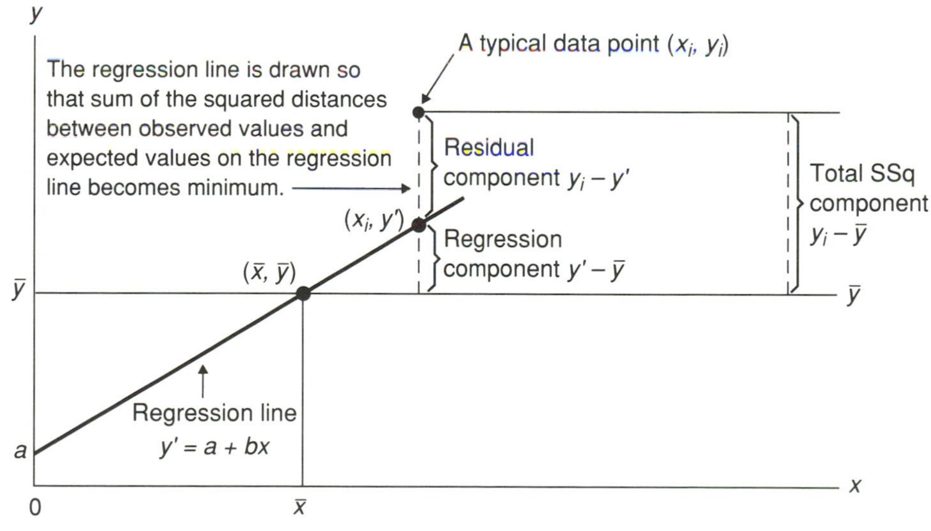

FIGURE 15.16 Components of the Sum of Squares

equation, a is the y-intercept (the value predicted for y when $x = 0$), and b is the **regression coefficient,** or the slope. x is the explanatory variable. Note that the regression equation is just the equation of a straight line. We call the response variable y' here instead of y to distinguish it from the actual measured values of y, the y-coordinates of the actual data points. For a given value of x_i we would have both a y_i value (the actual measured value of the variable) and a y' value, the y-coordinate of the point on the regression line for x_i.

Figure 15.17 is the scatter plot of the relationship between PREHR% and PREMAXHR that appeared in Figure 15.15, but with the regression line added.

The correlation coefficient for PREHR% and PREMAXHR is [1] _______. It is statistically significant. So it is reasonable to expect to predict heart rate percent from maximum heart rate. The regression equation is

$$\text{PREHR\%} = 24.8 + 0.417\ \text{PREMAXHR}$$

where PREHR% is y' and PREMAXHR is x. Use the equation to compute values in the following table.

PREMAXHR	Observed PREHR% (from FED)	PREHR% (read from Figure 15.17)	Predicted PREHR% (calculated from formula)	Difference between predicted and observed
168	96	2 ____________	3 ____________	8 ____________
180	102	4 ____________	5 ____________	9 ____________
195	105	6 ____________	7 ____________	10 ____________

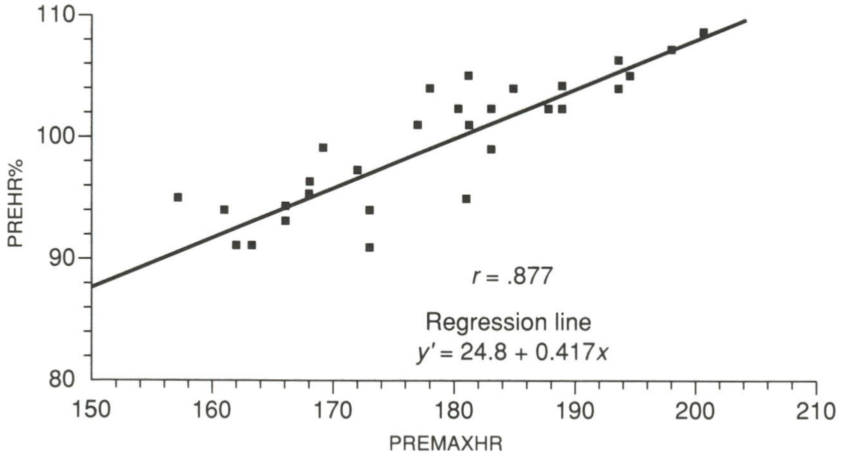

FIGURE 15.17 Scatter Plot and Regression Line for PREMAXHR VS. PREHR%

Computation of Slope, *b*, and Intercept, *a*

We now know the mechanics of computing y when the regression equation and x are given, but how is the equation arrived at? How do we find the slope b and intercept a? We will give the formulas that are used to do this, but keep in mind that computing from these formulas is tedious and time consuming. Computer software does the job for us quickly and easily so we would not calculate the least squares equation by hand for any meaningful data.

$$b = \frac{\Sigma(x_i - \bar{x})(y_i - \bar{y})}{\Sigma(x_i - \bar{x})^2} = r_{xy}\frac{s_y}{s_x} \qquad a = \bar{y} - b\bar{x}$$

where

r_{xy} = correlation coefficient of x and y
s_x = standard deviation of x
s_y = standard deviation of y

a is the intercept and b is the regression coefficient, or slope. What happens if s_y and s_x are equal? In that case, s_y/s_x will equal [1] _______. Therefore, b = [2] _______.

Suppose we transform the observations into corresponding z-scores and compute the correlation coefficient for the z-scores. It is an interesting fact that this correlation coefficient will be exactly the same number as r, the correlation coefficient for the original, untransformed data points. Since both sets of z-scores, those for the x-variable and those for the y-variable, have standard deviations of 1, the slope of the line relating the z-scores is [3] _______. Since both sets of z-scores have means of 0, it follows that the y-intercept for this line is at [4] _______. Thus, the equation of the line that relates the z-scores of the y-variable with the z-scores of the x-variable is $z_y = rz_x$. In general, this line is [5] the same as/different from from the regression line relating y to x.

If $r = 0$, then b will also be zero, because zero times any number is zero. A line with slope of 0 is horizontal, parallel to the x-axis. Such a line indicates that the explanatory variable, x, has no effect on the response variable, y. The regression line, being horizontal, predicts the same value for y in all cases; y is constant. In this situation, however, the scatter plot will be likely to have data points all over the place. As we would expect, the regression equation provides no help in predicting y from x. $r = 0$ tells us only that these variables do not have a linear relationship.

Table 15.2 compares the various formulas. Notice that the formulas for the correlation coefficient, r, and regression coefficient, b, are similar. The numerators are exactly the same. The correlation coefficient, r, has both s_x and s_y in the denominator, whereas the regression coefficient, b, has [6] _______ in the denominator.

TABLE 15.2 Comparison of the Formulas for Computing Correlation and Regression

Formula for Correlation	Formula for Regression
$r_{xy} = \frac{s_{xy}}{s_x s_y}$	$b = r_{xy}\frac{s_y}{s_x}$
$r = \frac{\Sigma\{(x_i - \bar{x})(y_i - \bar{y})\}}{\sqrt{\Sigma (x_i - \bar{x})^2 (y_i - \bar{y})^2}}$	$b = \frac{\Sigma (x_i - \bar{x})(y_i - \bar{y})}{\Sigma (x_i - \bar{x})^2}$
$r = \frac{\Sigma z_x * z_y}{n}$	$a = \bar{y} - b\bar{x}$

Figure 15.18 shows scatter plots and regression lines for data on height and weight for three groups of students. Circle the outlier in part (c). It corresponds to a student whose height is [1] ______ and whose weight is [2] ______. In Activity 15.7, predict your weight from your height using the equations provided in parts (a) or (b) and (c) of Figure 15.18.

ACTIVITY 15.7

Calculation of estimated weight and the difference between estimated and actual weight

	Regression equation	Your weight	Your height	Your actual weight	Estimated weight from equation	Difference between estimated and actual weight
Males	$y' = -304.8 + 6.8$ Ht.	176.3	______	______	______	______
Females	$y' = -4.3 + 2$ Ht.	120.9	______	______	______	______
Females (one outlier)	$y' = -56.5 + 2.8$ Ht.	121.1			______	______
Combined group	$y' = -273.6 + 6.2$ Ht.	140.5			______	______
Combined group (one outlier)	$y' = -330 + 7.1$ Ht.	141			______	______

The correlation coefficients given in parts (a), (b), and (c) of Figure 15.18 are statistically significant. When we eliminate the outlier, the correlation coefficient [3] increases/decreases. Eliminating the outlier increases the correlation coefficient of the combined group to .83. The correlation coefficient of the female

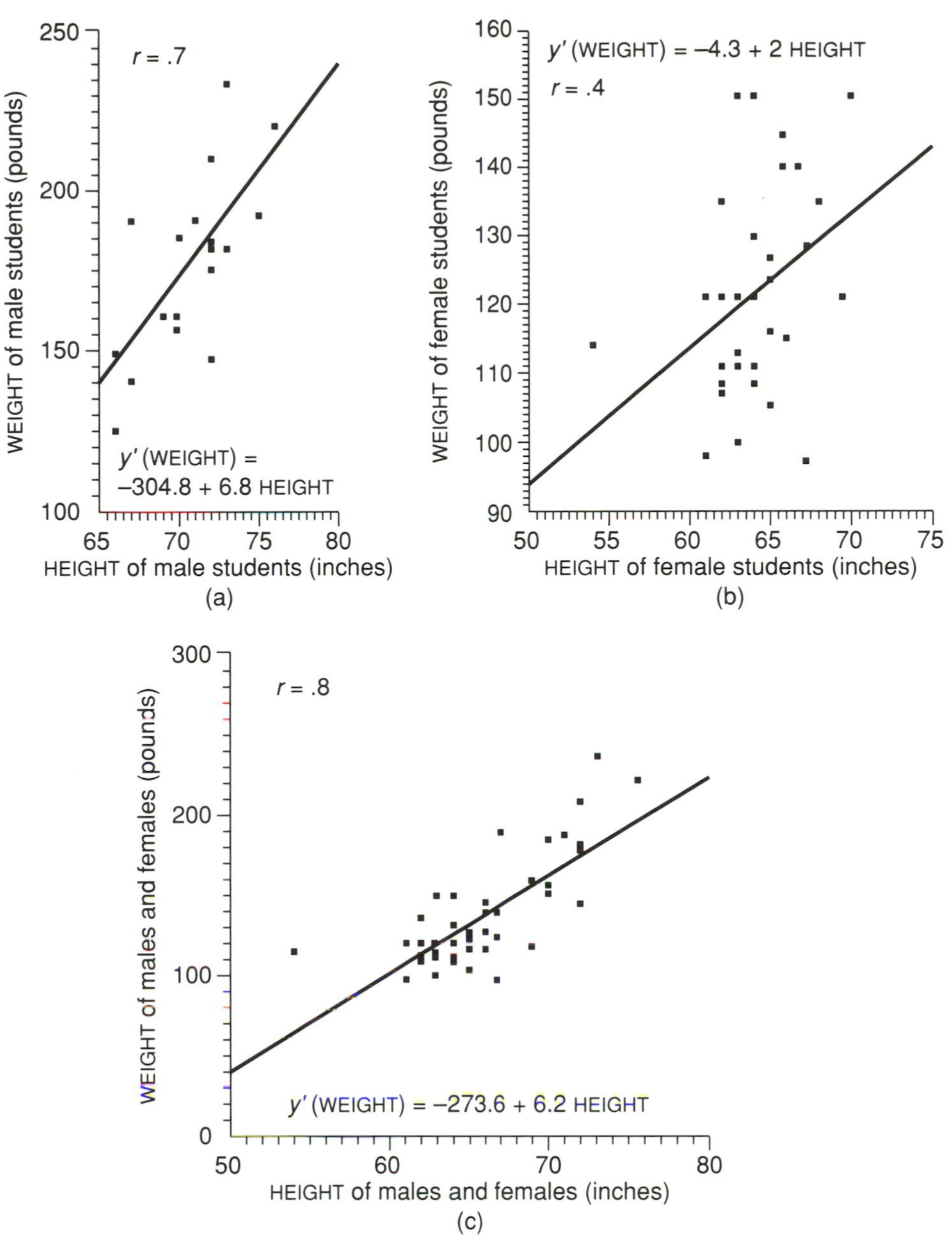

FIGURE 15.18 HEIGHT–WEIGHT Regression Lines for Students in Statistics Classes

group, after eliminating the outlier, is .410, in contrast to .371 before removing the outlier.

The equation for male weight predicted one of the author's weights better than the equation of the combined group (within 3 pounds). Even after the outlier was deleted from the combined group, the equation for male weight remained the best predictor. Which equation is the best predictor of your weight?

If we knew nothing about the regression equation and the relationship of weight and height, our best predictor of weight would be the mean of all of the weights (i.e., 140.5 pounds). From our knowledge of the relationship of weight and height we are able to use the regression equation and predict the weight more accurately, because this prediction uses added information, the height. Still, our prediction is likely to remain an approximation and will have a variance (residual component) due to factors not known to us (variation unexplained by height). In general, the weight predicted by the regression equation is different from the actual weight. The distance between the mean of the weight and the fitted point on the regression line indicates how much our prediction will improve if we know the height. This distance is called the variation (regression component) explained by height.

Analysis of Variance for Regression

Figure 15.19 shows the different components of variation of a point fitted by the regression line from a typical data point. The figure indicates these compo-

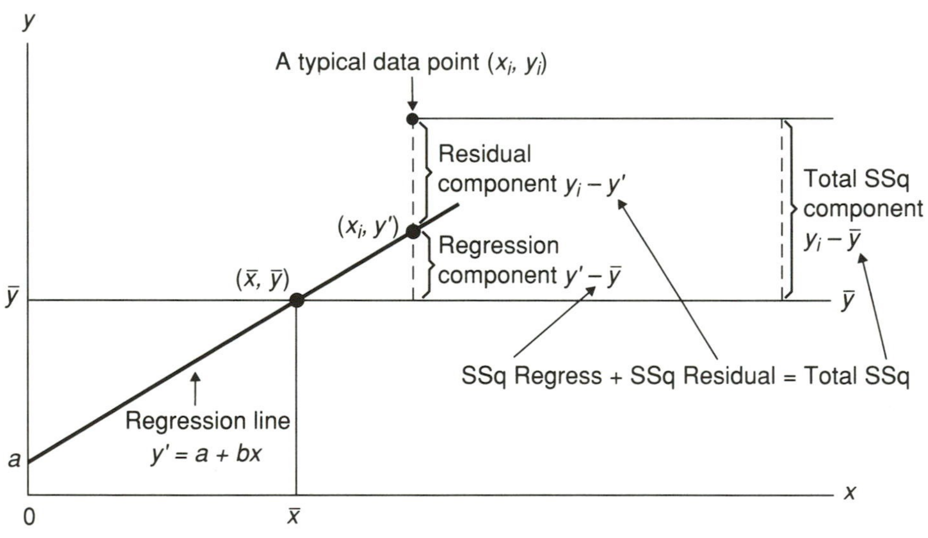

FIGURE 15.19 Components of the Analysis of Variance Table

nents for only one data point. Similar analysis can be made for each of the observed points in a data set, and this analysis will lead directly to the analysis of variance table. The components of variation indicated in Figure 15.19 are described as follows.

1. Total variation: Sum of squares of the distance between each data point y_i and the mean $\bar{y}$:

$$\sum (y_i - \bar{y})^2$$

2. Regression variation: Sum of squares of the distance between each fitted point on the line, y', and the mean, $\bar{y}$:

$$\sum (y_i' - \bar{y})^2$$

3. Residual variation (or error): Sum of squares of the distance between each data point y_i and each fitted point on the line y'; in other words, the variation that is still unexplained by the fitted line:

$$\sum (y_i - y_i')^2$$

It can be shown with the application of considerable algebra that

$$\sum (y_i - \bar{y})^2 = \sum (y_i' - \bar{y})^2 + \sum (y_i - y_i')^2$$

$$\text{Total SSq} = \text{SSq regress} + \text{SSq residual}$$

These sums of squares are incorporated into an analysis of variance table that is similar to the one we used when we tested three or more means. The difference is that here we have regression SSq instead of "between groups." The error here is residual SSq figured as the difference between data points and the fitted point on the line instead of "group means." One way of understanding this analysis is to think of the case in which there are a lot of points having the same x-coordinate but different y-coordinates. Take the mean of the ys for each x. Essentially this ANOVA is testing the hypothesis that all these means are equal. If they were all equal, the regression line would be horizontal, so the slope would be zero. Thus, the ANOVA is testing the hypothesis that $r = 0$. Table 15.3 summarizes the names. The analysis of variance table for PREHR% and PREMAXHR appears on page 558.

TABLE 15.3 Alternative Names for the Analysis of Variance Components

$\Sigma (y_i - \bar{y})^2 =$	$\Sigma (y_i' - \bar{y})^2$ +	$\Sigma (y_i - y_i')^2$
Total sum of squares	Regression sum of squares	Residual sum of squares
Total SSq	SSq regress	SSq Residual
Total	Regression	Error

Source of Variation	Degrees of Freedom, df	Sum of Squares, SS	Mean Sum of Squares, MS	MS Reg/MS Error, F	p-Value
Regression Error	1	782.33	782.33	93.16	0.000
(residual SSq)	28	235.14	8.40		
Total	29	1,017.47			
Note: MS = SS/df					

If we have no information about the relationship between x and y, the predicted value for y will be $\bar{y}$. When we make such a prediction the expected variation is $\Sigma\,(y_i - \bar{y})^2$. However, because we have an approximately linear relationship between x and y, we have specific y' values for each x. The explained variation based on the predictions is $\Sigma\,(y_i' - \bar{y})^2$. The proportion of the total (expected) variation that is the explained variation is $\Sigma\,(y_i' - \bar{y})^2 \div \Sigma\,(y_i - \bar{y})^2$. It turns out that this proportion is equal to r^2. We still have some variation that is not explained by x. This is $\Sigma\,(y_i - y_i')^2 \div \Sigma\,(y_i - \bar{y})^2$ which is equal to $1 - r^2$ and is the proportion of the variation unexplained by x.

Activity 15.8

Calculation of r^2 and $(1 - r^2)$

Use information from the analysis of variance table above to calculate the values below.

Explained variation	r^2	$\dfrac{\Sigma\,(y_i' - \bar{y})^2}{\Sigma\,(y_i - \bar{y})^2}$	$r^2 =$ [1] ____________
Unexplained variation	$1 - r^2$	$\dfrac{\Sigma\,(y_i - y_i')^2}{\Sigma\,(y_i - \bar{y})^2}$	$1 - r^2 =$ [2] __________
Standard deviation about regression	s	Square root of MS (error)	$s =$ [3] ____________

Testing for the Slope (Regression Coefficient)

As noted earlier, the regression equation for PREHR% and PREMAXHR is

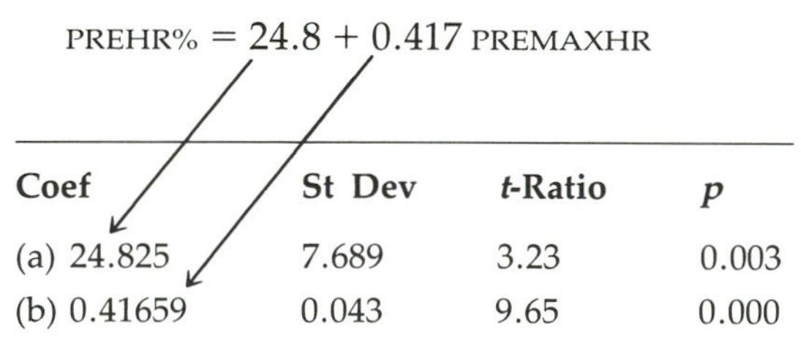

PREHR% = 24.8 + 0.417 PREMAXHR

Coef	St Dev	t-Ratio	p
(a) 24.825	7.689	3.23	0.003
(b) 0.41659	0.043	9.65	0.000

Suppose we wish to test the following hypothesis: Constant and regression coefficients for the population are 0. The test used here is a *t*-test. However, it is actually the same test as the one used above, because $F = t^2$. ($9.65^2 = 93.16$ but for round-off error.) The values of *p* associated with the *t*-ratios indicate that both constant and regression coefficient, *b*, are significantly different from 0.

Coefficient of Determination, Explained and Unexplained Variation

r is the correlation coefficient. r^2 is the percentage of variation in y explained by the explanatory variable, x. r^2 is also called the **coefficient of determination.** If we have no information about the relationship of x and y, the best predicted y we have for the population is $\bar{y}$. Because we fit a least squares line to the data and we know x, y can be predicted by y', the fitted point on the line. The proportion of the variation that is explained by the regression is r^2. Figure 15.20 illustrates these ideas.

When there is a lot of variation in the values of a variable, y, correlation and regression analysis are indicated. We may hit upon a variable that is highly correlated with y, but often this does not explain most of the variation in y. For example, if the correlation coefficient between x and y is $r = 0.8$, a fairly large correlation, the coefficient of determination is 0.64, which is significantly smaller. When r^2 is small even though r is significant, it is wise to look for other variables that may explain some more of the variation in y. For example, which of the FED variables have significant correlation with PREHR%? Each significant variable contributes something to the variation in PREHR%. We can determine which are significant by reading the correlation coefficients listed at the top of page 560.

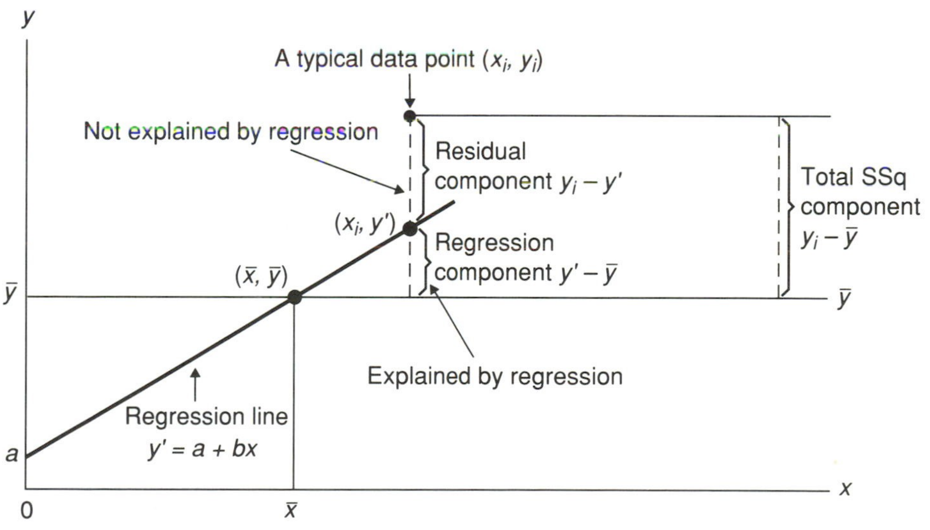

FIGURE 15.20 Variation Explained and Not Explained by Regression

	AGE	PREHR	PREMAXHR	PREVO2	PREVE	PREWAT	PRERQ	PRERESDIA
PREHR%	−.265	0.256	0.877	0.178	0.307	0.283	0.282	0.369

Which variables are significantly related to PREHR%? (Critical $r = .36$) [1] _______ _______. How much variation do each of the variables explain? r^2 for PREMAXHR = [2] _______; r^2 for PRERESDIA = [3] _______. The relationship of each of these variables to PREHR% is shown in Figure 15.21. Both variables together

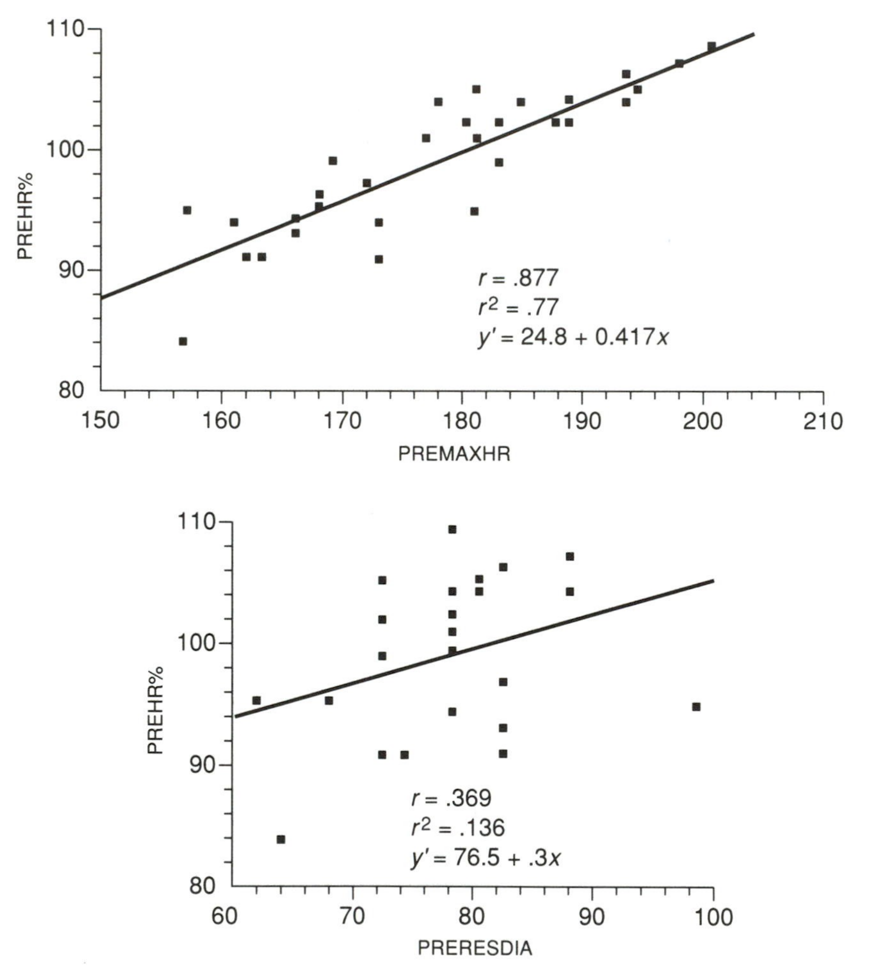

FIGURE 15.21 Variation in PREHR% Explained by Variables PREMAXHR and PRERESDIA

will explain [4] <u>less/more</u> of the variation in PREHR% than PREMAXHR alone. If we include both variables in something like a linear regression equation it will predict PREHR% better than either of the variables will by itself. This is included in the study of **multiple regression**, a full discussion of which is beyond the scope of this book. Computer software helps us find the multiple regression equation for these variables easily:

$$y' \text{ (PREHR\%)} = 13.1 + 0.408 \text{ PREMAXHR} + 0.165 \text{ PRERESDIA}$$

r for these two variables (called the multiple correlation coefficient) = 0.901. The proportion of the total variation explained by these two variables together is $r^2 =$ [5] _______.

ACTIVITY 15.9

Calculation of residuals and confidence intervals for the estimated values of PREHR% (y′)

The table on page 562 gives the observed x_i and y_i and the fit calculated from the regression equation shown in Figure 15.22. The standard deviation of the fit enables us to calculate the confidence interval for each fitted PREHR% using the same methods given in Chapter 14. The residual column gives residual $(y_i - y'_i)$; SD error is the standard deviation of the residual. Figure 15.22 gives the confidence belts for the points on the fitted line. Compute the missing residuals. (The last two have been calculated for you.)

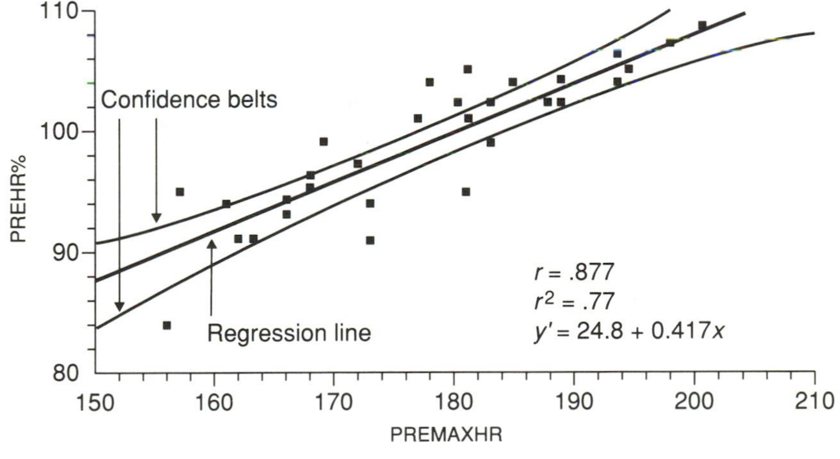

FIGURE 15.22 Regression Line for PREHR% and Predictor Variable PREMAXHR

Case number	x_i PREMAXHR	y_i PREHR%	y' fit	SD y'_i st. dev. fit	$y_i - y'_i$ residual	SD error st. resid.
1	169	99	95.228	0.650	1 ______	1.34
2	180	102	99.811	0.538	2 ______	0.77
3	163	91	92.729	0.827	3 ______	−0.62
4	195	105	106.060	0.914	4 ______	−0.39
5	183	99	101.061	0.576	5 ______	−0.73
6	189	104	103.560	0.719	6 ______	0.16
7	162	91	92.312	0.861	7 ______	−0.47
8	181	105	100.228	0.548	8 ______	1.68
9	173	91	96.895	0.567	−5.895	−2.07
10	168	95	94.812	0.676	0.188	0.07

The observations below were identified by a computer program as unusual observations. These are so identified because the standard deviations of the residual are larger than 2.

PREMAXHR	PREHR%	Fit	St dev. fit	Residual	Number of standard deviations from fit
173	91.000	96.895	0.567	−5.895	−2.07*R*
157	84.000	90.229	1.040	−6.229	−2.30*R*

R denotes an observation whose residual has a large standard deviation. The two observations have unusually low or high observed values compared to the fit. Locate these points on the graph in Figure 15.22 and circle them. These possibly do not belong to the population of observations. Further investigation of these cases would be warranted. The bad fit for these cases could indicate experimental error or an abnormal condition in the subjects.

ACTIVITY 15.10

Reading information from analysis of variance tables

The following computer output correlates HEIGHT and WEIGHT. Males and females are listed separately. For females:

Dependent var.: WEIGHT

n: 33

$r = .371$

$r^2 = .137$

Standard deviation about regression: 14.4

Variable	Coefficient	Std. Error	Std. Coef.	Tolerance	t	P (two-tailed)
Constant	−4.3	56.4	0.0	−0.1	0.9	
HEIGHT	2.0	0.9	0.4	1.0	2.2	0.0

Analysis of variance:

Source	Sum of Squares	df	Mean Square	F-Ratio	P
Regression	1,021.0	1	1,021.0	4.9	0.0
Residual	6,415.7	31	207.0		
Total	7,436.7	32			

For males:

Dependent var.: WEIGHT

n: 18

$r = .693$

$r^2 = .481$

Standard deviation about regression: 21.0

Variable	Coefficient	Std. Error	Std. Coef.	Tolerance	t	P (two-tailed)
Constant	−304.8	125.1	0.0		−2.4	0.0
HEIGHT	6.8	1.8	0.7	1.0	3.8	0.0

Analysis of variance:

Source	Sum of Squares	df	Mean square	F-Ratio	P
Regression	6,554.0	1	6,554.0	14.8	0.0
Residual	7,077.6	16	442.4		
Total	13,631.6	17			

Use information from the analysis of variance tables on page 563 above to calculate the indicated quantities.

	Female	Male	Comments
Explained variation	1 ________	2 ________	Coefficient of determination, r^2
Unexplained variation	3 ________	4 ________	$1 - r^2$
Standard deviation about regression	5 ________	6 ________	
$\Sigma (y_i - \bar{y})^2$	7 ________	8 ________	Total SSq
$\Sigma (y_i' - \bar{y})^2$	9 ________	10 ________	Regression SSq
$\Sigma (y_i - y_i')^2$	11 ________	12 ________	Residual
$\dfrac{\Sigma (y_i' - \bar{y})^2}{\Sigma (y_i - \bar{y})^2}$	13 ________	14 ________	Coefficient of determination, r^2
$\dfrac{\Sigma (y_i - y_i')^2}{\Sigma (y_i - \bar{y})^2}$	15 ________	16 ________	$1 - r^2$
Hypothesis: Regression coefficient = 0. Read p from the ANOVA table and reject or accept the hypothesis.	17 ________	18 ________	$\alpha = .05$ (two-tailed test)
Write the regression equation for each group.	19 ________	20 ________	

Examples from the Literature

Literature Example 15.1 is from the article entitled "Breastfeeding and Employment." Notice the title of the table. What two variables does this table correlate? $y =$ [1]______________ and $x =$ [2]______________. Consider problem scores at the same point in time. Look up the critical r in Table 15.1 for a one-sided test, $\alpha = .05$. For length of time after delivery = 6 weeks ($n = 18$), critical $r =$ [3]______ and observed $r = .36$. Therefore, the correlation [4]is/is not significant.

Notice the column for p-values for problem scores at the same point in time. All values but one are marked NS (not significant). r corresponding to 5 months is .27. Since $n = 51$, $\nu = 51 - 2 = 49$. There is no entry in Table 15.1 for $\nu = 49$. In order to err on the conservative side (i.e., to make it less likely to reject the null hypothesis), choose the next smaller entry, that is, $\nu = 45$. Look up the critical value of r, again, for a one-sided test with $\alpha = .05$. Critical $r =$ [5]_______. Therefore, correlation [6]is/is not significant.

If we used a two-sided test, what would be the critical values for $n = 18$ and 51? Critical $r(n = 18) =$ [7]_______; critical $r(n = 51;\ \nu = 45) =$ [8]_______.

The critical values of r for $n = 51$, $\nu = 49$ are approximately .233 (one-sided) and .273 (two-sided). Inference regarding significance in this case obviously depends upon whether the researcher uses a two-tailed or one-tailed test. Usually researchers use two-tailed tests unless there is good reason to use one-tailed tests. The authors of this article give the p-value, implying that they consider this r significant, so they seem to have used a one-tailed test. Although they did not say so in the excerpt, the reason for using a one-tailed test, in this case, would be that they did not expect negative correlation or did not consider any negative correlation meaningful. Notice that some of the observed values of r were negative, the most significant being -0.34, -0.36, and -0.47. However, these values were observed in groups with very small sample sizes. Notice that the one- and two-sided critical values are quite a bit larger for $n = 6$ and $n = 12$ ($\nu = 4$ and $\nu = 10$).

Consider problem scores at next point in time. Use the same logic as above to see whether the correlation coefficients are significant.

LITERATURE EXAMPLE 15.1

Breastfeeding and employment

Margaret H. Kearney, RNC, MS, Linda Cronenwett, RN, PhD, FAAN

Journal of Obstetric, Gynecologic, and Neonatal Nursing, vol. 20, no. 6 (November/December 1991)

Breastfeeding problems, outcomes, and satisfaction of married, well-educated first-time mothers who returned to work within six months postpartum were compared to those of mothers with the same characteristics who stayed at home. Mothers who planned to work after giving birth anticipated and experienced shorter durations of breastfeeding than did those who planned to remain at home. Breastfeeding experiences and satisfaction among working mothers differed little from the experiences and satisfaction of their nonworking counterparts; however, employment prior to two months postpartum exerted some negative effects on breastfeeding outcomes.

Table 3 Hours Worked per Week and Breastfeeding Problem Scores

Length of Time after Delivery	N*	Hours Worked per Week		Correlation with problem score at			
				Same point in time		Next point in time	
		Mean	SD	*r*	*p*	*r*	*p*
2 weeks	6	5.5	7.1	−0.34	NS	−0.47	NS
3 weeks	12	6.6	5.2	−0.36	NS	−0.17	NS
4 weeks	12	8.9	5.9	0.22	NS	0.07	NS
5 weeks	18	8.6	6.7	0.28	NS	0.47	0.03
6 weeks	18	8.1	5.3	0.36	NS	0.44	0.04
2 months	32	21.6	13.2	−0.06	NS	0.00	NS
3 months	49	27.1	12.1	0.16	NS	−0.05	NS
4 months	50	27.4	12.7	0.07	NS	0.18	NS
5 months	51	27.2	12.4	0.27	0.03	0.10	NS
6 months	48	28.6	13.2	0.09	NS	———	———

* Includes only those working mothers who were still breastfeeding.
——— indicates no next point in time in study.

Literature Example 15.2 is from the article entitled "Stress Process among Mothers of Infants: Preliminary Model Testing."

Notice the variables listed in the first column and first row in the table and the descriptive statistics at the bottom of each vertical column.

What are the descriptive statistics given at the bottom of each vertical column? [1]_______, _______. The top row lists by number the same variables given in the first column. Variable 6 in the top row is [2]_______.

The correlation coefficient for (HPLP) and self actualization = [3]_______. The critical r ($\alpha = .05$, two-tailed test) = [4]_______. r [5]is/is not significant. The asterisks (***) given in the table along with the r indicate that the probability of such r occurring if the hypothesis $\rho = 0$ is true is < [6]_______.

The author's explanation of the table notes that "None of the three stressor variables were significantly correlated with one another."

What are the values of r for variables 1 and 2? [7]_______

For variables 1 and 3? [8]_______

For variables 2 and 3? [9]_______

A table of critical values of r shows that for $n = 100$, r has to exceed .195 to be significant. Our table does not give n up to 173. The authors must have consulted larger tables or used computer software to compute the probabilities.

Read the rest of the explanation and relate it to the correlation coefficients in the table.

LITERATURE EXAMPLE 15.2

Stress process among mothers of infants: Preliminary model testing

Lorraine O. Walker

Nursing Research, vol. 38, no. 1 (January/February 1989)

Maternal employment, cesarean birth, and infant difficultness were used to test the mediating effect of perceived stress and the stress-buffering role of health practices on maternal identity. One hundred seventy-three mothers returned a parenting survey that focused on: stressors, perceived stress, health practices, maternal identity, and a demographic profile. Work status and infant difficultness were related to perceived stress. Neither had direct effects on maternal identity, but were related to it through the mediating effects of perceived stress. While health practices did not show buffering effects between stressors and perceived stress, these did contribute additively to the prediction of stress perception. Also, health practices contributed additively to the prediction of identity. Notable among the health practices predicting identity were self-actualizing expression, nutrition, interpersonal support, and stress management. These findings support a stress process model of parenting in which: (a) effects of stressors on maternal identity are mediated by perception of stress, and (b) health practices contribute positively and directly to maternal identity.

The intercorrelation of major variables representing constructs in the stress-process model is presented in Table 1. *None of the three stressor variables were significantly correlated with one another.* Since cesarean birth was not correlated with any other variables in the stress-process model, it was not included in any analyses that follow. Both work status and infant difficulties were modestly correlated with global perceived stress. In turn, global perceived stress, overall health-promotive lifestyle, and maternal identity variables were significantly correlated with one another. Both stressor variables were negatively correlated with several specific health-promotive practices, but a larger number of correlations for work status were statistically significant than for infant difficulties.

TABLE 1 Zero-Order Correlations and Descriptive Statistics for Variables in Stress Model

	1	2	3	4	5	6	7	8	9	10	11	12
Stressors												
1. Cesarean[a]	———											
2. Work status	−.04											
3. (INFDIFF)	.09	.01										
Mediator												
4. (PSS)	.02	.27**	.33***									
Moderators												
5. (HPLP)	.00	−.29***	−.16*	−.56***								
6. Self actualization	.01	−.21**	−.21**	−.57***	.84***							
7. Health responsibility	−.00	−.11	−.04	−.39***	.73***	.42***						
8. Exercise	.07	−.17*	−.03	−.26***	.55***	.25***	.41***					
9. Nutrition	−.03	−.29***	−.06	−.29***	.73***	.50***	.49***	.36***				
10. Interpersonal support	.01	−.23**	−.09	−.40***	.77***	.73***	.40***	.25***	.42***			
11. Stress management	−.06	−.23**	−.24***	−.44***	.74***	.57***	.47***	.30***	.49***	.50***		
Maternal Role												
12. (SD-Self)	.08	−.15	−.18*	−.53***	.48***	.54***	.22***	.16*	.33***	.38***	.39***	———
Scale Means	———	———	12.7	25.0	125.4	40.1	20.9	9.9	16.2	22.1	16.2	63.3
Scale SDs	———	———	3.4	8.0	20.2	7.0	5.1	3.8	4.0	3.9	3.4	8.3

Note: $N = 173$ in most cases. For work status, N = 148 because parttime employed mothers were omitted.
[a] Cesarean birth was coded 1 = yes, 0 = no. Work status was coded 1 = nonemployed, 2 = fulltime employed.
$* = p < .05$, $**p < .01$, $***p < .001$

Literature Example 15.3 is from the article entitled "Relationship Between Resistance to Insulin-Mediated Glucose Uptake, Urinary Uric Acid Clearance, and Plasma Uric Acid Concentration."

Notice Fig. 1, which shows two scatter plots.

Which of the scatter plots seems to show a curvilinear relationship?

[1] SSPG and Serum Uric Acid Level/ Insulin Response and Serum Uric Acid Level

Consider the title of Fig. 1.

What does $r = .61$; $p < .001$ indicate?

[2] ______________________________

____

____

Calculate the coefficient of determination, $r^2 =$ [3] ______

What does this indicate?

[4] ______________________________

____

____

Read the results section and look at the scatter plots in the light of the author's interpretation.

LITERATURE EXAMPLE 15.3

Relationship Between Resistance to Insulin-Mediated Glucose Uptake, Urinary Uric Acid Clearance, and Plasma Uric Acid Concentration

Francesco Facchini, MD; Y.-D. Ida Chen, PhD; Clarie B. Hollenbeck, PhD; Gerald M. Reaven, MD

Objective.—To define the relationship, if any, between insulin-mediated glucose disposal and serum uric acid.

Design.—Cross-sectional study of healthy volunteers.

Setting.—General Clinical Research Center, Stanford (Calif) University Medical Center.

Participants.—Thirty-six presumably healthy individuals, nondiabetic, without a history of gout.

Measurements.—Obesity (overall and regional), plasma glucose and insulin responses to a 75-g oral glucose load, fasting uric acid concentrations, plasma triglyceride and high-density lipoprotein–cholesterol concentrations, systolic and diastolic blood pressure, insulin-mediated glucose disposal, and urinary uric acid clearance.

Results.—Magnitude of insulin resistance and serum uric acid concentration were significantly related ($r = .69$; $P < .001$), and the relationship persisted when differences in age, sex, overall obesity, and abdominal obesity were taken into account ($r = .57$; $P < .001$). Insulin resistance was also inversely related to urinary uric acid clearance ($r = -.49$; $P < .002$), and, in addition, urinary uric acid clearance was inversely related to serum uric acid concentration ($r = -.61$; $P < .001$).

Conclusions.—Urinary uric acid clearance appears to decrease in proportion to increases in insulin resistance in normal volunteers, leading to an increase in serum uric acid concentration. Thus, it appears that modulation of serum uric concentration by insulin resistance is exerted at the level of the kidney.

(*JAMA*. 1991;266:3008-3011)

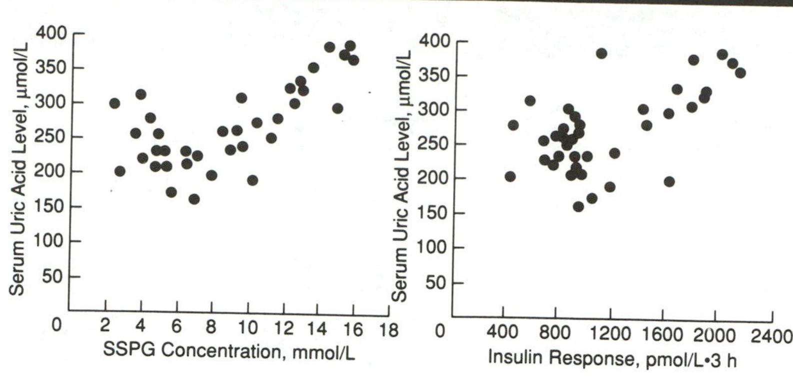

Fig 1.—Relationship between steady-state plasma glucose (SSPG) concentration during the insulin suppression test ($r=.69$; $P<.001$) (left panel) and plasma insulin response to oral glucose ($r=.61$; $P<.001$) (right panel) and serum uric acid concentration.

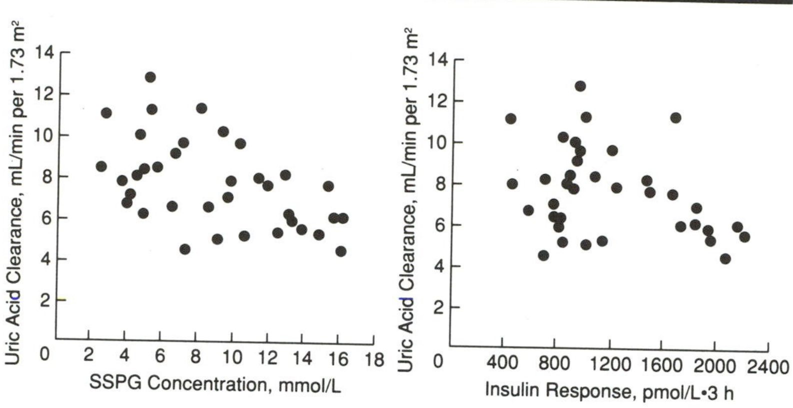

Fig 2.—Relationship between steady-state plasma glucose (SSPG) concentration during the insulin suppression test ($r=-.49$; $P<.002$) (left panel) and plasma insulin response to oral glucose ($r=-.33$; $P<.05$) (right panel) and uric acid clearance.

Literature Example 15.4 is from the article entitled "Relationships Between Physical Activity and Temporal-Distance Characteristics of Walking in Elderly Women."

Notice the scatter plot.

Is it reasonable to assume that there is a relationship between walking speed and cadence?

[1] Yes/No

$r = .91$

The percent of the variation in cadence accounted for by the walking speed r^2, is [2] _______.

What is the percent of variation not explained by walking speed?

$1 - r^2 =$ [3] _______

Regression equation

a (intercept) = [4]_______

b (regression coefficient) = [5] _______

Read the last sentence of the abstract.

What is the implication of this study?

[6] ______________________________

____

____

LITERATURE EXAMPLE 15.4

Relationships between physical activity and temporal-distance characteristics of walking in elderly women

Carol I. Leiper, Rebecca L. Craik

Physical Therapy, vol. 71, pp. 791–803

The purpose of this study was to investigate the relationships between physical activity and walking speed in women 64 years of age and over. Data were gathered from 81 nondisabled women ranging from 64.0 to 94.5 years of age. The women were categorized as sedentary, community active, or exercisers based on a combination of their living situation and level of daily activity. Subjects walked over a 3.84 m recording surface at five different paces, ranging from walking as slowly as possible to walking as quickly as possible. Actual walking speed and length of steps were measured. Stepping frequency and step length relative to leg length were derived measures. Mean walking speeds ranged from 0.43 m/s at the very slow pace to 1.42 m/s at the very fast pace. The walking speeds at the very slow pace were significantly different among the three physical activity groups. At the very slow pace women who exercised were able to walk significantly more slowly than the other women. The groups were not significantly different at any other pace. Normal walking speeds for all groups were slower than those previously reported for younger women, with the walking speed of the fastest pace of the elderly women being closer to the normal walking speed of younger women. The results of this study indicate that physical therapists need to utilize age-appropriate values as the standard when evaluating performance (Leiper CI, Craik RL. Relationships between physical activity and temporal-distance characteristics of walking in elderly women.)

Bivariate correlation tests were performed to determine the relationships among actual walking speed, cadence, and relative step length. Strong significant linear relationships were found for walking speed and cadence ($r = .92$, $n = 431$, $P < .001$), velocity and relative step length ($r = .90$, $n = 431$, $P < .001$), and relative step length and cadence ($r = .72$, $n = 431$, $P < .001$). These relationships and associated regression equations are illustrated in Figures 4 through 6.

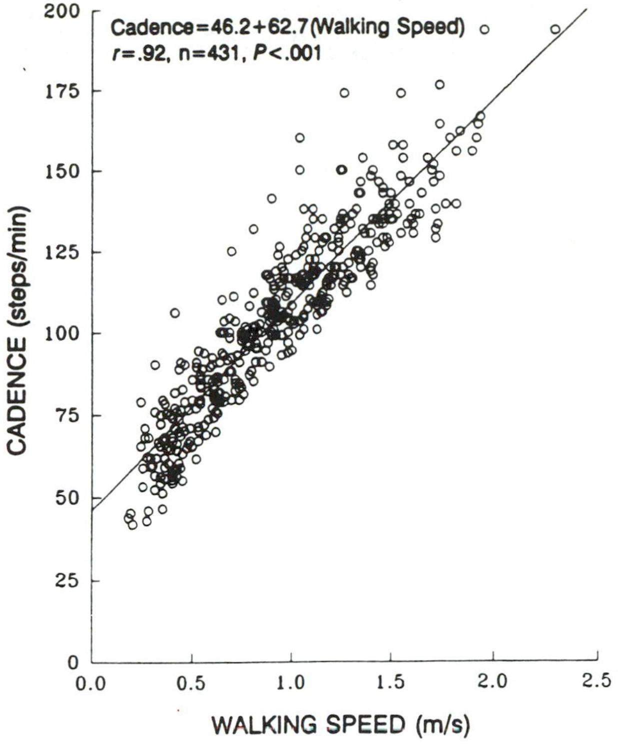

Relationship Between Walking Speed (in meters per second) and Cadence (in steps per minute) for Total Sample of Elderly Women (n = 81)

Literature Example 15.5 is from the article entitled "Carbon Monoxide in Indoor Ice Skating Rinks: Evaluation of Absorption by Adult Hockey Players."

Notice the regression equations for smokers and non-smokers.

The regression coefficients for smokers and non-smokers are [1] similar/not similar. The coefficients of determination for smokers and non-smokers are somewhat different. For non-smokers, [2] _______ % of the alveolar absorption is explained by environmental concentration, whereas for smokers, [3] _______ % is explained by this factor. A physiologist may be able to tell if this difference is attributable to smoking.

LITERATURE EXAMPLE 15.5

Carbon Monoxide in Indoor Ice Skating Rinks: Evaluation of Absorption by Adult Hockey Players

Benoît Lévesque, MD, Eric Dewailly, MD, MSc, Robert Lavoie, MD, Denis Prud'Homme, MD, MSc, and Sylvain Allaire, BSc

We evaluated alveolar carbon monoxide (CO) levels of 122 male, adult hockey players active in recreational leagues of the Quebec City region (Canada), before and after 10 weekly 90-minute games in 10 different rinks. We also determined exposure by quantifying the average CO level in the rink during the games. Other variables documented included age, pulmonary function, aerobic capacity, and smoking status. Environmental concentrations varied from 1.6 to 131.5 parts per million (ppm). We examined the absorption/exposure relationship using a simple linear regression model. In low CO exposure levels, physical exercise lowered the alveolar CO concentration. However, we noted that for each 10 ppm of CO in the ambient air, the players had adsorbed enough CO to raise their carboxyhemoglogin (COHb) levels by 1 percent. This relationship was true both for smokers and non-smokers. We suggest that an average environmental concentration of 20 ppm of CO for the duration of a hockey game (90 minutes) should be the reference limit not to be exceeded in indoor skating rinks. (*Am J Public Health* 1990; 80:594–598.)

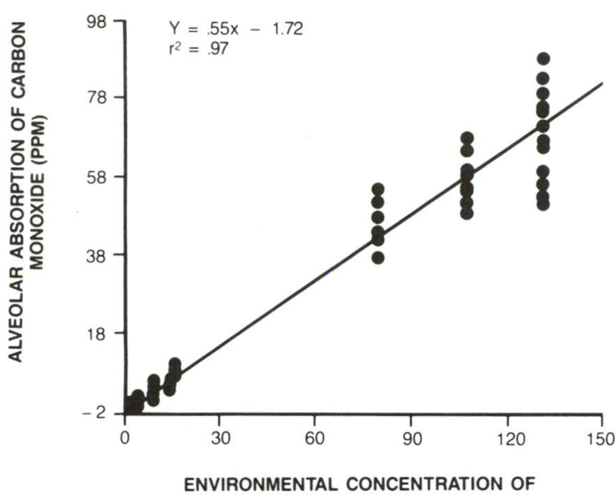

FIGURE 1—Relationship between Alveolar Absorption of Carbon Monoxide among Players during a 90-Minute Hockey Game based on Carbon Monoxide Concentrations in Indoor Skating Rinks (non-smokers).

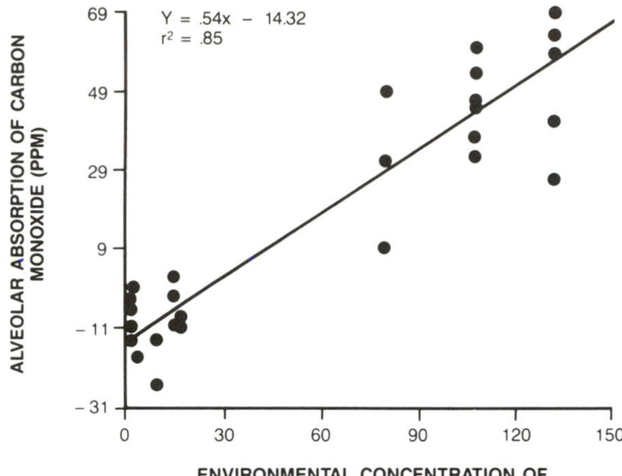

FIGURE 2—Relationship between Alveolar Absorption of Carbon Monoxide among Players during a 90-Minute Hockey Game Based on Carbon Monoxide Concentrations in Indoor Skating Rinks (smokers).

Literature Example 15.6 is from the article entitled "Effects of Mental Practice on Rate of Skill Acquisition."

In the abstract the author notes that "the experimental group's accuracy improved at a significantly greater rate than that of the control group."

The investigator obtained slopes for different variables and tested the hypothesis that each slope is 0. Which variable has a significant slope?

1 ________

What is the value of t for this slope?

2 ________

Look up the critical value of t from the t-table for $\alpha = .05$, two-tailed test: 3 ________ The probability given for the variable Accuracy is

4 ________.

Do you reject or accept the hypothesis that the slope is 0?

5 Reject/Accept

LITERATURE EXAMPLE 15.6

Effects of Mental Practice on Rate of Skill Acquisition

Joyce R Maring

The purpose of this study was to investigate the effectiveness of mental practice in increasing the rate of skill acquisition during a novel motor task. Twenty-six subjects were randomly assigned to two groups. The Control Group (n = 13) performed only physical practice; the Experimental Group (n = 13) performed both mental and physical practice. The task was to toss, by flexing the elbow, a Ping-Pong ball held in a cup on a forearm splint to a target. The biceps brachii muscle and the long and lateral heads of the triceps brachii muscle were monitored electromyographically to determine any changes occurring during skill acquisition. The Experimental Group's accuracy improved at a significantly greater rate than that of the Control Group. In addition, the Experimental Group demonstrated changes in timing variables that led to a more efficient movement. These changes included a decrease in time from the onset of muscle activity to peak activity and an increase in the time elapsed from the onset of agonist contraction to the onset of antagonist contraction. These results suggest that mental practice may be an important tool in facilitating the acquisition of a new motor skill. [Maring JR: Effects of mental practice on rate of skill acquisition. Phys Ther 70:165–172, 1990]

Table 2. *Results of* t *Test for Slopes of Dependent Variables of Control Group (n = 13)*

Variable	*t*	*df*	*p*[a]
Accuracy	4.35	12	.005
T1-biceps brachii muscle[b]	0.79	12	NS
T1-lateral head of triceps brachii muscle[b]	0.29	11	NS
T1-long head of triceps brachii muscle[b]	0.24	11	NS
T2-lateral head of triceps brachii muscle[c]	0.66	11	NS
T2-long head of triceps brachii muscle[c]	1.13	11	NS
Biceps brachii muscle[d]	0.59	12	NS
Lateral head of triceps brachii muscle[d]	0.17	11	NS
Long head of triceps brachii muscle[d]	1.86	11	.05

[a]Probability that slope is not different from zero.

[b]Onset of muscle activity to peak activity per trial.

[c]Onset of agonist to onset of antagonist per trial.

[d]Total integrated electromyographic activity expressed as percentage of maximal voluntary contraction per trial.

Self-Assessment

Tasks:	How well can you do it?					Page
	Poorly				*Very well*	
1. Interpret the correlation coefficient, r.	1	2	3	4	5	530
2. Interpret r^2, $(1 - r^2)$.	1	2	3	4	5	558
3. Formulate and test hypotheses about r using critical values of r and confidence intervals.	1	2	3	4	5	541
4. Estimate a confidence interval for a population correlation coefficient.	1	2	3	4	5	541
5. Predict y when x is known, on the basis of a line drawn using the least squares method.	1	2	3	4	5	550
6. Interpret an analysis of variance table for regression.	1	2	3	4	5	556
7. Interpret a regression equation.	1	2	3	4	5	551

Answers for Chapter 15

Answers for p. 529: 1) Chi-square 2) The correlation coefficient 3) Regression

Answers for p. 532: 1) Linear 2) Increases 3) Decrease 4) 0 5) Any number between −1 and +1 except ±1 6) Linear 7) Increases 8) Positive

Answers for p. 533: 1) Linear 2) −0.664 3) No

Answer for p. 535: 1) Multiply 2) Greater 3) Smaller 4) Smaller 5) Greater 6) Greater 7) Smaller

Answers for Activity 15.1: Two of the three negative products, rows 1 and 11, are in quadrant 2; the product in row 24 is in quadrant 4. The products in rows 3, 7, 9, 10, 15, 16, 17, 18, 19, 21, 25, and 28 are in quadrant 3. The rest are in quadrant 1.

Answer for p. 537: 1) $\frac{64.7}{12.47 * 5.92} = .877$ 2) 0.2

Answer for p. 538: 1) −0.1

Answers for Activity 15.2: 1) (.8, 1) 2) (−.4, +.4) 3) (−.1, .65) 4) (.15, .75)

Answers for Activity 15.3: 1) No 2) Reject 3) Yes 4) Accept 5) Yes 6) Accept 7) No 8) Reject

Answers for Activity 15.4: 1) Reject 2) Accept 3) Reject 4) Reject

Answers for p. 544: 1) ±2.048 2) −1.701 3) +1.701

Answers for Activity 15.5: 1) 9.8 2) Reject 3) 0 4) Accept 5) 2.1 6) Reject 7) 3.1 8) Reject

Answers for p. 548: 1) Refute 2) Refute 3) PREVO2

Answers for p. 552: (Answers have been rounded off.) 1) .877 2) 95 3) 95 4) 102 5) 100 6) 105 7) 106. 8) .0 9) 2 10) −1

Answers for p. 553: 1) 1 2) r_{xy} 3) 1 4) 0 5) Different from 6) The numerator of the variance of x or sum of squares for x

Answers for p. 554: 1) 54 inches 2) 114 pounds 3) Increases

Answers for Activity 15.8: 1) $r^2 = 76.9\%$ 2) $1 - r^2 = 23.1\%$ 3) $s = 2.898$

Answers for pp. 560–561: 1) PREMAXHR and PRERESDIA 2) .77 3) .136 4) More 5) 0.812

Answers for Activity 15.9: 1) 3.772 2) 2.189 3) −1.729 4) −1.060 5) −2.061 6) 0.440 7) −1.312 8) 4.772

Answers for Activity 15.10: 1) .137 2) .481 3) .863 4) .52 5) 14.4 6) 21 7) 7,436.7 8) 13,631.6 9) 1,021 10) 6,554 11) 6,415.7 12) 7,077.6 13) .137 14) .481 15) .863 16) .52 17) 0, reject 18) 0, reject 19) $y' = -4.3 + 2x$ 20) $y' = -304.8 + 6.8x$

Answers for Literature Example 15.1: 1) Problem scores 2) Hours worked per week 3) .400 4) Is not 5) .243 6) Is 7) .468 8) .288

Answers for Literature Example 15.2: 1) Scale means; scale SDs 2) Self actualization 3) .84 4) Table 15.1 does not give critical values for $n = 173$. The best we can say, based on Table 15.1, is that the critical r is less than .195 5) Is 6) .001 7) −.04 8) .09 9) .01

Answers for Literature Example 15.3: 1) SSPG and Serum Uric Acid Level 2) The correlation coefficient is .61 and the probability of such a correlation coefficient occurring if the hypothesis of $\rho = 0$ is true is $< .001$. 3) .3721 4) 37% of the variation in serum uric acid level can be explained by the insulin response. There must be other factors that also influence the acid level.

Answers for Literature 15.4: 1) Yes 2) 85% 3) 15% 4) 46.2 5) 62.7 6) Physical therapists need to use age-appropriate values when evaluating walking performance.

Answers for Literature Example 15.5: 1) Similar 2) 97 3) 85

Answers for Literature Example 15.6: 1) Accuracy 2) 4.35 3) 2.179 4) .005 5) Reject

CHAPTER

16

Statistics for Enumeration Data

Prelude

"Then one day they had the most terrible quarrel of all. King Azaz insisted that words were far more significant than numbers and hence his kingdom was truly the greater and the Mathemagician claimed that numbers were much more important than words and hence his kingdom was supreme. They discussed and debated and raved and ranted until they were on the verge of blows, when it was decided to submit the question to arbitration by the princesses.

"After days of careful consideration, in which all the evidence was weighed and all the witnesses heard, they made their decision:

"'Words and numbers are of equal value, for, in the cloak of knowledge, one is warp and the other woof. It is no more important to count the sands than it is to name the stars. Therefore, let both kingdoms live in peace.'"

— Norton Juster
The Phantom Toll Booth, pp. 76–77

Chapter Outline

Self-Assessment

Tasks:	How well can you do it? *Poorly*				*Very well*	Page
1. Explain the following terms:						
Enumeration data	1	2	3	4	5	583
Goodness of fit	1	2	3	4	5	586
2. Know when and how to use chi square to test each of the following types of hypotheses:						
There is no relationship between two categorical variables.	1	2	3	4	5	585
The distribution of different levels of x is the same for all categories of y (test for homogeneity).	1	2	3	4	5	586
The proportion of individuals in Group I having a certain characteristic is the same as the proportion in Group II having that characteristic.	1	2	3	4	5	586
The distribution of values for a variable obtained experimentally is the same as that predicted by a certain theoretical distribution (goodness of fit).	1	2	3	4	5	587
3. Know when and how to use each of the following measures of strength of association:						
Relative risk	1	2	3	4	5	606
Odds ratio	1	2	3	4	5	615
Phi coefficient (Φ)	1	2	3	4	5	603

Introduction

In this chapter we consider some ways of analyzing nominal and ordinal variables and ratio variables for which the data are given as frequencies of various categories. The unifying factor is that the values of the variable, whatever they might be, are grouped in categories, and the frequency (count) or percentage of data in each category is known. The variable might be dichotomous (e.g., yes or no answers), categorical (e.g., light, moderate, or strenuous exercise), or continuous (e.g., height, each category of which would be defined more precisely by cutoff values, that is, upper and lower limits).

χ^2 is often used to test categorical data in tables (see Chapter 11). The phi coefficient (Φ), based on χ^2, is used to measure strength of association. Dichotomous variables may also be analyzed in terms of relative risk and odds ratio (see Chapter 9). Table 16.1 and Figure 16.1 illustrate these four tests.

We end the chapter with a very brief discussion of some statistical tests that are exceptionally easy to implement and make almost no assumptions about the distribution of the population.

The chapter is organized by test. There are four steps in using each test: (1) deciding on the appropriate statistic; (2) computing the statistic; (3) looking up the statistic in a table; and (4) making a decision.

Examine the student questionnaire on page 584. Circle in the chart below all the variables that yield counts that can be presented in a table.

1	2	3	4	5	6	7	8	9	10	11	12	13	14
SEX	AGE	CLASS	HT	WT	PULSE 1	PULSE 2	SMOKE	ACTIVITY	STRESS	FRIENDS	TIMEOUT	HEALTH	TEMP

Data are frequently reported in table format. (Review formation of two-way and three-way tables, Chapter 5.) Some tables (one-way tables, for example) display values of the variables, thereby giving the distributions of those variables. Two-way tables, however, present more than just distributions of the variables. When a researcher constructs such a table, she does so to investigate the relationship between the two variables. Thus, a hypothesis is implicit in the table. We will first make explicit the type of hypothesis a given table poses, and then explicate the appropriate test for that kind of hypothesis.

TABLE 16.1 Some Tests and Techniques for Analyzing Enumeration Data

Test or Technique	Kind of Data
χ^2 (chi square)	Categorical data or enumeration data
Measures of strength of association: phi coefficient (Φ)	Categorical data or enumeration data
Relation of two variables in terms of risks:	
Relative risk	Yes-no data; mainly used with prospective studies (see Chapter 9 for examples)
Odds ratio	Yes-no data; mainly used in case-control studies (see Chapter 9 for examples)

STUDENT QUESTIONNAIRE

Please complete the following questionnaire by filling in the blank or circling the most appropriate answer.

1. Sex: (1) Female (2) Male
2. Age: _______
3. Class status:
   (1) Freshman (2) Sophomore (3) Junior (4) Senior (5) Graduate
4. Height in inches: _______
5. Weight in pounds: _______
6. Pulse (take your pulse for 15 seconds, multiply by 4, and record the number): _______
7. Run in place for one minute and then take your pulse again.

   Pulse after running in place for one minute: _______
8. Smoking cigarettes:
   (1) I never smoked. (2) I quit smoking. (3) I smoke. (4) Other
9. Daily physical activity level:
   (1) Minimal (2) Moderate (3) Vigorous
10. Daily stress level:
    (1) Minimal (2) Moderate (3) Heavy
11. I have at least three friends or relatives with whom I can discuss the positive and negative events of my day.
    (1) True (2) False
12. I take at least one five-minute "time out" per day to be quiet.
    (1) True (2) False
13. I would rate my overall health as:
    (1) Excellent (2) Good (3) Average (4) Below average
14. I generally keep the thermostat at my residence set at _______ °F.
15. What is your opinion about national health insurance?

    ____

    ____

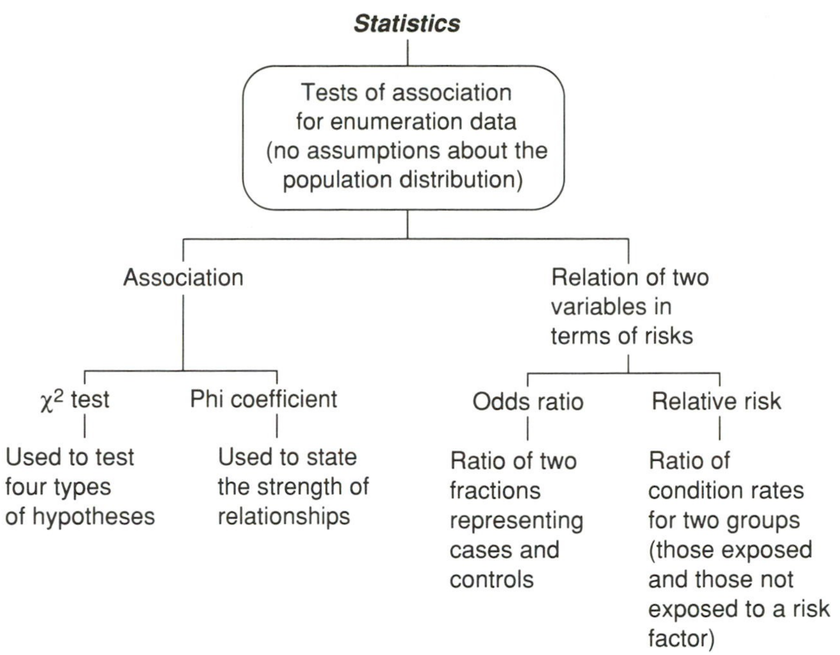

FIGURE 16.1 Decision Tree: Tests of Association for Enumeration Data

Chi-Square Statistic

Four Kinds of χ^2 Tests

Variables 1, 3, 8, 9, 10, 11, 12, and 13 from the student questionnaire are all categorical variables. For each of these we tabulate the frequency of responses in each category. Such data are often called **enumeration data**. A χ^2 test is used to test the following types of hypotheses about enumeration data:

1. There is no relationship between two categorical variables.
2. The distributions of different levels of x are the same for all categories of y (test for homogeneity).
3. The proportion of individuals in Group I having a certain characteristic is the same as the proportion in Group II having that characteristic.
4. The distribution of values for a variable obtained experimentally is the same as that predicted by a certain theoretical distribution (goodness of fit).

Activity 16.1 presents research questions and tables that represent each of the kinds of hypotheses mentioned above. For each type of hypothesis, you will formulate a research question and construct an appropriate table. (Review Chapter 12, Formulating Hypotheses, if necessary.)

ACTIVITY 16.1

Research questions and hypotheses concerning enumeration data in the student questionnaire

Hypothesis

There is no relationship between two categorical variables (SEX and ACTIVITY).

Research question

Is there a relationship between SEX and ACTIVITY?

Table

Rows: SEX Columns: ACTIVITY

	1	2	3	ALL
1	6	24	2	32
2	0	12	5	17
ALL	6	36	7	49

Write a parallel hypothesis about STRESS and HEALTH.

1 ______________

____

____

Write the corresponding question:

2 ______________

____

____

Sort questionnaire cards and create a table of the data.

3

The distributions of different levels of x (HEALTH) are the same for all categories of y (SEX).

Is the distribution of HEALTH the same for males and females?

Rows: SEX Columns: HEALTH

	1	2	3	4	ALL
1	5	21	6	1	33
2	4	11	3	0	18
ALL	9	32	9	1	51

Write a parallel hypothesis about ACTIVITY and HEALTH.

4 ______________

____

____

Write the corresponding question:

5 ______________

____

____

Sort questionnaire cards and create a table of the data.

6

The proportion of individuals in Group I (males) having a certain characteristic (have friends they can talk to, i.e., FRIENDS = TRUE) is the same as the proportion in Group II (females) having that characteristic.

Is the proportion of persons who have friends they can talk to the same for males (Group I) and females (Group II)?

Rows: SEX Columns: FRIENDS

		1	2	ALL
1	Number	29	4	33
	Proportion	.88	.12	1
2	Number	17	1	18
	Proportion	.94	.6	1
ALL	Number	46	5	51
	Proportion	.90	.10	1

Write a parallel hypothesis about TIMEOUT and HEALTH.

7 ______________

____

____

Write the corresponding question:

8 ______________

____

____

Sort questionnaire cards and create a table of the data.

9

The distribution of values for a variable (SEX) obtained experimentally is the same as predicted by a certain theoretical distribution (goodness of fit).

Theoretically we expect 50 percent females and 50 percent males in any given population.* Is the proportion of females in the sample .5?

	Observed		Expected	
SEX	Count	Proportion	Count	Proportion
1	33	.65	25.5	.5
2	18	.35	25.5	.5
$n =$	51		51	1.0

Write a parallel hypothesis about TIMEOUT.

10 ______________

____

____

Write the corresponding question:

11 ______________

____

____

Sort questionnaire cards and create a table of the data.

12

Assume that the theoretical distribution predicts a .5 proportion. (This proportion was selected simply for illustration.)

Hypothesis Testing Using the Chi-Square Statistic

All the questions in Activity 16.1 can be answered and the hypotheses tested using the χ^2 **statistic**. It is the appropriate statistic when the cells of tables contain counts, or enumeration data, or frequencies. The χ^2 statistic is defined by the following formula:

$$\chi^2 = \sum \frac{(O_i - E_i)^2}{E_i}$$

---

* The theoretical distribution in this example is called the uniform distribution. A variable is distributed uniformly when all categories have the same frequency. Thus, if there are two categories, each would contain 50 percent of the subjects; if there were four, each would contain 25 percent, etc.

Caution: You may not use χ^2 if:

1. any expected frequency is less than 1;
2. more than 20 percent of the expected frequencies are less than 5.

If your data have either of the above characteristics, the remedy is to combine categories. If this doesn't work, repeat the experiment with a larger sample.

ACTIVITY 16.2

Computation of χ^2

You practiced computing χ^2 in Chapter 7. As a review, calculate χ^2 using the data below:

$$\chi^2 = \sum \frac{(O_i - E_i)^2}{E_i}$$

Note that you need expected values to calculate χ^2

Data:

Observed, O_i	Expected, E_i	$\frac{(O_i - E_i)^2}{E_i}$
21	22.22	1 ______
12	10.78	2 ______
12	10.78	3 ______
4	5.22	4 ______

$$\chi^2 = \sum \frac{(O_i - E_i)^2}{E_i} = {}^{5}\_\_$$

Calculation of Expected Frequencies

To calculate expected frequency you must use your knowledge of probability. We will use the table of SEX by ACTIVITY to illustrate how to calculate expected values of females who engage in minimal exercise.

Rows: SEX Columns: ACTIVITY

	Minimal	Moderate	Vigorous	ALL
Female	6	24	2	32
Male	0	12	5	17
ALL	6	36	7	49

What is the probability of being a female in this sample (i.e., the proportion of females in the sample)?

(Total no. of females)/(No. of males + no. of females) = 32/49

What is the probability of exercising minimally (i.e., the proportion of the sample who exercise minimally)?

(No. who exercise minimally)/(Sample size) = 6/49

What is the probability of being a female and exercising minimally (assuming these variables are independent)?	[(32/49) * (6/49)] = .08
What is the expected number of females in the minimal exercise category?	(Relative proportion in the category) * Sample size = [(32/49) * (6/49)] * 49 = 3.92

Now calculate the expected frequency for females exercising moderately:

What is the probability of being a female in this sample?	32/49
What is the probability of exercising moderately?	1 ____________________
What is the probability of being a female and exercising moderately?	2 ______ * ______ = ______
What is the expected number of females in the moderate exercise category?	3 ____________________

Calculate the expected frequency for females exercising vigorously:

What is the probability of being a female in this sample?	4 ____________________
What is the probability of exercising vigorously?	5 ____________________
What is the probability of being a female and exercising vigorously?	6 ______ * ______ = ______
What is the expected number of females in the vigorous exercise category?	7 ____________________

Now follow the same steps for males. First, calculate the expected frequency for males exercising minimally.

What is the probability of being a male in this sample?	8 ____________________
What is the probability of exercising minimally?	9 ____________________
What is the probability of being a male and exercising minimally?	10 ____________________
What is the expected number of males in the minimum category?	11 ____________________

Calculate the expected frequency for males exercising moderately:

What is the probability of being a male in this sample?	12 ____________________

What is the probability of exercising moderately? 13 ______

What is the probability of being a male and exercising moderately? 14 ______

What is the expected number of males in the moderate category? 15 ______

Calculate the expected frequency for males exercising vigorously:

What is the probability of being a male in this sample? 16 ______

What is the probability of exercising vigorously? 17 ______

What is the probability of being a male and exercising vigorously? 18 ______

What is the expected number of males in the vigorous category? 19 ______

You can see the expected frequencies in the table.

Rows: SEX	**Columns:** ACTIVITY			
	Minimal	**Moderate**	**Vigorous**	**ALL**
Female				
Observed	6	24	2	32
Expected	3.92	23.51	4.57	32
$\frac{(O_i - E_i)^2}{E_i}$	20 ______	21 ______	22 ______	
Male				
Observed	0	12	5	17
Expected	2.08	12.49	2.43	17
$\frac{(O_i - E_i)^2}{E_i}$	23 ______	24 ______	25 ______	
ALL	6	36	7	49

Now use the formula to compute χ^2 (df = 2):

$$\chi^2 = \sum \frac{(O_i - E_i)^2}{E_i} = \text{ }^{26} \text{ \_\_\_\_\_\_}$$

Can we use χ^2? All the expected frequencies are more than 1. However, four out of six expected frequencies are below 5. This makes the use of χ^2 invalid, so we must collapse categories. If we combine data on minimal and moderate exercise we get a table like the following:

ACTIVITY: **Three levels**

Rows: SEX Columns: ACTIVITY

	Minimum	Moderate	Vigorous	ALL
Female				
Observed	6	24	2	32
Expected	3.92	23.51	4.57	32
$\frac{(O_i - E_i)^2}{E_i}$	1.106	0.010	1.446	
Male				
Observed	0	12	5	17
Expected	2.08	12.49	2.43	17
$\frac{(O_i - E_i)^2}{E_i}$	2.082	0.019	2.723	
	= 7.386			

ACTIVITY: **Collapsed to two levels**

	Min + Mod	Vigorous	Total
Female			
Observed	30	2	32
Expected	27.43	4.57	32
$\frac{(O_i - E_i)^2}{E_i}$	1 ________	2 ________	
Male			
Observed	12	5	17
Expected	14.57	2.43	17
$\frac{(O_i - E_i)^2}{E_i}$	3 ________	4 ________	
Total	42	7	49

Chi-square = 5 ______________

Degrees of freedom for χ^2 = (rows − 1) * (columns − 1)

For the collapsed table df = 6 ________

Notice that 50 percent of the expected frequencies in the collapsed table are below 5. So the χ^2 test is still not appropriate. The only other choice we have is to increase the sample size in future studies.*

* Note that combining the moderate and vigorous categories will not work, either, since entries in the vigorous column are less than 5. We therefore are assured that at least 50 percent of the entries would be less than 5 if we combined the moderate and vigorous categories, so we need do no further calculations.

ACTIVITY 16.3

Tables, hypotheses, and testing using the χ^2 statistic

We collected data on SEX and ACTIVITY from three more statistics classes to bring the sample size up to 150. Complete the calculations of χ^2 below.

Hypothesis	Table	χ^2 from table $\alpha = .05$	Calculated χ^2	Reject or accept the hypothesis
1. There is no relationship between two categorical variables (SEX and ACTIVITY).	Rows: SEX Columns: ACTIVITY	1 ______	11.3	2 R/A
Evaluate a similar hypothesis about ACTIVITY and STRESS: There is no relationship between STRESS and ACTIVITY.	Rows: STRESS Columns: ACTIVITY	3 ______	3.13	4 R/A
2. The distributions of different levels of x (HEALTH) are the same for all categories of y (SEX).	Rows: SEX Columns: HEALTH	5 ______	6.02	6 R/A

Rows: SEX Columns: ACTIVITY

	Min.	Mod.	Vig.	ALL
Female				
Observed	21	71	11	103
Expected	15.8	20.7	16.5	
Male				
Observed	2	32	13	47
Expected	7.2	32.3	7.5	
ALL	23	103	24	150

$\chi^2 = 11.297$ df = 2

Rows: STRESS Columns: ACTIVITY

	Min.	Mod.	Vig.	ALL
Min.	3	25	3	31
Mod.	15	60	15	90
Heavy	5	17	6	28
ALL	23	102	24	149

$\chi^2 = 3.126$ df = 4

Rows: SEX Columns: HEALTH

	Ex.	Good	Avg.	ALL
Female				
Observed	18	65	21	104
Expected	23.4	62.7	17.9	
Male				
Observed	16	26	5	47
Expected	10.6	28.3	8.1	
ALL	34	91	26	151

$\chi^2 = 6.019$ df = 2

Evaluate a similar hypothesis about HEALTH and ACTIVITY: The distributions of HEALTH are the same for each level of ACTIVITY.

Rows: ACTIVITY Columns: HEALTH [7] ______ 26 [8] R/A

	Ex.	Good	Avg.	ALL
Min.	3	13	7	23
Mod.	16	68	18	102
Vig.	15	9	0	24
ALL	34	90	25	149

$\chi^2 = 29.231$ df = 4

In this table the columns for "Good" and "Average" have been collapsed together because of the 0 in the Avg. column in the table above.

Rows: ACTIVITY Columns: HEALTH

	Ex.	Good + Avg.	ALL
1	3	20	23
2	16	86	102
3	15	9	24
ALL	34	115	149

$\chi^2 = 25.652$ df = 2
From table for df = 2 and $\alpha = .05$, $\chi^2 = 5.99$.
Note that both tables above give χ^2 larger than the χ^2 from the table.

3. The proportion of individuals in Group I (males) having a certain characteristic (FRIENDS) is the same as the proportion in Group II (females) having that characteristic.

Rows: SEX Columns: FRIENDS [9] ______ 0.27 [10] R/A

	True	False	ALL
Female			
Number	90	14	104
Proportion	.87	.13	
Male			
Number	40	8	48
Proportion	.83	.17	
ALL	130	22	152

$\chi^2 = 0.273$ df = 1

Evaluate a similar hypothesis about TIMEOUT and HEALTH: The proportion of persons who take TIMEOUT is the same for the different HEALTH groups.

Rows: TIMEOUT Columns: HEALTH [11] ______ 4.1 [12] R/A

	Ex.	Good	Avg.	ALL
True	28	63	14	105
False	6	28	10	44
ALL	34	91	24	149

$\chi^2 = 4.073$ df = 2

4. The distribution of values for a variable (SEX) obtained experimentally is the uniform distribution (goodness of fit).

SEX	Observed Count	Observed Prop.	Theoretical Count	Theoretical Prop.
Female	105	.69	77	.5
Male	48	.31	77	.5

$\chi^2 = 11$ df = 1

13 ____ 11

14 R/A

Evaluate a similar hypothesis about FRIENDS: The distribution of values for the variable FRIENDS in this sample is uniform, as expected on theoretical grounds.

FRIENDS	Observed Count	Observed Prop.	Theoretical Count	Theoretical Prop.
True	131	.86	76	.5
False	21	.14	76	.5

$\chi^2 = 46$ df = 1

15 ____ 46

16 R/A

Examples from the Literature

Literature Example 16.1 is from the article entitled "Medical and Psychosocial Factors Predictive of Psychotropic Drug Use in Elderly Patients." Read the abstract to learn about the study.

The main question investigated in this study was, Are the users of psychotropic drugs different from non-users with respect to various variables? What two kinds of statistical tests did the authors use?

1 ____________________

Notice the variables for which *t*-tests were performed. The hypotheses state that $\mu_1 = \mu_2$ or $\mu_1 - \mu_2 = 0$. Confidence intervals that cover $\mu_1 - \mu_2 = 0$ will lead us to accept the hypothesis. For which variables would you reject the hypothesis $\mu_1 - \mu_2 = 0$?

2 ____________________

When enumeration data or the corresponding percentages are involved, the test of choice is χ^2. Write the appropriate hypothesis for the variable Education vs. Users and Non-users.

3 ____________________

____

____

____

Write the value of χ^2 for df = 5 ($\alpha = .05$). Do you accept or reject the hypothesis?

4 ____________________

5 Accept/Reject

Would you reject any of the hypotheses tested with the χ^2 statistic?

6 Yes/No

LITERATURE EXAMPLE 16.1

Medical and Psychosocial Factors Predictive of Psychotropic Drug Use in Elderly Patients

L. Douglas Ried, PhD, Dale B. Christensen, PhD, and Andy Stergachis, PhD

The purpose of this study was to investigate medical and psychosocial factors that may be used to identify patients at risk of psychotropic drug use. Population-based surveys were completed by 278 elderly health maintenance organization (HMO) patients in August 1984. Physical and mental health status and social support were measured in the survey. Automated prescription records from the year prior to and the year after the survey were linked to data from the survey. Patients received 737 prescriptions for psychotropic drugs during the two-year period under study. Doxepin (20.2 percent), flurazepam (15.2 percent), and diazepam (14.8 percent) were dispensed most frequently. Nearly 30 percent of the patients received a prescription for at least one psychotropic drug during the two-year period, and 14 percent received at least one prescription during both years. Three significant predictors of subsequent psychotropic drug use were: prior use (odds ratio = 17.2, 95% CI = 6.25, 47.33), the number of physical impairments (OR = 1.73, 95% CI = 1.05, 2.84), and the respondent's rating on the Alameda Health Scale (OR = 1.65, 95% CI = 0.99, 2.75). Patients' self-reported mental health status and sociodemographic characteristics were not significant predictors of subsequent use. (*Am J Public Health* 1990; 80:1349–1353.)

TABLE 2—Comparisons of Users and Non-users of Psychotropic Drugs in the Year after the Survey

Variables	Users (N = 59)	Non-users (N = 219)	Test Statistic	Difference	95% CI
Age	72.9	74.9	2.17[a]	1.98	0.09, 3.87
Percent female	66.1	54.8	1.98[b]	0.16	
Percent married	59.3	64.8	0.39[b]	0.52	
Percent living alone	27.1	28.3	0.00[b]	0.99	
Education (percent)			2.83[c]		
<8th grade	15.8	11.7			
9–11th grade	24.6	20.4			
High school graduate	24.6	26.3			
13–15 years	21.1	23.0			
College graduate	8.8	7.5			
Graduate school	5.3	11.3			
Income (percent)			3.12[c]		
<$10,000	26.5	23.5			
$10–$15,000	32.7	25.7			
$15–$25,000	22.4	29.9			
$25–$35,000	14.3	11.8			
$35–$50,000	2.0	5.9			
$50,000+	2.0	3.2			
Number of daily activities interfered with	1.04	0.61	2.09[a]	0.43	−0.03, 0.89
Number of medical conditions	4.32	3.42	2.13[a]	0.90	0.03, 1.79
Alameda scale	3.01	3.39	1.55[a]	0.38	−0.13, 0.88
Emotional ties	9.57	9.69	0.32[a]	0.12	−0.65, 0.99
Positive effect	43.89	46.77	1.91[a]	2.88	−0.21, 5.99
Depressive symptomatology	7.18	6.76	0.85[a]	0.42	−0.41,1.25
Self-reported health status	2.23	1.95	2.56[a]	0.28	0.02, 0.52

[a]Independent t-test
[b]*chi*-square (df = 2)
[c]*chi*-square (df = 5)

Table 2 shows the results of the comparisons on each of the sociodemographic and health status predictor variables during the subsequent year between users and non-users of psychotropic drugs. The average user was nearly two years younger than the average nonuser. Psychotropic users reported that their physical condition interfered with more of their daily activities, had more self-reported medical conditions, had a less positive outlook on life, and, according to their own reports, were in somewhat poorer health. It is interesting to note that while self-reported health status of users was significantly lower than that of non-users, it was still in the "good" to "fair" range.

Literature Example 16.2 is from the article entitled "Patterns of Medical Employment: A Survey of Imbalances in Urban Mexico." Read the abstract to learn about the study.

Write a hypothesis for the variable in Table 4.	1 ______________________ ______________________ ______________________
On the basis of the reported *p*, do you reject or accept the hypothesis?	2 Reject/Accept
Write a hypothesis for variables in Table 5.	3 ______________________ ______________________ ______________________
Look up χ^2 in the table for df = 3 and $\alpha = .05$.	4 ______________________
On the basis of the calculated $\chi^2 =$ 17.22, do you reject or accept the hypothesis?	5 Reject/Accept
Read the results written by the authors. Do you agree?	6 Yes/No

Do you think the results would be much different in the United States?

LITERATURE EXAMPLE 16.2

Patterns of Medical Employment: A Survey of Imbalances in Urban Mexico

Julio Frenk, MD, MPH, PhD, Javier Alagon, DPhil, Gustavo Nigenda, MA, Alejandro Muñoz-delRio, BSc, Cecilia Robledo, BA, Luis A. Vaqúez-Segovia, BA, and Catalin Ramírez-Cuadra, BA

This article quantifies the magnitude and correlates of the major imbalances affecting the employment of physicians in the urban areas of Mexico. Since the early 1970s the country has experienced a rapid increase in the supply of doctors, which its health system was unable to absorb fully. In 1986, we conducted a survey in the 16 most important cities based on a probability sample of households where someone with an MD degree lived. A total of 604 physicians were interviewed for a response rate of 97 percent.

The unemployment rate was 7 percent of potentially active physicians; 11 percent held a nonmedical job, and another 11 percent exhibited low productivity and/or income. All in all, we project that 23,500 physicians in these cities were either unemployed or underemployed.

This medical employment pattern was analyzed against five independent variables: generation (i.e. the year in which the physician started medical school), gender, social origin, medical school quality, and specialty. Apart from generation, type of specialty exhibited the strongest correlation with the employment situation of a physician.

The results suggest that higher education and health care in Mexico may be producing rather than correcting social inequalities. Policy alternatives are discussed to restore a balance between the training of physicians, their gainful employment, and the health needs of the population. (*Am J Public Health* 1991; 81:23–29)

In the case of gender, there were no significant differences between men and women in the quality of the medical school attended, but their respective opportunities for postgraduate training presented sharp contrasts. Thus, when specialty is examined by sex, we find that more than two-thirds of female doctors have no specialty, compared to less than half of their male colleagues (Table 4). Among those who do complete a residency, the proportions of men and women are about equal in family medicine. However, males dominate in the basic specialties and even more so in the subspecialties, where their proportion is twice that of females.

Social origin could also affect employment indirectly, through a relationship that is mediated by medical school. As shown in Table 5, the association between social origin and medical school is fairly strong. Thus, 40 percent of physicians from the lower social origin attended an inadequate medical school, compared with 44 percent of doctors from the higher social origin who went to a good school.

TABLE 4—Type of Specialty by Gender

Type of Specialty	Gender: Female %	Male %
Without specialty	69	49
Family medicine	9	8
Basic specialties	11	20
Subspecialties	11	23
Total	100	100
N	100	391

$X^2 = 14.56$ $p = 0.0022$ Gamma = 0.36

TABLE 5—Quality of the Medical School Attended, by Social Origin of the Physician

Medical School Quality	Social Origin: Low %	Middle %	High %
Inadequate	40	28	25
Average	36	34	31
Good	24	38	44
Total	100	100	100
N	197	180	114

$X^2 = 17.22$ $p = 0.0018$ Gamma = 0.24

Literature Example 16.3 is from the article entitled "The Impact of HIV-Related Illness on Employment." Read the abstract to learn about the study.

The Data Analysis section refers to the "proportional hazard" model. What are the maximum and minimum for the proportional hazards?

1 Minimum _______;

maximum _______

A proportional hazard of 1 indicates that a subject's working status will be the same for each interval after onset of HIV-related symptoms. The confidence intervals indicate where the population proportional hazard may lie. Those variables whose confidence intervals do not cover 1 are likely to be significantly different from 1, hence working status is unlikely to be the same. What are the variables whose confidence intervals do not cover 1?

2 ________________________________

____

____

____

A χ^2 test was used, and the p-values are given in the last column of the table. Use $\alpha = .05$ to test the hypothesis that there is no relationship between "stopping working" and a given characteristic. Which characteristics are associated with "stopping working"? In other words, you would reject the null hypothesis associated with which characteristics?

3 ________________________________

____

____

Notice that the confidence interval test and the χ^2 test yielded the same characteristics.

LITERATURE EXAMPLE 16.3

The Impact of HIV-Related Illness on Employment

Edward H. Yelin, PhD, Ruth M. Greenblatt, MD, Harry Hollander, MD, and James R. McMaster, MHSA

We used structured telephone interviews to determine the extent of work loss following onset of symptoms, the interval between onset of symptoms and cessation of work, and the risk factors for work loss among 193 persons with symptoms of human immunodeficiency virus (HIV)-related illness attending the AIDS Clinic at the University of California, San Francisco, between October 1, 1988, and September 30, 1989. Estimates of the duration of time between onset of HIV-related symptoms and work loss derive from the life table method of Kaplan and Meier. A Cox proportional hazards model is used to estimate the effect of risk factors on the probability of withdrawing from work in each time interval. Eighty-six percent of the respondents worked prior to onset of the first symptom of HIV-related illness; 40 percent were working at the time of the most recent interview, a mean of 958 days later. The total number of hours worked declined by 59 percent during this time. Kaplan-Meier analysis indicates that 50 percent who worked prior to onset of HIV-related illness stopped working within two years and all had stopped within 10 years after onset of the first symptom. (*Am J Public Health* 1991; 81:79–84)

Data Analysis

A Cox proportional hazards model[31] is used to estimate the effect of the respondents' demographic and medical characteristics, as well as of their job characteristics on their probability of withdrawing from work in each time interval.

In Table 5, we present estimates of the proportional hazard of work loss associated with characteristics of the respondents and of the jobs they held at the time they first noticed symptoms of HIV-related illness. A proportional hazard less (greater) than 1.0 indicates that the person with the risk factor is less (more) likely to stop working in each time interval after onset. Persons with a diagnosis of AIDS stopped working at much higher rates than those who were diagnosed with ARC or who were asymptomatic. Age, a history of male homosexuality or intravenous (IV) drug use, union status, and job tenure did not affect work status. Members of minorities were slightly, though not significantly, more likely to stop working than non-Hispanic whites. The nature of work at the job held at the first onset of symptoms, however, did affect subsequent work status. The proportional hazard of work loss associated with an increment of *each* of the 11 physical requirements of the job was 1.09, indicating that persons with very physically demanding work would have much higher rates of work loss than those in sedentary occupations. Conversely, the proportional hazard of work loss associated with an increment of *each* of the nine measures of discretion in work activities was .93, indicating that persons who controlled the pace and scheduling of work activities were much less likely to leave employment.

TABLE 5—Impact of Medical and Social Characteristics of Respondents and of Their Jobs on Subsequent Work Status

Characteristic	Proportional Hazard (95% Confidence Interval)	p-value[a]
Age/year	0.99 (0.96, 1.03)	.94
Minority race	1.71 (0.94, 3.13)	.07
Acquired immunodeficiency syndrome	2.73 (1.02, 7.35)	.04
Aids-related complex	1.21 (0.42, 3.52)	.72
Male homosexual	0.61 (0.24, 1.59)	.30
Intravenous drug user	0.30 (0.03, 2.92)	.29
Physical requirements of jobs[b]	1.09 (1.01, 1.17)	.03
Discretion over work[c]	0.93 (0.87, 0.99)	.02
Member of labor union	0.76 (0.40, 1.43)	.38
Tenure on job: years	0.99 (0.99, 1.01)	.99

[a] *p*-value by chi-square test.

[b] Count of physical requirements of jobs, including walking, using stairs, standing or sitting for long periods, stooping, reaching, using eyes for reading or inspection, concentrating or memorizing, using computers, using fingers to grasp, lifting 25 lbs.

[c] Count of discretion over work activities, including pace of work, work procedures, what is produced, time of breaks, sick leave, time of arrival and departure.

Literature Example 16.4 is from the article entitled "How Women's Adopted Low-Fat Diets Affect Their Husbands." Read the abstract to learn about study.

What tests were done to get the information in Tables 3 and 4?	1 ______________________ ______________________
Write a hypothesis regarding high-fat foods, using symbols.	2 ______________________ ______________________
Do you reject or accept all the hypotheses for high-fat foods? Why?	3 Reject/Accept ______________________
Do you reject or accept all the hypotheses for other foods? Why?	4 Reject/Accept ______________________
Refer to Table 4. For which weight is the difference in the means of control and intervention groups significant?	5 ______________________

Read the authors' results (page 602) with a view to learning how to write about research results.

LITERATURE EXAMPLE 16.4

How Women's Adopted Low-Fat Diets Affect Their Husbands

Ann L. Shattuck, MPH, MS, RD, Emily White, PhD, and Alan R. Kristal, DrPH

Objectives. One way of promoting a reduction in dietary fat intake is by changing the diet of family members. This study investigated the long-term effects of a low-fat dietary intervention on husbands of women who participated in the Women's Health Trial (WHT).

Methods. An average of 12 months after the end of the WHT, a randomly selected sample of participants' husbands was sent dietary and health questionnaires as part of a follow-up study of the maintenance of the low-fat diet among WHT participants.

Results. We found an absolute difference in fat intake between groups of 4 percentage points (32.9% energy from fat among intervention husbands [n = 188] vs 36.9% among control husbands [n = 180]). The wife's attitude and fat intake were among the most important predictors of her husband's fat intake, indicating that the effect of the WHT intervention on the husbands of participants was more likely due to their acceptance of lower-fat foods being served at home than to overt actions by the men.

Conclusions. Our results suggest that a dietary intervention aimed at women can have an effect on their husbands and may be a cost-effective approach to healthy dietary change for both women and men. (*Am J Public Health.* 1992;82:1244–1250)

TABLE 3—Reported Change in Consumption of Selected Foods by Women's Health Trial Intervention and Control Husbands at Follow-up

	Intervention			Control		
	n[a]	Eat Less, %		n[a]	Eat Less, %	*P* Value[b]
High-fat foods						
Whole milk	112	67.0		110	43.6	.001
Butter	128	78.1		121	52.9	.001
Margarine	172	37.2		173	17.3	.001[c]
Eggs	176	77.8		174	54.6	.001
Cheese	182	52.8		174	25.9	.001
Red meat	184	71.7		178	56.7	.005
Cakes, cookies, ice cream	177	48.6		177	32.8	.002[c]
	n[a]	Eat More, %		n[a]	Eat More, %	*P* Value[b]
Other foods						
Fish	180	62.8		176	55.1	.228
Fruits	185	46.5		178	42.1	.398
Vegetables	185	41.1		178	32.0	.101
Whole wheat breads	166	45.2		165	35.2	.165[c]
Whole grains	161	47.8		164	40.2	.329[c]
	n	Mean	SD	n	Mean SD	
Total change score[d]	185	6.0	2.8	180	4.4 2.8	.001[e]

[a]Number of men who consumed the foods.
[b]The intervention–control group difference was based on Fisher's Exact Test (two-tail) unless otherwise noted.
[c]Chi-square test.
[d]The score range was 1–12, increasing with number of dietary changes made in the recommended direction (e.g., lower fat, higher fiber).
[e]Student's *t* test.

TABLE 4—Self-reported Mean Body Weight and Weight Change of Husbands of Women's Health Trial Participants

	Intervention (n = 185)		Control (n = 178)		
Weight, lb	Mean	SD	Mean	SD	*P* Value for Difference
Baseline	187.3	29.9	192.1	31.2	.136
Follow-up	185.3	29.0	191.8	32.0	.044
Weight change	−2.1	9.0	−0.4	8.3	.063

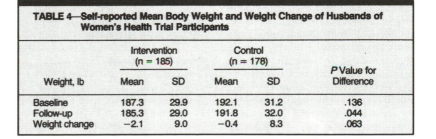

The husbands were asked about how their consumption of selected foods and food groups had changed since their wives entered the WHT (Table 3). Significantly more intervention than control husbands reported eating less of each high-fat food (all *P*s < .01), but there were no significant differences between groups in increased consumption of foods promoted as healthy. The decreased consumption of foods high in both saturated and polyunsaturated fat is consistent with the WHT goal of lowering total fat. The summary dietary change score differed substantially between intervention and control group husbands ($P < .001$), indicating that intervention husbands made significantly more dietary changes in the recommended direction (e.g., decreasing high-fat foods) than did the control husbands.

Mean recalled body weight at the time their wives entered WHT did not differ between intervention and control group husbands (Table 4). At follow-up, however, intervention husbands had a significantly lower mean reported weight than did control husbands ($P < .05$). Consistent with the lower estimated energy intake by the intervention husbands, weight change between the two time periods was greater in the intervention group than in the control group ($P = .06$).

Strength of Association

Three measures of strength of association are frequently used in health sciences. They are **phi coefficient, relative risk,** and **odds ratio.** The phi coefficient is used to measure the **strength of association** of variables in a two-by-two table and is based on χ^2. Relative risk helps us estimate risks of certain events. For example, not all persons who are exposed to a causative agent will contract the illness. For this reason it is necessary to estimate the risk of acquiring a particular disease when exposed to the causative agent. Odds ratios can be used, in some cases, as estimates of relative risk. They are also used in case-control (retrospective) studies as a measure of the odds of a subject's having a particular condition when the person engaged in a behavior that may contribute to the condition. Table 16.2 summarizes information about these three tests.

Phi Coefficient

χ^2 provides a way of testing whether two variables are associated or independent of each other but gives no indication of the strength of the association. Also, the magnitude of χ^2 is influenced by the sample size. A typical way of adjusting for the sample size, n, is to divide by it. If we divide χ^2 by n and take the square root of this number, the resulting statistic is Φ, which is an index

TABLE 16.2

Measures of Strength of Association	Comment	Formula
Phi coefficient (Φ)	Based on χ^2 statistic; varies from 0 to 1	$\Phi = \sqrt{\frac{\chi^2}{n}}$
Relative risk (RR)	Used in clinical studies; a ratio of disease rates for dichotomous exposure groups	$RR = \frac{a/(a+b)}{c/(c+d)}$
Odds ratio (OR)	Used in case-control studies; not based on disease rates; also used as an estimate for relative risk	$OR = \frac{a/b}{c/d}$

of the strength of the association. Φ varies from 0, indicating no association, to 1, indicating perfect association. Calculate Φ for the following data on health status:

	HEALTH		
SEX	**1**	**2**	**Total**
Female	8	11	19
Male	2	11	13
Total	10	22	32

$\chi^2 = 2.565$

$$\Phi = \sqrt{\frac{\chi^2}{n}} = \_\_\_\_\_\_$$

Example from the Literature

Literature Example 16.5 is from the article entitled "How Valid Are Mammography Self-Reports?" Read the abstract to learn about the study.

What is the Φ coefficient given in the table? __________

Since Φ is close to 1, we infer that the association between reporting a mammogram and actually having had one is strong. Read the results to see how the authors report the results.

LITERATURE EXAMPLE 16.5

How Valid Are Mammography Self-Reports?

EUNICE S. KING, RN, PHD, BARBARA K. RIMER, MPH, DRPH, BRUCE TROCK, PHD, ANDREW BALSHEM, BA, AND PAUL ENGSTROM, MD

Abstract: We compared mammography reports in medical records to self-reports obtained during a 1989 telephone interview survey for a sample of 100 women members of a health maintenance organization (HMO) who indicated they had mammograms within the past year and 100 who said they had not had mammograms within the past year. Of the women reporting they had not had mammograms within the past year, none had mammogram reports in the HMO data center. Of the 100 women reporting they had mammograms within the past year, 94 had confirmatory radiology records. (*Am J Public Health* 1990; 80:1386–1388.)

Women's self-reports of having obtained mammograms within the past year correlated highly with the presence of actual mammogram reports (Table 2). Of the 99 women reporting they had *not* had mammograms within the past year, none had a mammogram report in the database. Of the 100 women reporting they "had a mammogram" within the past year, we were able to validate self-reports on 94 women. Mammogram reports on 50 of these women were obtained through the database; the remaining 44 were validated through phone calls to radiology centers or to the women's physicians. Although *all* of the remaining six women had obtained mammograms, they were not done within the past year. Five of the women had mammograms between 13 and 18 months prior to the survey interview; the other woman had a mammogram two years prior to the survey. These six women did not share any particular sociodemographic characteristics when examined by age, employment, race, education, and marital status. Thus, ultimately, we were able to account for *every* one of the women who reported that they had mammograms.

TABLE 2—Comparison of Women's Self-reports with Actual Mammogram Reports

	Radiology Reports		
Self-reports	Yes n	No n	Total n
Had mammogram in preceding year	94	6	100
Did not have mammogram in preceding year	0	99	99
Total	94	105	199

Phi coefficient=.94 (95% CI = 0.93, 9.95), p<.001.

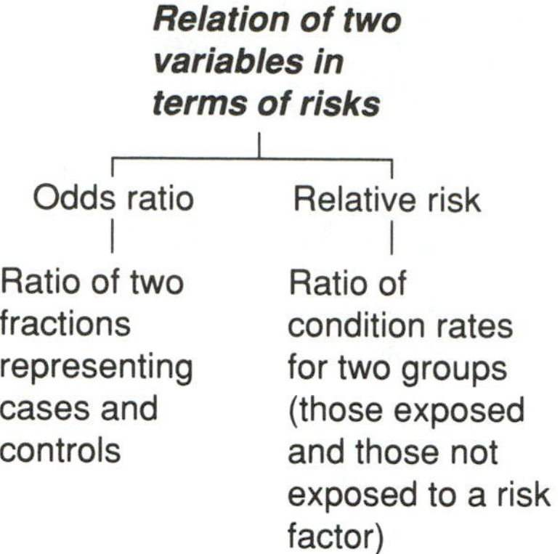

FIGURE 16.2

Relative Risk and Odds Ratio

Relative Risk

Two measures also used as indexes of strength of association are relative risk and odds ratio. Relative risk (RR) is used to give rates of any condition for groups exposed and not exposed to a hazard or procedure—typically, disease rates for groups exposed and not exposed to a specific suspected health hazard. Odds ratio (OR), or **relative odds**, is used to evaluate data gathered in case-control studies, which track the number of individuals manifesting and not manifesting a condition of interest. See Figure 16.2 (and Chapter 9.)

Relative risk is the ratio of disease (or any condition) rates for dichotomous exposure groups, and it is given by the following formula:

$$\mathrm{RR} = \frac{a/(a+b)}{c/(c+d)}$$

where a, b, c, and d are the respective numbers in the categories depicted in Table 16.3.

None of the variables in the SQD and FED data sets lend themselves to this kind of analysis, so we will use the data from a journal article. Literature Example 16.6 is from the article entitled "Predictors of Vaccination Behavior among Persons Ages 65 Years and Older." The data from the article are presented in the following table. The numbers in the table are the proportions of subjects in each category.

TABLE 16.3

	Condition	
Risk Factor	**Present**	**Absent**
Present	a	b
Absent	c	d

Exposed or Not Exposed to the Crucial Condition	Event Whose Risk We Are Assessing		
	Vaccinated	**Not Vaccinated**	**Total**
Visited doctor	.36 (*a*)	.64 (*b*)	1.00
Didn't visit doctor	.18 (*c*)	.82 (*d*)	1.00

$$RR = \frac{a/(a+b)}{c/(c+d)} = \_\_\_\_\_\_$$

Note: In computing RR it makes no difference whether absolute numbers or proportions are used. In fact, the computations are easier using proportions because both denominators are 1. Therefore, when the values of a, b, c, and d are given as proportions, we may use the following simpler formula:

$$RR = \frac{a}{c}$$

The authors of Literature Example 16.6 stated, "Persons who had had a medical examination within the preceding year were twice as likely to have received influenza vaccination than persons who had not had a checkup within the preceding year." How large RR must be to be significant is not a statistical question but a question of the researchers' judgment. Now read the abstract to learn about the study.

Computation of confidence intervals for relative risk is complicated. However, the concept is similar to the concept of confidence intervals of means. We test the hypothesis the same way. If the confidence interval includes the number 1, we accept the hypothesis that the risk of having the condition is the same whether the risk factor is present or not. What are the minimum and maximum relative risks given in the table?

1 Minimum RR = ______;

Maximum RR = ______

List the factors whose confidence intervals do not cover RR = 1.

2 ____________________

____

____

Read the authors' description of the results. Which factor did they think was the most important predictor of vaccination status, and why?

3 ____________________

____

____

LITERATURE EXAMPLE 16.6

Predictors of Vaccination Behavior among Persons Ages 65 Years and Older

Paul A. Stehr-Green, DrPH, MPH, Mary Ann Sprauer, MD, MPH, Walter W. Williams, MD, MPH, and Kevin M. Sullivan, MPH, MHA

Abstract: We estimated influenza vaccination coverage of 32 percent among persons 65 years of age and older from the 1987 Behavioral Risk Factor Surveillance System survey. Race other than White, obesity, lack of seatbelt use, and current smoking were associated with decreased likelihood of having been vaccinated. Controlling for these factors, the best predictor of having received influenza vaccination was having had a medical checkup within the last year (Odds Ratio = 2.40, 95% confidence interval = 1.84, 3.14). (*Am J Public Health* 1990; 80:1127–1129.)

When health "risk-taking" behaviors were examined, persons who were obese (RR = 0.86, 95% CI = 0.76, 0.97), persons who did not regularly use seatbelts (RR = 0.89, 95% CI = 0.80, 0.99), and current smokers (RR = 0.80, 95% CI = 0.68, 0.93) were less likely to have received the influenza vaccine (Table 1). There were no differences in vaccination coverage among older persons who were users of smokeless tobacco products, at risk for acute or chronic alcohol consumption problems (including self-reported drinking and driving), or who had a sedentary lifestyle.

The general characteristics that were most strongly associated with whether persons over 64 years old received an influenza vaccination during the preceding year were those related to knowledge and utilization of other health care services (Table 1). Persons who had had a medical examination within the preceding year were twice as likely to have received influenza vaccination than persons who had not had a checkup within the preceding year. Persons who had ever been told by a health professional that they had high blood pressure were somewhat more likely to have received influenza vaccination; however, among these persons there was no increased likelihood of vaccination in those currently taking antihypertensive medications nor among those who reportedly were still hypertensive. Among women, knowledge and utilization of mammography for breast cancer screening was associated with a greater likelihood of receiving influenza vaccination.

When influences of all demographic and behavioral factors were examined simultaneously, those that were previously shown not to be associated were again not correlated with receipt of influenza vaccination. Estimates of relative risk for vaccination associated with race, obesity, lack of seatbelt use, current smoking, or having been diagnosed with hypertension were essentially unchanged. Having had a medical checkup within the last year remained most strongly associated with having received influenza vaccination (Odds Ratio = 2.40, 95% CI = 1.84, 3.15).

Discussion

Because a recommendation by a health care provider appears to be one of the most important factors influencing a person's decision to be vaccinated,[13] greater attention among providers to implementing recommended immunization practices could help improve influenza vaccine coverage.

TABLE 1—Relative Risks of Receipt of Influenza Vaccine by Demographic and Behavioral Risk Factors among Persons 65 Years of Age and Older[a]—Behavioral Risk Factor Surveillance System, 1987

Risk Factors	% with Risk Factor	% Vaccinated in Risk Group[b]	Relative Risk[c]	95% CI[d]
Demographic Characteristics				
Male sex	33.6	34.1	1.09	(0.98, 1.22)
White race	89.2	33.8	1.63	(1.38, 1.93)
High school graduate	56.5	32.1	0.98	(0.88, 1.09)
College graduate	12.6	36.3	1.14	(0.98, 1.33)
Low income (< $10,000/yr)	37.5	31.9	0.98	(0.87, 1.10)
Employed outside home	9.1	29.8	0.91	(0.75, 1.11)
Health Risk-taking Behaviors				
Obesity	28.8	29.4	0.86	(0.76, 0.97)
Lack of seatbelt use	36.8	29.6	0.89	(0.80, 0.99)
Current smoking	13.3	26.5	0.80	(0.68, 0.93)
Use of smokeless tobacco	3.6	31.2	0.96	(0.72, 1.28)
Acute drinking problem	2.1	39.3	1.22	(0.83, 1.78)
Chronic drinking problem	2.7	25.4	0.78	(0.56, 1.08)
Drinking and driving	0.2	17.5	0.54	(0.25, 1.15)
Sedentary lifestyle	67.5	31.9	0.96	(0.86, 1.08)
Utilization of Health Care				
Medical exam in last year	80.0	35.9	2.00	(1.73, 2.31)
Ever told had hypertension	45.6	30.0	1.18	(1.06, 1.31)
(Women) Know of mammography	85.6	32.1	1.26	(1.04, 1.53)
Ever had mammogram	44.7	36.7	1.31	(1.14, 1.50)
Mammogram in last year	60.2	40.3	1.27	(1.03, 1.55)

[a]Total sample size was 9,799 persons; however, due to incomplete responses for some questions, the effective sample size for individual analysis varied.

[b]Data for each respondent were weighted by a factor that was inversely related to that respondent's probability of selection in order to compensate for sources of variation in selection probabilities.

[c]Ratio of the proportion of persons with the specified demographic or behavioral characteristic who received an influenza vaccination within the last year to the proportion of persons without the specified characteristic who received the vaccination.

[d]95% confidence intervals taking into account the complex sample design.

Literature Example 16.7 is from the article entitled "Perforating Eye Injury in Allegheny County, Pennsylvania." Read the abstract to learn about the study.

What are the minimum and maximum RR for B/W?

1 Minimum _______;

maximum _______

For which age group does the confidence interval include a RR of 1? What do you infer from this observation?

2 ______________________________

____

____

____

Which age group has the highest risk of perforating eye injury?

3 ______________________________

Comment on RR for B/W across age groups.

4 ______________________________

____

____

Usually authors refer to other investigations and compare their findings with those of others. How do the authors' findings compare with those from the Swedish study?

5 ______________________________

____

____

____

LITERATURE EXAMPLE 16.7

Perforating Eye Injury in Allegheny County, Pennsylvania

Deborah Landen, MD, MPH, David Baker, MD, MPH, Ronald LaPorte, PhD, and Richard A. Thoft, MD

Abstract: From 1980 through 1986, acute perforating eye injury (ICD codes 871.0–871.9) was diagnosed in 345 residents of Allegheny County, Pennsylvania. The mean incidence rate was 3.49 per 100,000 person years. There was no significant change in incidence over the seven-year period. The largest number of injuries occurred among individuals working with tools, of which 47 percent were occupational. Males had a 6.5-fold risk of injury relative to females. Blacks had a risk of 2.2 times that of Whites, mainly due to an excess of assaultive injuries. Individuals who had had recent ocular surgery accounted for 4.6 percent of cases overall, and for 31.6 percent of cases in those over age 60. (*Am J Public Health* 1990; 80:1120–1122.)

There was only a slight difference in incidence between Whites and Blacks for those under age 20. For those over age 20, the incidence in Whites was reduced, while for Blacks it increased, reaching a peak among those ages 20–59, and then declined in those over age 60 (Table 1).

TABLE 1—Relative Risk of Perforating Eye Injury, Black:White and Male:Female, by Age, Allegheny County, PA 1980–86

Age	RR B/W	(95% CI)	RR M/F	(95% CI)
0–19	1.30	(.82, 2.07)	7.71	(6.68, 8.74)
20–39	2.77	(1.89, 4.07)	11.18	(10, 12.36)
40–59	2.93	(1.64, 5.23)	4.88	(3.21, 6.55)
60+	2.17	(1.09, 3.25)	2.53	(1.25, 3.81)
Total (Age-Adjusted)	2.18	(1.53, 3.09)	6.55	(5.98, 7.12)

Discussion

Comparison with the Swedish study of perforating eye injuries (6.6 per 100,000 for men and 1.2 for women) shows that our sex-specific incidence rates of 5.7 and 1.5, respectively are quite similar; they also showed no consistent trend over time.

Young males experienced the highest risk of injury in both studies. We found an increased risk among Blacks. However, this excess risk may represent lower socioeconomic status of Blacks. Individuals living in low income census tracts had a higher risk of injury; mean income for Blacks in Allegheny County is substantially lower than that for Whites.[11]

Literature Example 16.8 is from the article entitled "A Measles Outbreak at a College with a Prematriculation Immunization Requirement." Read the abstract to learn about the study.

The research question asks whether time since vaccination is a risk factor for vaccine failure. The bars in Figure 2 can give the impression that those who were vaccinated not more than four years ago had a lesser attack rate. The confidence interval bars allow us to test the hypothesis that attack rates in all age groups are similar. Do you reject or accept the hypothesis?

1 Reject/Accept

What did the researchers conclude about the risk of measles attack when living in a dormitory?

2 ______________________________

____

LITERATURE EXAMPLE 16.8

A Measles Outbreak at a College with a Prematriculation Immunization Requirement

Bradley S. Hersh, MD, MPH, Lauri E. Markowitz, MD, Richard E. Hoffman, MD, MPH, Daniel R. Hoff, PA, Mary J. Doran, Jessica C. Fleishman, Stephen R. Preblud, MD, and Walter A. Orenstein, MD

Background. In early 1988 an outbreak of 84 measles cases occurred at a college in Colorado in which over 98 percent of students had documentation of adequate measles immunity (physician diagnosed measles, receipt of live measles vaccine on or after the first birthday, or serologic evidence of immunity) due to an immunization requirement in effect since 1986.

Methods. To examine potential risk factors for measles vaccine failure, we conducted a retrospective cohort study among students living in campus dormitories using student health service vaccination records.

Results. Overall, 70 (83 percent) cases had been vaccinated at ≥12 months of age. Students living in campus dormitories were at increased risk for measles compared to students living off-campus (RR = 3.0, 95% CI = 2.0, 4.7). Students vaccinated at 12–14 months of age were at increased risk compared to those vaccinated at ≥15 months (RR = 3.1, 95% CI = 1.7, 5.7). Time since vaccination was not a risk factor for vaccine failure. Measles vaccine effectiveness was calculated to be 94% (95% CI = 86, 98) for vaccination at ≥15 months.

Conclusions. As in secondary schools, measles outbreaks can occur among highly vaccinated college populations. Implementation of recent recommendations to require two doses of measles vaccine for college entrants should help reduce measles outbreaks in college populations. (*Am J Public Health* 1991;81:360–364)

The overall attack rate among Fort Lewis College students was 2.4 percent. There were no differences in attack rates by race. Attack rates by housing status, sex, and class are summarized in Table 2. Students living in campus dormitories were at higher risk for measles than students living off-campus. When the analysis was limited to students living in campus dormitories, differences in attack rates by class are no longer observed.

Students vaccinated before 1980 were not at significantly increased risk for measles compared with those vaccinated in 1980 or after. No trend for increasing attack rate with increasing time since vaccination was observed among students vaccinated once at 15 months of age or older (Figure 2).

The investigation of this outbreak found measles vaccine to be over 90 percent effective for vaccination at 15 months or greater. Other investigations have found similar results.

TABLE 2—Risk Factors for Measles, Attack Rates by Student Characteristics, Fort Lewis College

Characteristic	Students	Cases	Attack Rate	Relative Risk	95% CI
Housing					
Dormitories	1,278	53	4.1%	3.0	(2.0–4.7)
Off-campus	2,277	31	1.4%	1.0	referent
Sex					
Male	1,922	46	2.4%	1.0	(0.7–1.6)
Female	1,633	38	2.3%	1.0	referent
Class					
Freshman	1,347	46	3.4%	4.3	(1.9–10.1)
Sophomore	842	19	2.3%	2.9	(1.2–7.2)
Junior	603	13	2.1%	2.7	(1.1–7.2)
Senior	763	6	0.8%	1.0	referent
Class*					
Freshman	759	34	4.5%	2.5	(0.3–17.7)
Sophomore	352	13	3.7%	2.0	(0.3–15.2)
Junior	112	5	4.5%	2.5	(0.3–20.5)
Senior	55	1	1.8%	1.0	referent

*Among students living in campus dormitories.

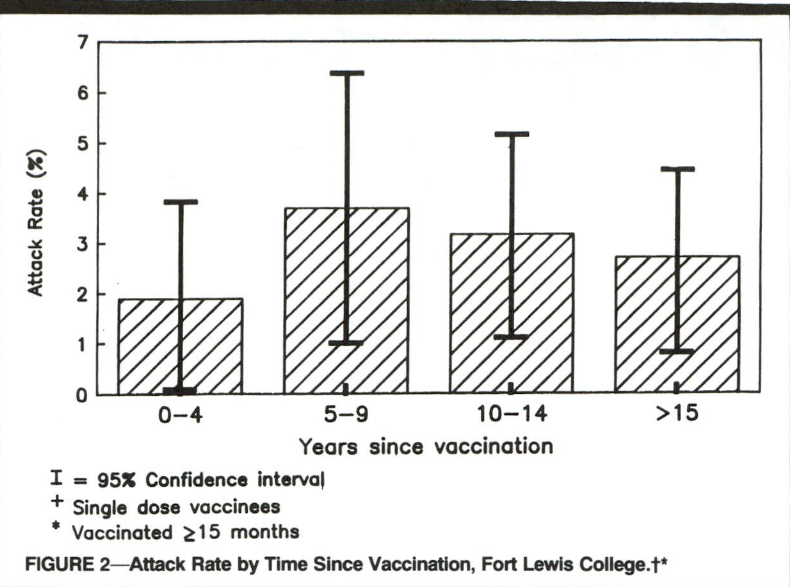

FIGURE 2—Attack Rate by Time Since Vaccination, Fort Lewis College.†*

Odds Ratio

Odds ratio is defined as the ratio of two fractions: a/b, representing risk factor present, and c/d, representing risk factor absent. The odds ratio is given by the following formula:

$$\text{OR} = \frac{a/b}{c/d}$$

where a, b, c, and d are as defined in Table 16.4.

The odds ratio is widely used to evaluate the results of **case-control** studies. In such studies subjects are assigned to case or control groups based on whether they have or do not have the illness (or whatever condition is designated as the response variable). The odds ratio may also be used in prospective studies where a and c are relatively small in comparison to b and d. In this context, odds ratio may be thought of as an approximation to the relative risk.

None of the variables in SQD and FED lend themselves to analysis using the odds ratio, so we use an example from the literature to make the computation explicit. Literature Example 16.9 describes a case-control study. Use the values in the following table to calculate the odds ratio for the study described in Literature Example 16.9.

Suspected Risk Factor: Eating Green Salad		
Eat Green Salad?	**Cases**	**Controls**
Yes	32(a)	11(b)
No	6(c)	24(d)

$$\text{OR} = \frac{a/b}{c/d} = \underline{\qquad\qquad}$$

TABLE 16.4

Explanatory Variable	**Response Variable**	
(Risk factor)	**Positive (cases)**	**Negative (controls)**
Present	a	b
Absent	c	d

LITERATURE EXAMPLE 16.9

A Multifocal Outbreak of Hepatitis A Traced to Commercially Distributed Lettuce

LISA S. ROSENBLUM, MD, MPH, IRENE R. MIRKIN, MD, DAVID T. ALLEN, MD, SUSAN SAFFORD, MD, AND STEPHEN C. HADLER, MD

Abstract: From February 1 through March 20, 1988, 202 cases of hepatitis A were reported in and around Jefferson County, Kentucky. The epidemic curve indicated a common-source exposure. However, there was no apparent single source of exposure from a restaurant, or community gathering; nor was there a geographic clustering by residence. Cases were mainly adults 20–59 years old (89 percent); 51 percent were female. A case-control study using neighborhood controls found that factors associated with hepatitis A were: having eaten downtown (odds rátio [OR] = 4.0) and having dined at any one of three restaurants (OR = 21.0). Case-control studies of patrons of two of these restaurants found that eating green salad was strongly associated with acquiring hepatitis A: OR = 11.6 and OR = 4.4. The three implicated restaurants accounted for 71 percent of the cases. All three restaurants were supplied by the same fresh produce distributor; however, investigation suggested that contamination most likely occurred prior to local distribution. This outbreak of hepatitis A is the first in the United States apparently associated with fresh produce contaminated before distribution to restaurants, and raises important public health issues regarding the regulation of fresh produce. (*Am J Public Health* 1990; 80:1075–1080.)

Below are the data from Literature Example 16.6 for which you computed the relative risk (page 607). Calculate the odds ratio for these data.

Exposed or Not Exposed to the Crucial Condition	Event Whose Risk We Are Assessing: Vaccinated	Not Vaccinated	Total
Visited doctor	.36 (*a*)	.64 (*b*)	1.00
Didn't visit doctor	.18 (*c*)	.82 (*d*)	1.00

$$\text{OR} = \frac{a/b}{c/d} = \_\_\_\_\_\_$$

Literature Example 16.6 is interesting because the authors refer to both relative risk and odds ratio. The authors say this about odds ratio: "Having had a medical checkup within the last year remained most strongly associated with having received influenza vaccination (Odds Ratio = 2.40 . . .)." Other factors were studied, and "medical checkup" was more strongly associated than the others. The odds ratio derives its significance from a comparison of the odds ratios of the various factors.

The following quote from the article refers to the relative risk of 2 for the same data: "Persons who had had a medical examination within the preceding year were twice as likely to have received influenza vaccination than persons who had not had a checkup within the preceding year." Notice the difference in terminology. You can't say "twice as likely" when odds ratio = 2.

Examples from the Literature

Literature Example 16.10 is from the article entitled "Risk Factors for Syphilis: Cocaine Use and Prostitution." Read the abstract to learn about the study.

What kind of study is this? 1 ____________________

In case-control studies, the odds ratio is used to estimate the odds of being a case when one engages in a high-risk behavior. Which factor in this study carries the greatest odds ratio? 2 ____________________

For which factor does the confidence interval include the (adjusted) odds ratio of 1? 3 ____________________

Read the authors' interpretation of the risk factors. (Note that the concept of adjusting is beyond the scope of this book.)

LITERATURE EXAMPLE 16.10

Risk Factors for Syphilis: Cocaine Use and Prostitution

ROBERT T. ROLFS, MD, MARTIN GOLDBERG, AND ROBERT G. SHARRAR, MD, MSc

Abstract: In Philadelphia, a large increase in syphilis among minority group heterosexuals began in 1986 and preceded similar increases elsewhere in the United States. To determine reasons for this increase, we conducted a case-control study in the metropolitan sexually transmitted diseases clinic during 1987 and 1988. Cocaine use (odds ratio [OR] 3.1; 95% confidence interval [95% CI] = 1.5, 6.5 among men; OR 5.8; 95% CI = 1.5, 33 among women) and exchange of drugs for sex (OR 3.5; 95% CI = 1.4, 8.7 among men) were risk factors for syphilis. Although cocaine users reported more sexual partners and more frequently reported sex with prostitutes, cocaine use remained a risk factor after adjustment for these behaviors. These data suggest that sexual behavior or another factor, such as availability or utilization of health care, among cocaine users leads to increased risk of syphilis in this population. Increases in cocaine use may be partly responsible for recent increases in syphilis incidence in the United States. (*Am J Public Health* 1990; 80:853–857.)

Use of cocaine during the three months before the interview was significantly more common among both male and female cases (Tables 2 and 3). Among men, having given a woman drugs in exchange for sex, having had sex with a woman met at a "crack house" or a place where drugs are sold or used ("crack house" sex), and sex with a woman on the same day she was met (first day sex) were also risk factors for syphilis. Twenty-six of 27 men, for whom the specific drug exchanged for sex was recorded, reported trading cocaine for sex. These associations persisted after adjustment for age, race, education, location of residence within the city, and whether the patient presented voluntarily to the city STD clinic for care, using logistic regression (Table 2). Adjustment for the number of sexual partners reported in the last three months and five years also did not change the risk estimates.

TABLE 2—Risk Factors for Syphilis among Heterosexual Men in Philadelphia, 1987–88

Risk Behavior[a]	Cases (n = 95)		Controls (n = 126)		Odds Ratio (95% CI)	
	n	(%)	n	(%)	Crude	Adjusted[b]
No cocaine use	50	(53)	93	(74)	1.0	
Cocaine use	45	(47)	33	(26)	2.5 (1.4,4.5)	3.1 (1.5,6.5)
Cocaine, by route of use[c]						
Intravenous	10	(11)	1	(1)	18.6 (2.3, 150)	24.1 (2.0, 295)
Smoked	15	(16)	13	(10)	2.2 (0.9,5.3)	2.2 (0.8,6.3)
Nasal and unknown	20	(21)	19	(15)	2.0 (0.9,4.3)	2.7 (1.1,6.9)
Cocaine use, other than intravenous	35	(37)	32	(25)	2.0 (1.1,3.8)	2.5 (1.2,5.4)
Sex with prostitute	26	(27)	22	(17)	1.8 (0.9,4.7)	2.4 (1.1,5.4)
Sex with prostitute by what was exchanged[d]						
Drugs	21	(22)	14	(11)	2.3 (1.1,4.7)	3.5 (1.4,8.7)
Money	15	(16)	11	(9)	2.1 (0.9,4.7)	2.9 (1.03,8.5)
"Crack house" sex[e]	23	(25)	16	(13)	2.2 (1.1,4.5)	2.6 (1.1,6.2)
First day sex[f]	51	(54)	50	(40)	1.8 (1.03,3.0)	2.2 (1.1,4.4)

a) All behaviors refer to the 3 months prior to the interview.
b) Adjusted for age, race, education, geographic residence within the city, and whether patient presented voluntarily to STD clinic for care.
c) Route of cocaine use is classified hierarchically as follows: any intravenous, smoked but not intravenous, and nasal or unspecified route, but not intravenous or smoked.
d) Exchange of drugs and money are not mutually exclusive categories.
e) Sex with a woman met at a "crack house" or a place where drugs are sold or used.
f) Sex with a woman on the same day that woman was met.

Literature Example 16.11 is from the article entitled "Cigarette, Alcohol, and Coffee Consumption and Prematurity." Read the abstract to learn about the study.

Notice how the OR increased for every ten cigarettes consumed in all three categories: Low birth weight, Low birth weight for gestational age (LBWGA), and Preterm birth. Are all the ORs statistically significant?

[1] Yes/No

Summarize the results of the study.

[2] ______________________________

____

____

____

____

LITERATURE EXAMPLE 16.11

Cigarette, Alcohol, and Coffee Consumption and Prematurity

Alison D. McDonald, MD, Ben G. Armstrong, PhD, and Margaret Sloan, BA

We analyzed data from a survey of occupational and other factors in pregnancy to assess the effects of cigarette, alcohol, and coffee consumption on pregnancy outcome. The risk of low birth weight for gestational age was found to increase substantially with smoking. Occasional consumers of alcohol had a slightly reduced risk relative to total abstainers. In more frequent drinkers, there was a small increase in risk. Risk increased slightly with coffee consumption. (*Am J Public Health.* 1992;82:87–90)

Results

Smoking

The risk of all three measures of prematurity, especially LBWGA, was higher among smokers than nonsmokers, and increased with number of cigarettes smoked ($P < .001$) (Table 2). For every 10 cigarettes smoked per day, the risk of LBWGA increased by a factor of 1.51 (95% CI, 1.44–1.57). Smoking accounted for 39% of cases of LBWGA, 35% of cases of low birth weight, and 11% of preterm births.

We investigated risk according to changes in smoking habits only for low birth weight. Women who smoked before but not during the first trimester had no excess risk (Table 3). Risk was also consistently reduced in women who cut down their consumption. Cessation before the second trimester brought the risk close to that for nonsmokers (Table 4). It is less clear that continued but reduced consumption was associated with a reduction in risk.

Mean placental weight varied little with smoking. For nonsmokers it was 626.6 g; for smokers of 1 to 9 cigarettes daily, 617.8 g; for 10 to 19 cigarettes, 618.7 g; and for 20 or more, 614.8 g. Accounting for birth weight, the placentas of smokers were a little heavier than those of nonsmokers; for example, the adjusted mean weight for heavy smokers was 30 g more than for nonsmokers.

Alcohol

For all three outcomes, light consumers of alcohol had a small but statistically significant reduction in risk relative to total abstainers (Table 5). Risk then appears to increase with alcohol intake, especially for low birth weight ($P = .002$ for trend). We examined risk of low birth weight by type of alcohol. After controlling for number of drinks, we found that risk was higher by a factor of 1.24 (95% CI, 1.00–1.53) for beer drinkers, and by a factor of 1.38 (95% CI, 0.96–1.99) for drinkers of spirits, than for wine drinkers.

Coffee

Trends of increasing risk with coffee consumption were consistent for low birth weight ($P = .02$) and LBWGA ($P = .01$), reaching a peak at a factor of about 1.4 in women consuming 10 or more cups per day (Table 6). Risk of LBWGA increased by a factor of 1.04 (95% CI, 1.01–1.06) per cup of coffee per day. If these associations were causal, coffee would account for 6% of cases of LBWGA, and 4% of cases of low birth weight.

Interactions

Interactions between smoking, alcohol, coffee, and other variables were seldom substantial but often statistically significant ($P < .05$). For LBWGA, the risk of smoking was apparently greater for French-speaking women than for others by a factor of 1.55 (95% CI, 1.17–2.05); that of alcohol consumption was 1.35 times greater for mothers over 30 (95% CI, 1.08–1.69); that of coffee consumption was 1.27 times greater for alcohol consumers than for abstainers (95% CI, 1.02–1.60).

TABLE 2—Risk of Prematurity by Cigarette Consumption

	No. Pregnancies	% Premature	OR	95% CI	Change[a]
Low birth weight					
Nonsmoker	26 089	3.6	1.00		73.4 g
< 10 per day	3 147	6.1	1.64	1.39–1.93	−37.1 g
10–19 per day	5 042	9.2	2.39	2.12–2.71	−138.3 g
20+ per day	6 167	11.2	2.85	2.53–3.21	−178.7 g
LBWGA					
Nonsmoker	26 089	3.0	1.00		59.8 g
< 10 per day	3 147	6.3	1.97	1.68–2.33	−29.5 g
10–19 per day	5 042	8.6	2.58	2.27–2.93	−107.2 g
20+ per day	6 167	10.8	3.19	2.82–3.60	−150.4 g
Preterm birth					
Nonsmoker	26 089	6.0	1.00		−0.08 wk
< 10 per day	3 147	7.4	1.22	1.05–1.41	−0.08 wk
10–19 per day	5 042	9.1	1.43	1.27–1.60	−0.20 wk
20+ per day	6 167	8.9	1.33	1.18–1.49	−0.16 wk

Note. OR = odds ratio, estimated by logistic regression with all factors, including age, pregnancy order, previous spontaneous abortion, previous low-birth-weight infant, prepregnancy weight, ethnic group, education employment at start of pregnancy, and alcohol and coffee consumption, in the model. CI = confidenct interval. LBWGA = low birth weight for gestational age.
[a]Mean, adjusted for other risk factors by analysis of variance.

Literature Example 16.12 is from the article entitled "Body Fatness and Risk for Elevated Blood Pressure, Total Cholesterol, and Serum Lipoprotein Ratios in Children and Adolescents." The article uses the following abbreviations: SBP = systolic blood pressure; DBP = diastolic blood pressure; TCHOL = total cholesterol; LDL-C = low-density lipoprotein fraction of cholesterol; HDL-C = high-density lipoprotein fraction of cholesterol; VLDL-C = very low-density lipoprotein fraction of cholesterol; LR-1 = LDL-C/HDL-C; LR-2 = (VLDL-C + LDL-C)/HDL-C. Read the abstract to learn about the study.

Note that both figures are three-dimensional graphs. Which two variables are depicted on the horizontal axes?	1 ______________________ ______________________
What type of variable is % fat?	2 Nominal/Ordinal/Interval/Ratio
What type of variable are the other variables on the horizontal axes?	3 Nominal/Ordinal/Interval/Ratio
For men, what are the maximum and minimum odds ratios associated with SBP for fat levels higher than 10%?	4 Minimum = _______; maximum = _______
An asterisk by a ratio indicates that the confidence interval for that odds ratio does not include 1, which signifies that the odds ratio is statistically significant. At what % fat do the odds ratios for LR-1, LR-2, SBP, and DBP on the horizontal axes become significant?	5 ______________________ ______________________ ______________________ ______________________
There are no significant odds ratios up to 25% fat level when we consider TCHOL for males. Is this true for females?	6 Yes/No

Read the results part of the abstract and study both graphs. Can you think of better ways of displaying these data?

LITERATURE EXAMPLE 16.12

Body Fatness and Risk for Elevated Blood Pressure, Total Cholesterol, and Serum Lipoprotein Ratios in Children and Adolescents

Daniel P. Williams, MS, Scott B. Going, PhD, Timothy G. Lohman, PhD, David W. Harsha, PhD, Sathanur R. Srinivasan, PhD, Larry S. Webber, PhD, and Gerald S. Berenson, MD

Background. Recent studies have shown considerable variation in body fatness among children and adolescents defined as obese by a percentile rank for skinfold thickness.

Methods. We examined the relationship between percent body fat and risk for elevated blood pressure, serum total cholesterol, and serum lipoprotein ratios in a biracial sample of 3320 children and adolescents aged 5 to 18 years. Equations developed specifically for children using the sum of subscapular (S) and triceps (T) skinfolds were used to estimate percent fat. The S/T ratio provided an index of trunkal fat patterning.

Results. Significant overrepresentation (>20%) of the uppermost quintile (UQ) for cardiovascular disease (CVD) risk factors was evident at or above 25% fat in males (32.2% to 37.3% in UQ) and at or above 30% fat in females (26.6% to 45.4% in UQ), even after adjusting for age, race, fasting status, and trunkal fat patterning.

Conclusions. These data support the concept of body fatness standards in White and Black children and adolescents as significant predictors of CVD risk factors. Potential applications of these obesity standards include epidemiologic surveys, pediatric health screenings, and youth fitness tests. (*Am J Public Health.* 1992;82:358–363)

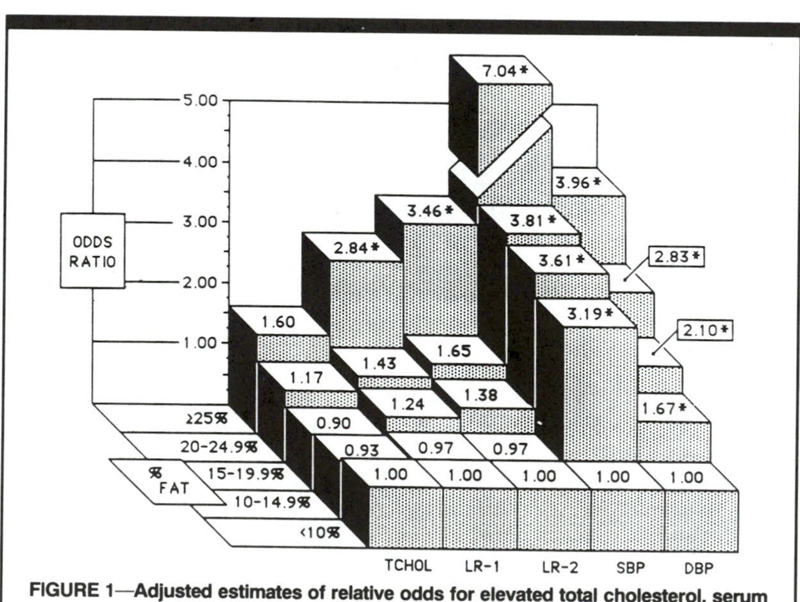

FIGURE 1—Adjusted estimates of relative odds for elevated total cholesterol, serum lipoprotein ratios, and blood pressure in males (*95% CI does not include 1.00). Abbreviations are the same as in Table 2.

Multivariate-based estimates of the relative odds for elevated BP, total cholesterol, and lipoprotein ratios, adjusted for potential confounding variables at increasing levels of percent body fat, are presented in Figure 1 for males and Figure 2 for females. After accounting for potential confounding by age, race, trunkal fat patterning, and fasting status, males with body fat ≥25.0% were 2.8 (1.7–4.8) to 7.0 (3.6–13.6) times as likely as those in the leanest group (<10% fat) to have elevated systolic BP, diastolic BP, LDL-C/HDL-C ratios, and VLDL-C + LDL-C/HDL ratios (Figure 1). In females, the 30% to 34.9% fat group was 2.7 (1.6–4.5) to 3.8 (2.3–6.1) times more likely than the leanest group (<20% fat) to be in the uppermost quintile for these CVD risk factor variables; the likelihood for elevated BP and lipoprotein ratios in the female ≥35.0% fat group ranged from 3.3 (1.8–5.9) to 4.9 (2.8–8.7) (Figure 2).

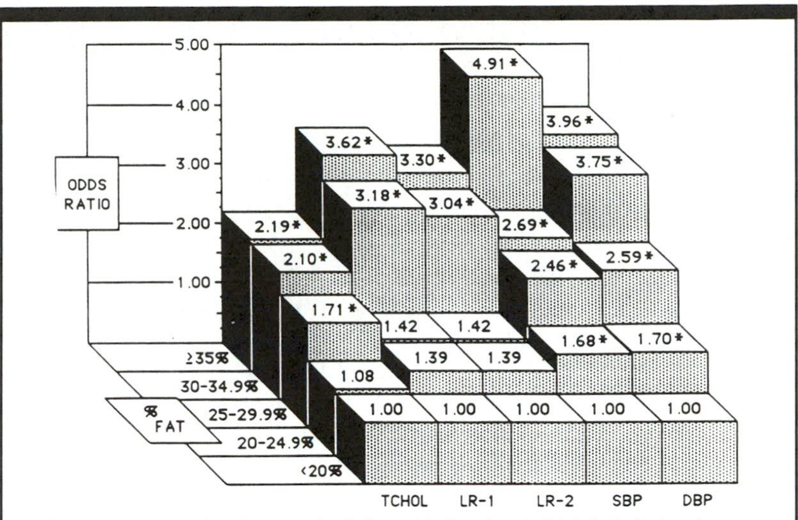

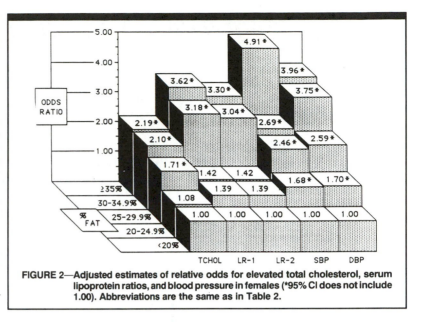

FIGURE 2—Adjusted estimates of relative odds for elevated total cholesterol, serum lipoprotein ratios, and blood pressure in females (*95% CI does not include 1.00). Abbreviations are the same as in Table 2.

Distribution-Free Tests

One category of tests that we have not yet mentioned is **distribution-free tests**, so named because they make no assumptions about the distribution of the population from which the samples are drawn. (t-tests, for example, assume that the samples were drawn from populations that are normally distributed, so t-tests are not distribution-free.) Distribution-free tests are also called nonparametric tests by some statisticians, but we prefer to reserve this term for tests that are truly not about population parameters (such as goodness of fit and randomness tests). Since distribution-free tests are frequently used, we will describe a few of them and give some invented examples. We will not present examples from the literature because although we have seen many references to these tests in journal articles, we have been unable to find any in which details are given.

Sign Tests

Sign tests basically test whether a population's median has a certain value or whether two populations have the same median. Like tests of means, sign tests can be either one-sided or two-sided.

The sign test is exceptionally easy to implement. You just note whether the sample values are larger or smaller than the hypothesized median, recording a plus (+) if they are larger or a minus (−) if they are smaller. If you observe a sufficiently small number of either plus or minus signs, you reject the null hypothesis of equality. For instance, for a one-sided test for which the alternative hypothesis is population median $< M_0$, you would reject the null hypothesis (population median $= M_0$) if you get too few plus signs. For a two-sided hypothesis, you reject the null hypothesis if either plus or minus signs are too few.

How many plus or minus signs count as "too few" can be determined from appropriate tables of critical values for the sign test. For example, such tables tell us that the critical two-tailed number of plus or minus signs at the 5 percent level of significance when $n = 10$ is 1; when $n = 15$ the critical number is 3; when $n = 20$ the critical number is 5; when $n = 30$ the critical number is 9; when $n = 40$ the critical number is 13; and when $n = 50$ the critical number is 17. An example will illustrate.

Data: 1, 1, 2, 5, 6, 7, 7, 8, 9, 11
H_0: Median = 10
H_a: Median $\neq$ 10

There are nine minuses and one plus; since the smaller of these two numbers is 1, and 1 is the critical number for $n = 10$, we conclude that there is significant evidence to reject the null hypothesis at the 5 percent level.

What do we do if a sample value is exactly equal to the hypothesized median? In that case, neither a plus nor minus is assigned. Instead, the sample size is reduced by one—the sample value is essentially treated as non-existent. In the above example, had there also been a 10 in the data set, we would have ignored the 10 and arrived at the same conclusion.

A similar test can be used to test hypotheses that a treatment caused no effect or that two different treatments had the same effect. In the case of pre-treatment and post-treatment data, for instance, we test the hypothesis that the "post" value minus the "pre" value has a median value of zero. This kind of test can be used even when the variable, though continuous, has been measured on an ordinal scale. (For example, we might investigate whether the pulse rate of a person who meditates after exercising returns to the resting pulse rate quicker than that of a person who exercises but does not meditate. If we did not record the actual times it took for pulse rates to return to normal but only whether it was more, less, or the same amount of time under the two conditions, we'd be measuring a continuous variable on an ordinal scale.)

Although easy to perform, a sign test is generally not powerful, and for large sample sizes it is less efficient than a *t*-test. The reason for this is that the sign test uses very little information. Differences are noted only as plus or minus; a minuscule difference counts just as much as an exceptionally large difference.

Rank Tests

Rank tests are more powerful than sign tests because they make use of the size of the deviation from the hypothesized median. For these tests differences are computed and ranked first, second, third, etc., according to size. The simplest of these tests is the **Wilcoxon Signed-Ranks test**.

Wilcoxon Signed-Ranks Test

Assume that the data consist of the numbers 11, 13, 17, 19, 20, 24, 28, 28, 28, 29, and 30, and the hypothesized median is 20. The differences are −9, −7, −3, −1, 0, 4, 8, 8, 8, 9, and 10. As with the sign test, we omit the 0 difference. We rank the remaining differences in order of size from smallest to largest, disregarding plus or minus signs. Thus, −1 is the smallest difference so it is ranked 1, −3 is next so it is ranked 2, 4 is next so it is ranked 3, etc. Duplicate differences are given the average of the ranks they would have had if they'd been different. Finally, we attach the plus or minus sign of the difference to the rank. Following is a table that gives the result for our contrived data set.

x	$x - 20$	Rank	Signed Rank
11	−9	(8 + 9)/2 = 8.5	−8.5
13	−7	4	−4
17	−3	2	−2
19	−1	1	−1
20	0	Omit	
24	4	3	+3
28	8	(5 + 6 + 7)/3 = 6	+6
28	8	6	+6
28	8	6	+6
29	9	8.5	+8.5
30	10	10	+10

The positive ranks are summed to obtain T_+ and the negative ranks are summed to obtain T_-. If one or both of these numbers are sufficiently small (based on tables) the null hypothesis is rejected. For our example, $T_+ = 39.5$ and $T_- = 15.5$. For a two-sided test at the 5 percent level with $n = 10$, we find that we can reject the null hypothesis if the smaller of T_+ and T_- is less than 8. Since this is not the case in our example, we would have no significant statistical evidence to reject the hypothesis that the median is 20.

The **Wilcoxon Matched-Pairs Signed-Ranks test** is similar. The pre sample values are subtracted from the post sample values, the values are ranked, and signs are assigned exactly as above. The hypothesis being tested is that the median of the difference is zero.

Mann-Whitney *U* Test

The **Mann-Whitney *U* test** is very similar to the Wilcoxon Signed-Ranks test. It is based on a similar, though less general, test devised by Wilcoxon. This test is used when samples are independent, for example, when they come from an experimental group and a control group composed of different subjects. (The Wilcoxon Matched-Pairs test discussed above is used for dependent samples.) The test assumes that if the two samples come from different populations, the populations differ in location, not shape. The null hypothesis for this test states that the two samples come from the same population. The alternative hypothesis would be that they come from different populations (two-sided) or that the members of one population tend to be larger than the members of the other (one-sided).

In the U test, the two samples are pooled and ranked according to size. The ranks of the separated samples are then summed. As with the signed-ranks test, if the same number appears more than once, the ranks that these numbers would have had if they'd been different are averaged, and this average becomes the rank for each number.

The statistic U that is to be tested (the so-called **test statistic**) is the smaller of U_1 and U_2 as computed from the following formulas:

$$U_1 = mn + \frac{m(m + 1)}{2} - S_1$$

$$U_2 = mn + \frac{n(n + 1)}{2} - S_2$$

where m is the size of sample 1, n is the size of sample 2, and S_1 and S_2 are the sums of the ranks of the respective samples.

ACTIVITY 16.4

Computations of the test statistic for the U test

Consider the following data:

Experimental group: 135, 162, 164, 170, 180, 225, 312
Control group: 122, 160, 210, 210, 233, 312

At the 5 percent level, the critical U for a one-sided test ($m = 7$, $n = 6$) is 8. Fill in the missing values or letters (E = experimental, C = control) below.

Data	Group	Rank
122	C	1
135	E	2
160	C	3
1 ______	2 ______	3 ______
4 ______	5 ______	6 ______
7 ______	8 ______	9 ______
10 ______	11 ______	12 ______
13 ______	14 ______	15 ______
210	C	16 ______
225	E	10
233	C	11
312	E	12.5
312	E	12.5

$S_1 =$ 17 ______
$S_2 =$ 18 ______
$U_1 =$ 19 ______
$U_2 =$ 20 ______
$U =$ 21 ______

We would [22]accept/reject the null hypothesis that both samples are from the same population (versus the hypothesis that the first population members tend to be larger).

Kruskal-Wallis *H* Test

The **Kruskal-Wallis *H* Test** is used when there are more than two groups. It is a generalization of the *U* test discussed above. It tests a hypothesis about whether *k* samples originate from a common population. As with the *U* test, the data are pooled and ranked, and the ranks of each group are summed. The test statistic *H* is defined by the following formula:

$$H = \left(\frac{12}{n(n+1)} \sum_{i=1}^{k} \frac{S_i^2}{n_i} \right) - 3(n+1)$$

where n_i is the size of the ith sample, k is the number of samples, n is the sum of the n_i (the size of the pooled sample), and S_i is the sum of the ranks for the ith group. Again, tables can be found for this statistic. Also, when all n_i are greater than 5 and k is greater than 4, H has a χ^2 distribution with df $= k - 1$.

Resampling Methods

Resampling comprises a family of techniques that have been gaining popularity. They are especially useful for collecting information about a statistic when little or nothing is known about its distribution. To implement one of these methods, a computer is used to draw a huge number of samples from the given sample. Resampling methods seem to give something for nothing, since the original data are sampled over and over. It has been shown theoretically that the distributions so obtained accurately represent the actual sampling distributions. We will briefly describe two examples of resampling.

Bootstrapping

Bootstrapping is useful for estimating parameters such as the mean of a population, when the actual sampling distribution of the estimate is unobtainable. This is a very powerful method for constructing confidence intervals in situations where it is impossible to collect more data.

In this method we start off with a sample of size *n*. We then treat that sample as if it were a population, and we draw samples from it. The trick is that we draw samples of size *n* (the same *n*) from the original sample, a lot of them. Won't we just get the same sample over and over? No, because we draw samples with replacement, i.e., we choose a number from our original sample, replace that number, choose another number from the sample, etc. For example, if our original sample consisted of the numbers 1, 2, 3, some of the samples drawn could be 1, 1, 2, or 2, 1, 2, or even 3, 3, 3.

If we wanted the 95 percent confidence interval for the mean, for example, we could then throw out the largest 2.5 percent and the smallest 2.5 percent of the sample means to obtain the middle 95 percent of the bootstrap sampling distribution. The width of this is essentially the width of the desired confidence interval, although it should be centered at the mean of the original sample.

(Some details of this process have been left out here; the confidence interval is not computed in exactly this way. This description should give you the flavor of the method, however.)

Permutation Tests

One type of **permutation test** is used to test whether two independent samples come from the same population. To test this we test whether the difference of the two means is zero.

Assume we have available a sample of size m (perhaps from an experimental group) and a second independent sample of size n (perhaps from the control group). We would pool the two samples to obtain a "population" of size $m + n$. We would then draw samples of size m from this population (without replacement), compute their means, and contrast each with the mean of the n elements that were not included in the first sample. Thus, if our two original samples were 1, 2, 3 and 3, 4, we might get as our first "resample" 4, 3, 2, with a sample mean of 3. The remaining elements 1, 3 have a mean of 2. Computing the difference, we get 1.

We continue this process until we have a very large number of differences. We reject the null hypothesis that the two original samples come from the same population at the 5 percent level, for example, if the difference of means from the original two samples falls within the range of the top 5 percent of the resampled differences of means.

A permutation test can be used in place of a t-test for difference of means. The t-test assumes that the populations from which the samples are drawn are normal, but the permutation test can be applied when little or nothing is known about the distributions or when they are actually known to be non-normal.

Self-Assessment

Tasks:	**How well can you do it?**					**Page**
	Poorly				*Very well*	
1. Explain the following terms:						
Enumeration data	1	2	3	4	5	583
Goodness of fit	1	2	3	4	5	586
2. Know when and how to use chi square to test each of the following types of hypotheses:						
There is no relationship between two categorical variables.	1	2	3	4	5	585
The distribution of different levels of x is the same for all categories of y (test for homogeneity).	1	2	3	4	5	586
The proportion of individuals in Group I having a certain	1	2	3	4	5	586

characteristic is the same as the proportion in Group II having that characteristic.						
The distribution of values for a variable obtained experimentally is the same as that predicted by a certain theoretical distribution (goodness of fit).	1	2	3	4	5	587
3. Know when and how to use each of the following measures of strength of association:						
Relative risk	1	2	3	4	5	606
Odds ratio	1	2	3	4	5	615
Phi coefficient (Φ)	1	2	3	4	5	603

Answers for Chapter 16

Answers for p. 583: The following variables yield counts: 1, SEX; 3, CLASS; 8, SMOKE; 9, ACTIVITY; 10, STRESS; 11, FRIENDS; 12, TIMEOUT; 13, HEALTH. If you wish to form tables and use the χ^2 statistic to test for association you need to categorize variables 2, 4, 5, 6, 7, and 14.

Answers for Activity 16.1: 1) There is no relationship between STRESS and HEALTH. 2) Is there a relationship between STRESS and HEALTH? 4) The distributions of different levels of HEALTH are the same for all levels of ACTIVITY. 5) Are the distributions of HEALTH the same for each level of ACTIVITY? 7) The proportion of persons who take TIMEOUT is the same for the different HEALTH groups. 8) Are the proportions of persons who take TIMEOUT the same for the different HEALTH groups? 10) The distribution of TIMEOUT groups in this sample is 3:4:3, as expected on theoretical grounds. 11) Is the distribution of ACTIVITY groups in this sample 3:4:3, as expected on theoretical grounds?

Answers to questions 3, 6, 9, and 12 are the tables shown below.

3)

Rows: STRESS Columns: HEALTH

	1	2	3	4	ALL
1	2	6	3	0	11
2	5	23	3	0	31
3	1	3	3	1	8
ALL	8	32	9	1	50

6)

Rows: ACTIVITY Columns: HEALTH

	1	2	3	4	ALL
1	0	6	0	0	6
2	4	23	8	1	36
3	5	2	0	0	7
ALL	9	31	8	1	49

9)

Rows: TIMEOUT Columns: HEALTH

	1	2	3	4	ALL
1	8	21	3	1	33
2	1	11	4	0	16
ALL	9	32	7	1	49

12)

	Observed		Expected	
TIMEOUT	Count	Proportion	Count	Proportion
1	33	.67	24.5	.5
2	16	.33	24.5	.5
$n = 49$				

Answers for Activity 16.2: 1) .067 2) .138 3) .138 4) .284 5) .627

Answers for pp. 589–590: 1) 36/49 2) 32/49 * 36/49 = .48 3) [32/49 * 36/49] * 49 = 23.52 4) 32/49 5) 7/49 6) 32/49 * 7/49 = .093 7) [32/49 * 7/49] * 49 = 4.57 8) 17/49 9) 6/49 10) [17/49 * 6/49] = .042 11) [17/49 * 6/49)] * 49 = 2.08 12) 17/49 13) 36/49 14) .255 15) 12.49 16) 17/49 17) 7/49 18) .0496 19) 2.43 20) 1.106 21) .010 22) 1.446 23) 2.082 24) .019 25) 2.723 26) 7.386

Answers for p. 591: 1) .241 2) 1.445 3) .453 4) 2.718 5) 4.857 6) 1

Answers for Activity 16.3: 1) 5.99 2) Reject 3) 9.5 4) Accept 5) 5.99 6) Reject 7) 9.5 8) Reject 9) 3.84 10) Accept 11) 5.99 12) Accept 13) 3.8 14) Reject 15) 3.8 16) Reject

Answers for Literature Example 16.1: 1) Independent *t*-test; chi-square test 2) Age; Number of medical conditions; Self-reported health status 3) There is no relationship between education and use of psychotropic drugs. 4) 11.07 5) Accept 6) No

Answers for Literature Example 16.2: 1) There is no relationship between gender and the types of speciality. 2) Reject 3) There is no relationship between social origin and quality of medical school attended. 4) 7.81 5) Reject 6) Based on their findings, we certainly agree. Both *p*-values are extremely small, indicating that it would be highly unlikely to have gotten these values were the hypotheses true.

Answers for Literature Example 16.3: 1) 0.30; 2.73 2) Acquired immunodeficiency syndrome; Physical requirements of jobs; Discretion over work 3) Acquired immunodeficiency syndrome; Physical requirements of jobs; Discretion over work

Answers for Literature Example 16.4: 1) Fisher's Exact Test (which is beyond the scope of this book); *t*-test; chi-square test 2) Intervention eat less, % = Control eat less, % 3) Reject, because all the *p*-values are less than 0.05. 4) Accept, because all the *p*-values are more than 0.05. 5) Follow-up weight

Answer for p. 604: $\Phi = 0.28$

Answer for Literature Example 16.5: $\Phi = .94$

Answer for p. 607: RR = 2

Answers for Literature Example 16.6: 1) 0.54; 2.0 2) Obesity; Lack of seat belt use; Current smoking; all factors listed under "Utilization of health care" 3) Medical exam in last year; they thought so because RR = 2.

Answers for Literature Example 16.7: 1) 1.30; 2.93 2) 0–19; the difference between blacks and whites in risk of perforating eye injury is not significant for that age group. 3) 20–39 4) There was only a slight difference in incidence between whites and blacks for those under age 20. 5) The sex-specific incidence rates of perforating eye injury in this study are quite similar to those found in the Swedish study.

Answers for Literature Example 16.8: 1) Accept 2) Students living in dormitories were at increased risk for measles.

Answer for Literature Example 16.9: OR = 11.6

Answer for p. 617: OR = 2.4

Answers for Literature Example 16.10: 1) Case-control study 2) Intravenous cocaine use 3) Smoked cocaine

Answers for Literature Example 16.11: 1) Yes, none of the C.I.s includes 1. 2) The risk of all three measures of prematurity, especially LBWGA, was higher among smokers than nonsmokers and increased with the number of cigarettes smoked.

Answers for Literature Example 16.12: 1) % fat, cholesterol components, and blood pressure 2) Ratio 3) Nominal 4) 3.19; 7.04 5) LR-1 and LR-2 are significant at 25% level for men and 30% level for women; SBP is significant at 10% level for men and 20% level for women; DBP is significant at 20% level for both men and women. 6) No; odds ratios become significant at fat percent levels above 29.9%.

Answers for Activity 16.4: 1) 162 2) E 3) 4 4) 164 5) E 6) 5 7) 170 8) E 9) 6 10) 180 11) E 12) 7 13) 210 14) C 15) 8.5 16) 8.5 17) 46.5 18) 44.5 19) 23.5 20) 18.5 21) 18.5 22) Accept

CHAPTER

17

Designs Using Analysis of Variance

Prelude

The dusty wagon was lined with shelves full of curious boxes and jars of a kind found in old apothecary shops. It looked as though it hadn't been swept out in years. Bits and pieces of equipment lay strewn all over the floor, and at the rear was a heavy wooden table covered with books, bottles, and bric-a-brac. . . .

Sitting at the table, busily mixing and measuring, was the man who had invited them in. He was wearing a long white coat with a stethoscope around his neck and a small round mirror attached to his forehead, and the only really noticeable things about him were his tiny mustache and his enormous ears, each of which was fully as large as his head.

"Are you a doctor?" asked Milo, trying to feel as well as possible.

"I am **KAKOFONOUS A. DISCHORD, DOCTOR OF DISSONANCE**," roared the man, and, as he spoke, several small explosions and a grinding crash were heard.

"What does the 'A' stand for?" stammered the nervous bug, too frightened to move.

"**AS LOUD AS POSSIBLE**," bellowed the doctor, and two screeches and a bump accompanied his response. "Now, step a little closer and stick out your tongues."

"Just as I suspected," he continued, opening a large dusty book and thumbing through the pages. "You're suffering from a severe lack of noise."

He began to jump around the wagon, snatching bottles from the shelves until he had a large assortment in various colors and sizes collected at one end of the table. All were neatly labeled: Loud Cries, Soft Cries, Bangs, Bongs, Smashes, Crashes, Swishes, Swooshes, Snaps and Crackles, Whistles and Gongs, Squeeks, Squawks, and Miscellaneous Uproar. After pouring a little of each into a large glass beaker, he stirred the mixture thoroughly with a wooden spoon, watching intently as it smoked and steamed and boiled and bubbled.

— Norton Juster
The Phantom Toll Booth, pp. 135–137

Chapter Outline

Self-Assessment

Tasks:	How well can you do it?					Page
1. Explain the following terms:	*Poorly*				*Very well*	
Factor	1	2	3	4	5	634
Level	1	2	3	4	5	635
Fixed effects model	1	2	3	4	5	635
Random effects model	1	2	3	4	5	635
Block design	1	2	3	4	5	635
Latin square design	1	2	3	4	5	653
Repeated measures design	1	2	3	4	5	655
2. Understand computations for the following designs:						
Completely randomized design with one factor	1	2	3	4	5	645
Randomized block design	1	2	3	4	5	646
Completely random design with two factors; measurements on one unit per cell	1	2	3	4	5	649
Completely random design with two factors; measurements on several units per cell	1	2	3	4	5	651

Introduction

In Chapter 14 we discussed analysis of variance for experiments and studies with one experimental variable. In this chapter we consider similar designs, some of which involve more than one experimental variable.

Explanatory (experimental) variables are called **factors.** Thus, a factor can be thought of as a criterion for classifying or distinguishing different experimental units. In this chapter we will only consider factors that are measured on the nominal, ordinal, or interval scales; variables that are measured on a ratio scale will be assumed to have been broken into categories (intervals). For example, height might be broken into the following categories: 4 to 5 feet, more than 5 feet but less than 5.5 feet, more than 5.5 feet but less than 6 feet, more than 6 feet. The categories that a factor is split into are called **levels.** The **response variable** is the effect that we are interested in measuring as the factors vary from level to level. In this context, experimental subjects (humans, animals, or things used for experimentation) are often called **units.** In this chapter we will not distinguish between experiments, in which there is an intervention by the researcher, and studies, in which there is no intervention. When we speak of a **treatment** we mean a specific combination of levels for the various factors. Thus, for example, referring to the FED data set, we might consider AGE and PARTICIP to be factors and PREVO2 to be the response variable. Levels for AGE might be "under 50" and "50 or older"; levels for PARTICIP are "control," "low-intensity exercise," and "high-intensity exercise." The treatment, then, for unit (or case) number 3 in the FED would be 50 or older and control; the value of PREVO2, the response variable, is 29.6 in this case. Table 17.1 gives examples of factors and levels.

The purpose of the designs introduced in this chapter is to test the hypothesis that the various combinations of treatments do not result in statistically significant differences in the response variable. When such a hypothesis is rejected, refinements of the theory (which are beyond the scope of this elementary treatment) allow a researcher to determine which combinations of levels are significantly different from which other combinations of levels. In this chapter we will only discuss tests that detect significant evidence for such differences.

Some Design Considerations

Experimental designs may assess the effects of one, two, three, or more factors on the response variable. If the levels of these factors are chosen (fixed) by the experimenter, the model is called a **fixed effects model** (as in examples 1, 2, 3a, or 5 in Table 17.1). A **random effects model** is one in which factors are sampled from many available levels (as in examples 3b and 4 in Table 17.1). The results of fixed effects models apply only to the population consisting of those levels, whereas the results of random effects models can be generalized to the larger population from which the sample was drawn. Thus, the statistical results of example 3a in Table 17.1 would apply only to the Great Lakes, but the results of example 3b could be generalized to all lakes in the continental United States bigger than an acre in area.

In some of the designs that we will study, units from various samples are matched, a procedure called **blocking.** If in example 4 from Table 17.1 all male subjects were grouped into one block and all females into another, the design would be a **block design.** Designs that do not involve blocking are called **completely random designs.** Some other common criteria for blocking are age, weight, time period, location, and observer gathering the data.

Designs are said to be one-way, two-way, etc., depending on how many factors there are. Thus, an experimental design to study the factor and response variable in example 2 in Table 17.1 would be a one-way, completely random design with fixed effects. If, in addition to the chemical intervention, subjects

TABLE 17.1 Examples of Factors and Levels

Factor, or Criterion for Classifying	Examples of Levels	Response Variable	Experimental Unit
1. Different interventions	a) Placebo, old drug, new drug b) Control, low-intensity exercise, high-intensity exercise	a) Cholesterol level b)Variables such as change in VO2, change in Fat%, etc.	a) Human subjects b) Human subjects, such as those in FED
2. Quantity; for example, different doses of a drug	0.1 mg, 0.01 mg, 0.001 mg	Blood pressure	Animals such as laboratory rats
3. Name; for example, names of different lakes	a) The Great Lakes: Huron, Erie, Superior, Michigan, Ontario b) A random sample of twenty-five lakes chosen from a list of all lakes in the continental United States that are bigger than an acre in area	Phosphorous content (no intervention is applied; the study assesses the phosphorous content of the water from each lake)	One liter of water from each lake
4. Kind of chemical or substance; for example, analgesic drugs	From a random sample of five from all available over-the-counter analgesic drugs, each drug is considered a level.	Minutes required for a drug to remove headache	Humans with headache
5. Other variables collected concurrently with experimental data could be used for classifying; for example, activity in student data	Minimal, moderate, vigorous	Pulse	Students in classes that participated in SQD

were subjected to varying levels of stress (chosen by the experimenter), the design would be a two-way, completely random design with fixed effects. If subjects were also blocked by sex, the design would be a two-way, randomized block design with fixed effects. With designs having two or more factors there can be one or many observations for each combination of levels of the factors. **Balanced designs** are designs in which there are the same number of observations for each combination of levels. For pedagogical simplicity, we will consider only balanced designs.

Block Designs

Blocking should not be an altogether new concept, since we actually introduced it in Chapter 14, though we did not use the term. When dependent observations are paired and the difference of the means is tested, basically a block design is being employed. It certainly makes sense to pair the weight of a subject taken before an intervention (such as high-intensity exercise for a semester) with the weight of the same subject after that intervention. Pairing in this way controls for the fact that the participants in the experiment start out with different weights. Similarly, in cases where a good bit of judgment on the part of an observer is required in estimating a response variable, we might group together all the data collected by each observer in its own block. In this way, we would control for the differences in judgment of the various observers.

When a researcher expects to find much variation between the units regardless of treatment, some criterion for blocking is needed. The criterion for blocking is chosen in the hope that there will be noticeable differences between units in different blocks. The computations used in analyzing a block design tend to minimize the effect of the variation from block to block; hence block designs tend to control for variability between blocks. As you will see, a one-way design with blocking is actually a two-way design, since the criterion for blocking is another variable. The computations for one-way block designs are virtually identical to the computations for two-way designs. Blocking can also be done on designs with more than one factor. A two-way design with blocking on one of the factors would be analyzed much like a three-way design.

One particular class of block designs are those in which each subject is treated as a block. These designs are sometimes used when it makes sense to apply all the treatments to every subject. For example, in example 2 in Table 17.1, if we gave each rat 0.1 mg of a drug on Monday of the first week, 0.01 mg on Monday of the second week, and 0.001 mg on Monday of the third week, we would be using this kind of design. In these designs each subject is said to act as its own control. Each treatment instance within each block is treated as a different unit.

Figure 17.1 illustrates the particular designs that we will discuss in detail in this chapter. There are many other possible designs, most of which are variations on these basic types. We present these designs to introduce you to the basic principles involved, not to provide you with a broad knowledge of a large number of possible designs and their related analyses. Each of these designs can be either a fixed effects design or a random effects design. If there

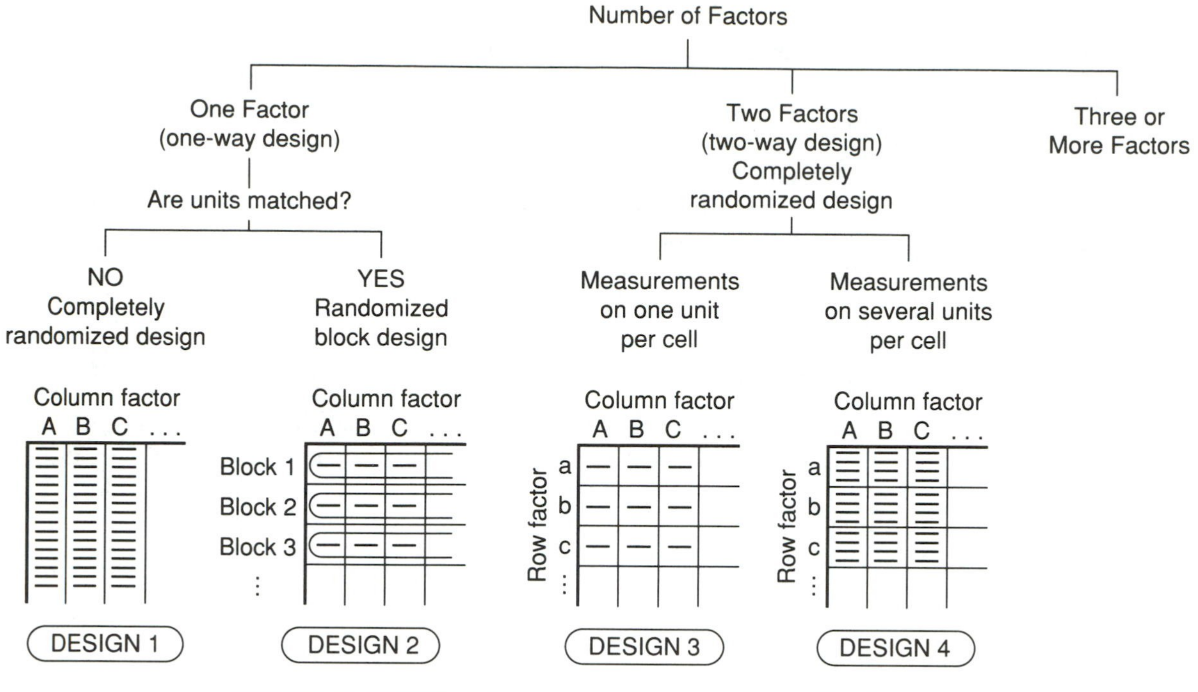

FIGURE 17.1 Experimental Designs for One or More Factors

are two or more factors the design can be mixed, with some of the factors being fixed effect and some random.

ACTIVITY 17.1

Identifying examples

Consult Table 17.1 and Figure 17.1 to decide which design is used in each of the following experiments.

Description of experiment	**Design**
Subjects were randomly assigned to three groups: control with no exercise, light-intensity exercise for a semester, and high-intensity exercise for a semester. The purpose of the experiment was to determine whether exercise had any effect on percent fat, which was measured at the beginning and end of the semester. The response variable was POSTFAT − PREFAT.	What are the factors? 1 ____________________ What are the levels for each factor? 2 ____________________ ____________________ What is the response variable? 3 ____________________ What are the experimental units? 4 ____________________

"The experimental and comparison group subjects were treated with either mobilization and extension (a treatment matched to the category) or a flexion exercise regimen (an unmatched treatment). Outcome was assessed with a modified Oswestry Low Back Pain Questionnaire administered initially and at 3 and 5 days after initiation of treatment." Anthony Delitto et al., *Physical Therapy* 73, number 4, April 1993.

What is the design?

5 ______________________________

What are the factors?

6 ______________________________

What are the levels for each factor?

7 ______________________________

____

What is the response variable?

8 ______________________________

What are the experimental units?

9 ______________________________

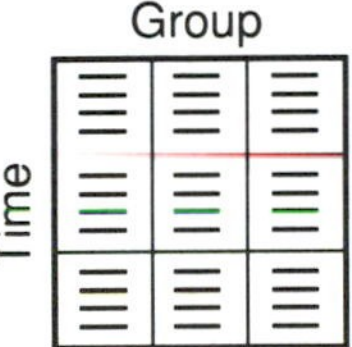

"Subjects in the experimental group ($n = 6$) trained to apply specific forces of 1, 5, 10, 15, 20 and 25 kiloponds using bathroom scales. They practiced for 10 minutes per day for 30 days. Their ability to produce these forces on command was measured using a force platform as they applied posteroanterior passive accessory intervertebral joint movements to the lumbar spine of the healthy subjects. This testing was done prior to training (pretest), immediately after training (posttest), and 1 month following cessation of training (retention test). The control group subjects ($n = 6$) had no training with scales but were also students of the postgraduate

What is the design?

10 ______________________________

What are the factors?

11 ______________________________

What are the levels for each factor?

12 ______________________________

____

____

What is the response variable?

13 ______________________________

What are the experimental units?

14 ______________________________

What is the design?

15 ______________________________

manipulative physical therapy course." Jenny Keating et al., *Physical Therapy* 73, number 1, January 1993.

"The subjects were measured with goniometer for dorsiflexion and STJ range of motion (ROM). The ROMs for each subject's casted and noncasted legs were compared before and after treatment with TCC for neuropathic plantar ulcers by use of a 2X2 repeated-measures analysis of variance design." Jay E. Diamond et al., *Physical Therapy* 73, number 5, May 1993.

What are the factors?

16 ______________________________

What are the levels for each factor?

17 ______________________________

____

What is the response variable?

18 ______________________________

What are the experimental units?

19 ______________________________

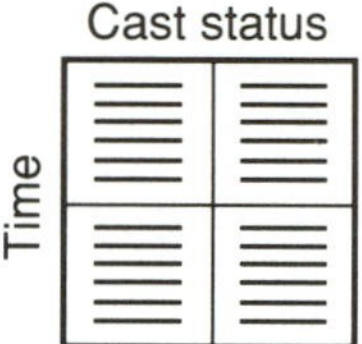

What is the design?

20 ______________________________

"Eight subjects in each of three kinetic feedback groups performed an isometric elbow extension task in an attempt to minimize error between their effort and a force template over a 5-second period. Feedback was provided (1) concurrently with and after each attempt (concurrent feedback) (2) after each attempt (100% feedback), or (3) after every other attempt (50% feedback). Immediate and delayed (48-hour) retention tests were used to compare task error among the three feedback groups for acquisition, immediate retention, and delayed retention trials." Darl W. Vander Linden, *Physical Therapy* 73, number 2, February 1993.

What are the factors?

21 ______________________________

What are the levels for each factor?

22 ______________________________

____

____

What is the response variable?

23 ______________________________

What are the experimental units?

24 ______________________________

Groups

Time

What is the design?

25 ______________________________

General Linear Models

In a **general linear model** with one factor, we assume that each observation's deviation from the mean is a sum of the effect of the factor under study (which we call the **column factor effect**) and the effect of other (usually unknown) variables (which we call the **random effect,** or **error**). That is, Observation = Mean + Column factor effect + random effect. Mathematically, the model is expressed as follows:

$$X_{ij} = \mu + (\mu_j - \mu) + (X_{ij} - \mu_j)$$

where X_{ij} is the value of the response variable for the ith unit in the jth level of the factor, μ is the theoretical mean of the population that would consist of all possible observations, and μ_j represents the mean of the response variable over the population that consists of all units at the jth level of the factor. Each level of the factor is a different population. μ is the mean of the population consisting of the aggregate of all units at any level of the factor. Note that the formula is true algebraically. μ_j and μ cancel out, leaving only X_{ij} on the right-hand side of the equation. $(\mu_j - \mu)$ represents the variation of the mean of the jth level from the overall mean. This term measures the average change in the variable due to the jth level of the treatment. $\mu_j - \mu$ is called the column effect. $(X_{ij} - \mu_j)$ is the variation in the observation that is not caused by the treatment. This term takes into account all other (usually unknown) factors that cause the individual units to vary. This variation is often called the random variation.

Mathematical Assumptions

The following mathematical assumptions allow us to develop the formula for the general linear model into a method for doing useful computations.

1. The samples are independent random samples from the populations of interest.
2. Each population is normally distributed.
3. Each population has the same variance.

(These are actually the assumptions used for the fixed effects model; the random effects model has slightly different assumptions. In practice, however, both models are treated the same; the computations for both are identical. These assumptions are rarely checked in practice. However, it should be noted that the model has been shown to be highly sensitive to the third assumption when the design is unbalanced, i.e., when there are unequal numbers of units in the samples. In this case, the results of testing with analysis of variance are very unreliable. The normality assumption has been found not to be critical. The test is not very sensitive to violations of symmetry, provided all the samples are skewed in the same direction. Visual inspection of the data is usually sufficient to judge whether the samples are roughly normal.)

From Population to Sample

In general we do not know the means of the various populations, so we use the unbiased estimators, the sample means. The formula for the general linear model translates into

$$X_{ij} = \bar{X}_{GM} + (\bar{X}_j - \bar{X}_{GM}) + (X_{ij} - \bar{X}_j)$$

where $\bar{X}_{GM}$ is the grand mean (the mean of all observations) and $\bar{X}_j$ is the mean of the observations in the jth column (all observations at the jth level of the factor). $\bar{X}_j - \bar{X}_{GM}$ estimates the column effect (the effect due to the jth level of the factor) and $X_{ij} - \bar{X}_j$ estimates the random effect (the effect due to all other causes).

Two or More Factors

When there are two factors a similar partitioning of the observations takes place, except that the value of the response variable would be the sum of the grand mean plus row effects as well as column effects and random effects. For more factors the model can be extended to take in additional factors. Additional terms are also considered in these multi-factor models to account for interactions between the various factors. Interactions occur when the factors are not independent.

Activity 17.2 generalizes the above formulas to the specific designs that we are considering. Fill in the appropriate formulas for the sample estimates.

ACTIVITY 17.2

Mathematical ANOVA models

Mathematical Model

Design	Response Variable	Mean	Column effect	Row (or block) effect	Interaction	Random error
Design 1: Completely randomized design, (random or fixed effects)	$X_{ij} =$	μ Estimated by 1 ______	$+ (\mu_{col} - \mu)$ Estimated by 2 ______	None	None	$+ (X_{ij} - \mu_{col})$ Estimated by 3 ______
Design 2: Randomized block design	$X_{ij} =$	μ Estimated by 4 ______	$+ (\mu_{col} - \mu)$ Estimated by 5 ______	$+ (\mu_{block} - \mu)$ Estimated by 6 ______		$+ \{X_{ij} - [\mu + (\mu_{col} - \mu) + (\mu_{block} - \mu)]\}$ Estimated by 7 ______ ______ ______ ______

Design 3: Two-way completely randomized design, one observation per cell, no interaction	$X_{ij} =$	μ Estimated by 8 ______	$+ (\mu_{col} - \mu)$ Estimated by 9 ______	$+ (\mu_{row} - \mu)$ Estimated by 10 ______		$+ \{X_{ij} - [\mu + (\mu_{col} - \mu) + (\mu_{row} - \mu)]\}$ Estimated by 11 ______ ______ ______ ______
Design 4: Two-way completely randomized design, observations on several units per cell	$X_{ijk} =$	μ Estimated by 12 ______	$+ (\mu_{col} - \mu)$ Estimated by 13 ______	$+ (\mu_{row} - \mu)$ Estimated by 14 ______	$+ (\mu_{cell} - \mu_{col} - \mu_{row} + \mu)$ Estimated by 15 ______	$+ (X_{ijk} - \mu_{cell})$ Estimated by 16 ______

Notation becomes complicated for these models, because no really obvious way has been devised to indicate exactly which of the many means is being referred to. Furthermore, knowing which terms are being added in each summation can be confusing, and this becomes more confusing as the number of factors increases. We hope that Table 17.2 will help familiarize you with our notation. Working through the exercises that follow will probably be most helpful in understanding the formulas.

The formulas that are actually used to calculate the various means are derived from the formula for the general linear model using a good bit of

Table 17.2 Explanation of Notations for Means

Notation for Particular Sample Mean	Explanation	Corresponding Population Mean
$\bar{X}_{GM}$	Indicates grand mean (mean of all data in the table)	μ
$\bar{X}_{col}$	Indicates mean of levels in each column	μ_{col}
$\bar{X}_{row}$	Indicates mean of levels in each row	μ_{row}
Measure of Effect for Sample		**Corresponding Population Effect**
$(\bar{X}_{col} - \bar{X}_{GM})$	Column effect (effect of a specific column on the particular X_{ij} under consideration)	$(\mu_{col} - \mu)$
$(\bar{X}_{row} - \bar{X}_{GM})$	Row effect (effect of a specific row)	$(\mu_{row} - \mu)$

algebra. Basically, the grand mean is subtracted from both sides of the equation and both sides are squared. The resulting equation is summed over all possible entries in the table, some mathematical magic is applied, and what results is a formula involving elements that look a lot like numerators of variances. For example, the formula for the one-factor case (Design 1) is as follows:

$$\Sigma(X_{ij} - \bar{X}_{\text{GM}})^2 = n\Sigma(\bar{X}_{\text{col}} - \bar{X}_{\text{GM}})^2 + \Sigma(X_{ij} - \bar{X}_{\text{col}})^2$$

The summation on the left side of the equation is taken over all possible combinations of i and j, that is, over all observations. On the right side of the equation, n is the number of observations in each column, and the first summation is taken over all columns. That is, we compute the mean of the first column, subtract the grand mean from it, and square the result; then we compute the mean of the second column, subtract the grand mean, square the result, and add it to the previous number. We continue in this way adding squares of the differences until we have added that value for the last column. The final summation on the right side of the equation is taken over all combinations of i and j. The resulting formula is the formula from Chapter 14 that says that the total sum of squares is the between sum of squares plus the within sum of squares.

Computing Sums of Squares and the Analysis of Variance Table

We will give an invented example for each design that we have introduced and lead you through the computations that precede a test of the hypothesis that the various combinations of levels produce statistically significant different values of the response variable. The purpose is not for you to learn how to perform these calculations, however. Rather, it is to help you develop a feel for the differences and similarities between the various models and acquire some familiarity with what is being computed. In actual practice, computer software is usually used to perform the calculations. If you must perform the computations by hand, there are formulas other than the ones we have given that make the computations much simpler.

Completely Randomized Designs

Consider design 1. You have already performed computations on data from a design of this type in Chapter 14, so this example will be familiar. You may want to review Activity 14.5 (page 468). Recall that sum of squares, abbreviated SSq, is the numerator of the variance formula,

$$s^2 = \frac{\Sigma(x_i - \bar{x})^2}{n - 1}$$

ACTIVITY 17.3

One-way completely randomized design

Data:

A	B	C
3	5	8
3	4	8
4	5	7
2	6	8
3	7	9
3	3	2

Design 1

Calculations:

First compute means for each column, $\bar{X}_{col}$:

Level A mean = $\bar{X}_A$ = [1] _______

Level B mean = $\bar{X}_B$ = [2] _______

Level C mean = $\bar{X}_C$ = [3] _______

Grand Mean, $\bar{X}_{GM}$ = [4] _______

Sums of squares (SSq):

Total SSQ = $\Sigma (X_{ij} - \bar{X}_{GM})^2$ = [5] _______ (This summation is taken over all combinations of i and j, that is, over every entry in the data table.)

Within SSq for level A = $\Sigma (X_{colAi} - \bar{X}_{colA})^2$ = [6] _______, where X_{colAi} indicates each observation in column A.

Within SSq for level B = $\Sigma (X_{colBi} - \bar{X}_{colB})^2$ = [7] _______, where X_{colBi} indicates each observation in column B.

Within SSq for level C = $\Sigma (X_{colCi} - \bar{X}_{colC})^2$ = [8] _______, where X_{colCi} indicates each observation in column C.

Within SSq = sum of within SSq for levels A, B, C =

[9] _______

In the following formula, n is the number of observations in each column. We assume that each sample is the same size.

Between SSq = $n\{(\bar{X}_A - \bar{X}_{GM})^2 + (\bar{X}_B - \bar{X}_{GM})^2 + (\bar{X}_C - \bar{X}_{GM})^2\}$ = 6{[10] _______ + [11] _______ + [12] _______} =

[13] _______

Total SSq = Between SSq + Within SSq =

[14] _______ = [15] _______ + [16] _______

The following analysis of variance table is the result of the above computations. It was generated by computer. The probability shown in the table cannot be derived from the tables given in this book. From the F-table in this book you find that for $p = .05$ with numerator df = 2 and denominator df = 15, $F = 3.68$.

Source	df	SSq	Mean SSq	F	Probability
Between Col Levels	2	48	24		
Within Col Levels	15	44	2.933	8.18	.004
Total SSq	17	92			

Mean SSq is SSq/df. F is the mean SSq for the treatments (between) divided by mean SSq for error (within). If the means of the three levels are equal the probability of getting this large an F is .004, so we would reject the hypothesis that the treatments are the same, and we would conclude that the different treatments produce different results. The test gives significant evidence ($p =$ .004) that at least two of the three populations from which the column samples were drawn have different means. If it were true that the means of these populations were equal, only 0.4 percent of the possible samples that could be drawn would result in sample means at least as different as the ones in our samples.

Randomized Block Designs

When using the designs that we have been considering, it is sometimes possible to increase the power of the test by blocking, that is, by matching up some of the observations. Blocking essentially changes a one-way design into a two-way design. (Actually, blocking can be used on a design with any number of factors, resulting in a design with one more factor. We illustrate with the one-way case for simplicity.)

The computations in Activity 17.3 involved estimates of the column effect that were labeled between SSq and within SSq. This terminology is not quite applicable to the block design or to designs involving more than one factor. In these designs what corresponds to the between SSq is a contribution from columns, called the column factor SSq, and a contribution from rows, called the row factor SSq, or block factor SSq. We will refer to the within SSq as the residual, or **random error**. This error term is rather complicated but basically it can be computed by subtracting the sum of the row and column factor SSqs from the total SSq. The formula for the random error SSq in the block design case is Random Error SSq = Total SSq − (column factor SSq + block factor SSq), where total SSq is $\Sigma (X_{ij} - \bar{X}_{GM})^2$, as in Activity 17.3. Column factor SSq = $b \Sigma (\bar{X}_{col} - \bar{X}_{GM})^2$, where b is the number of blocks. Block factor SSq = $c \Sigma (\bar{X}_{block} - \bar{X}_{GM})^2$, with c being the number of columns.

ACTIVITY 17.4

Randomized block design

Data:

	Column Factor A	B	C
Block 1	4	5	6
Block 2	4	5	9
Block 3	2	3	4
Block 4	5	7	9

Design 2

Computations:
Means for each level of the column factor, $\bar{X}_{col}$:

Level A mean = $\bar{X}_A$ = 1 _______

Level B mean = $\bar{X}_B$ = 2 _______

Level C mean = $\bar{X}_C$ = 3 _______
Means for each level of the block factor, $\bar{X}_{block}$:

Block 1 mean = $\bar{X}_{block1}$ = 4 _______

Block 2 mean = $\bar{X}_{block2}$ = 5 _______

Block 3 mean = $\bar{X}_{block3}$ = 6 _______

Block 4 mean = $\bar{X}_{block4}$ = 7 _______

Grand Mean, $\bar{X}_{GM}$ = 8 _______

Sums of squares (SSq):
Total SSq = $\Sigma (X_{ij} - \bar{X}_{GM})^2$ =
9 _______

Factor level contributions:
Level A contribution = $(\bar{X}_A - \bar{X}_{GM})^2$ =
10 _______

Level B contribution = $(\bar{X}_B - \bar{X}_{GM})^2$ =
11 _______

Level C contribution = $(\bar{X}_C - \bar{X}_{GM})^2$ =
12 _______

Column factor SSq = no. of blocks * sum of contributions of levels A, B, C =

$b \Sigma (\bar{X}_{col} - \bar{X}_{GM})^2$ = 13 _______

(*b* is the number of blocks.)

Block contributions:
Block 1 contribution = $(\bar{X}_{block1} - \bar{X}_{GM})^2$ =
14 _______

Block 2 contribution = $(\bar{X}_{block2} - \bar{X}_{GM})^2$ =
15 _______

Block 3 contribution = $(\bar{X}_{block3} - \bar{X}_{GM})^2$ =
16 _______

Block 4 contribution = $(\bar{X}_{block4} - \bar{X}_{GM})^2$ =
17 _______

Block factor SSq = no. of columns * sum of contributions of blocks 1, 2, 3, 4 =

$c \Sigma (\bar{X}_{block} - \bar{X}_{GM})^2 =$ 18 ________

Random error SSq:
Total SSq − (Column SSq + Block SSq) = Random error SSq
19 ________ − (20 ________ + 21 ________) =
22 ________

We now have all the elements that go into the analysis of variance table. The mean sum of squares (MS) for each term is calculated by dividing the SSq for that term by the df for that term. Degrees of freedom for blocks = $b - 1$; df for columns = $c - 1$. F is computed by dividing the mean sum of squares between columns by the mean sum of squares for the error,

$$F = \frac{\text{MS(between columns)}}{\text{MS(error)}}$$

Given this information, you should be able to fill in the table in Activity 17.5.

Activity 17.5

Analysis of variance for one-way block design

Source	df	SSq	Mean SSq	F
Between blocks	3	1 ________	2 ________	7 ________ (computed)
Between columns	2	3 ________	4 ________	8 ________ (table)
Error	6	5 ________	6 ________	
Total	11	52.25		

Since the computed value of F in Activity 17.5 is larger than the value from the table, we would reject the hypothesis that the means of the column populations are equal. This test has given us very significant evidence that there really is a difference between the column treatment levels.

The result of introducing the block factor was to remove the between blocks MS from the error term, thereby decreasing MS (error), the divisor of the F ratio. Decreasing the divisor increases the fraction, so F is likely to be larger when computed this way than it would be if the data were analyzed as a one-way design without blocking. Note that F may not be larger, however, because the degrees of freedom computed this way are different, and hence the divisors in the MS formulas are different. Blocking works to make the test more powerful (likely to give a larger F, thereby making it more likely that the null hypothesis will be rejected) when the researcher has good reason to believe that there is variation between blocks.

Two-Way Designs

Calculations for two-way designs are very similar to the calculations for block designs. Only the names of the various factors change. Instead of block SSq we have row factor SSq (which is computed exactly the same way). The reason for the similarity is that a one-way block design, as mentioned, is actually a two-way design with the blocking criterion as the second factor. The major difference between a one-way block design and a two-way design arises in the computation of F and the method of interpretation.

ACTIVITY 17.6

Two-way design, one measurement per cell

Data:

Row Factor	Column Factor A	B	C
a	7	9	11
b	6	10	11
c	3	8	4
d	4	5	6

Design 3

Computations:

Means for each level of the column factor: $\bar{X}_{col}$

Level A mean = $\bar{X}_{colA}$ = [1] _______

Level B mean = $\bar{X}_{colB}$ = [2] _______

Level C mean = $\bar{X}_{colC}$ = [3] _______

Means for each level of the row factor: $\bar{X}_{row}$

Row a mean = $\bar{X}_{rowa}$ = [4] _______

Row b mean = $\bar{X}_{rowb}$ = [5] _______

Row c mean = $\bar{X}_{rowc}$ = [6] _______

Row d mean = $\bar{X}_{rowd}$ = [7] _______

Grand mean, $\bar{X}_{GM}$ = [8] _______

Sums of squares (SSq):

Total SSq = $\sum (X_{ij} - X_{GM})^2$ = [9] _______

Column factor contributions

Level A contribution = $(\bar{X}_{colA} - \bar{X}_{GM})^2$ = [10] _______

Level B contribution = $(\bar{X}_{colB} - \bar{X}_{GM})^2$ = [11] _______

Level C contribution = $(\bar{X}_{colC} - \bar{X}_{GM})^2$ = [12] _______

Column factor SSq = no. of rows * sum of contributions of levels A, B, C =

$r \sum (\bar{X}_{col} - \bar{X}_{GM})^2$ = [13] _______

(r is the number of rows.)

Row factor contributions:

Row a contribution = $(\bar{X}_{rowa} - \bar{X}_{GM})^2 =$ 14 ______

Row b contribution = $(\bar{X}_{rowb} - \bar{X}_{GM})^2 =$ 15 ______

Row c contribution = $(\bar{X}_{rowc} - \bar{X}_{GM})^2 =$ 16 ______

Row d contribution = $(\bar{X}_{rowd} - \bar{X}_{GM})^2 =$ 17 ______

Row factor SSq = no. of columns * sum of SSq of rows a, b, c, d $= c\sum(\bar{X}_{row} - \bar{X}_{GM})^2 =$ 18 ______

Total SSq − (Column SSq + Row SSq) = Random error

19 ______ − (20 ______ + 21 ______) =

22 ______

The analysis of variance table looks much like the one for the block design. However, F is computed for both the row factor and the column factor. As before, F is the between MS divided by the error MS. Verify the following values.

Source	df	SS	MS	*F* (computed)	*F* (table)	*p*
Between rows	3	48	16	6.86	4.76	.0229
Between columns	2	24	12	5.143	5.14	.050007
Error	6	14	2.33			
Total	11	86				

Since both Fs are bigger than the values given in the table, we would reject the hypotheses that the row means are equal and that the column means are equal. We would conclude that there is statistically significant evidence that both row and column treatments have an effect. The evidence that the row factor is significant is much stronger than the evidence that the column factor is significant.

Activity 17.6 points out why using .05 as a universal standard for rejection of null hypotheses is a bad idea. Would you reject or accept the null hypothesis for columns in this example? To four decimal places, the p-value is exactly .05. The evidence against this null hypothesis is significant—not as significant as it would be if the p-value were less than .05, but more significant than it would be if the p-value were greater. Why draw a hard and fast line at .05? The current trend of reporting p-values allows the reader to judge how significant the evidence is against the null hypothesis.

Activity 17.7

Two-way design with several observations per cell

Data:

Row Factor	A	B	C
1	2	3	5
	3	4	3
	2	2	3
2	1	4	4
	3	2	4
	2	4	3
3	1	2	2
	1	4	1
	1	2	1
4	2	2	2
	1	2	3
	1	1	1

(Column Factor: A, B, C)

Three units for each factor combination

Design 4

Note that if cell sizes are not the same, the design is unbalanced and the various values are computed from different formulas than the ones used here.

Computations:

The means of each cell have already been computed and are shown below.

	A	B	C
1	2.33	3	3.67
2	2	3.33	3.67
3	1	2.67	1.33
4	1.33	1.67	2

The following means can be found by averaging the means of the cells in each column.

Means for each level of the column factor: $\bar{X}_{col}$

Level A mean = $\bar{X}_{colA}$ = 1.667

Level B mean = $\bar{X}_{colB}$ = [1] _______

Level C mean = $\bar{X}_{colC}$ = 2.667

Grand Mean, $\bar{X}_{GM}$ = [2] _______

The following means can be found by averaging the means of the cells over each row.

Means for each level of the row factor: $\bar{X}_{row}$

Row 1 mean = $\bar{X}_{row1}$ = 3.00

Row 2 mean = $\bar{X}_{row2}$ = [3] _______

Row 3 mean = $\bar{X}_{row3}$ = [4] _______

Row 4 mean = $\bar{X}_{row4}$ = 1.667

Sums of squares (SSq):

Total SSq $= \Sigma (X_{ij} - \bar{X}_{GM})^2$

[5] _______

Factor level contributions:

Level A contribution $= (\bar{X}_{colA} - \bar{X}_{GM})^2 = .4444$

Level B contribution $= (\bar{X}_{colB} - \bar{X}_{GM})^2 =$ [6] _______

Level C contribution $= (\bar{X}_{colC} - \bar{X}_{GM})^2 =$ [7] _______

Column factor SSq = (no. rows) * (no. entries in each cell) * sum of contributions of levels A, B, C = [8] _______

Row factor contributions:

Row 1 contribution = $(\bar{X}_{row1} - \bar{X}_{GM})^2$ = .4444

Row 2 contribution = $(\bar{X}_{row2} - \bar{X}_{GM})^2$ = [9] _______

Row 3 contribution = $(\bar{X}_{row3} - \bar{X}_{GM})^2$ = [10] _______

Row 4 contribution = $(\bar{X}_{row4} - \bar{X}_{GM})^2$ = .4444

Row factor SSq = (no. columns) * (no. entries in each cell) * sum of contributions of levels 1, 2, 3, 4 = [11] _______

Interaction contribution:

With this design we introduce a term not used in the designs that we have previously studied: the interaction SSq, or row * column SSq. It is important to evaluate this SSq because there may be contributions to the total SSq caused by combinations of the factors. The interaction SSq can be computed by multiplying the number of entries per cell by the sum of all terms of the form (cell mean − row mean − column mean + grand mean)^2, where row and column means are the means of the row and column in which the cell appears.

Interaction contribution A, 1 = 0

Interaction contribution B, 1 = .1111

Interaction contribution C, 1 = [12] _______

Interaction contribution A, 2 = .1111

Interaction contribution B, 2 = [13] _______

Interaction contribution C, 2 = .1111

Interaction contribution A, 3 = [14] _______

Interaction contribution B, 3 = .4444

Interaction contribution C, 3 = [15] _______

Interaction contribution A, 4 = .1111

Interaction contribution B, 4 = .1111

Interaction contribution C, 4 = [16] _______

Interaction SSq = 3 * sum of the above contributions = [17] _______

[18] _______ − ([19] _______ + [20] _______ + [21] _______) = [22] _______

Total SSq − (Col SSq + Row SSq + Int SSq) = Random error

The analysis of variance table is as follows:

Source	df	SS	MS	F	p
Row	3	16.0000	5.3333	7.38	0.001
Column	2	8.0000	4.0000	5.54	0.011
Row * Column	6	4.6667	0.7778	1.08	0.404
Error	24	17.3333	0.7222		
Total	35	46.0000			

Interpretation We test the null hypothesis that the various means are equal by first considering the row * column MS. If the p-value for this interaction term is small, we may reject the null hypothesis without further consideration. However, if it is large, as it is in this case, we test the row and column MS. Since these are both small we would conclude that levels of both the row and column factors are likely to affect the response variable.

Latin Square Designs

You have probably noticed that in going from one-factor designs to two-factor designs the number of observations required increased markedly, even for the designs with only one observation per cell. If we have four levels of each factor and n subjects in each group, we would need $4n$ subjects if there were one factor, $16n$ subjects for two factors, and $64n$ subjects for three factors. Latin square designs provide us with a method for testing three factors using only $16n$ subjects instead of $64n$. These designs are therefore very economical, but they can only be used in situations in which each factor has the same number of levels and the effects are additive (i.e., there are no interaction effects between the factors).

Suppose there are three factors with four levels each. Assume that the levels of factor I are a, b, c, d, the levels of factor II are 1, 2, 3, 4, and the levels of factor III are A, B, C, D. We prepare a table for factors I and II and then assign levels of factor III in such a way that each treatment appears exactly once in each row and exactly once in each column. Figure 17.2 illustrates one possible assignment. Thus, for example, the subjects getting treatment b of the first factor and 4 of the second factor would get treatment C of the third factor.

There are many possible arrangements for the third factor, and often they are chosen randomly from a table of Latin square configurations. This is necessary when other considerations suggest that there might conceivably be a bias introduced if the third-factor level were not assigned randomly.

Analysis proceeds much as in the two-factor case already discussed, except that SSq for factor three must also be computed. The four different cells that received treatment A would be averaged to get the mean for level A; the mean for levels B, C, and D would be computed similarly. The grand mean would be subtracted from each of these third-factor means and the result squared. These numbers would be added together to get the third-factor sum of squares. The degrees of freedom for each factor are the same: one less than the number of levels. We assumed no interaction, so there is no interaction term. df for the total is the number of subjects minus one and df for the error term is figured by subtraction, as in the previous cases.

For four factors the design would be similar: the levels of the fourth factor would be assigned to the cells the way the third factor was assigned in the example above. These arrangements of four factors are called Graeco-Latin square designs. The procedure can be generalized to more than four factors in a similar fashion.

When it is not possible to assume that there are no interactions between the factors or when the factors do not all have the same number of levels, using more subjects is unavoidable. Sometimes it is not possible to have the same number of subjects in every cell (in fact, some combinations may not have any subjects). Techniques for analyzing such unbalanced designs are beyond the scope of this book.

Data for a three-way non-Latin square ANOVA could be arranged in layouts similar to the ones given for two-way data with a separate grid for each level of the third factor. One of the examples in the next section employs such a design, and we have provided an example of the layout there.

	a	b	c	d
1	A	B	C	D
2	B	D	A	C
3	C	A	D	B
4	D	C	B	A

FIGURE 17.2

Examples from the Literature

Each of the following Literature Examples employs what is known as a **repeated measures design**, which is somewhat different from the designs we have discussed. Repeated measures designs are used when each level of at least one of the factors is applied to every unit. That is, multiple observations of each unit are made, at least one for each treatment. This design is desirable because it controls for the variability between different experimental units. It is a kind of block design with blocking by experimental unit.

A repeated measures design has the advantage of getting a lot of information from a relatively few units. The design must be applied carefully, however. The danger is that one treatment may affect the results of the next treatment. If we were testing medications for lowering blood pressure, the apparent effect of the second treatment could actually be a delayed effect from the first treatment. Also, there could be synergistic effects, whereby a later treatment is enhanced or inhibited by an earlier treatment. Therefore, care must be taken to ensure that the treatment levels are truly independent.

The main difference in analysis of repeated measures designs can be seen in the analysis of variance table. You will notice that there are two error terms. One factor is divided by the one error term, the other factor and the interaction are divided by the other. (Similar analyses are performed on so-called split-plot designs in which there are two factors but the researcher desires more precise information on one of the factors than on the other.) For further information on these and other ANOVA designs, we recommend *Statistics in Research: Basic Concepts and Techniques for Research Workers*, by Bernard Ostle and Linda C. Malone (Ames, IA: Iowa State University Press, 1988).

Literature Example 17.1 is from the article entitled "Evidence for Use of an Extension-Mobilization Category in Acute Low Back Syndrome: A Prescriptive Validation Pilot Study." Answer the following questions about the article.

What are the two factors and what are the levels for each?	[1] Factor 1: _______; levels: _________ Factor 2: _______; levels: _________
What is the response variable?	[2] Response variable: ______________
What is the design for this experiment? Describe and diagram the design.	[3] ________________________________
Refer to Table 3. Are the factors and interaction of factors significant at the 5 percent level?	[4] Groups and period interaction: Yes/No Treatment groups: Yes/No Treatment periods: Yes/No
What extra information does Figure 2 furnish, and how does it augment the results in the ANOVA table?	[5] ________________________________
Read the results excerpt. Do the table and figure give all the information the researchers needed for their inferences?	[6] Yes/no

LITERATURE EXAMPLE 17.1

Evidence for Use of an Extension-Mobilization Category in Acute Low Back Syndrome: A Prescriptive Validation Pilot Study

Background and Purpose. *The prescriptive validity of a treatment-oriented extension-mobilization category for patients with low back syndrome (LBS) was examined.* ***Subjects.*** *Of a total of 39 patients with LBS referred for physical therapy, 24 patients (14 male, 10 female), aged 14 to 50 years ($\bar{X}$=31.3, SD=11.6), were classified as having signs and symptoms indicating treatment with an extension-mobilization approach. The remaining subjects were dismissed from the study. Patients in the extension-mobilization category were randomly assigned to either an experimental (treatment) group (n=14) or a comparison group (n=10).* ***Methods.*** *The experimental and comparison group subjects were treated with either mobilization and extension (a treatment matched to the category) or a flexion exercise regimen (an unmatched treatment). Outcome was assessed with a modified Oswestry Low Back Pain Questionnaire administered initially and at 3 and 5 days after initiation of treatment. Data were analyzed with a 2×3 (treatment group×treatment period) analysis of variance.* ***Results.*** *The subjects' rate of improvement, as indicated by the Oswestry questionnaire scores, was dependent on the treatment group to which they were assigned. Subjects treated with extension and mobilization positively responded at a faster rate than did those treated with a flexion-oriented program.* ***Conclusion and Discussion.*** *This study illustrates that a priori classification of selected patients with LBS into a treatment category of extension and mobilization and subsequently treating the patients accordingly with specified interventions can be an effective approach to conservative management of selected patients. [Delitto A, Cibulka MT, Erhard RE, et al. Evidence for use of an extension-mobilization category in acute low back syndrome: a prescriptive validation pilot study. Phys Ther. 1993;73:216–228.]*

Anthony Delitto
Michael T Cibulka
Richard E Erhard
Richard W Bowling
Janet A Tenhula

Data Analysis

Oswestry questionnaire data were analyzed with a 2×3 (treatment group×treatment period) analysis of variance (ANOVA), with treatment group a between-group factor and treatment period a within-group factor.[32]

The ANOVA results are summarized in Table 3. These results indicate main effects of treatment group as well as treatment period. The significant main effects were precluded by a significant treatment group×treatment period interaction ($F[1,22]=4.91$, $P<.05$). Thus, the rate of improvement, as indicated by the Oswestry questionnaire scores, was dependent on the treatment group to which the patient was assigned. This result is illustrated in Figure 2.

Table 3. *Analysis of Variance Summary*

	df	SS	MS	*F*	*P*
Treatment group	1	3815	3815	4.54	.045
Error	22	18488	840		
Treatment period	2	2950	1475	28.32	.000
Treatment group×treatment period	2	512	256	4.91	.012
Error	44	2291	52		

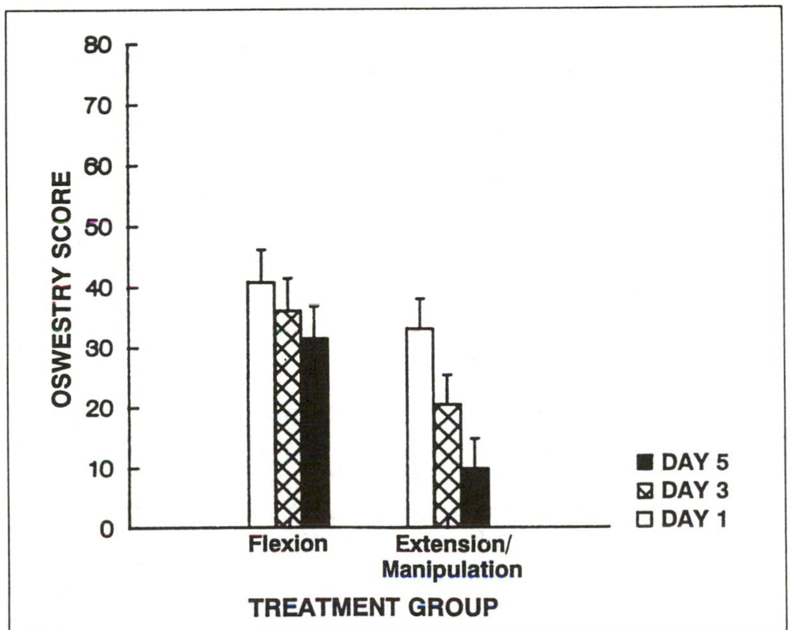

Figure 2. *Between-group comparison of Oswestry questionnaire scores initially and at 3- and 5-day follow-ups. (Error bars represent standard deviation.)*

Literature Example 17.2 is from the article entitled "The Effect of Training on Physical Therapists' Ability to Apply Specified Forces of Palpation." Read the excerpts and answer the questions below.

What are the three factors and what are the factor levels?

[1] Factor 1: ____________________

levels: ____________________

Factor 2: ____________________

levels: ____________________

Factor 3: ____________________

levels: ____________________

What is the response variable?

[2] Response variable: ____________________

____

Two three-way analyses of variance were performed with repeated measures. We might arrange the data for each in a pair of tables like the one at right:

Experimental group

		Time	
		Pre	Post
Force	1		
	5		
	10		
	15		
	20		

Control group

		Time	
		Pre	Post
Force	1		
	5		
	10		
	15		
	20		

Refer to figures 2A and 2B. Read the excerpts from the procedure section for definitions of PSE and PRE. Read the results excerpt, paying special attention to how ANOVA was used.

Did the reseachers conclude that the groups were significantly different on PSE as measured by pretest and posttest? [3] Yes/No

Were they significantly different on PRE? Yes/No

Did the researchers conclude that the groups were significantly different on PSE as measured by pretest and retention test? Yes/No

Why do you think they included the graphs in Figure 2?

____

____

Refer to Figure 3 and the results for PSE.

[4] Did the ANOVA reveal that PSE differences were significant for different forces? Yes/No

Why do you think the researchers included Figure 3?

____

____

LITERATURE EXAMPLE 17.2

The Effect of Training on Physical Therapists' Ability to Apply Specified Forces of Palpation

Jenny Keating
Thomas A Matyas
Timothy M Bach

Background and purpose. *The aim of this study was to evaluate whether postgraduate physical therapy students studying manipulation could learn to accurately produce specific forces during palpation of an intervertebral joint.* ***Subjects.*** *The 12 subjects (7 female, 5 male), aged 26 to 36 years ($\bar{X}$=29.5, SD=2.9), had each completed a 4-year degree course in physical therapy and had worked between 3 and 10 years in clinical practice. All subjects were enrolled in a 12-month postgraduate manipulative therapy diploma course.* ***Methods.*** *Subjects in the experimental group (n=6) trained to apply specific forces of 1, 5, 10, 15, 20, and 25 kiloponds using bathroom scales. They practiced for 10 minutes per day for 30 days. Their ability to produce these forces on command was measured using a force platform as they applied posteroanterior passive accessory intervertebral joint movements to the lumbar spine of the healthy subjects. This testing was done prior to training (pretest), immediately after training (posttest), and 1 month following cessation of training (retention test). The control group subjects (n=6) had no training with scales but were also students of the postgraduate manipulative physical therapy course.* ***Results.*** *In comparison with the control group, the experimentally trained group showed reduced error in force production both immediately after training and 1 month later. This improvement was significant for the retention test. For the retention test, the experimental group subjects were also tested on the trained task (ie, their ability to apply specific forces to the scales). They developed higher levels of accuracy than did the control group.* ***Conclusion and Discussion.*** *Experimental training, therefore, was an effective addition to normal training, suggesting that therapists can learn to quantify applied forces, with significant implications for communication and evaluation of joint behavior. [Keating J, Matyas TA, Bach TM. The effect of training on physical therapists' ability to apply specified forces of palpation. Phys Ther. 1993;73:38–46.]*

Procedure

In order to allocate subjects to either the control group or the experimental group, the data collected at the pretest were analyzed to determine two kinds of error: percentage systematic error (PSE) and percentage random error (PRE) (see "Data Analysis"). The PSE represents the difference between the requested force and the applied force. A low PSE score would indicate that the subject was able to apply forces that approximated the requested force. The PRE represents the difference between repeated attempts to produce the same force and is independent of the requested force.

Data Analysis

Performance of subjects was assessed by calculating the mean PSE and mean PRE scores for each force at the pretest, posttest, and retention test. Two three-way analyses of variance

(ANOVAs) for repeated measures on the last two factors (test phase and force) were performed on the mean PSE scores. The first analysis examined the effect of training. The factors in this analysis were group (control, experimental), test phase (pretest, posttest), and force (1, 5, 10, 15, 20, and 25 kp). The second analysis examined retention of the training effects. The factors were group (control, experimental), test phase (pretest, retention test), and force (1, 5, 10, 15, 20, and 25 kp). Two additional ANOVAs, using the same designs, were conducted to analyze the PRE scores.

Results

Force Platform Data

Figure 2A presents the average PSE scores for the control and experimental groups at the pretest, posttest, and retention test. The PSE, averaged over the six forces, decreased between the pretest and the posttest for the experimental group, whereas it increased for the control group. These trends, however, were not significant. The graph further shows that this downward trend in PSE scores for the experimental group continued between the posttest and the retention test, whereas only a very slight decrease in PSE scores was suggested for the control group between the same tests. The second ANOVA confirmed that the difference in the trends exhibited by the two groups from the pretest to the retention test was significant ($F[1/10]=5.05, P<.05$). This finding indicates that, when tested during palpation of the lumbar spine, the ability of the trained subjects to produce the requested force improved significantly more than that of the untrained subjects in the period between the pretest and the retention test.

Although the PSE scores shown in Figure 3 suggest that the improvement in the trained subjects did not occur uniformly at all forces tested (eg, the differences between groups appear to be more pronounced in the lower forces, with virtually no apparent differences at 20 and 25 kp), neither of the ANOVAs revealed this difference to be statistically significant.

Figure 2B shows the average PRE scores for the control and experimental groups for the pretest, posttest, and retention test. Little change in PRE scores appears to have occurred in either group at the different test phases, and the ANOVAs also failed to detect significant variations.

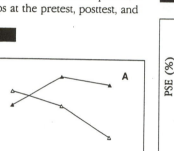

Figure 2. *(A) Average percentage systematic error (PSE) scores for control group (solid triangles) and experimental group (open triangles) at pretest, posttest, and retention test of performance on the lumbar spine. (B) Average percentage random error (PRE) scores for control group (solid triangles) and experimental group (open triangles) at pretest, posttest, and retention test of performance on the lumbar spine.*

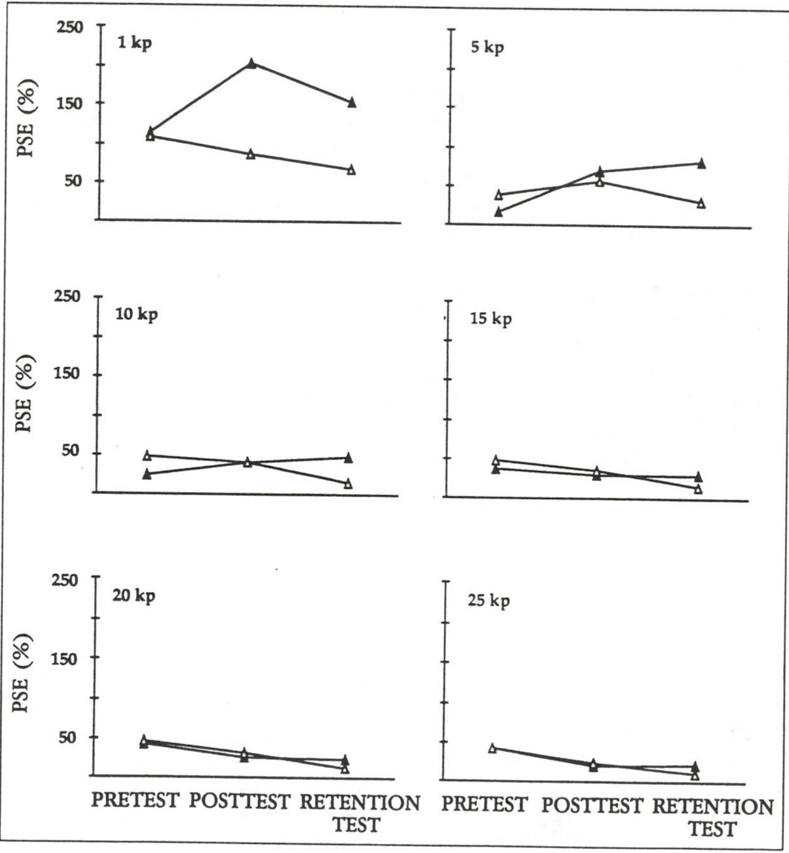

Figure 3. *Percentage systematic error (PSE) score at each force for control group (solid triangles) and experimental group (open triangles) at pretest, posttest, and retention test of performance on the lumbar spine.*

Literature Example 17.3 is from an article entitled "Effect of Total Contact Cast Immobilization on Subtalar and Talocrural Joint Motion in Patients with Diabetes Mellitus." Read the excerpt and answer the questions below.

What are the two factors and what are the factor levels?

1 Factor 1: ____________________

levels: ____________________

Factor 2: ____________________

levels: ____________________

What are the response variables?

2 Response variables: ____________

____

____

What is the design for this experiment? In the space to the right, draw a diagram showing the factors and layout of the data and give the name of the design.

3 ____________________

Read Table 2. Look at casted and noncasted means for dorsiflexion. What do you notice? Can you tell whether the difference is significant?

4 ____________________

____

____

What other statistical tools did the authors use to highlight the results for bilateral eversion range of motion? How does marking standard deviation on the bar graph help?

5 ____________________

____

____

____

Look at Table 5. For total inversion and eversion, was time a significant factor at the 5 percent level?

6 Yes/no

At what level was casting significant?

7 ____________________

Look at Table 6. What factors are significant for dorsiflexion at the 5 percent level?

8 ____________________

Look at Table 2. Note the means and interpret them in the light of the results from the analysis of variance table.

9 ____________________

____

____

LITERATURE EXAMPLE 17.3

Effect of Total Contact Cast Immobilization on Subtalar and Talocrural Joint Motion in Patients with Diabetes Mellitus

Jay E Diamond
Michael J Mueller
Anthony Delitto

Background and Purpose. *The purpose of this study was to determine the effect of total contact casting (TCC) on dorsiflexion at the talocrural joint (TCJ) and motion (inversion/eversion) at the subtalar joint (STJ).* ***Subjects.*** *Thirty-seven patients (29 men, 8 women), ranging in age from 32 to 79 years ($\bar{X}$=54, SD=11), with diabetes mellitus and a unilateral plantar ulceration participated in the study.* ***Methods.*** *The subjects were measured with a goniometer for dorsiflexion and STJ range of motion (ROM). The ROMs for each subject's casted and noncasted legs were compared before and after treatment with TCC for neuropathic plantar ulcers by use of a 2×2 repeated-measures analysis of variance design.* ***Results.*** *Mean time of immobilization in TCC (healing time) was 42 days (SD=43, range=8–119). The results indicated (1) ROM was unchanged at the STJ, but dorsiflexion decreased slightly (1°) on both the casted and noncasted sides following the last cast removal, and (2) ROM was less on the ulcerated side prior to casting compared with the nonulcerated side.* ***Conclusion and Discussion.*** *We believe the beneficial effects (healing of wounds) outweigh the minimal detrimental effects (decreased dorsiflexion) of treatment with TCC. [Diamond JE, Mueller MJ, Delitto A. Effect of total contact cast immobilization on subtalar and talocrural joint motion in patients with diabetes mellitus. Phys Ther. 1993;73:310–315.]*

Table 2. *Means, Standard Deviations, and Ranges for Subtalar Joint and Talocrural Joint Inversion, Eversion, and Dorsiflexion (in Degrees)*

	Ulcerated (Casted) Leg						Nonulcerated (Noncasted) Leg					
	Precast			Postcast			Precast			Postcast		
Motion	$\bar{X}$	SD	Range	$\bar{X}$	SD	Range	$\bar{X}$	SD	Range	$\bar{X}$	SD	Range
Inversion	22.7	6.8	10–40	22.0	6.3	9–35	25.6	8.9	10–60	24.2	8.2	12–55
Eversion	3.6	4.2	−6–12	3.7	4.0	−5–12	4.8	3.5	−5–10	5.0	3.8	−5–12
Total inversion and eversion	26.3	7.6	13–44	25.6	7.4	12–42	30.4	8.8	10–55	29.2	8.3	11–50
Dorsiflexion	3.4	4.8	−8–15	2.4	4.9	−10–11	5.2	4.3	−5–15	4.2	3.3	−2–12

Results

Table 2 contains the means and standard deviations of all goniometric measures before and after casting, and those results are displayed graphically in Figures 1 through 4. Tables 3 through 6 contain the ANOVA results and indicate which of these findings were significantly different. The ANOVA results indicate the following: (1) a significant difference between precast and postcast measurements for dorsiflexion but not for any STJ motion (ie, dorsiflexion decreased by approximately 1° on both sides [casted and noncasted], whereas STJ motion remained unchanged); (2) a significant difference between casted (ulcerated) and noncasted (nonulcerated) extremity for inversion, total ROM, and dorsiflexion (ie, the casted foot demonstrated less ROM than did the noncasted foot before and after casting); and (3) no significant interaction between the variables (ie, casting did not affect ROM of the casted leg differently than in the noncasted leg following treatment with TCC) (Tabs. 3–6).

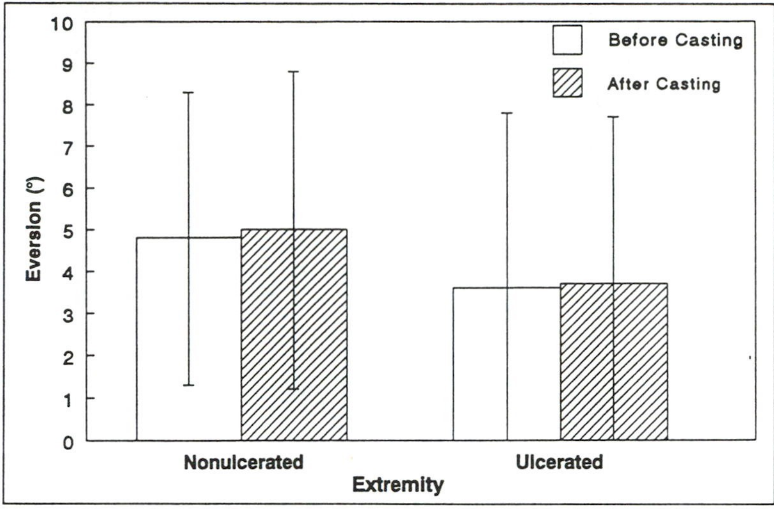

Figure 2. *Means and standard deviations of bilateral eversion range of motion.*

Table 5. *Results of Analysis of Variance of Casted Versus Noncasted Extremity Over Time for Total Inversion and Eversion*

Source	*df*	SS	MS	*F*	*P*[a]
Pretreatment × posttreatment (A)	1	34.06	34.06	1.91	.176
Error	36	642.19	17.84		
Casted × noncasted (B)	1	533.52	533.52	9.73	<.004
Error	36	1973.73	54.83		
A × B	1	2.44	2.44	0.24	.630
Error	36	372.81	10.36		

[a]Alpha level significant at .05.

Table 6. *Results of Analysis of Variance of Casted Versus Noncasted Extremity Over Time for Dorsiflexion*

Source	*df*	SS	MS	*F*	*P*[a]
Pretreatment × posttreatment (A)	1	39.03	39.03	6.22	<.017
Error	36	225.97	6.28		
Casted × noncasted (B)	1	121.32	121.32	18.22	<.000
Error	36	239.68	6.66		
A × B	1	0.00	0.00	0.00	1.00
Error	36	110.00	3.06		

[a]Alpha level significant at .05.

Self-Assessment

Tasks:	How well can you do it?					Page
1. Explain the following terms:	*Poorly*				*Very well*	
Factor	1	2	3	4	5	634
Level	1	2	3	4	5	635
Fixed effects model	1	2	3	4	5	635
Random effects model	1	2	3	4	5	635
Block design	1	2	3	4	5	635
Latin square design	1	2	3	4	5	653
Repeated measures design	1	2	3	4	5	655
2. Understand computations for the following designs:						
Completely randomized design with one factor	1	2	3	4	5	645
Randomized block design	1	2	3	4	5	646
Completely random design with two factors; measurements on one unit per cell	1	2	3	4	5	649
Completely random design with two factors; measurements on several units per cell	1	2	3	4	5	651

Answers for Chapter 17

Answers for Activity 17.1: 1) Amount of exercise 2) No exercise, light-intensity exercise, high-intensity exercise 3) POSTFAT − PREFAT 4) Humans 5) One-way, completely randomized design with fixed effects (Design 1) 6) Type of group, time 7) Unmatched and matched; initial, 3-day, and 5-day 8) Back pain as measured with Oswestry questionnaire 9) Humans 10) Two-way, completely randomized design 11) Type of groups, time, forces 12) Control and experimental groups; before training, after training, 1 month after training; 1, 5, 10, 15, and 25 kiloponds 13) Ability to produce specific forces (probably measured as difference between force produced and target force) 14) Humans 15) Three-way, probably completely randomized design with fixed effects (The excerpt does not give enough information to be sure of this.) 16) Casting status, time 17) Casted and noncasted; before and after treatment 18) ROM 19) Humans 20) Two-way completely randomized, repeated measures (a term explained later in the chapter) 21) Groups, time 22) Concurrent feedback, 100% feedback, 50% feedback; immediate and delayed (48 hours) tests 23) Error between subjects' effort and force template 24) Humans 25) Two-way, completely randomized design

Answers for Activity 17.2: 1) $\bar{X}_{GM}$ 2) $(\bar{X}_{col} - \bar{X}_{GM})$ 3) $(X_{ij} - \bar{X}_{col})$ 4) $\bar{X}_{GM}$ 5) $(\bar{X}_{col} - \bar{X}_{GM})$ 6) $(\bar{X}_{block} - \bar{X}_{GM})$ 7) $\{X_{ij} - [\bar{X}_{GM} + (\bar{X}_{col} - \bar{X}_{GM}) + (\bar{X}_{block} - \bar{X}_{GM})]\}$ 8) $\bar{X}_{GM}$ 9) $(\bar{X}_{col} - \bar{X}_{GM})$ 10) $(\bar{X}_{row} - \bar{X}_{GM})$ 11) $\{X_{ij} - [\bar{X}_{GM} + (\bar{X}_{col} - \bar{X}_{GM}) + (\bar{X}_{row} - \bar{X}_{GM})]\}$ 12) $\bar{X}_{GM}$ 13) $(\bar{X}_{col} - \bar{X}_{GM})$ 14) $(\bar{X}_{row} - \bar{X}_{GM})$ 15) $(\bar{X}_{cell} - \bar{X}_{col} - \bar{X}_{row} + \bar{X}_{GM})$ 16) $(X_{ijk} - \bar{X}_{cell})$

Answers for Activity 17.3: 1) 3 2) 5 3) 7 4) 5 5) 92 6) 2 7) 10 8) 32 9) 44 10) 4 11) 0 12) 4 13) 48 14) 92 15) 48 16) 44

Answers for Activity 17.4: 1) 3.75 2) 5 3) 7 4) 5 5) 6 6) 3 7) 7 8) 5.25 9) 52.25 10) 2.25 11) .0625 12) 3.0625 13) 21.5 14) .0625 15) .5625 16) 5.0625 17) 3.0625 18) 26.25 19) 52.25 20) 21.5 21) 26.25 22) 4.5

Answers for Activity 17.5 1) 26.25 2) 8.75 3) 21.50 4) 10.75 5) 4.50 6) 0.75 7) 14.33 8) 5.14

Answers for Activity 17.6: 1) 5 2) 8 3) 8 4) 9 5) 9 6) 5 7) 5 8) 7 9) 86 10) 16 11) 4 12) 4 13) 24 14) 12 15) 12 16) 12 17) 12 18) 48 19) 86 20) 24 21) 48 22) 14

Answers for Activity 17.7: 1) 2.6667 2) 2.3333 3) 3 4) 1.6667 5) 46 6) .1111 7) .1111 8) 8 9) .4444 10) .4444 11) 16 12) .1111 13) 0 14) 0 15) .4444 16) 0 17) 4.6667 18) 46 19) 8 20) 16 21) 4.6667 22) 17.3333

Answers for Literature Example 17.1: 1) Type of group; experimental (matched) and comparison (unmatched); time; initial, 3-day, 5-day 2) Back pain as measured with modified Oswestry Low Back Pain Questionnaire 3) The diagram should have three rows labeled Init, 3-day, and 5-day and two columns labeled Experimental and Comparison. Rows and columns can be switched; it doesn't matter which factor is the row factor and which is the column factor. The design is a two-way completely random (fixed effects), repeated measures design. (The tip-off that it's repeated measures is that there are two error rows in Table 3.) 4) Yes; yes; yes 5) The figure shows that mean scores decreased noticeably more rapidly for the experimental group over time. 6) Yes

Answers for Literature Example 17.2: 1) Groups; experimental and control; times; pretest, posttest, and retention test; forces; 1, 5, 10, 15, 20, and 25 kiloponds 2) Actual force applied as measured on a force platform 3) No; no; yes; Figure 2 points out that the control group members declined in ability whereas the experimental group members maintained the skill gained. We believe that the mention of "trends" in the data is not warranted, however, because the initial part of the trend is not found to be statistically significant by the ANOVA. For this reason, it can be argued, these graphs are misleading. 4) No; we think that the authors of this article are trying to illustrate a trend (that differences between the groups were more pronounced at the lower forces) that their ANOVA did not support. We think inclusion of this figure is also misleading.

Answers for Literature Example 17.3: 1) Status of casting; casted and noncasted; time; before cast and after cast 2) Range of motion for bilateral inversion, bilateral eversion, bilateral dorsiflexion

3)

Time	Cast status: Cast	Cast status: No cast
Before		
After		

Several measurements in each cell

Two-way analysis of variance with repeated measures 4) Means for the casted legs went from 3.4 (pretest) to 2.4 (posttest), a difference of 1 unit. For the uncasted leg the difference is also 1 unit, but since Precast was 5.2 this is a smaller percentage change. We cannot tell if the difference is significant until we have done a statistical analysis. 5) Bar graphs, tables, and analysis of variance; a reader has no idea how the observations are distributed if the lines indicating standard deviations are absent. These lines help put the differences in the heights of bars in perspective. 6) No 7) We don't know for sure since we only know p is smaller than .004, but we don't know how much smaller. We do know that this result would be significant at any level greater than .004, so it is extremely unlikely that the means for the casted and noncasted legs are the same. 8) Pretreatment and posttreatment as well as casted and noncasted 9) In light of this table and the significance of the results, the differences we noted in dorsiflexion in Table 2 take on a new meaning. Read the Results section.

APPENDIX A

Data Sets

Faculty Exercise Data (FED)

Data for each case have been arranged so that they can be copied and pasted on a 3 × 5 card. Cards can then be sorted various ways. For the FED, $n = 30$.

Case 1

AGE	PREHR	PREMAXHR	PREVO2	PREVE
61	71	169	29.5	142.8
PREWAT	PREHR%	PRERQ	PREFAT%	PREWT
224	99	1.28	36.33	202
PREHT	PRERESHR	PRRESSYS	PRRESDIA	PRERPP
72.00	77	152	78	19,608
POSTNHR	POSTMAXHR	POSTVO2	POSTVEE	POSTWAT
____	____	____	____	____
POSTHR%	POSTRQ	POSTFAT	POSTWT	POSTHT
____	____	____	____	____
POSTREHR	POSSYS	POSDIA	POSTRPP	PARTICIP
____	____	____	____	1

Case 2

AGE	PREHR	PREMAXHR	PREVO2	PREVE
52	85	180	28.7	142.4
PREWAT	PREHR%	PRERQ	PREFAT%	PREWT
246	102	1.18	31.49	207
PREHT	PRERESHR	PRRESSYS	PRRESDIA	PRERPP
71.25	____	____	____	____
POSTNHR	POSTMAXHR	POSTVO2	POSTVEE	POSTWAT
____	____	____	____	____
POSTHR%	POSTRQ	POSTFAT	POSTWT	POSTHT
____	____	____	____	____
POSTREHR	POSSYS	POSDIA	POSTRPP	PARTICIP
____	____	____	____	1

Case 3

AGE	PREHR	PREMAXHR	PREVO2	PREVE
57	79	163	29.6	103.2
PREWAT	PREHR%	PRERQ	PREFAT%	PREWT
185	91	1.27	31.74	160
PREHT	PRERESHR	PRRESSYS	PRRESDIA	PRERPP
67.00	78	118	74	27,930
POSTNHR	POSTMAXHR	POSTVO2	POSTVEE	POSTWAT
81	162	29.4	103.6	185
POSTHR%	POSTRQ	POSTFAT	POSTWT	POSTHT
93	1.27	27.02	161	67.00
POSTREHR	POSSYS	POSDIA	POSTRPP	PARTICIP
97	110	70	24,960	2

Case 4

AGE	PREHR	PREMAXHR	PREVO2	PREVE
43	67	195	34.2	104.1
PREWAT	PREHR%	PRERQ	PREFAT%	PREWT
205	105	1.20	31.77	160
PREHT	PRERESHR	PRRESSYS	PRRESDIA	PRERPP
71.00	88	108	72	24,016
POSTNHR	POSTMAXHR	POSTVO2	POSTVEE	POSTWAT
115	195	40.4	117.6	246
POSTHR%	POSTRQ	POSTFAT	POSTWT	POSTHT
107	1.13	30.77	161	71.00
POSTREHR	POSSYS	POSDIA	POSTRPP	PARTICIP
84	110	58	17,550	4

Case 5

AGE	PREHR	PREMAXHR	PREVO2	PREVE
48	81	183	40.7	157.4
PREWAT	PREHR%	PRERQ	PREFAT%	PREWT
246	99	1.31	23.61	158
PREHT	PRERESHR	PRRESSYS	PRRESDIA	PRERPP
65.50	69	108	72	15,478
POSTNHR	POSTMAXHR	POSTVO2	POSTVEE	POSTWAT
94	182	49.9	195.9	308
POSTHR%	POSTRQ	POSTFAT	POSTWT	POSTHT
102	1.23	23.46	158	65.50
POSTREHR	POSSYS	POSDIA	POSTRPP	PARTICIP
68	102	68	13,938	4

Case 6

AGE	PREHR	PREMAXHR	PREVO2	PREVE
44	93	189	27.6	141.9
PREWAT	PREHR%	PRERQ	PREFAT%	PREWT
246	104	1.19	37.00	225
PREHT	PRERESHR	PRRESSYS	PRRESDIA	PRERPP
72.75	75	132	88	21,840
POSTNHR	POSTMAXHR	POSTVO2	POSTVEE	POSTWAT
94	194	33.2	158.0	286
POSTHR%	POSTRQ	POSTFAT	POSTWT	POSTHT
107	1.19	37.36	218	72.75
POSTREHR	POSSYS	POSDIA	POSTRPP	PARTICIP
69	122	80	20,230	3

Case 7

AGE	PREHR	PREMAXHR	PREVO2	PREVE
61	87	162	24.6	90.2
PREWAT	PREHR%	PRERQ	PREFAT%	PREWT
164	91	1.18	28.87	164
PREHT	PRERESHR	PRRESSYS	PRRESDIA	PRERPP
67.25	80	138	82	23,460
POSTNHR	POSTMAXHR	POSTVO2	POSTVEE	POSTWAT
82	170	28.4	104.0	205
POSTHR%	POSTRQ	POSTFAT	POSTWT	POSTHT
100	1.24	29.57	167	67.25
POSTREHR	POSSYS	POSDIA	POSTRPP	PARTICIP
74	112	60	22,420	3

Case 8

AGE	PREHR	PREMAXHR	PREVO2	PREVE
58	87	181	23.3	98.6
PREWAT	PREHR%	PRERQ	PREFAT%	PREWT
185	105	1.15	37.36	207
PREHT	PRERESHR	PRRESSYS	PRRESDIA	PRERPP
69.50	86	126	80	24,300
POSTNHR	POSTMAXHR	POSTVO2	POSTVEE	POSTWAT
____	____	____	____	____
POSTHR%	POSTRQ	POSTFAT	POSTWT	POSTHT
____	____	____	____	____
POSTREHR	POSSYS	POSDIA	POSTRPP	PARTICIP
____	____	____	____	1

Case 9

AGE	PREHR	PREMAXHR	PREVO2	PREVE
43	88	173	36.5	139.8
PREWAT	PREHR%	PRERQ	PREFAT%	PREWT
246	91	1.13	26.33	180
PREHT	PRERESHR	PRRESSYS	PRRESDIA	PRERPP
69.25	95	132	72	21,844
POSTNHR	POSTMAXHR	POSTVO2	POSTVEE	POSTWAT
70	172	42.5	144.2	288
POSTHR%	POSTRQ	POSTFAT	POSTWT	POSTHT
94	1.17	25.66	186	69.25
POSTREHR	POSSYS	POSDIA	POSTRPP	PARTICIP
74	110	58	15,732	3

Case 10

AGE	PREHR	PREMAXHR	PREVO2	PREVE
51	95	168	23.3	175.8
PREWAT	PREHR%	PRERQ	PREFAT%	PREWT
227	95	1.12	43.97	312
PREHT	PRERESHR	PRRESSYS	PRRESDIA	PRERPP
75.00	95	132	98	20,700
POSTNHR	POSTMAXHR	POSTVO2	POSTVEE	POSTWAT
95	165	24.3	166.1	246
POSTHR%	POSTRQ	POSTFAT	POSTWT	POSTHT
93	1.18	42.18	319	75.00
POSTREHR	POSSYS	POSDIA	POSTRPP	PARTICIP
105	142	100	22,464	2

Case 11

AGE	PREHR	PREMAXHR	PREVO2	PREVE
57	96	177	25.7	95.0
PREWAT	PREHR%	PRERQ	PREFAT%	PREWT
144	101	1.41	22.80	135
PREHT	PRERESHR	PRRESSYS	PRRESDIA	PRERPP
67.75	98	118	78	32,680
POSTNHR	POSTMAXHR	POSTVO2	POSTVEE	POSTWAT
92	169	29.0	89.0	164
POSTHR%	POSTRQ	POSTFAT	POSTWT	POSTHT
98	1.29	26.26	137	67.75
POSTREHR	POSSYS	POSDIA	POSTRPP	PARTICIP
87	112	58	22,032	3

Case 12

AGE	PREHR	PREMAXHR	PREVO2	PREVE
41	108	189	28.6	108.7
PREWAT	PREHR%	PRERQ	PREFAT%	PREWT
183	102	1.27	24.71	146
PREHT	PRERESHR	PRRESSYS	PRRESDIA	PRERPP
67.00	95	102	72	21,016
POSTNHR	POSTMAXHR	POSTVO2	POSTVEE	POSTWAT
108	197	39.7	134.9	226
POSTHR%	POSTRQ	POSTFAT	POSTWT	POSTHT
108	1.30	28.10	153	67.00
POSTREHR	POSSYS	POSDIA	POSTRPP	PARTICIP
85	104	62	17,424	4

Case 13

AGE	PREHR	PREMAXHR	PREVO2	PREVE
40	122	194	36.1	183.3
PREWAT	PREHR%	PRERQ	PREFAT%	PREWT
264	106	1.38	36.88	210
PREHT	PRERESHR	PRRESSYS	PRRESDIA	PRERPP
71.00	82	118	82	17,936
POSTNHR	POSTMAXHR	POSTVO2	POSTVEE	POSTWAT
97	193	35.9	179.4	246
POSTHR%	POSTRQ	POSTFAT	POSTWT	POSTHT
105	1.33	33.47	212	71.00
POSTREHR	POSSYS	POSDIA	POSTRPP	PARTICIP
86	112	62	21,060	2

Case 14

AGE	PREHR	PREMAXHR	PREVO2	PREVE
62	90	178	28.7	113.0
PREWAT	PREHR%	PRERQ	PREFAT%	PREWT
246	104	1.12	35.57	233
PREHT	PRERESHR	PRRESSYS	PRRESDIA	PRERPP
75.00	113	148	88	23,970
POSTNHR	POSTMAXHR	POSTVO2	POSTVEE	POSTWAT
85	169	29.3	121.7	246
POSTHR%	POSTRQ	POSTFAT	POSTWT	POSTHT
101	1.11	34.91	233	75.00
POSTREHR	POSSYS	POSDIA	POSTRPP	PARTICIP
86	142	88	16,660	2

Case 15

AGE	PREHR	PREMAXHR	PREVO2	PREVE
47	82	166	23.0	125.1
PREWAT	PREHR%	PRERQ	PREFAT%	PREWT
205	93	1.32	34.42	255
PREHT	PRERESHR	PRRESSYS	PRRESDIA	PRERPP
71.00	86	120	82	17,024
POSTNHR	POSTMAXHR	POSTVO2	POSTVEE	POSTWAT
87	171	25.9	138.0	288
POSTHR%	POSTRQ	POSTFAT	POSTWT	POSTHT
96	1.18	35.56	263	71.00
POSTREHR	POSSYS	POSDIA	POSTRPP	PARTICIP
80	108	80	15,680	4

Case 16

AGE	PREHR	PREMAXHR	PREVO2	PREVE
54	83	168	30.3	103.3
PREWAT	PREHR%	PRERQ	PREFAT%	PREWT
223	96	1.07	31.62	187
PREHT	PRERESHR	PRRESSYS	PRRESDIA	PRERPP
73.00	———	———	———	———
POSTNHR	POSTMAXHR	POSTVO2	POSTVEE	POSTWAT
———	———	———	———	———
POSTHR%	POSTRQ	POSTFAT	POSTWT	POSTHT
———	———	———	———	———
POSTREHR	POSSYS	POSDIA	POSTRPP	PARTICIP
———	———	———	———	1

Case 17

AGE	PREHR	PREMAXHR	PREVO2	PREVE
40	93	173	27.0	162.4
PREWAT	PREHR%	PRERQ	PREFAT%	PREWT
246	94	1.27	40.67	249
PREHT	PRERESHR	PRRESSYS	PRRESDIA	PRERPP
72.00	80	112	78	17,820
POSTNHR	POSTMAXHR	POSTVO2	POSTVEE	POSTWAT
87	165	37.2	199.2	329
POSTHR%	POSTRQ	POSTFAT	POSTWT	POSTHT
92	1.16	35.40	242	72.00
POSTREHR	POSSYS	POSDIA	POSTRPP	PARTICIP
64	110	68	12,610	4

Case 18

AGE	PREHR	PREMAXHR	PREVO2	PREVE
49	83	157	26.0	101.0
PREWAT	PREHR%	PRERQ	PREFAT%	PREWT
205	84	1.13	35.24	210
PREHT	PRERESHR	PRRESSYS	PRRESDIA	PRERPP
72.75	83	118	64	18,848
POSTNHR	POSTMAXHR	POSTVO2	POSTVEE	POSTWAT
98	166	29.5	132.9	226
POSTHR%	POSTRQ	POSTFAT	POSTWT	POSTHT
102	1.19	33.92	206	72.75
POSTREHR	POSSYS	POSDIA	POSTRPP	PARTICIP
89	104	62	19,500	2

Case 19

AGE	PREHR	PREMAXHR	PREVO2	PREVE
61	70	161	32.8	153.5
PREWAT	PREHR%	PRERQ	PREFAT%	PREWT
226	94	1.24	29.07	188
PREHT	PRERESHR	PRRESSYS	PRRESDIA	PRERPP
72.50	—	—	—	—
POSTNHR	POSTMAXHR	POSTVO2	POSTVEE	POSTWAT
—	—	—	—	—
POSTHR%	POSTRQ	POSTFAT	POSTWT	POSTHT
—	—	—	—	—
POSTREHR	POSSYS	POSDIA	POSTRPP	PARTICIP
—	—	—	—	1

Case 20

AGE	PREHR	PREMAXHR	PREVO2	PREVE
41	78	198	33.3	166.6
PREWAT	PREHR%	PRERQ	PREFAT%	PREWT
224	107	1.43	36.41	188
PREHT	PRERESHR	PRRESSYS	PRRESDIA	PRERPP
71.25	87	152	88	27,636
POSTNHR	POSTMAXHR	POSTVO2	POSTVEE	POSTWAT
83	200	44.2	158.5	288
POSTHR%	POSTRQ	POSTFAT	POSTWT	POSTHT
107	1.30	35.32	184	71.25
POSTREHR	POSSYS	POSDIA	POSTRPP	PARTICIP
71	130	80	16,692	4

Case 21

AGE	PREHR	PREMAXHR	PREVO2	PREVE
57	78	157	26.9	115.9
PREWAT	PREHR%	PRERQ	PREFAT%	PREWT
164	95	1.30	36.55	170
PREHT	PRERESHR	PRRESSYS	PRRESDIA	PRERPP
64.50	64	124	68	19,320
POSTNHR	POSTMAXHR	POSTVO2	POSTVEE	POSTWAT
69	165	36.1	127.6	205
POSTHR%	POSTRQ	POSTFAT	POSTWT	POSTHT
96	1.26	29.16	156	64.50
POSTREHR	POSSYS	POSDIA	POSTRPP	PARTICIP
68	108	58	16,380	3

Case 22

AGE	PREHR	PREMAXHR	PREVO2	PREVE
44	82	201	29.4	146.9
PREWAT	PREHR%	PRERQ	PREFAT%	PREWT
246	109	1.26	34.42	214
PREHT	PRERESHR	PRRESSYS	PRRESDIA	PRERPP
71.50	89	122	78	18,900
POSTNHR	POSTMAXHR	POSTVO2	POSTVEE	POSTWAT
97	202	32.7	162.9	265
POSTHR%	POSTRQ	POSTFAT	POSTWT	POSTHT
111	1.23	37.71	224	71.50
POSTREHR	POSSYS	POSDIA	POSTRPP	PARTICIP
91	112	68	24,310	2

Case 23

AGE	PREHR	PREMAXHR	PREVO2	PREVE
52	80	185	37.6	139.5
PREWAT	PREHR%	PRERQ	PREFAT%	PREWT
246	104	1.20	25.42	178
PREHT	PRERESHR	PRRESSYS	PRRESDIA	PRERPP
70.50	80	118	78	16,188
POSTNHR	POSTMAXHR	POSTVO2	POSTVEE	POSTWAT
86	190	39.4	149.0	288
POSTHR%	POSTRQ	POSTFAT	POSTWT	POSTHT
108	1.22	27.51	182	70.50
POSTREHR	POSSYS	POSDIA	POSTRPP	PARTICIP
69	110	60	16,872	4

Case 24

AGE	PREHR	PREMAXHR	PREVO2	PREVE
40	90	181	39.1	114.5
PREWAT	PREHR%	PRERQ	PREFAT%	PREWT
226	95	1.22	31.62	162
PREHT	PRERESHR	PRRESSYS	PRRESDIA	PRERPP
65.00	82	110	62	18,080
POSTNHR	POSTMAXHR	POSTVO2	POSTVEE	POSTWAT
62	178	48.0	161.9	288
POSTHR%	POSTRQ	POSTFAT	POSTWT	POSTHT
97	1.30	19.30	159	65.00
POSTREHR	POSSYS	POSDIA	POSTRPP	PARTICIP
59	100	58	13,596	3

Case 25

AGE	PREHR	PREMAXHR	PREVO2	PREVE
52	91	172	30.5	68.6
PREWAT	PREHR%	PRERQ	PREFAT%	PREWT
144	97	1.27	19.42	122
PREHT	PRERESHR	PRRESSYS	PRRESDIA	PRERPP
64.50	76	132	82	31,310
POSTNHR	POSTMAXHR	POSTVO2	POSTVEE	POSTWAT
70	166	32.7	81.6	156
POSTHR%	POSTRQ	POSTFAT	POSTWT	POSTHT
94	1.23	18.19	124	64.50
POSTREHR	POSSYS	POSDIA	POSTRPP	PARTICIP
73	132	82	31,482	2

Case 26

AGE	PREHR	PREMAXHR	PREVO2	PREVE
41	77	188	40.8	169.7
PREWAT	PREHR%	PRERQ	PREFAT%	PREWT
267	102	1.25	38.41	180
PREHT	PRERESHR	PRRESSYS	PRRESDIA	PRERPP
70.00	83	130	78	21,590
POSTNHR	POSTMAXHR	POSTVO2	POSTVEE	POSTWAT
73	183	44.8	174.2	320
POSTHR%	POSTRQ	POSTFAT	POSTWT	POSTHT
97	1.17	25.45	179	70.00
POSTREHR	POSSYS	POSDIA	POSTRPP	PARTICIP
81	112	72	16,100	4

Case 27

AGE	PREHR	PREMAXHR	PREVO2	PREVE
46	117	183	21.4	154.5
PREWAT	PREHR%	PRERQ	PREFAT%	PREWT
226	102	1.19	24.23	312
PREHT	PRERESHR	PRRESSYS	PRRESDIA	PRERPP
73.50	107	152	90	23,876
POSTNHR	POSTMAXHR	POSTVO2	POSTVEE	POSTWAT
105	185	24.4	152.7	267
POSTHR%	POSTRQ	POSTFAT	POSTWT	POSTHT
103	1.09	43.50	310	73.50
POSTREHR	POSSYS	POSDIA	POSTRPP	PARTICIP
82	112	72	18,486	3

Case 28

AGE	PREHR	PREMAXHR	PREVO2	PREVE
52	72	166	40.8	135.1
PREWAT	PREHR%	PRERQ	PREFAT%	PREWT
267	94	1.19	25.73	177
PREHT	PRERESHR	PRRESSYS	PRRESDIA	PRERPP
72.50	74	112	78	13,248
POSTNHR	POSTMAXHR	POSTVO2	POSTVEE	POSTWAT
77	175	45.4	150.8	306
POSTHR%	POSTRQ	POSTFAT	POSTWT	POSTHT
93	1.13	21.24	175	72.50
POSTREHR	POSSYS	POSDIA	POSTRPP	PARTICIP
60	100	62	10,560	3

Case 29

AGE	PREHR	PREMAXHR	PREVO2	PREVE
40	97	194	48.0	199.2
PREWAT	PREHR%	PRERQ	PREFAT%	PREWT
326	104	1.22	21.73	175
PREHT	PRERESHR	PRRESSYS	PRRESDIA	PRERPP
70.50	85	108	80	15,180
POSTNHR	POSTMAXHR	POSTVO2	POSTVEE	POSTWAT
91	186	61.3	226.3	390
POSTHR%	POSTRQ	POSTFAT	POSTWT	POSTHT
101	1.20	18.00	176	70.50
POSTREHR	POSSYS	POSDIA	POSTRPP	PARTICIP
74	112	60	13,910	4

Case 30

AGE	PREHR	PREMAXHR	PREVO2	PREVE
48	80	181	30.0	104.6
PREWAT	PREHR%	PRERQ	PREFAT%	PREWT
205	101	1.36	33.98	163
PREHT	PRERESHR	PRRESSYS	PRRESDIA	PRERPP
66.00	———	———	———	———
POSTNHR	POSTMAXHR	POSTVO2	POSTVEE	POSTWAT
———	———	———	———	———
POSTHR%	POSTRQ	POSTFAT	POSTWT	POSTHT
———	———	———	———	———
POSTREHR	POSSYS	POSDIA	POSTRPP	PARTICIP
———	———	———	———	1

Student Questionnaire Data (SQD)

Data for each case have been arranged so that they can be copied and pasted on a 3 × 5 card. Cards can then be sorted various ways. For the SQD, $n = 51$.

Case	SEX	AGE	CLASS	HEIGHT	WEIGHT
1	1	22	4	69	120
	PULSE1	PULSE2	SMOKE	ACTIVITY	STRESS
	64	72	1	2	1
	FRIENDS	TIMEOUT	HEALTH	TEMP	
	1	1	1	75	

Case	SEX	AGE	CLASS	HEIGHT	WEIGHT
2	2	25	4	66	148
	PULSE1	PULSE2	SMOKE	ACTIVITY	STRESS
	72	88	1	2	2
	FRIENDS	TIMEOUT	HEALTH	TEMP	
	1	1	1	70	

Case	SEX	AGE	CLASS	HEIGHT	WEIGHT
3	1	33	4	67	128
	PULSE1	PULSE2	SMOKE	ACTIVITY	STRESS
	72	104	4	2	3
	FRIENDS	TIMEOUT	HEALTH	TEMP	
	1	2	3	70	

Case	SEX	AGE	CLASS	HEIGHT	WEIGHT
4	1	33	4	54	114
	PULSE1	PULSE2	SMOKE	ACTIVITY	STRESS
	58	82	1	2	3
	FRIENDS	TIMEOUT	HEALTH	TEMP	
	1	2	2	70	

Case	SEX	AGE	CLASS	HEIGHT	WEIGHT
5	1	21	3	65	126
	PULSE1	PULSE2	SMOKE	ACTIVITY	STRESS
	88	120	1	2	1
	FRIENDS	TIMEOUT	HEALTH	TEMP	
	1	1	3	75	

Case	SEX	AGE	CLASS	HEIGHT	WEIGHT
6	1	23	3	62	108
	PULSE1	PULSE2	SMOKE	ACTIVITY	STRESS
	76	88	1	2	2
	FRIENDS	TIMEOUT	HEALTH	TEMP	
	1	1	2	68	

Case	SEX	AGE	CLASS	HEIGHT	WEIGHT
7	2	22	3	71	190
	PULSE1	PULSE2	SMOKE	ACTIVITY	STRESS
	68	92	1	2	2
	FRIENDS	TIMEOUT	HEALTH	TEMP	
	1	1	2	——	

Case	SEX	AGE	CLASS	HEIGHT	WEIGHT
8	2	23	3	73	180
	PULSE1	PULSE2	SMOKE	ACTIVITY	STRESS
	68	76	1	2	2
	FRIENDS	TIMEOUT	HEALTH	TEMP	
	1	1	2	——	

Case	SEX	AGE	CLASS	HEIGHT	WEIGHT
9	1	24	4	70	150
	PULSE1	PULSE2	SMOKE	ACTIVITY	STRESS
	84	84	1	2	1
	FRIENDS	TIMEOUT	HEALTH	TEMP	
	1	2	3	——	

Case	SEX	AGE	CLASS	HEIGHT	WEIGHT
10	1	21	3	65	105
	PULSE1	PULSE2	SMOKE	ACTIVITY	STRESS
	80	100	1	2	1
	FRIENDS	TIMEOUT	HEALTH	TEMP	
	1	2	2	75	

Case					
Case 11	SEX 1	AGE 21	CLASS 4	HEIGHT 62	WEIGHT 107
	PULSE1 68	PULSE2 80	SMOKE 1	ACTIVITY 1	STRESS 2
	FRIENDS 1	TIMEOUT 1	HEALTH 2	TEMP 70	
Case 12	SEX 1	AGE 22	CLASS 4	HEIGHT 64	WEIGHT 98
	PULSE1 60	PULSE2 80	SMOKE 1	ACTIVITY 2	STRESS 2
	FRIENDS 1	TIMEOUT 2	HEALTH 3	TEMP 80	
Case 13	SEX 1	AGE 24	CLASS 4	HEIGHT 64	WEIGHT 108
	PULSE1 84	PULSE2 116	SMOKE 1	ACTIVITY 2	STRESS 2
	FRIENDS 1	TIMEOUT 1	HEALTH 2	TEMP ——	
Case 14	SEX 1	AGE 32	CLASS 4	HEIGHT 62	WEIGHT 108
	PULSE1 88	PULSE2 100	SMOKE 1	ACTIVITY 3	STRESS 2
	FRIENDS 1	TIMEOUT 1	HEALTH 1	TEMP ——	
Case 15	SEX 2	AGE 26	CLASS 4	HEIGHT 67	WEIGHT 190
	PULSE1 92	PULSE2 120	SMOKE 1	ACTIVITY 2	STRESS 3
	FRIENDS 1	TIMEOUT 1	HEALTH 3	TEMP 75	
Case 16	SEX 2	AGE 21	CLASS ——	HEIGHT 70	WEIGHT 155
	PULSE1 56	PULSE2 60	SMOKE 4	ACTIVITY 2	STRESS 2
	FRIENDS 1	TIMEOUT ——	HEALTH 3	TEMP ——	

Case	SEX	AGE	CLASS	HEIGHT	WEIGHT
17	1	23	5	61	120
	PULSE1	**PULSE2**	**SMOKE**	**ACTIVITY**	**STRESS**
	72	108	2	2	2
	FRIENDS	**TIMEOUT**	**HEALTH**	**TEMP**	
	1	1	2	68	

Case	SEX	AGE	CLASS	HEIGHT	WEIGHT
18	2	23	4	75	192
	PULSE1	**PULSE2**	**SMOKE**	**ACTIVITY**	**STRESS**
	84	92	1	———	2
	FRIENDS	**TIMEOUT**	**HEALTH**	**TEMP**	
	1	2	2	70	

Case	SEX	AGE	CLASS	HEIGHT	WEIGHT
19	1	21	4	66	140
	PULSE1	**PULSE2**	**SMOKE**	**ACTIVITY**	**STRESS**
	88	116	1	1	2
	FRIENDS	**TIMEOUT**	**HEALTH**	**TEMP**	
	1	2	2	70	

Case	SEX	AGE	CLASS	HEIGHT	WEIGHT
20	1	21	4	63	110
	PULSE1	**PULSE2**	**SMOKE**	**ACTIVITY**	**STRESS**
	92	104	4	2	2
	FRIENDS	**TIMEOUT**	**HEALTH**	**TEMP**	
	1	1	4	———	

Case	SEX	AGE	CLASS	HEIGHT	WEIGHT
21	2	21	3	76	220
	PULSE1	**PULSE2**	**SMOKE**	**ACTIVITY**	**STRESS**
	72	80	3	3	1
	FRIENDS	**TIMEOUT**	**HEALTH**	**TEMP**	
	1	1	2	60	

Case	SEX	AGE	CLASS	HEIGHT	WEIGHT
22	1	26	4	68	135
	PULSE1	**PULSE2**	**SMOKE**	**ACTIVITY**	**STRESS**
	68	88	1	2	1
	FRIENDS	**TIMEOUT**	**HEALTH**	**TEMP**	
	1	1	2	70	

Case 23

SEX	AGE	CLASS	HEIGHT	WEIGHT
1	25	4	62	110

PULSE1	PULSE2	SMOKE	ACTIVITY	STRESS
68	88	1	2	2

FRIENDS	TIMEOUT	HEALTH	TEMP
2	1	1	74

Case 24

SEX	AGE	CLASS	HEIGHT	WEIGHT
1	22	3	66	115

PULSE1	PULSE2	SMOKE	ACTIVITY	STRESS
72	100	1	2	2

FRIENDS	TIMEOUT	HEALTH	TEMP
2	2	2	75

Case 25

SEX	AGE	CLASS	HEIGHT	WEIGHT
2	30	4	69	160

PULSE1	PULSE2	SMOKE	ACTIVITY	STRESS
72	88	1	2	2

FRIENDS	TIMEOUT	HEALTH	TEMP
1	1	2	70

Case 26

SEX	AGE	CLASS	HEIGHT	WEIGHT
2	29	4	70	160

PULSE1	PULSE2	SMOKE	ACTIVITY	STRESS
88	116	2	2	2

FRIENDS	TIMEOUT	HEALTH	TEMP
1	1	2	65

Case 27

SEX	AGE	CLASS	HEIGHT	WEIGHT
1	38	5	65	123

PULSE1	PULSE2	SMOKE	ACTIVITY	STRESS
48	54	3	3	3

FRIENDS	TIMEOUT	HEALTH	TEMP
1	1	1	———

Case	SEX	AGE	CLASS	HEIGHT	WEIGHT
28	2	23	5	72	180
	PULSE1	PULSE2	SMOKE	ACTIVITY	STRESS
	76	84	1	3	1
	FRIENDS	TIMEOUT	HEALTH	TEMP	
	1	1	1	68	

Case	SEX	AGE	CLASS	HEIGHT	WEIGHT
29	2	22	3	72	175
	PULSE1	PULSE2	SMOKE	ACTIVITY	STRESS
	96	124	3	2	2
	FRIENDS	TIMEOUT	HEALTH	TEMP	
	1	1	2	70	

Case	SEX	AGE	CLASS	HEIGHT	WEIGHT
30	1	20	3	64	120
	PULSE1	PULSE2	SMOKE	ACTIVITY	STRESS
	80	92	1	2	2
	FRIENDS	TIMEOUT	HEALTH	TEMP	
	1	1	2	68	

Case	SEX	AGE	CLASS	HEIGHT	WEIGHT
31	2	28	4	72	210
	PULSE1	PULSE2	SMOKE	ACTIVITY	STRESS
	66	88	1	3	2
	FRIENDS	TIMEOUT	HEALTH	TEMP	
	1	2	1	———	

Case	SEX	AGE	CLASS	HEIGHT	WEIGHT
32	1	33	3	62	135
	PULSE1	PULSE2	SMOKE	ACTIVITY	STRESS
	60	———	2	1	3
	FRIENDS	TIMEOUT	HEALTH	TEMP	
	2	2	2	70	

Case	SEX	AGE	CLASS	HEIGHT	WEIGHT
33	1	38	4	63	150
	PULSE1	PULSE2	SMOKE	ACTIVITY	STRESS
	68	84	2	2	2
	FRIENDS	TIMEOUT	HEALTH	TEMP	
	1	1	2	65	

Case	SEX	AGE	CLASS	HEIGHT	WEIGHT
34	1	21	3	62	120
	PULSE1	PULSE2	SMOKE	ACTIVITY	STRESS
	72	108	1	2	2
	FRIENDS	TIMEOUT	HEALTH	TEMP	
	1	1	2	75	

Case	SEX	AGE	CLASS	HEIGHT	WEIGHT
35	1	23	4	64	120
	PULSE1	PULSE2	SMOKE	ACTIVITY	STRESS
	104	116	4	———	2
	FRIENDS	TIMEOUT	HEALTH	TEMP	
	1	1	3	———	

Case	SEX	AGE	CLASS	HEIGHT	WEIGHT
36	1	24	4	63	100
	PULSE1	PULSE2	SMOKE	ACTIVITY	STRESS
	74	104	1	1	2
	FRIENDS	TIMEOUT	HEALTH	TEMP	
	1	2	2	70	

Case	SEX	AGE	CLASS	HEIGHT	WEIGHT
37	1	20	3	63	112
	PULSE1	PULSE2	SMOKE	ACTIVITY	STRESS
	84	116	1	1	2
	FRIENDS	TIMEOUT	HEALTH	TEMP	
	1	2	2	68	

Case	SEX	AGE	CLASS	HEIGHT	WEIGHT
38	2	20	3	72	147
	PULSE1	PULSE2	SMOKE	ACTIVITY	STRESS
	72	116	2	3	2
	FRIENDS	TIMEOUT	HEALTH	TEMP	
	1	1	1	———	

Case	SEX	AGE	CLASS	HEIGHT	WEIGHT
39	1	20	3	67	97
	PULSE1	PULSE2	SMOKE	ACTIVITY	STRESS
	76	96	1	2	2
	FRIENDS	TIMEOUT	HEALTH	TEMP	
	1	1	2	68	

Case	SEX	AGE	CLASS	HEIGHT	WEIGHT
40	2	25	3	66	125
	PULSE1	PULSE2	SMOKE	ACTIVITY	STRESS
	80	120	1	2	2
	FRIENDS	TIMEOUT	HEALTH	TEMP	
	1	1	2	68	

Case	SEX	AGE	CLASS	HEIGHT	WEIGHT
41	2	21	3	73	233
	PULSE1	PULSE2	SMOKE	ACTIVITY	STRESS
	96	136	1	2	1
	FRIENDS	TIMEOUT	HEALTH	TEMP	
	2	———	3	68	

Case	SEX	AGE	CLASS	HEIGHT	WEIGHT
42	1	30	5	66	145
	PULSE1	PULSE2	SMOKE	ACTIVITY	STRESS
	72	84	1	2	3
	FRIENDS	TIMEOUT	HEALTH	TEMP	
	1	1	2	70	

Case	SEX	AGE	CLASS	HEIGHT	WEIGHT
43	1	61	5	64	150
	PULSE1	**PULSE2**	**SMOKE**	**ACTIVITY**	**STRESS**
	60	100	1	2	———
	FRIENDS	**TIMEOUT**	**HEALTH**	**TEMP**	
	1	1	1	50	

Case	SEX	AGE	CLASS	HEIGHT	WEIGHT
44	1	20	2	63	120
	PULSE1	**PULSE2**	**SMOKE**	**ACTIVITY**	**STRESS**
	88	100	1	2	1
	FRIENDS	**TIMEOUT**	**HEALTH**	**TEMP**	
	1	2	2	70	

Case	SEX	AGE	CLASS	HEIGHT	WEIGHT
45	1	23	3	64	110
	PULSE1	**PULSE2**	**SMOKE**	**ACTIVITY**	**STRESS**
	80	128	1	2	1
	FRIENDS	**TIMEOUT**	**HEALTH**	**TEMP**	
	1	1	2	70	

Case	SEX	AGE	CLASS	HEIGHT	WEIGHT
46	2	24	4	70	186
	PULSE1	**PULSE2**	**SMOKE**	**ACTIVITY**	**STRESS**
	72	100	1	2	1
	FRIENDS	**TIMEOUT**	**HEALTH**	**TEMP**	
	1	1	2	70	

Case	SEX	AGE	CLASS	HEIGHT	WEIGHT
47	2	21	4	67	140
	PULSE1	**PULSE2**	**SMOKE**	**ACTIVITY**	**STRESS**
	84	104	1	2	2
	FRIENDS	**TIMEOUT**	**HEALTH**	**TEMP**	
	1	2	2	68	

Case	SEX	AGE	CLASS	HEIGHT	WEIGHT
48	1	30	4	65	116
	PULSE1	PULSE2	SMOKE	ACTIVITY	STRESS
	72	112	1	1	2
	FRIENDS	TIMEOUT	HEALTH	TEMP	
	1	1	2	70	

Case	SEX	AGE	CLASS	HEIGHT	WEIGHT
49	2	21	4	72	182
	PULSE1	PULSE2	SMOKE	ACTIVITY	STRESS
	84	100	2	3	2
	FRIENDS	TIMEOUT	HEALTH	TEMP	
	1	2	2	68	

Case	SEX	AGE	CLASS	HEIGHT	WEIGHT
50	1	23	4	64	130
	PULSE1	PULSE2	SMOKE	ACTIVITY	STRESS
	72	88	1	2	2
	FRIENDS	TIMEOUT	HEALTH	TEMP	
	1	1	2	72	

Case	SEX	AGE	CLASS	HEIGHT	WEIGHT
51	1	24	3	67	140
	PULSE1	PULSE2	SMOKE	ACTIVITY	STRESS
	80	104	1	2	3
	FRIENDS	TIMEOUT	HEALTH	TEMP	
	2	2	3	___	

APPENDIX B

Statistical Tables

TABLE B-1 Cumulative standard normal distribution, or z

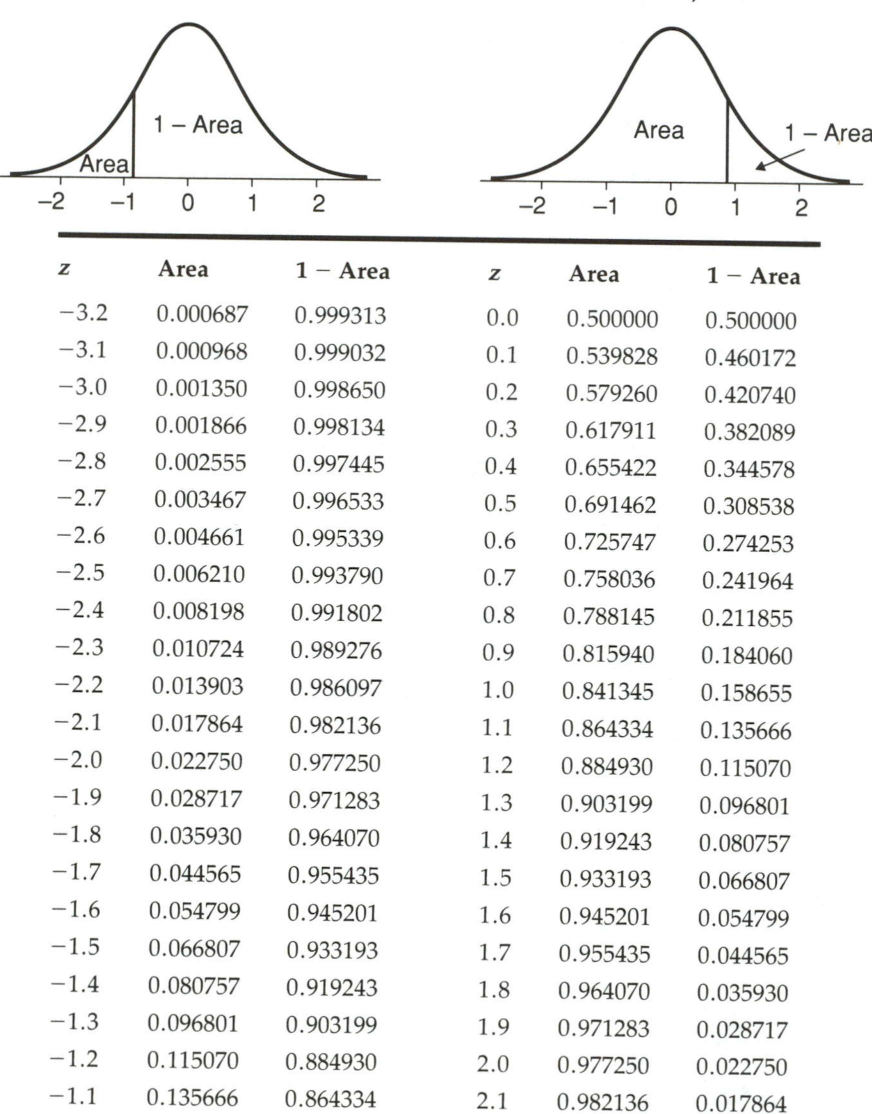

z	Area	1 − Area	z	Area	1 − Area
−3.2	0.000687	0.999313	0.0	0.500000	0.500000
−3.1	0.000968	0.999032	0.1	0.539828	0.460172
−3.0	0.001350	0.998650	0.2	0.579260	0.420740
−2.9	0.001866	0.998134	0.3	0.617911	0.382089
−2.8	0.002555	0.997445	0.4	0.655422	0.344578
−2.7	0.003467	0.996533	0.5	0.691462	0.308538
−2.6	0.004661	0.995339	0.6	0.725747	0.274253
−2.5	0.006210	0.993790	0.7	0.758036	0.241964
−2.4	0.008198	0.991802	0.8	0.788145	0.211855
−2.3	0.010724	0.989276	0.9	0.815940	0.184060
−2.2	0.013903	0.986097	1.0	0.841345	0.158655
−2.1	0.017864	0.982136	1.1	0.864334	0.135666
−2.0	0.022750	0.977250	1.2	0.884930	0.115070
−1.9	0.028717	0.971283	1.3	0.903199	0.096801
−1.8	0.035930	0.964070	1.4	0.919243	0.080757
−1.7	0.044565	0.955435	1.5	0.933193	0.066807
−1.6	0.054799	0.945201	1.6	0.945201	0.054799
−1.5	0.066807	0.933193	1.7	0.955435	0.044565
−1.4	0.080757	0.919243	1.8	0.964070	0.035930
−1.3	0.096801	0.903199	1.9	0.971283	0.028717
−1.2	0.115070	0.884930	2.0	0.977250	0.022750
−1.1	0.135666	0.864334	2.1	0.982136	0.017864

TABLE B-1 *Continued*

z	Area	1 − Area	z	Area	1 − Area
−1.0	0.158655	0.841345	2.2	0.986097	0.013903
−0.9	0.184060	0.815940	2.3	0.989276	0.010724
−0.8	0.211855	0.788145	2.4	0.991802	0.008198
−0.7	0.241964	0.758036	2.5	0.993790	0.006210
−0.6	0.274253	0.725747	2.6	0.995339	0.004661
−0.5	0.308538	0.691462	2.7	0.996533	0.003467
−0.4	0.344578	0.655422	2.8	0.997445	0.002555
−0.3	0.382089	0.617911	2.9	0.998134	0.001866
−0.2	0.420740	0.579260	3.0	0.998650	0.001350
−0.1	0.460172	0.539828	3.1	0.999032	0.000968
0.0	0.500000	0.500000	3.2	0.999313	0.000687

TABLE B-2 Normal Distribution: Central Area (incremented *z*-scores)

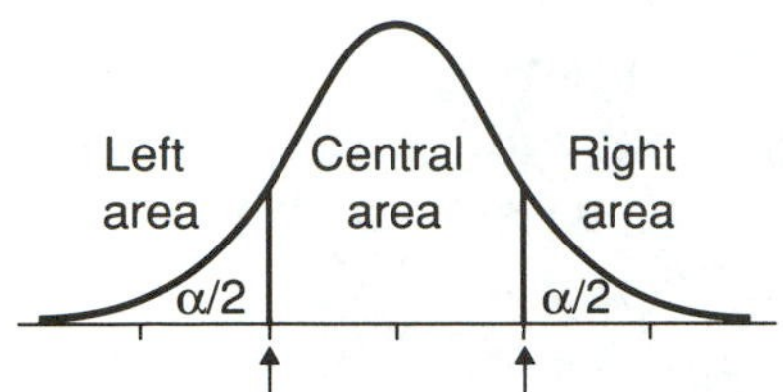

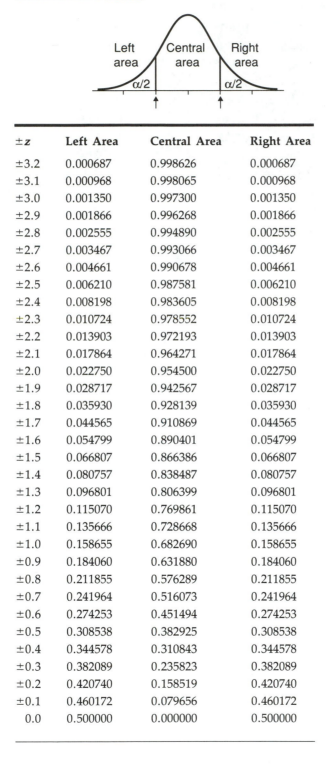

$\pm z$	Left Area	Central Area	Right Area
±3.2	0.000687	0.998626	0.000687
±3.1	0.000968	0.998065	0.000968
±3.0	0.001350	0.997300	0.001350
±2.9	0.001866	0.996268	0.001866
±2.8	0.002555	0.994890	0.002555
±2.7	0.003467	0.993066	0.003467
±2.6	0.004661	0.990678	0.004661
±2.5	0.006210	0.987581	0.006210
±2.4	0.008198	0.983605	0.008198
±2.3	0.010724	0.978552	0.010724
±2.2	0.013903	0.972193	0.013903
±2.1	0.017864	0.964271	0.017864
±2.0	0.022750	0.954500	0.022750
±1.9	0.028717	0.942567	0.028717
±1.8	0.035930	0.928139	0.035930
±1.7	0.044565	0.910869	0.044565
±1.6	0.054799	0.890401	0.054799
±1.5	0.066807	0.866386	0.066807
±1.4	0.080757	0.838487	0.080757
±1.3	0.096801	0.806399	0.096801
±1.2	0.115070	0.769861	0.115070
±1.1	0.135666	0.728668	0.135666
±1.0	0.158655	0.682690	0.158655
±0.9	0.184060	0.631880	0.184060
±0.8	0.211855	0.576289	0.211855
±0.7	0.241964	0.516073	0.241964
±0.6	0.274253	0.451494	0.274253
±0.5	0.308538	0.382925	0.308538
±0.4	0.344578	0.310843	0.344578
±0.3	0.382089	0.235823	0.382089
±0.2	0.420740	0.158519	0.420740
±0.1	0.460172	0.079656	0.460172
0.0	0.500000	0.000000	0.500000

TABLE B-3 Critical Values of the *t*-Distributions for One-Tailed Tests

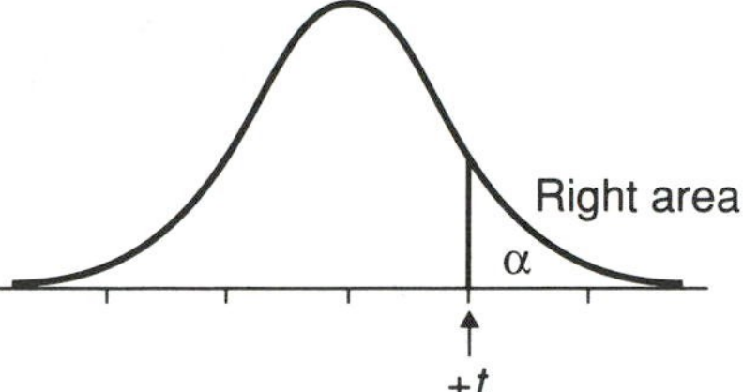

The values given are critical *t*-values for one-tailed *t* tests on the right side.

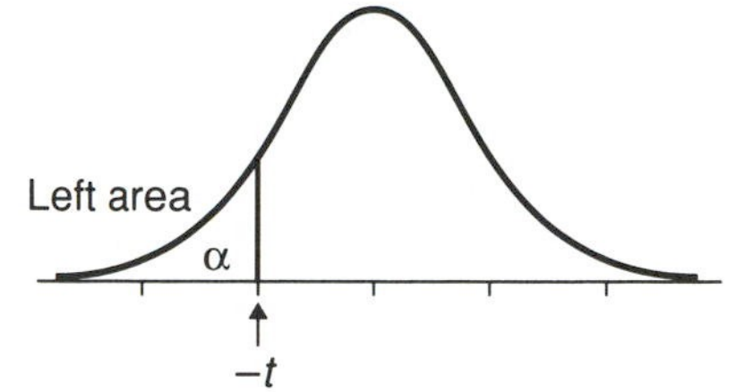

The values given are critical *t*-values for one-tailed *t* tests on the left side.

df	$\alpha = .05$	$\alpha = .01$	df	$\alpha = .05$	$\alpha = .01$
5	2.015	3.365	5	−2.015	−3.365
6	1.943	3.143	6	−1.943	−3.143
7	1.895	2.993	7	−1.895	−2.993
8	1.860	2.896	8	−1.860	−2.896
9	1.833	2.821	9	−1.833	−2.821
10	1.812	2.764	10	−1.812	−2.764
11	1.796	2.718	11	−1.796	−2.718
12	1.782	2.681	12	−1.782	−2.681
13	1.771	2.650	13	−1.771	−2.650
14	1.761	2.624	14	−1.761	−2.624
15	1.753	2.602	15	−1.753	−2.602
16	1.746	2.583	16	−1.746	−2.583
17	1.740	2.567	17	−1.740	−2.567
18	1.734	2.552	18	−1.734	−2.552
19	1.729	2.539	19	−1.729	−2.539
20	1.725	2.528	20	−1.725	−2.528
21	1.721	2.518	21	−1.721	−2.518
22	1.717	2.508	22	−1.717	−2.508
23	1.714	2.500	23	−1.714	−2.500
24	1.711	2.492	24	−1.711	−2.492
25	1.708	2.485	25	−1.708	−2.485
26	1.706	2.479	26	−1.706	−2.479
27	1.703	2.473	27	−1.703	−2.473
28	1.701	2.467	28	−1.701	−2.467
29	1.699	2.462	29	−1.699	−2.462
30	1.697	2.457	30	−1.697	−2.457
40	1.684	2.423	40	−1.684	−2.423
60	1.671	2.390	60	−1.671	−2.390
120	1.658	2.358	120	−1.658	−2.358
∞	1.645	2.326	∞	−1.645	−2.326

At df = ∞, the *t*-distribution and the standard normal distribution are exactly the same.

At df = ∞, the *t*-distribution and the standard normal distribution are exactly the same.

TABLE B-4 Critical Values of the *t*-Distribution for Two-Tailed Tests and Confidence Intervals

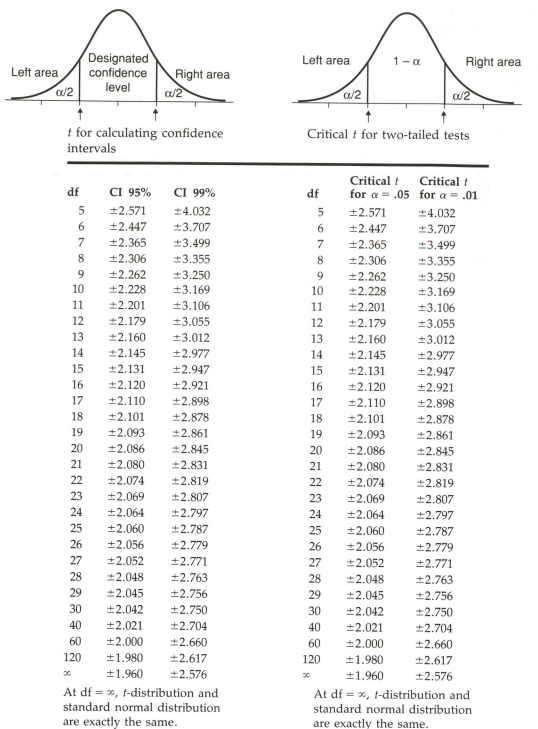

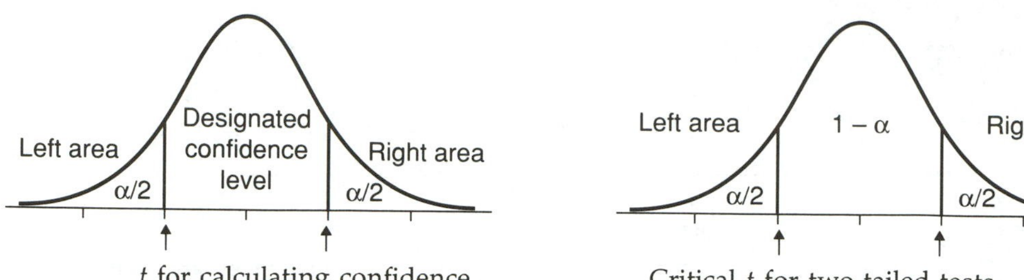

df	CI 95%	CI 99%	df	Critical t for $\alpha = .05$	Critical t for $\alpha = .01$
5	±2.571	±4.032	5	±2.571	±4.032
6	±2.447	±3.707	6	±2.447	±3.707
7	±2.365	±3.499	7	±2.365	±3.499
8	±2.306	±3.355	8	±2.306	±3.355
9	±2.262	±3.250	9	±2.262	±3.250
10	±2.228	±3.169	10	±2.228	±3.169
11	±2.201	±3.106	11	±2.201	±3.106
12	±2.179	±3.055	12	±2.179	±3.055
13	±2.160	±3.012	13	±2.160	±3.012
14	±2.145	±2.977	14	±2.145	±2.977
15	±2.131	±2.947	15	±2.131	±2.947
16	±2.120	±2.921	16	±2.120	±2.921
17	±2.110	±2.898	17	±2.110	±2.898
18	±2.101	±2.878	18	±2.101	±2.878
19	±2.093	±2.861	19	±2.093	±2.861
20	±2.086	±2.845	20	±2.086	±2.845
21	±2.080	±2.831	21	±2.080	±2.831
22	±2.074	±2.819	22	±2.074	±2.819
23	±2.069	±2.807	23	±2.069	±2.807
24	±2.064	±2.797	24	±2.064	±2.797
25	±2.060	±2.787	25	±2.060	±2.787
26	±2.056	±2.779	26	±2.056	±2.779
27	±2.052	±2.771	27	±2.052	±2.771
28	±2.048	±2.763	28	±2.048	±2.763
29	±2.045	±2.756	29	±2.045	±2.756
30	±2.042	±2.750	30	±2.042	±2.750
40	±2.021	±2.704	40	±2.021	±2.704
60	±2.000	±2.660	60	±2.000	±2.660
120	±1.980	±2.617	120	±1.980	±2.617
∞	±1.960	±2.576	∞	±1.960	±2.576

At df = ∞, *t*-distribution and standard normal distribution are exactly the same.

At df = ∞, *t*-distribution and standard normal distribution are exactly the same.

TABLE B-5 Critical χ^2 for Chosen Tail Areas or Percentiles

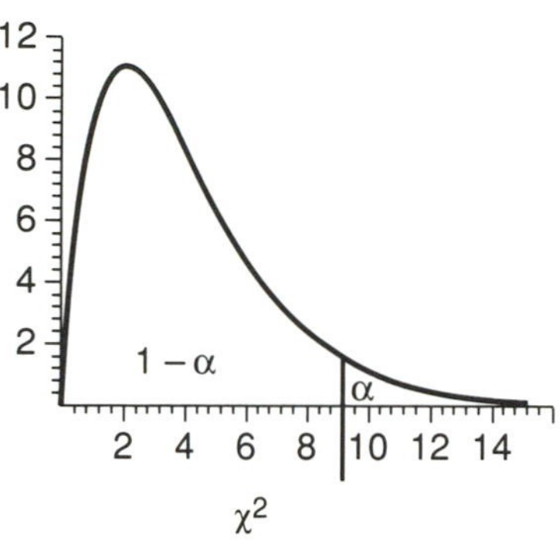

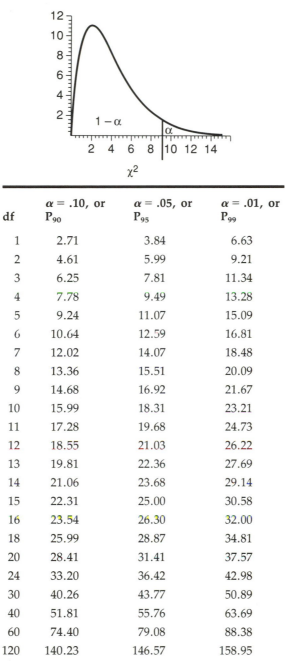

df	α = .10, or P_{90}	α = .05, or P_{95}	α = .01, or P_{99}
1	2.71	3.84	6.63
2	4.61	5.99	9.21
3	6.25	7.81	11.34
4	7.78	9.49	13.28
5	9.24	11.07	15.09
6	10.64	12.59	16.81
7	12.02	14.07	18.48
8	13.36	15.51	20.09
9	14.68	16.92	21.67
10	15.99	18.31	23.21
11	17.28	19.68	24.73
12	18.55	21.03	26.22
13	19.81	22.36	27.69
14	21.06	23.68	29.14
15	22.31	25.00	30.58
16	23.54	26.30	32.00
18	25.99	28.87	34.81
20	28.41	31.41	37.57
24	33.20	36.42	42.98
30	40.26	43.77	50.89
40	51.81	55.76	63.69
60	74.40	79.08	88.38
120	140.23	146.57	158.95

Table B-6 *F* **Distribution, Upper 5 Percent Values for Various Degrees of Freedom**

df for Denominator	df for Numerator																		
	1	2	3	4	5	6	7	8	9	10	12	15	20	24	30	40	60	120	∞
1	161	200	216	225	230	234	237	239	241	242	244	246	248	249	250	251	252	253	254
2	18.5	19.0	19.2	19.2	19.3	19.3	19.4	19.4	19.4	19.4	19.4	19.4	19.4	19.5	19.5	19.5	19.5	19.5	19.5
3	10.1	9.55	9.28	9.12	9.01	8.94	8.89	8.85	8.81	8.79	8.74	8.70	8.66	8.64	8.62	8.59	8.57	8.55	8.53
4	7.71	6.94	6.59	6.39	6.26	6.16	6.09	6.04	6.00	5.96	5.91	5.86	5.80	5.77	5.75	5.72	5.69	5.66	5.63
5	6.61	5.79	5.41	5.19	5.05	4.95	4.88	4.82	4.77	4.74	4.68	4.62	4.56	4.53	4.50	4.46	4.43	4.40	4.37
6	5.99	5.14	4.76	4.53	4.39	4.28	4.21	4.15	4.10	4.06	4.00	3.94	3.87	3.84	3.81	3.77	3.74	3.70	3.67
7	5.59	4.74	4.35	4.12	3.97	3.87	3.79	3.73	3.68	3.64	3.57	3.51	3.44	3.41	3.38	3.34	3.30	3.27	3.23
8	5.32	4.46	4.07	3.84	3.69	3.58	3.50	3.44	3.39	3.35	3.28	3.22	3.15	3.12	3.08	3.04	3.01	2.97	2.93
9	5.12	4.26	3.86	3.63	3.48	3.37	3.29	3.23	3.18	3.14	3.07	3.01	2.94	2.90	2.86	2.83	2.79	2.75	2.71
10	4.96	4.10	3.71	3.48	3.33	3.22	3.14	3.07	3.02	2.98	2.91	2.85	2.77	2.74	2.70	2.66	2.62	2.58	2.54
11	4.84	3.98	3.59	3.36	3.20	3.09	3.01	2.95	2.90	2.85	2.79	2.72	2.65	2.61	2.57	2.53	2.49	2.45	2.40
12	4.75	3.89	3.49	3.26	3.11	3.00	2.91	2.85	2.80	2.75	2.69	2.62	2.54	2.51	2.47	2.43	2.38	2.34	2.30
13	4.67	3.81	3.41	3.18	3.03	2.92	2.83	2.77	2.71	2.67	2.60	2.53	2.46	2.42	2.38	2.34	2.30	2.25	2.21
14	5.60	3.74	3.34	3.11	2.96	2.85	2.76	2.70	2.65	2.60	2.53	2.46	2.39	2.35	2.31	2.27	2.22	2.18	2.13
15	4.54	3.68	3.29	3.06	2.90	2.79	2.71	2.64	2.59	2.54	2.48	2.40	2.33	2.29	2.25	2.20	2.16	2.11	2.07
16	4.49	3.63	3.24	3.01	2.85	2.74	2.66	2.59	2.54	2.49	2.42	2.35	2.28	2.24	2.19	2.15	2.11	2.06	2.01
17	4.45	3.59	3.20	2.96	2.81	2.70	2.61	2.55	2.49	2.45	2.38	2.31	2.23	2.19	2.15	2.10	2.06	2.01	1.96
18	4.41	3.55	3.16	2.93	2.77	2.66	2.58	2.51	2.46	2.41	2.34	2.27	2.19	2.15	2.11	2.06	2.02	1.97	1.92
19	4.38	3.52	3.13	2.90	2.74	2.63	2.54	2.48	2.42	2.38	2.31	2.23	2.16	2.11	2.07	2.03	1.98	1.93	1.88
20	4.35	3.49	3.10	2.87	2.71	2.60	2.51	2.45	2.39	2.35	2.28	2.20	2.12	2.08	2.04	1.99	1.95	1.90	1.84
21	4.32	3.47	3.07	2.84	2.68	2.57	2.49	2.42	2.37	2.32	2.25	2.18	2.10	2.05	2.01	1.96	1.92	1.87	1.81
22	4.30	3.44	3.05	2.82	2.66	2.55	2.46	2.40	2.34	2.30	2.23	2.15	2.07	2.03	1.98	1.94	1.89	1.84	1.78
23	4.28	3.42	3.03	2.80	2.64	2.53	2.44	2.37	2.32	2.27	2.20	2.13	2.05	2.01	1.96	1.91	1.86	1.81	1.76
24	4.26	3.40	3.01	2.78	2.62	2.51	2.42	2.36	2.30	2.25	2.18	2.11	2.03	1.98	1.94	1.89	1.84	1.79	1.73
25	4.24	3.39	2.99	2.76	2.60	2.49	2.40	2.34	2.28	2.24	2.16	2.09	2.01	1.96	1.92	1.87	1.82	1.77	1.71
30	4.17	3.32	2.92	2.69	2.53	2.42	2.33	2.27	2.21	2.16	2.09	2.01	1.93	1.89	1.84	1.79	1.74	1.68	1.62
40	4.08	3.23	2.84	2.61	2.45	2.34	2.25	2.18	2.12	2.08	2.00	1.92	1.84	1.79	1.74	1.69	1.64	1.58	1.51
60	4.00	3.15	2.76	2.53	2.37	2.25	2.17	2.10	2.04	1.99	1.92	1.84	1.75	1.70	1.65	1.59	1.53	1.47	1.39
120	3.92	3.07	2.68	2.45	2.29	2.18	2.09	2.02	1.96	1.91	1.83	1.75	1.66	1.61	1.55	1.50	1.43	1.35	1.25
∞	3.84	3.00	2.60	2.37	2.21	2.10	2.01	1.94	1.88	1.83	1.75	1.67	1.57	1.52	1.46	1.39	1.32	1.22	1.00

TABLE B-7 z, t, F, and χ^2 Distribution Values of Variance Ratio F Exceeded in 5 Percent of Random Samples

df of Denominator	t	df of Numerator 1	2	3	4	5	6	7	8
6	2.45	5.99	5.14	4.76	4.53	4.39	4.28	4.21	4.15
7	2.365	5.59	4.74	4.35	4.12	3.97	3.87	3.79	3.73
8	2.31	5.32	4.46	4.07	3.84	3.69	3.58	3.50	3.44
9	2.26	5.12	4.26	3.86	3.63	3.48	3.37	3.29	3.23
10	2.23	4.96	4.10	3.71	3.48	3.33	3.22	3.14	3.07
11	2.20	4.84	3.98	3.59	3.36	3.20	3.09	3.01	2.95
12	2.18	4.75	3.88	3.49	3.26	3.11	3.00	2.91	2.85
13	2.16	4.67	3.80	3.41	3.18	3.02	2.92	2.83	2.77
14	2.145	4.60	3.74	3.34	3.11	2.96	2.85	2.76	2.70
15	2.13	4.54	3.68	3.29	3.06	2.90	2.79	2.71	2.64
16	2.12	4.49	3.63	3.24	3.01	2.85	2.74	2.66	2.59
17	2.11	4.45	3.59	3.20	2.96	2.81	2.70	2.61	2.55
18	2.10	4.41	3.55	3.16	2.93	2.77	2.66	2.58	2.51
19	2.09	4.38	3.52	3.13	2.90	2.74	2.63	2.54	2.48
20	2.086	4.35	3.49	3.10	2.87	2.71	2.60	2.51	2.45
25	2.06	4.24	3.38	2.99	2.76	2.60	2.49	2.40	2.34
30	2.04	4.17	3.32	2.92	2.69	2.53	2.42	2.34	2.27
40	2.02	4.08	3.23	2.84	2.61	2.45	2.34	2.25	2.18
60	2.00	4.00	3.15	2.76	2.52	2.37	2.25	2.17	2.10
100	1.98	3.94	3.09	2.70	2.46	2.30	2.19	2.10	2.03
∞	1.96	3.84	3.00	2.605	2.37	2.21	2.10	2.01	1.94
	χ^2	3.84	5.99	7.815	9.49	11.07	12.59	14.07	15.51

TABLE B-8 Percentage Points, Distribution of the Correlation Coefficient, When $\rho = 0$
$P[r \leq \text{tabular value}] = 1 - \alpha$

ν	$\alpha = 0.05$ $2\alpha = 0.1$	0.025 0.05	0.01 0.02	0.005 0.01	0.0025 0.005	0.0005 0.001	ν	$\alpha = 0.05$ $2\alpha = 0.1$	0.025 0.05	0.01 0.02	0.005 0.01	0.0025 0.005	0.0005 0.001
1	0.9877	0.9^2692	0.9^3507	0.9^3877	0.9^4692	0.9^6877	16	0.400	0.468	0.543	0.590	0.631	0.708
2	0.9000	0.9500	0.9800	0.9^2000	0.9^2500	0.9^4000	17	0.389	0.456	0.529	0.575	0.616	0.693
3	0.805	0.878	0.9343	0.9587	0.9740	0.9^2114	18	0.378	0.444	0.516	0.561	0.602	0.679
4	0.729	0.811	0.882	0.9172	0.9417	0.9741	19	0.369	0.433	0.503	0.549	0.589	0.665
5	0.669	0.754	0.833	0.875	0.9056	0.9509	20	0.360	0.423	0.492	0.537	0.576	0.652
6	0.621	0.707	0.789	0.834	0.870	0.9249	25	0.323	0.381	0.445	0.487	0.524	0.597
7	0.582	0.666	0.750	0.798	0.836	0.898	30	0.296	0.349	0.409	0.449	0.434	0.554
8	0.549	0.632	0.715	0.765	0.805	0.872	35	0.275	0.325	0.381	0.418	0.452	0.519
9	0.521	0.602	0.685	0.735	0.776	0.847	40	0.257	0.304	0.358	0.393	0.425	0.490
10	0.497	0.576	0.658	0.708	0.750	0.823	45	0.243	0.288	0.338	0.372	0.403	0.465
11	0.476	0.553	0.634	0.684	0.726	0.801	50	0.231	0.273	0.322	0.354	0.384	0.443
12	0.457	0.532	0.612	0.661	0.703	0.780	60	0.211	0.250	0.295	0.325	0.352	0.408
13	0.441	0.514	0.592	0.641	0.683	0.760	70	0.195	0.232	0.274	0.302	0.327	0.380
14	0.426	0.497	0.574	0.623	0.664	0.742	80	0.183	0.217	0.257	0.283	0.307	0.357
15	0.412	0.482	0.558	0.606	0.647	0.725	90	0.173	0.205	0.242	0.267	0.290	0.338
							100	0.164	0.195	0.230	0.254	0.276	0.321

$\alpha = 1 - F(r$, given ν and assuming $\rho = 0)$ is the upper-tail area of the distribution of r appropriate for use in a single-tail test. For a two-tail test, 2α must be used. If r is calculated from n paired observations, enter the table with $\nu = n - 2$. For partial correlations enter with $\nu = n - k - 2$, where k is the number of variables held constant.

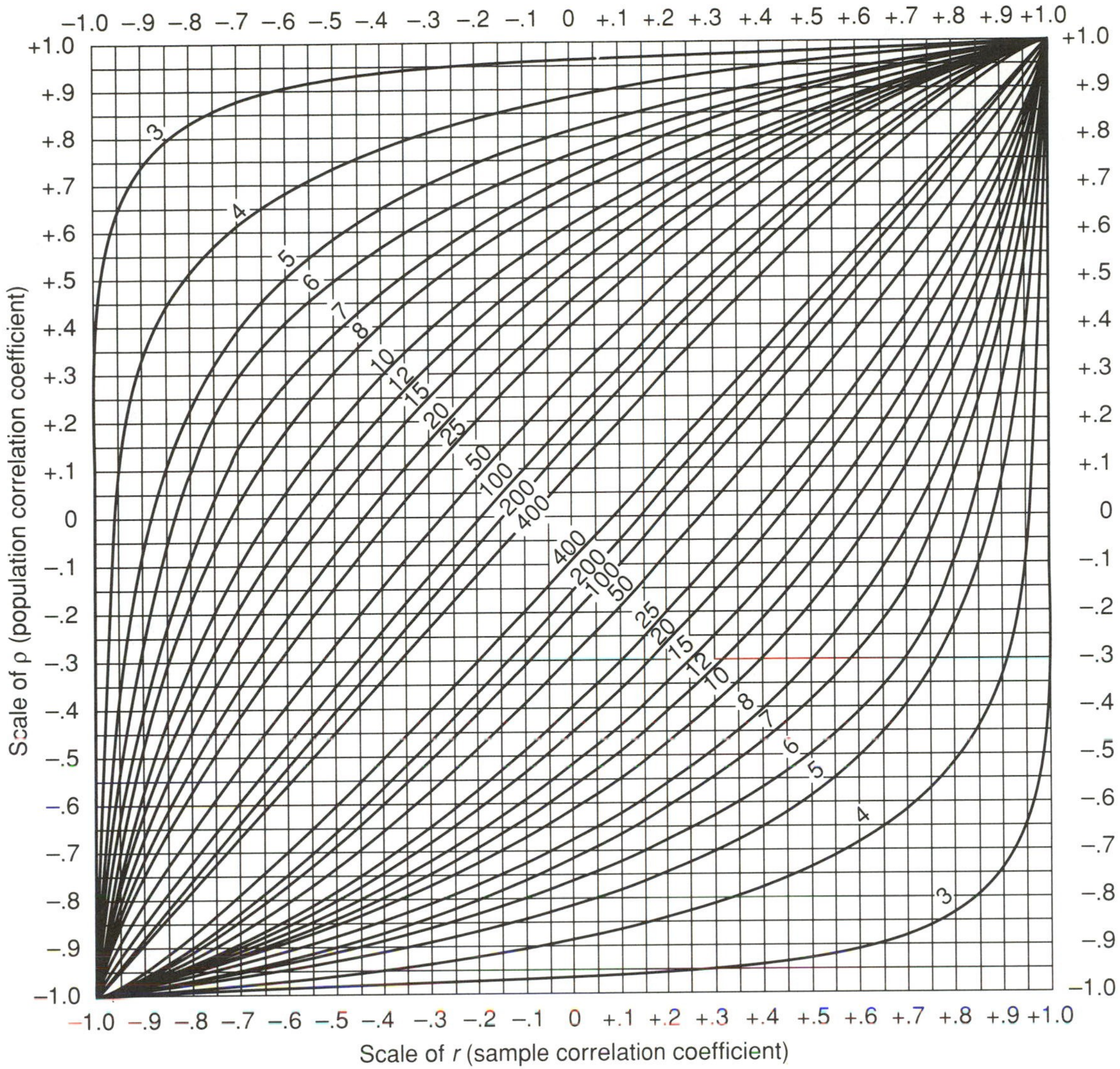

FIGURE B-1 Confidence Belts for the Correlation Coefficient ($1 - \alpha = .95$)

The numbers on the curves indicate sample size. The chart can also be used to determine upper and lower 2.5 percent significance points for r, given ρ.

Index

"Progress Toward Meeting the 1990 Nutrition Objectives for the Nation: Nutrition Services and Data Collection in State/Territorial Health Agencies," by M. Kaufman, J. Heimendinger, S. Foester, and M. A. Carrol; 1987, 77:299–303.

"Adolescent Smoking, Weight Changes, and Binge-Purge Behavior: Associations with Secondary Amenorrhea," by J. Johnson and A. H. Whitaker; 1992, 82:47–54.

"The Effect of Legal Drinking Age on Fatal Injuries of Adolescents and Young Adults," by N. E. Jones, C. F. Pieper, and L. S. Robertson; 1992, 82:112–115.

"Vaginal Douching among Women of Reproductive Age in the United States: 1988," by S. O. Aral, W. D. Mosher, and W. Cates, Jr.; 1992, 82:210–214.

"Knowledge and Attitudes about AIDS among Corporate and Public Service Employees," by J. K. Barr, J. M. Waring, and L. J. Warshaw; 1992, 82:225–228.

"Employment Status and Heart Disease Risk Factors in Middle-Aged Women: The Rancho Bernardo Study," by D. Kritz-Silverstein, D. L. Wingard, and E. Barrett-Connor; 1992, 82:215–219.

"Attitudes about Infertility Intervention among Fertile and Infertile Couples," by L. J. Halman, A. Abbey, and F. M. Andrews; 1992, 82:191–194.

"The Quality of Life in the Year before Death," by H. R. Lentzner, E. R. Pamuk, E. P. Rhodenhiser, R. Rothenberg, and E. Powell-Griner; 1992, 82:1093–1098.

"HIV Antibody Testing among Adults in the United States: Data from 1988 NHIS," by A. M. Hardy and D. A. Dawson; 1990, 80:586–589.

"The Economic Impact of Injuries: A Major Source of Medical Costs," by L. C. Harlan, W. R. Harlan, and P. E. Parsons; 1990, 80:453–459.

"Maternal Recall Error of Child Vaccination Status in a Developing Nation," by J. J. Valadez and L. H. Weld; 1991, 82:120–123.

"Patient 'Dumping' Post-COBRA," by A. L. Kellermann and B. B. Hackman; 1990, 80:864–867.

"The Lowest Birth-Weight Infants and the US Infant Mortality Rate: NCHS 1983 Linked Birth/Infant Death Data," by M. D. Overpeck, H. J. Hoffman, and K. Prager; 1992, 82:441–444.

"The Differential Effect of Traditional Risk Factors on Infant Birthweight among Blacks and Whites in Chicago," by J. W. Collins, Jr. and R. J. David; 1990, 80:679–681.

"Field Comparison of Several Commercially Available Radon Detectors," by R. W. Field and B. C. Kross; 1990, 80:926–930.

"Cancer Rates after the Three Mile Island Nuclear Accident and Proximity of Residence to the Plant," by M. C. Hatch, S. Wallenstein, J. Beyea, J. W. Nieves, and M. Susser; 1991, 81:719–724.

"Increase in Condom Sales following AIDS Education and Publicity, United States," by J. S. Moran, H. R. Janes, T. A. Peterman, and K. M. Stone; 1990, 80:607–608.

"Environmental Tobacco Smoke Exposure during Infancy," by B. A. Chilmonczyk, G. J. Knight, G. E. Palomaki, A. J. Pulkkinen, J. Williams, and J. E. Haddow; 1990, 80: 1205–1208.

"Decreased Access to Medical Care for Girls in Punjab, India: The Roles of Age, Religion, and Distance," by B. E. Booth and M. Verma; 1992, 82:1155–1157.

"Does Maternal Tobacco Smoking Modify the Effect of Alcohol on Fetal Growth?" by J. Olsen, A. da Costa Pereira, and S. F. Olsen; 1991, 81:69–73.

"Birth Weight and Perinatal Mortality: The Effect of Gestational Age," by A. J. Wilcox, and R. Skjoerven; 1992, 82:378–382.

"Performance of the Reflotron in Massachusetts' Model System for Blood Cholesterol Screening Program," by S. Havas, R. Bishop, L. Koumjian, J. Reisman, and S. Wozenski; 1992, 82:458–461.

"Weight Gain Prevention and Smoking Cessation: Cautionary Findings," by S. M. Hall, C. D. Tunstall, K. L. Vila, and J. Duffy, 1992, 82:799–803.

"Effects on Serum Lipids of Adding Instant Oats to Usual American Diets," by L. Van Horn, A. Moag-Stahlberg. K. Liu, C. Ballew, K. Ruth, R. Hughes, and J. Stamler; 1991, 81:183–188.

"Relationship between Body Mass Indices and Measures of Body Adiposity," by D. A. Revicki and R. G. Israel; 1986, 76:992–994.

"The Behavioral Risk Factor Survey and the Stanford Five-City Project Survey: A Comparison of Cardiovascular Risk Behavior Estimates," by C. Jackson, D. E. Jatulis, and S. P. Fortmann; 1992, 82:412–416.

"Patient, Provider and Hospital Characteristics Associated with Inappropriate Hospitalization," by A. L. Siu, W. G. Manning, and B. Benjamin; 1990, 80:1253–1256.

"Crop Duster Aviation Mechanics: High Risk for Pesticide Poisoning," by R. McConnell, A. F. P. Anton, and R. Magnotti; 1990, 80:1236–1239.

"Gender Differences in Cigarette Smoking and Quitting in a Cohort of Young Adults," by P. L. Pirie, D. M. Murray, and R. V. Luepker; 1991, 81:324–327.

"Home Apnea Monitoring and Disruptions in Family Life: A Multidimensional Controlled Study," by E. Ahmann, L. Wulff, and R. G. Meny; 1992, 82:719–722.

"Work-Site Nutrition Intervention and Employees' Dietary Habits: The Treatwell Program," by G. Sorensen, D. M. Morris, M. K. Hunt, J. R. Hebert, D. R. Harris, A. Stoddard, and J. K. Ockene; 1992, 82:877–880.

"Carbon Monoxide in Indoor Ice Skating Rinks: Evaluation of Absorption by Adult Hockey Players," by B. Levesque, E. Dewailly, R. Lavoie, D. Prud'Homme, and S. Allaire; 1990, 80:594–598.

"Medical and Psychosocial Factors Predictive of Psychotropic Drug Use in Elderly Patients," by L. D. Ried, D. B. Christensen, and A. Stergachis; 1990, 80:1349–1353.

"Patterns of Medical Employment: A Survey of Imbalances in Urban Mexico," by J. Frenk, J. Alagon, G. Nigenda, A. Munoz-delRio, C. Robledo, L. A. Vaquez-Segovia, and C. Ramirez-Cuadra; 1991, 81:23–29.

"The Impact of HIV-Related Illness on Employment," by E. H. Yelin, R. M. Greenblatt, H. Hollander, and J. R. McMaster; 1991, 81:79–84.

"How Women's Adopted Low-Fat Diets Affect Their Husbands," by A. L. Shattuck, E. White, and A. R. Kristal; 1992, 82:1244–1250.

"How Valid Are Mammography Self-Reports?" by E. S. King, B. K. Rimer, B. Trock, A. Balshem, and P. Engstrom; 1990, 80:1386–1388.

"Predictors of Vaccination Behavior among Persons Ages 65 Years and Older," by P. A. Stehr-Green, M. A. Sprauer, W. W. Williams, and K. M. Sullivan; 1990, 80:1127–1129.

"Perforating Eye Injury in Allegheny County, Pennsylvania," by D. Landen, D. Baker, R. LaPorte, and R. A. Thoft; 1990, 80:1120–1122.

"A Measles Outbreak at a College with a Prematriculation Immunization Requirement," by B. S. Hersh, L. E. Markowitz, R. E. Hoffman, D. R. Hoff, M. J. Doran, J. C. Fleishman, S. R. Preblud, and W. A. Orenstein; 1991, 81:360–364.

"A Multifocal Outbreak of Hepatitis A Traced to Commercially Distributed Lettuce," by L. S. Rosenblum, I. R. Mirkin, D. T. Allen, S. Safford, and S. C. Hadler; 1990, 80:1075–1080.

"Risk Factors for Syphilis: Cocaine Use and Prostitution," by R. T. Rolfs, M. Goldberg, and R. G. Sharrar; 1990, 80:853–857.

"Cigarette, Alcohol, and Coffee Consumption and Prematurity," by A. D. McDonald, B. G. Armstrong, and M. Sloan; 1992, 82:87–90.

"Body Fatness and Risk for Elevated Blood Pressure, Total Cholesterol, and Serum

Lipoprotein Ratios in Children and Adolescents," D. P. Williams, S. B. Going, T. G. Lohman, D. W. Harsha, S. R. Srinivasan, L. S. Webber, and G. S. Berenson; 1992, 82:358–363.

To the American Medical Association for permission to reprint excerpts from the following articles from the *Journal of the American Medical Association*:

"Bicycle-Associated Head Injuries and Deaths in the United States from 1984 Through 1988: How Many Are Preventable?" by J. J. Sacks, P. Holmgreen, S. M. Smith, and D. M. Sosin; 1991, 266:3016–3018. Copyright 1991, American Medical Association.

"How Strictly Do Dialysis Patients Want Their Advance Directives Followed?" by A. Sehgal, A. Galbraith, M. Chesney, P. Schoenfeld, G. Charles, and B. Lo; 1992, 267:59–63. Copyright 1992, American Medical Association.

"No-Smoking Laws in the United States: An Analysis of State and City Actions to Limit Smoking in Public Places and Workplaces," by N. A. Rigotti and C. L. Pashos; 1991, 266:3162–3167. Copyright 1991, American Medical Association.

"Predictors of Physicians' Smoking Cessation Advice," by E. Frank, M. A. Winkleby, D. G. Altman, B. Rockhill, and S. P. Fortmann; 1991, 266:3139–3144. Copyright 1991, American Medical Association.

"Prevalence of Migraine Headache in the United States: Relation to Age, Income, Race, and Other Sociodemographic Factors," by W. F. Stewart, R. B. Lipton, D. D. Celentano, and M. L. Reed; 1992, 267:64–69. Copyright 1992, American Medical Association.

"RJR Nabisco's Cartoon Camel Promotes Camel Cigarettes to Children," by J. R. DiFranza, J. W. Richards, Jr., P. M. Paulman, N. Wolf-Gillespie, C. Fletcher, R. D. Jaffe, and D. Murray; 1991, 266:3149–3153. Copyright 1991, American Medical Association.

"Active Enforcement of Cigarette Control Laws in the Prevention of Cigarette Sales to Minors," by L. A. Jason, P. Y. Ji, M. D. Anes, and S. H. Birkhead; 1991, 266:3159–3161. Copyright 1991, American Medical Association.

"Ventricular Arrhythmias in Patients Undergoing Noncardiac Surgery," by B. O'Kelly, W. S. Browner, B. Massie, J. Tubau, L. Ngo, D. T. Mangano, for the Study of Perioperative Ischemia Research Group; 1992, 268:217–221. Copyright 1992, American Medical Association.

"Relationship Between Resistance to Insulin-Mediated Glucose Uptake, Urinary Uric Acid Clearance, and Plasma Uric Acid Concentration," by F. Facchini, Y.-D. I. Chen, C. B. Hollenbeck, and G. M. Reaven; 1991, 266:3008–3011. Copyright 1991, American Medical Association.

"Transdermal Nicotine for Smoking Cessation," by the Transdermal Nicotine Study Group; 1991, 266:3133–3138. Copyright 1991, American Medical Association.

"The Presence and Accessibility of Firearms in the Homes of Adolescent Suicides: A Case-Control Study," by D. A. Brent, J. A. Perper, C. J. Allman, G. M. Moritz, M. E. Wartella, and J. P. Zelenak; 1991, 266:2989–2995. Copyright 1991, American Medical Association.

"Incidence and Risk Factors for Gout in White Men," by R. Roubenoff, M. J. Klag, L. A. Mead, K.-Y. Liang, A. J. Seidler, and M. C. Hochberg; 1991, 266:3004–3007. Copyright 1991, American Medical Association.

"Safety-Belt and Helmet Use Among High School Students—United States, 1990," CDC; 1992, 41:111–114. Copyright 1992, American Medical Association.

"Brand Logo Recognition by Children Aged 3 to 6 Years: Mickey Mouse and Old Joe the Camel," by P. M. Fischer, M. P. Schwartz, J. W. Richards, Jr., A. O. Goldstein, and T. H. Rojas; 1991, 266:3145–3148. Copyright 1991, American Medical Association.

"Concordance for Dyslipidemic Hypertension in Male Twins," by J. V. Selby, B. Newman, J. Quiroga, J. C. Christian, M. A. Austin, and R. R. Fabsitz; 1991, 265:2079–2084. Copyright 1991, American Medical Association.

"Aspirin Effects on Mortality and Morbidity in Patients With Diabetes Mellitus: Early Treatment Diabetic Retinopathy Study Report 14," by ETDRS Investigators; 1992, 268:1292–1300. Copyright 1992, American Medical Association.

"Comparisons of Hospital Care for Patients With AIDS and Other HIV-Related Conditions," by D. P. Andrulis, V. B. Weslowski, E. Hintz, and A. W. Spolarich; 1992, 267:2482–2486. Copyright 1992, American Medical Association.

"Isolated Systolic Hypertension and Subclinical Cardiovascular Disease in the Elderly: Initial Findings From the Cardiovascular Health Study," by B. M. Psaty, C. D. Furberg, L. H. Kuller, N. O. Borhani, P. M. Rautaharju, D. H. O'Leary, D. E. Bild, J. Robbins, L. P. Fried, and C. Reid; 1992, 268:1287–1291. Copyright 1992, American Medical Association.

To the American Society for Clinical Nutrition for permission to reprint excerpts from the following articles from the *American Journal of Clinical Nutrition*:

"Self-Reported Weight and Height," by M. L. Rowland; 1990, 52:1125–1133.

"Calcium Supplementation on Normotensive and Hypertensive Pregnant Women," by K. B. Knight and R. E. Keith; 1992, 55:891–895.

An excerpt from the article "Determining Sample Size for Research Activities," by R. V. Krejcie and D. W. Morgan; 1970, 30:607–610 is reprinted by permission of Educational and Psychological Measurement from *Educational and Psychological Measurement*.

To the *Journal of Obstetric, Gynecologic, and Neonatal Nursing* for permission to reprint excerpts from the following articles:

"Breastfeeding and Employment," by M. H. Kearney and L. Cronenwett; 1991, 20(6):471–480.

"Infant Responses to Saline Instillations and Endotracheal Suctioning," by D. R. Shorten, P. J. Byrne, and R. L. Jones; 1991, 20(6):464–469.

To the *Journal of Environmental Health* for permission to reprint excerpts from the following articles:

"A Graduate Certificate Program: Education and Training for Hazardous Waste Specialists," by R. Powitz, J. McMicking, and R. Kummler; 1990, 52(4):230–231.

"Measuring Airborne Asbestos: The Significance of Small Fibers Detected under TEM," by B. E. Lippy and J. A. Boggs; 1989, 52(3):157–160.

Excerpts from the article "Use of Nonionic Contrast Media in a High-Risk Outpatient Population," by J. V. Zelch; 1990, 19(11):16–20 are reprinted by permission of Anderson Publishing Ltd. from *Applied Radiology*.

Excerpts from the article "Stress Process Among Mothers of Infants: Preliminary Model Testing," by L. O. Walker; 1989, 38(1) are reprinted with permisison of *Nursing Research*.

Excerpts from the article "Continuing Education: Do Mandates Matter?" by T. J. Edwards III; 1992, 63(3) are reprinted with permission of *Radiologic Technology*, the official journal of the American Society of Radiologic Technologists, © 1992.

Excerpts from the article "Self-Reported Eating Disorders of Black, Low-Income Adolescents: Behavior, Body Weight Perceptions, and Methods of Dieting," by M. Balentine, K. Stitt, J. Bonner, and L. Clark are reprinted with permission of the American School Health Association, Kent, Ohio, from *Journal of School Health*, 1991, 61(9):392–396.

To the American Physical Therapy Association for permission to reprint excerpts from the following articles from *Physical Therapy*:

"Canadian Therapists' Priorities for Clinical Research: A Delphi Study," by C. Miles-Tapping, A. Dyck, S. Brunham, E. Simpson, and L. Barber; 1990, 70:448–454.

"Effect of Acupuncture-Point Stimulation on Diastolic Blood Pressure in Hypertensive Subjects: A Preliminary Study," by T. Williams, K. Mueller, and M. W. Cornwall; 1991, 71:523–529.

"Effects of an Exercise Program on Sick Leave Due to Back Pain," by K. M. Kellett, D. A. Kellett, and L. A. Nordholm; 1991, 71:283–293.

"Exercise Effect on Electromyographic Activity of the Vastus Medialis Oblique and Vastus Lateralis Muscles," by W. P. Hanten and S. S. Schulthies; 1990, 70:561–565.

"Effect of Total Contact Cast Immobilization on Subtalar and Talocrural Joint Motion in Patients with Diabetes Mellitus," by J. E. Diamond, M. J. Mueller, and A. Delitto; 1993, 73:310–315.

"Effect of Feedback on Learning a Vertebral Joint Mobilization Skill," by M. Lee, A. Moseley, and K. Refshauge; 1990, 70:97–104.

"Reduction of Chronic Posttraumatic Hand Edema: A Comparison of High Voltage Pulsed Current, Intermittent Pneumatic Compression, and Placebo Treatments," by J. W. Griffin, L. S. Newsome, S. W. Straika, and P. E. Wright; 1990, 70:279–286.

"Progressive Exercise Testing in Closed Head-Injured Subjects: Comparison of Exercise Apparatus in Assessment of a Physical Conditioning Program," by M. Hunter, J. Tomberlin, C. Kirkikis, and S. T. Kuna; 1990, 70:363–371.

"Effects of Mental Practice on Rate of Skill Acquisition," by J. R. Maring; 1990, 70:165–172.

"Influence of High Voltage Pulsed Direct Current on Edema Formation Following Impact Injury," by J. A. Bettany, D. R. Fish, and F. C. Mendel; 1990, 70:219–224.

"Relationships between Physical Activity and Temporal-Distance Characteristics of Walking in Elderly Women," by C. I. Leiper and R. L. Craik; 1991, 71:791–803.

"Evidence for Use of an Extension-Mobilization Category in Acute Low Back Syndrome: A Prescriptive Validation Pilot Study," by A. Delitto, M. T. Cibulka, R. E. Erhard, R. W. Bowling, and J. A. Tenhula; 1993, 73:216–228.

"The Effect of Training on Physical Therapists' Ability to Apply Specified Forces of Palpation," by J. Keating, T. A. Matyas, and T. M. Bach; 1993, 73:38–46.

To Boxwood Press for permission to reprint the table entitled "Value of Variance-ratio F exceeded in 5% of random samples" and Figure 46, "The relationship between four fundamental sampling distributions from a basic analysis of variation: Numbers from experiments" by C. C. Li, 1959.

To CRC Press for permission to reprint the table entitled "Percentage Points, Distribution of the Correlation Coefficient, when $\rho = 0$" from *CRC Standard Probability and Statistics Tables and Formulae*, W. H. Beyer, ed, © 1991 by CRC Press, Inc.

To McGraw-Hill for permission to reprint Tables A-12e and A-5 and the graph showing confidence limits for the population correlation coefficient from *Introduction to Statistical Analysis*, by W. Dixon and F. Massey, 4th edition, 1983.